Paul Jarvis

Adaptation of Plants to Water and High Temperature Stress

ADAPTATION OF PLANTS TO WATER AND HIGH TEMPERATURE STRESS

EDITED BY

NEIL C. TURNER

Division of Plant Industry, CSIRO
Canberra, Australia

PAUL J. KRAMER

Duke University
Durham, North Carolina

A WILEY-INTERSCIENCE PUBLICATION

JOHN WILEY & SONS

New York · Chichester · Brisbane · Toronto

Library of Congress Cataloging in Publication Data:

Main entry under title:

Adaptation of plants to water and high temperature stress:

   "Proceedings of [a] seminar held from November 6 to
10, 1978, at the Carnegie Institution of Washington,
Department of Plant Biology, Stanford, California."
   "A Wiley-Interscience publication."
   Includes index.
   1. Plant–water relationships—Congresses. 2. Crops
and water—Congresses. 3. Plants, Effect of heat on—
Congresses. 4. Plants, Effect of stress on—Congresses.
5. Adaptation (Biology)—Congresses. 6. Plants,
Effect of drought on—Congresses. 7. Plants—Drought
resistance—Congresses. I. Turner, Neil C., 1940– II. Kramer,
Paul Jackson, 1904–

QK870.A3   581.5   79-24428
ISBN 0-471-05372-4

Printed in the United States of America

10 9 8 7 6 5 4 3 2 1

Contributors

Paul A. Armond
Department of Plant Biology
Carnegie Institution of Washington
Stanford, California

D. Aspinall
Department of Plant Physiology
Waite Agricultural Research Institute
University of Adelaide
Adelaide, South Australia
Australia

Murray R. Badger
Department of Plant Biology
Carnegie Institution of Washington
Stanford, California

E. W. R. Barlow
School of Biological Sciences
Macquarie University
Sydney, New South Wales
Australia

J. E. Begg
Division of Plant Industry
CSIRO
Canberra
Australian Capital Territory, Australia

Joseph A. Berry
Department of Plant Biology
Carnegie Institution of Washington
Stanford, California

F. Bidinger
International Crops Research Institute
   for the Semi-Arid Tropics
Hyderabad, Andhra Pradesh, India

Olle Björkman
Department of Plant Biology
Carnegie Institution of Washington
Stanford, California

A. Blum
Agricultural Research Organization
The Volcani Center
Bet Dagan, Israel

J. S. Boyer
USDA-SEA-AR
Departments of Botany and Agronomy
University of Illinois
Urbana, Illinois

C. J. Brady
Plant Physiology Unit
Division of Food Research
CSIRO
Sydney, New South Wales
Australia

J. M. Cutler
Department of Agronomy
Cornell University
Ithaca, New York

James Ehleringer
Department of Biology
University of Utah
Salt Lake City, Utah

R. A. Fischer
Division of Plant Industry
CSIRO
Canberra
Australian Capital Territory, Australia

D. F. Gaff
Department of Botany
Monash University
Melbourne, Victoria, Australia

Andrew D. Hanson
MSU–DOE Plant Research
    Laboratory
Michigan State University
East Lansing, Michigan

Kuni Ishihara
Faculty of Agriculture
Tokyo University of
    Agriculture and Technology
Tokyo, Japan

Paul G. Jarvis
Department of Forestry
    and Natural Resources
University of Edinburgh
Edinburgh, U.K.

Douglas A. Johnson
USDA-SEA-AR
Crop Research Laboratory
Utah State University
Logan, Utah

H. G. Jones
East Malling Research Station
East Malling
Kent, U.K.

Madeleine M. Jones
Department of Environmental Biology
Research School of Biological
    Sciences
Australian National University
Canberra
Australian Capital Territory, Australia

Wayne R. Jordan
Texas Agricultural Experiment Station
Temple, Texas

Paul J. Kramer
Department of Botany
Duke University
Durham, North Carolina

J. Kummerow
Department of Botany
San Diego State University
San Diego, California

A. N. Lahiri
Division of Soil-Water-Plant
    Relationship
Central Arid Zone Research Institute
Jodhpur, Rajasthan, India

J. Levitt
Department of Plant Biology
Carnegie Institution of Washington
Stanford, California

M. M. Ludlow
Division of Tropical Crops and
    Pastures
CSIRO
Brisbane, Queensland, Australia

J. R. McWilliam
Department of Agronomy and Soil
    Science
University of New England
Armidale, New South Wales, Australia

Fred R. Miller
Texas Agricultural Experiment Station
Texas A&M University
College Station, Texas

H. A. Mooney
Department of Biological Sciences
Stanford University
Stanford, California

J. M. Morgan
New South Wales Department of
    Agriculture
Agricultural Research Centre
Tamworth, New South Wales
Australia

R. E. Munns
Department of Agronomy
Institute of Agriculture
University of Western Australia
Perth, Western Australia, Australia

Park S. Nobel
Department of Biology and Division of
  Environmental Biology of the
  Laboratory of Nuclear Medicine and
  Radiation Biology
University of California
Los Angeles, California

C. B. Osmond
Department of Environmental Biology
Research School of Biological
  Sciences
Australian National University
Canberra
Australian Capital Territory, Australia

J. C. O'Toole
International Rice Research Institute
Los Baños, The Philippines

Carl S. Pike
Department of Biology
Franklin and Marshall College
Lancaster, Pennsylvania

S. B. Powles
Department of Environmental Biology
Research School of Biological
  Sciences
Australian National University
Canberra
Australian Capital Territory, Australia

John K. Raison
Plant Physiology Unit
Division of Food Research
CSIRO
Sydney, New South Wales
Australia

Joe T. Ritchie
USDA-SEA-AR
Grassland, Soil and Water Research
  Laboratory
Temple, Texas

P. L. Steponkus
Department of Agronomy
Cornell University
Ithaca, New York

Cecil R. Stewart
Department of Botany and Plant
  Pathology
Iowa State University
Ames, Iowa

Howard M. Taylor
USDA-SEA-AR
Soil and Water Management Unit
Iowa State University
Ames, Iowa

Tadayoshi Tazaki
Faculty of Science
Toho University
Chiba, Japan

J. A. Teeri
Barnes Laboratory
University of Chicago
Chicago, Illinois

Neil C. Turner
Division of Plant Industry
CSIRO
Canberra
Australian Capital Territory, Australia

Tadahiro Ushijima
Faculty of Agriculture
Tokyo University of
  Agriculture and Technology
Tokyo, Japan

H. H. Wiebe
Department of Biology
Utah State University
Logan, Utah

K. Winter
Department of Environmental Biology
Research School of Biological
  Sciences
Australian National University
Canberra
Australian Capital Territory, Australia

Preface

Water is the earth's most abundant compound, yet the lack of water is the most important single factor limiting plant productivity and crop yields throughout the world. It is therefore not surprising that the effects of water stress on physiological processes, productivity, and yield of plants have been widely studied and reported. Moreover, since water stress in nature is usually accompanied by above-average temperatures, studies of high temperature stress have received considerable, though less, attention.

In recent years interest has focused not simply on the response of plants to water and high temperature stress, but also on the importance of morphological and physiological mechanisms of adaptation to water and high temperature stress. The provisions of the United States–Australia Cooperative Science Agreement enabled the organization of a workshop/seminar to review progress in this field. This volume reports the proceedings of this seminar, held from November 6–10, 1978, at the Carnegie Institution of Washington, Department of Plant Biology, Stanford, California. The workshop was unique in that it brought together scientists from a wide range of disciplines—agronomy, biochemistry, biophysics, crop physiology, ecology, ecophysiology, forestry, horticulture, plant breeding, plant physiology, and soil science—to discuss the adaptation of plants to water and high temperature stress. Furthermore, the workshop convened scientists working on plants from a range of ecosystems—native, agricultural, and forest—in which objectives in the face of drought may differ.

A second feature of the workshop was that it attempted to identify mechanisms of adaptation to water and high temperature stress, and also to understand the interaction and integration of these mechanisms on productivity and yield. The significance of these mechanisms in breeding and selection for drought tolerance was evaluated.

We are indebted to the government of the United States, through the National Science Foundation and the U.S. Department of Agriculture, and the government of Australia, through the Department of Science, for help in arranging this workshop and for financial assistance that allowed six participants from Australia and fourteen participants from the United States to attend. In addition, many institutions in both countries contributed to the success of the workshop by providing funds for participants. Participation by scientists from outside the United States and Australia was made possible by funds from the Japanese Society for the Promotion of Science, the Ford Foundation, the International Crops Research Institute for the Semi-Arid

Tropics, the Agricultural Research Council in the United Kingdom, the University of Edinburgh, the Agricultural Research Organization in Israel, and the International Rice Research Institute.

We are also grateful to Dr. L. T. Evans of the CSIRO Division of Plant Industry, whose initial advice and encouragement helped to launch this workshop, and to Drs. J. E. Begg and R. A. Fischer of the same division for their assistance in organizing the workshop. We are indebted to the Carnegie Institution of Washington, Department of Plant Biology, at Stanford, for support, practical help, and provision of facilities for the workshop: in particular we are grateful for the local arrangements made by Dr. H. A. Mooney of Stanford University and Dr. O. Björkman and F. Nicholson of the Carnegie Institution's Plant Biology Department. Finally, we express our thanks to Mrs. Robyn Long for typing assistance and to M. J. Long for editorial assistance.

NEIL C. TURNER

PAUL J. KRAMER

Canberra, Australia
Durham, North Carolina
June 1980

Contents

INTRODUCTION | I

Water is an essential component of plant life. It comprises approximately 85 to 90% of the total fresh weight in physiologically active herbaceous plants. If the water content in most species falls much below this level, many of the physiological activities of the plant are impaired. Yet the amount of water in the plant comprises only a small fraction of the water that passes through the plant during its lifetime and is lost to the atmosphere in transpiration. Therefore the relationship between plants and water has interested scientists for generations and has led to a series of books on this theme, such as *The Plant in Relation to Water* (1), *Die Hydratur der Pflanze und ihre physiologisch-ökologische Bedeutung* (2), *Plant and Soil Water Relationships* (3), *The Water Relations of Plants* (4), *Plant-Water Relationships* (5), *Plant and Soil Water Relationships: A Mod-*

ern Synthesis (6), *Water and Plant Life: Problems and Modern Approaches* (7), and the five volumes of *Water Deficits and Plant Growth* (8). In addition, several reviews over the past decade have discussed aspects of plant water relations (9–20).

Rather than reviewing yet again the responses of plants to water stress, this book focuses attention on the responses or mechanisms that enable a plant to survive water deficits. This ability to adapt or acclimate to environmental variation has received much less attention than the responses to environmental variables. Some discussion of the adaptation of plants to water and temperature stresses is contained in *Physiological Adaptation to the Environment* (21) and *Plant Structure, Function and Adaptation* (22) and also in recent reviews by Arnon (23), Turner (24), and Jones et al. (25). This book treats in detail the physiological and morphological adaptations to water stress. Since water stress is usually, but not always (26), accompanied by high temperatures, adaptation to high temperature is also included. Adaptation to low temperature stress is omitted not because it lacks importance, but to limit the scope of this book. Readers are referred to the treatise by Levitt (27) and to the proceedings of two recent workshops (28, 29) for further details on cold temperature stress.

A little more than half the chapters in this book are devoted to identifying and evaluating the physiological and morphological mechanisms of adaptation to water and high temperature stress by both ecologists and crop physiologists. Part V considers the interaction of these mechanisms and discusses their integrated effect, as judged by productivity and yield. The four chapters in Part VI look into the variability among cultivars and the possibilities of breeding for and selection for mechanisms of adaptation to water and high temperature stress.

As mentioned in the Preface, this book reports the proceedings of a workshop/seminar on the subject. Limitations of space make it impossible to incorporate the results of the many formal and informal discussions arising in such a workshop. Authors had an opportunity to revise their manuscripts in light of discussions or questions raised at the workshop, and many accepted this opportunity. Additionally, the final chapter incorporates the summaries of each session prepared by the chairman. These are personal summaries and do not necessarily reflect a consensus of opinion at the workshop. Indeed, chairmen were chosen for their ability to contribute to the subject, not because of their inclination to summarize every detail.

The water status of soil or plant tissue can be characterized in several ways (30). This book describes the state of water by its chemical potential within the system relative to that of pure free water, that is, water containing no solutes and bound by no forces. This gives a measure of the capacity of the water at any point to do work compared to the work capacity of free water. It is defined by the water potential (31). The total water potential ψ in the soil or plant consists of several mutually independent components: the osmotic potential π, arising from dissolved solutes in the water, the turgor potential or pressure potential P, arising from turgor and hydrostatic forces in the system, and the matric potential τ, arising from capillary forces at the water-air interfaces.* The water potential is measured in units of energy per unit volume: this is dimensionally equivalent to pressure, and the unit of bars is used throughout this book. Although there is a move toward the use of SI units (32), the unit of bars is retained here because it is understood by a wide range of readers [1 bar = 1.0×10^{6} dynes/cm^{2} = 1.0×10^{5} newtons/m^{2} = 0.987 atmosphere = 1017 cm water = 75.0 cm Hg = 14.50 lb/in^{2} = 10^{5} pascals = 0.1 megapascal (MPa) = 100 joules/kg].

Alternatively, the water status of plant tissue can be described in terms of the water content relative to the saturated water content. Now called the relative water content (RWC), and earlier named relative turgidity, it is

$$RWC = \frac{\text{fresh weight} - \text{dry weight}}{\text{saturated weight} - \text{dry weight}} \times 100$$

A numerically different, but similarly measured, parameter is the water saturation deficit (WSD), obtained as follows:

$$WSD = \frac{\text{saturated weight} - \text{fresh weight}}{\text{saturated weight} - \text{dry weight}} \times 100$$

that is, WSD = 100 − RWC.

Both systems of defining water status are used in

*Symbols of water status are standardized in this book as follows:

ψ = total water potential
π = osmotic or solute potential
P = turgor or pressure potential
τ = matric potential
RWC = relative water content
WSD = water saturation deficit

The component of the plant or soil to which the water status refers is identified by the subscript.

this book. At extreme stresses such as those experienced by seeds or "resurrection plants," values of water potential and relative water content have little meaning, and the water status of plant tissue is best defined in terms of the equivalent relative humidities: this is used in Chapter 14, where the ability of plants to withstand extreme water stress is discussed.

This introductory section emphasizes the importance of drought, stress research, and the adaptability of plants. Chapter 1 defines the scope of the book and defines many of the terms used in subsequent chapters; Chapter 2 brings out the significance for food production of research into the effects of stress in crop plants.

REFERENCES

1. N. A. Maximov, *The Plant in Relation to Water*, Allen and Unwin, London, 1929.
2. H. Walter, *Die Hydratur der Pflanze und ihre physiologisch-ökologische Bedeutung (Untersuchungen über den osmotischen Wert)*, Fischer, Jena, 1931.
3. P. J. Kramer, *Plant and Soil Water Relationships*, McGraw-Hill, New York–Toronto–London, 1949.
4. A. J. Rutter and F. H. Whitehead, Eds., *The Water Relations of Plants*, Blackwell, London, and Wiley, New York, 1963.
5. R. O. Slatyer, *Plant-Water Relationships*, Academic Press, London–New York, 1967.
6. P. J. Kramer, *Plant and Soil Water Relationships: A Modern Synthesis*, McGraw-Hill, New York–St. Louis–San Francisco–London–Sydney–Toronto–Mexico–Panama, 1969.
7. O. L. Lange, L. Kappen, and E.-D. Schulze, Eds., *Water and Plant Life: Problems and Modern Approaches*, Springer-Verlag, Berlin–Heidelberg–New York, 1976.
8. T. T. Kozlowski, Ed., *Water Deficits and Plant Growth*, Vols. 1–5, Academic Press, New York–San Francisco–London, 1968–1978.
9. J. Dainty, in M. B. Wilkins, Ed., *The Physiology of Plant Growth and Development*, McGraw-Hill, London–New York–Toronto–Sydney–Mexico–Johannesburg–Panama, 1969, p. 421.

10. R. O. Slatyer, in J. D. Eastin, F. A. Haskins, C. Y. Sullivan, and C. H. M. Van Bavel, Eds., *Physiological Aspects of Crop Yield,* Am. Soc. Agron. and Crop Sci. Soc. Am., Madison, Wisconsin, 1969, p. 53.

11. P. E. Weatherley, *Adv. Bot. Res.,* **3,** 171 (1970).

12. T. C. Hsiao, *Annu. Rev. Plant Physiol.,* **24,** 519 (1973).

13. R. O. Slatyer, in R. O. Slatyer, Ed., *Plant Response to Climatic Factors. Proceedings of the Uppsala Symposium, 1970,* UNESCO, Paris, 1973, p. 177.

14. P. J. Kramer, *Plant Physiol.,* **54,** 463 (1974).

15. C. T. Gates, *J. Aust. Inst. Agric. Sci.,* **40,** 121 (1974).

16. P. G. Jarvis, in D. A. de Vries and N. H. Afgan, Eds., *Heat and Mass Transfer in the Biosphere. I. Transfer Processes in Plant Environment,* Scripta, Washington, DC, 1975, p. 369.

17. J. S. Boyer and H. G. McPherson, *Adv. Agron.,* **27,** 1 (1975).

18. J. E. Begg and N. C. Turner, *Adv. Agron.,* **28,** 161 (1976).

19. N. C. Turner and J. E. Begg, in J. R. Wilson, Ed., *Plant Relations in Pastures,* CSIRO, Melbourne, Australia, 1978, p. 50.

20. N. C. Turner and G. J. Burch, in I. D. Teare, Ed., *Crop-water Relations,* Interscience, New York–London, 1980 (in press).

21. F. J. Vernberg, Ed., *Physiological Adaptation to the Environment,* Intext, New York, 1975.

22. M. A. Hall, Ed., *Plant Structure, Function and Adaptation,* Macmillan, London–Basingstoke, 1976.

23. I. Arnon, in U. S. Gupta, Ed., *Physiological Aspects of Dryland Farming,* Oxford and IBH, New Delhi–Bombay–Calcutta, 1975, p. 3.

24. N. C. Turner, in H. Mussell and R. C. Staples, Eds., *Stress Physiology in Crop Plants,* Interscience, New York–London, 1979, p. 343.

25. M. M. Jones, N. C. Turner, and C. B. Osmond, in L. G. Paleg and D. Aspinall, Eds., *The Physiology and Biochemistry of Drought Resistance,* Academic Press, New York–San Francisco–London, 1980 (in press).

26. R. A. Fischer and N. C. Turner, *Annu. Rev. Plant Physiol.,* **29,** 277 (1978).

27. J. Levitt, *Responses of Plants to Environmental Stresses,* Academic Press, New York–London, 1972.

28. P. H. Li and A. Sakai, Eds., *Plant Cold Hardiness and Freezing Stress: Mechanisms and Crop Implications,* Academic Press, New York–San Francisco–London, 1978.

29. J. M. Lyons, J. K. Raison, and D. Graham, Eds., *Low Temperature Stress in Crop Plants: The Role of the Membrane,* Academic Press, New York–San Francisco–London, 1980.

30. B. Slavík, *Methods of Studying Plant-Water Relations,* Springer-Verlag, Berlin–Heidelberg–New York, and Academia, Prague, 1974.

31. R. O. Slatyer and S. A. Taylor, *Nature (London),* **187,** 922 (1960).

32. L. D. Incoll, S. P. Long, and M. R. Ashmore, *Curr. Adv. Plant Sci.,* **9,** 331 (1977).

1 | Drought, Stress, and the Origin of Adaptations

PAUL J. KRAMER

Department of Botany, Duke University, Durham, North Carolina

1. INTRODUCTION

Before discussing the modifications or adaptations in structure and function that enable some plants to survive water and high temperature stress better than others, let us review the general mechanism by which environmental stresses affect plant growth. For example, drought causes plant water deficits that reduce cell turgor, causing closure of stomata and reduction in cell enlargement, thereby reducing both the leaf surface area and the rate of photosynthesis per unit of leaf area. If the plant water deficit becomes more severe, the photosynthetic machinery is damaged, further reducing the rate of photosynthesis per unit of leaf area. High temperature increases the rate of water loss and the use of food in respiration. It also sometimes damages the photosynthetic machinery, as pointed out in Chapter 15.

An environmental factor such as water deficit or high temperature, or a change in genotype, can affect plant growth and yield only by affecting the physiological processes and conditions in plants (Figure 1.1). Thus to know why certain species and varieties survive or even thrive in habitats where others fail requires a better understanding of how their physiological processes are affected by various environmental factors. Furthermore, if a plant breeder produces a higher yielding variety, it is because a more efficient

7

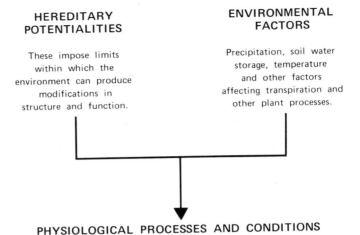

HEREDITARY
POTENTIALITIES

These impose limits
within which the
environment can produce
modifications in
structure and function.

ENVIRONMENTAL
FACTORS

Precipitation, soil water
storage, temperature
and other factors
affecting transpiration and
other plant processes.

PHYSIOLOGICAL PROCESSES AND CONDITIONS

Water absorption, ascent of sap, transpiration.
Plant water balance as reflected in water potential,
turgor, stomatal opening, and cell enlargement.
Photosynthesis, carbohydrate and nitrogen metabolism,
and other metabolic processes.

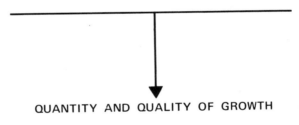

QUANTITY AND QUALITY OF GROWTH

Size of cells, organs, and plants.
Root--shoot ratio.
Succulence, kinds and amounts of
various compounds accumulated.
Economic yield.

Figure 1.1. How the quantity and quality of plant growth is controlled by an organism's hereditary potentialities and its environment, operating through its physiological processes and conditions.

combination of physiological processes has been arrived at for a particular environment.

2. THE IMPORTANCE OF WATER AND HEAT STRESS

Crop plants rarely attain their full genetic potential for yield because of the limitations imposed by the environment, especially unfavorable temperatures and lack of water. About one-third of the world's potentially arable land suffers from an inadequate supply of water, and on most of the remainder crop yields are periodically reduced by drought. Plant water deficits affect every aspect of plant growth (1) and the worldwide losses in yield from water stress probably exceed the losses from all other causes combined. The importance of water in relation to plant growth is discussed in detail in Part V.

High temperature stress receives less attention than water stress, except in relation to vegetation growing in hot deserts. However the manner in which high temperature affects plants is probably better understood than are some of the effects of water stress.

2.1. Heat Stress

The effects of heat stress are often confounded with those of water stress because drought usually is accompanied by high temperature, which increases the rate of transpiration and hastens the occurrence of injurious dehydration. There also may be depletion of carbohydrates because the maximum rate of respiration usually occurs at a higher temperature than the maximum rate of photosynthesis. Furthermore, the cooling effects of transpiration are reduced in water-stressed plants, resulting in increased leaf temperatures, which decrease apparent photosynthesis and may disturb nitrogen and lipid metabolism and injure cell membranes. Some Soviet investigators claim that ammonia is released at high temperatures, causing injury. Direct heat injury is less common but often occurs to stems of seedlings at the soil surface and to plants in the hottest deserts. The effects of high temperature were reviewed by Levitt (2) and are discussed in more detail in Part IV.

Low Temperature Stress. The extensive research on cold and chilling injury lies outside of the scope of this book. However low soil temperature often reduces water absorption and causes water stress, resulting in injury that sometimes is mistakenly attributed to low temperature. This was observed in cotton by St. John and Christiansen (3, p. 259).

2.2. Temperature Perturbations

There is much discussion concerning the probable effects on plant growth and crop yield of a small increase or decrease in the global temperature. This subject is important in the long term, but of more immediate concern in agriculture are the short periods of abnormally high or low temperatures that often occur during the growing season. There is evidence that a week or less of abnormally high or low temperature can measurably affect growth and yield. For example, Powell and Huffman (4) reported that 6 days at above or below normal temperature during seed development reduces seed size of sorghum, and Akpan and Bean (5) concluded that year-to-year variations in temperature affect the yield and quality of seed of pasture grasses. It also has been shown that the temperature regime in which the seed was developed affects the growth of tobacco seedlings (6). It therefore seems that more research is needed on the effects of short periods of temperature stress at various stages of growth because these incidents are more immediately important than possible long-term shifts in the world temperature. Fischer (Chapter 21, Section 6.2) also points to effects of short periods of atmospheric stress on seed filling in grain crops and concludes that more research is needed to determine the significance of these stresses.

3. MEASUREMENT OF WATER STRESS

Progress in all branches of science has always depended on the development of new methods and new instrumentation. This was particularly true in the field of plant water relations, where progress was hindered for many years by lack of convenient methods of measuring stress.

The need for quantitative measurements of water stress was appreciated by ecologists and physiologists early in this century, but the only method available was to measure the osmotic potential of expressed sap. This method provided useful information, including evidence of daily and seasonal changes in osmotic potential and a decrease in osmotic potential in plants subjected to water stress (7; 8, pp. 39–45). However by 1940 fewer measurements of osmotic potential were being made. One reason for this was the uncertainty about whether expressed sap provided reliable samples. In addition, it was increasingly appreciated that water movement is controlled by differences in water potential rather than by differences in osmotic potential. Unfortunately, no convenient method of measuring water potential was then available; hence the value of several decades of research on plant water relations was reduced by lack of quantitative measurements of plant water stress.

Usable thermocouple psychrometers became available by 1960, and shortly afterwards the pressure chamber was introduced. Now we can measure the water potential of soil, roots in the soil, and attached leaves, as well as detached leaves and twigs. The availability of equipment for measuring water potential may have led to overemphasis on this variable, which is only one of three terms in the equation for cell and tissue water potential ψ:

$$\psi = \pi + P \tag{1.1}$$

where π = osmotic potential

P = turgor or pressure potential

It is generally agreed that water movement is controlled by the water potential and cell enlargement by the turgor or pressure potential. Now there is increasing interest in the possibility that reduced turgor is the factor directly affecting metabolic processes in stressed plants (1, p. 563; Chapter 7, this book). Furthermore, there is strong interest in the importance of a decrease in osmotic potential or osmotic adjustment as an adaptive mechanism to water stress (Chapter 7). Thus to fully understand what is occurring in plants subjected to water deficits it is necessary to measure both water potential and osmotic potential and to calculate the turgor potential. Water potential usually is measured by the thermocouple psychrometer or pressure chamber (9, 10). However the validity of measurements of osmotic potential made on frozen tissue or expressed sap is sometimes questioned because of the possibility of dilution by cell wall water. One way of avoiding this error is use of the pressure-volume technique described by Tyree and Hammel (11).

Water content, calculated on either a dry or a fresh weight basis, is an unsatisfactory measure of water stress, but field water content as a percentage of water content at saturation is sometimes useful. This is calculated as follows:

$$\text{relative water content (RWC)} = \frac{\text{fresh weight} - \text{dry weight}}{\text{saturated weight} - \text{dry weight}} \times 100 \ (1.2)$$

The water saturation deficit (WSD), another expression of the same value, is calculated in another way; that is, WSD = 100 − RWC. It is claimed by some workers that plants tolerant of desiccation show a smaller decrease in water potential for a given decrease in water content than those that are less tolerant (see Ref. 12). Methods of measuring water deficit or RWC are described by Barrs (9) and Slavík (10).

4. THE TERMINOLOGY OF DROUGHT TOLERANCE

Since this book deals with adaptations to water and temperature stress, it is imperative that some of the terms be discussed, and it is particularly important to differentiate between ''drought'' and ''plant water stress.''

4.1. Drought

Drought is a meteorological and environmental event (13), defined somewhat loosely as absence of rainfall for a period of time long enough to cause depletion of soil moisture and damage to plants. The length of time without rain that is necessary to cause injury depends on the kind of plant, the water-holding characteristics of the soil in which it is growing, and the atmospheric conditions that affect the rates of evaporation and transpiration. Drought may be permanent, as in desert areas; seasonal, as in areas with well-defined wet and dry seasons; or unpredictable, as in many humid climates. Agricultural drought was defined by Van Bavel and Verlinden (14) as the condition that exists when there is insufficient water available to a crop. They made this more quantitative by estimating the amount of readily available water stored in the root zone, the average rate of loss by evaporation, and the average replenishment by precipitation during each month of the year. From these data were calculated the probable number of drought days, that is, days when soil moisture might be limiting, for plants growing in soil with various water storage capacities in the root zone.

4.2. Stress

It is difficult to define "stress" precisely. In engineering and the physical sciences, the term is usually defined as the force applied per unit area. For example, pressure on a girder produces strain in it. However in biology stress is usually described more loosely as any factor that disturbs the normal functioning of an organism. Levitt (2, also Chapter 28), attempted to apply the physical terminology to living organisms, but it seems unlikely that plant water "stress" will be replaced by plant water "strain."

In the commonly used terminology, "drought" is an environmental stress of sufficient duration to produce a plant water deficit or stress, which in turn causes disturbance of physiological processes. The degree of plant water stress or deficit depends on the extent to which water potential and cell turgor are reduced below their optimum values. Quantitative measurement of water stress was discussed earlier in this chapter (Section 3).

Although plant water stress always accompanies drought, it may occur in the absence of drought, either because of excessive transpiration or because water absorption is inhibited by cold soil, an excess of salt in the soil solution, deficient aeration, or injury to root systems. Most plants are subjected to transient, midday water deficits in hot, sunny weather, even when growing in moist soil. However this book is concerned primarily with stress periods of longer duration, caused by the decreasing availability of soil moisture.

4.3. Drought Resistance or Tolerance

The term "drought resistance" has long been used with reference to the ability of plants to survive drought. However it is an unsatisfactory and even ambiguous term, which I prefer to replace with "drought tolerance." Although "drought" refers to a meteorological phenomenon, it is sometimes misapplied to plant phenomena. An example of this confusion occurs in the following statement from a recent paper: "Growth drought tolerance, defined as the plant drought that is just sufficient to halt the increase of seedling leaf area, varied among seedlings from about −20 to −40 bars. Growth drought avoidance, defined as the difference between plant drought and the drought of the shoot environment, . . ." (15). These are thoroughly confusing statements, because there is no plant drought, but only plant water stress. Also, how can a drought be confined to the shoot environment?

To avoid this kind of confusion, I propose to classify the adaptations by which plants survive in regions subject to drought in two major categories: drought escape and drought tolerance. I prefer use of "avoidance" rather than "escape" because the former more accurately describes the actual situation, but "escape" is used here to prevent confusion with Levitt's terminology (2), which is used in some chapters of this book. The adaptations contributing to drought tolerance can be subdivided into categories of dehydration postponement and dehydration tolerance.

Drought Escape. Drought escape is characteristic of only a few plants, such as desert ephemerals and some plants growing in areas possessing well-defined wet and dry seasons. Drought-escaping plants complete their life cycle, or at least their reproductive cycle, before the dry season begins and are seldom severely stressed. Early maturity often is important when droughts occur late in the growing season. For example, each day by which wheat in Kansas and Nebraska matures earlier than the Kharkof variety results in an increase in yield of nearly 60 kg/ha (16). In mediterranean climates tolerance of low temperature may be an important adaptation for drought escape because it permits growth to start early enough in the season to ensure completion of the life cycle before water becomes limiting.

Drought Tolerance. It is impossible for plants of humid climates to escape random droughts, which are characteristic of much of the temperate zone, and some of them possess other adaptations that increase their tolerance of drought. These can be classified as those that postpone dehydration and those that increase tolerance of dehydration.

Dehydration Postponement. This occurs by means of morphological or physiological modifications that reduce transpiration or increase absorption. Thick cuticle, leaf rolling, responsive stomata, and deep root systems all increase the ability of plants to endure droughts for considerable periods of time without becoming severely dehydrated. Water storage occurs in a few succulents and other plants with large storage organs, but it is negligible factor in crop plants. Among cultivated crops, only pineapple has the thick cuticle and the stomatal behavior characteristic of succulent plants with crassulacean acid metabolism (CAM), and as a result it has a transpiration ratio much lower than that of any other cultivated crop. Dehydration postponement of many crops and wild plants depends on the efficiency of their root systems, a topic discussed in Chapters 5, 6, and 25.

The value of responsive stomata and large root systems is sometimes questioned. For example, prompt closure of stomata in moderately stressed leaves may be advantageous in regions where droughts are of short duration, but detrimental where droughts last for a long time, since photosynthesis is reduced by cutting off the supply of carbon dioxide before the photosynthetic machinery has been inhibited by water stress. It also has been argued

14   PAUL J. KRAMER

that a high root resistance is advantageous where plants must live on stored moisture, because if less water is absorbed early in the season, more will be available later when the crop is maturing. Many of these special situations were discussed by Begg and Turner (17).

"Dehydration postponement" is equivalent to the term "drought avoidance" as used by Levitt (2) for plants that maintain a high water potential when exposed to an external water stress. This usage is unfortunate because such plants do not avoid drought; rather, they possess various adaptations that enable them to tolerate it. As stated earlier, drought is a meteorological event, and the only plants to avoid it are those that complete their life cycle before drought occurs.

Dehydration Tolerance. The degree of dehydration without permanent injury varies widely, depending on the process under consideration, the stage of development, the duration of stress, and the kind of plant. Readers are referred to Hsiao (1) and Begg and Turner (17) for detailed discussions of factors affecting the degree of injury caused by dehydration.

The reaction of plants to dehydration can be considered from two standpoints, *survival* and *yield*. Survival of severe dehydration is of greater importance with respect to native vegetation than for crop plants because if crops are severely stressed they usually have little economic value. However dehydration tolerance may be more important in crop plants than generally is supposed, because some plants such as sorghum and western wheat grass can tolerate considerable water stress and recover when a drought ends.

The physiological basis for differences among species in tolerance of dehydration must be sought largely at the molecular level, perhaps chiefly in membrane structure and enzyme activity. For example, the cellular fine structure of maize is injured by water stress more than that of sorghum (18). Vieira da Silva (19) also found that the fine structure of cells of drought susceptible cotton is injured more by water stress than that of drought tolerant types. These and other reports suggest that tolerance of moderate dehydration may be of significance in crop plants.

It should be emphasized that the three categories of adaptation are not mutually exclusive because one kind of plant can possess more than one category of adaptation. For example, some varieties of sorghum exhibit early maturity, and sorghums in general possess extensive root systems and some degree of protoplasmic tolerance of dehydration. One of the major problems in the field of stress physiology is to determine what adaptations or categories of adaptation are most important in respect to the survival of plants of particular kinds in specific habitats.

The concept of water stress tolerance is too broad and general to serve as a good basis for plant breeding. The purpose of this book is to identify some of the specific adaptations to water and heat stress that can be used both in ecology and in plant breeding.

5. OTHER PROBLEMS OF TERMINOLOGY

Discussions of the manner in which plants are adapted to special environmental conditions often contain terminology that is of questionable validity both philosophically and scientifically.

5.1. Strategy

Consider the common misuse of the word "strategy." Misapplication of "strategy" and "tactics" in biology is exemplified by the title of a paper by Harper and Ogden, "The Reproductive Strategy of High Plants" (20). Another paper states, "We may usefully view a plant as an intricate control system in which responses to stress are strategies directed towards achieving certain goals" (21, p. 375). Earlier in the same paper we find: "It is the control of leaf area and morphology which is often the most powerful means a mesophytic plant has of influencing its fate." Since "strategy" refers to a plan of action designed to attain some desired end, this is pure teleology. How does a plant know what it should do when confronted by water stress or some other environmental crisis? Scientists can develop research strategies to solve their problems, but plants cannot develop strategies because they do not possess the power to make intelligent decisions. Although terms such as "strategy" and "tactics" may attract reader attention, they are philosophically objectionable when used with respect to the behavior of plants.

5.2. The Concept of Adaptation

Although the theme of this book is adaptation to stress, the concept of "adaptation" is difficult to define because it is used both with respect to the evolutionary origin of a character and with respect to the contribution of a character to the fitness of an organism to survive in its present environment. We are concerned with the latter usage. We therefore describe adaptations as *heritable* modifications in structures or functions that increase the probability of an organism surviving and reproducing in a particular environment, or both. However it often is difficult to determine whether a particular modification in a character is beneficial (i.e., has adaptive value) in a particular environment. It is tempting to assume that if a modification in a character survives, it must be beneficial; but Williams (22) warned that the presence of a character is not proof that it is currently essential 'or even beneficial to survival in the environment in which the organism now lives.

It also should be emphasized that success of an organism in a particular environment rarely depends on possession of a single adaptive character. According to Bradshaw (23), fitness or adaptation of an organism to an environment depends on possession of an optimum combination of characters that minimizes the deleterious effects and maximizes the advantageous effects. Thus plant breeders are faced with the difficult task of producing genotypes with an optimum combination of adaptive characters rather than the simpler task of producing genotypes with a single adaptive character. There may be exceptions to this generalization, however, as when success depends on tolerance of a single factor such as excess aluminum.

"Adaptation" is sometimes used carelessly, in a manner implying that plants and animals can undergo modifications *in order to* become better adapted to an environment. There is an important, but sometimes neglected, difference between stating that plants have acquired adaptations in structure and function that enable them to survive water stress and stating that plants acquire adaptations to survive water stress, as implied in the quotations in Section 5.1. The concept of purposeful adaptation in nature was widely held by theologians and some biologists in the eighteenth and nineteenth centuries. However there is no scientific basis for such a view, and no justification for the teleological terminology that seems to be coming back into use.

During the evolution of the plant kingdom, innumerable modifications in structures and processes have occurred as a result of random mutations and recombinations. Most of these were deleterious and disappeared, but a few were beneficial because they enabled the plants possessing them to survive and reproduce more successfully, and these were preserved by natural selection. As a result, plants growing in increasingly dry habitats accumulated various modifications of characters with adaptive values, such as thick cuticle, extensive root systems, low osmotic potential, and tolerance of dehydration that increased the probability of their survival. However the modifications or adaptations that enable plants to survive droughts and live in dry habitats *did not originate for those purposes*. They originated from random mutations and recombinations that probably occurred in plants growing in both moist and dry habitats, but usually were not preserved in moist habitats because they had no survival value. The fact that a certain kind of modification would be beneficial in a changing environment has no bearing on whether it will appear. Many organisms have become extinct because the modifications that would have enabled them to become adapted to a new set of environmental conditions failed to develop.

5.3. Stability of Adaptive Characters

The expression of all characters is controlled by the interaction of heredity and environment, as shown in Figure 1.1. However some are more responsive to the environment than others. If a character is unresponsive and remains relatively unchanged over a wide range of environments, it is described as lacking in plasticity or as phenotypically constant. Examples are the water storage tissue and thick cuticle of succulents, and the low osmotic potential and dehydration tolerance of many xerophytes. If a character is responsive to the environment, it is regarded as phenotypically flexible or phenotypically plastic. Examples of phenotypic plasticity are the differences between sun and shade leaves, the reversible decrease in osmotic potential observed in some mesophytes when water stressed (Chapter 7, Section 6.1), and the switch from the C_3 pathway of photosynthesis to CAM carbon metabolism found in some succulents when subjected to water stress (Chapter 10, Section 6).

Various terms have been used to categorize the two extremes in phenotypic plasticity (see Ref. 23). Recently Fischer and Turner (24) proposed to classify characters with low phenotypic plasticity as "constitutive" and those with a high degree of plasticity as "facultative." They regarded these terms as synonymous with the concepts of strategy and tactics used by Harper and Ogden (20). The validity of such a classification seems questionable to me because it suggests qualitative differences in plasticity or flexibility, when in fact there are only differences in degree of plasticity.

5.4. Acclimation

"Acclimation" refers to the *nonheritable* modification of characters caused by exposure of organisms to new climatic conditions, such as warmer, cooler, or drier weather. It depends on the occurrence of temporary phenotypic modifications caused by the changing environment. The extent of acclimation that can occur varies widely, depending on the plasticity of the species. For example, the summer annual *Tidestromia oblongifolia* has a very high optimum temperature for photosynthesis and cannot adjust to a lower temperature. In contrast, the optimum temperature for photosynthesis of the evergreen perennial creosote bush (*Larrea divaricata*), is about 10°C lower in the winter than in summer (Chapter 18, Section 2). *Larrea* likewise undergoes considerable acclimation to water stress, the photosynthetic apparatus of plants grown in Death Valley remaining uninhibited at a level of water stress that severely inhibits photosynthesis of plants grown in a humid environment (25).

The difference between adaptation and acclimation is illustrated by the adaptation of photosynthesis in *Eucalyptus* to the temperatures characteristic of various elevations, on which is superimposed short-term acclimation to seasonal variations in temperature (26).

5.5. Hardening

The process of "hardening" appears to be the equivalent of acclimation, and it also depends on phenotypic modifications. It is generally found that plants that have been subjected to several cycles of mild water stress suffer less injury from drought than do plants not stressed previously. Plants are commonly "hardened" before transplanting by exposure to full sun, by decreased frequency of watering, and even by undercutting and loosening (wrenching) the root systems. These treatments usually result in temporary water stress, reduced leaf size, thicker cuticle, and sometimes a larger root/shoot ratio. There also are evidences of protoplasmic changes, as indicated by the differences in reaction of photosynthesis of previously stressed

18 PAUL J. KRAMER

and unstressed creosote bush to water stress, mentioned in Section 5.4. Some Soviet physiologists claim that treatment of seed by wetting and redrying before sowing will increase the tolerance of water stress by plants grown from the treated seed (27).

There is an extensive body of literature on the hardening of plants to increase their tolerance of low temperature, but the subject is outside the scope of this book. There also is some evidence that plants can be hardened or acclimated to high temperature, and this is discussed in Part IV.

5.6. Water Use Efficiency

The "efficiency" with which water is used refers in general to the amount of water used per unit of plant material produced. Like the term itself, however, this has been expressed in various ways. At one time "water requirement" designated the amount of water transpired per unit of dry matter produced by a crop. Later this term was replaced by "transpiration ratio," since there is no specific amount of water *required* to grow a crop. For example, water requirement or transpiration ratio for corn varied from 250 to 400 and for alfalfa from 660 to over 1000, over a period of 7 years at Akron, Colorado (8, p. 499). These data were based on transpiration rates of plants grown in containers.

In recent years the term "water use efficiency" (WUE) has come into general use, but it is employed in at least two ways. Physiologists sometimes calculate WUE in units such as mg CO_2/g H_2O, or occasionally as mol CO_2/mol H_2O. This is the ratio of carbon fixed in photosynthesis to water lost by transpiration. This use of the term is discussed in detail by Fischer and Turner (24) and it is used in this book by Nobel (Chapter 4).

Agronomists usually define *water use efficiency* in terms of the ratio of dry matter produced or crop yield, to water used in transpiration and evaporation. Thus:

$$\text{WUE} = \frac{\text{dry matter or crop yield}}{\text{evapotranspiration}} \tag{1.3}$$

This places emphasis on the amount of water used per unit of economic product, such as grain, forage, fruit, or wood. It also accounts for the loss in crop production of a substantial fraction of the water by evaporation from the soil. The numerator of this equation, yield, can be greatly modified by cultural practices such as early planting, high plant density, and fertilization. These can materially increase the ratio of dry matter produced to water used by a crop, although they usually have little effect on the amount of CO_2 fixed in photosynthesis per unit of water lost by transpiration. Cultural and plant breeding programs designed to increase the ratio of dry matter produced to water used by crops are discussed by various writers in the special issue of *Agricultural Meterology* subsequently published as an edited book by Stone (28) and the book *Plant Environment and Efficient Water Use,* edited by Pierre et al. (29).

6. OBJECTIVES AND PROBLEMS

In the breeding and selection of crop plants, artificial selection has replaced natural selection, and the rate of accumulation of modifications or adaptations that result in increased yield in a particular environment can be tremendously increased. To be successful, however, plant breeders need to know what kinds of modification have adaptive value. Such concepts as drought or heat tolerance are too general to form a satisfactory basis for a breeding program. Several chapters in this book identify and evaluate some of the specific morphological and physiological adaptations that contribute to tolerance of water and heat stress. This information should be equally valuable to ecologists, crop physiologists, and plant breeders. For example, the relative importance of morphological adaptations such as deep, extensively branched roots and thick cuticle compared to the physiological adaptations such as responsive stomata, regulation of osmotic potential, protoplasmic tolerance of dehydration, and C_4 carbon metabolism, need to be evaluated so that the information can be used by plant breeders in crop improvement programs and by ecologists in explaining the distribution of natural vegetation.

Because of the complexity of these problems, we may need to reconsider our research methods. For example, in controlled environment research we need to develop water stress more slowly, in the same manner as it develops in the field (17). We also need to apply stress at various stages of development, since the effects are often different when plants are stressed in the vegetative stage and in the reproductive stage. Furthermore we need to consider possible differences in behavior between individual plants in a controlled environment and plants in a crop or community out of doors. We also should make more measurements of physiological processes, chemical composition, and morphological development. Growth analysis seems desirable to learn how the growth of the various organs is modified. Such large-scale research requires a team effort.

More interdisciplinary research is necessary at every step, from identifying the problems, through research on them, to application of the results of the research to explaining plant distribution and increasing crop production. The problems of adaptation are too broad for solution by practitioners of any one discipline. Usually scientists with field experience, such as ecologists and agronomists, are required to recognize the problems at the environmental level, but physiologists can then assist in identifying and solving them at the process level. Finally ecologists, agronomists, and physiologists can join with plant breeders in finding solutions at the whole plant and crop level.

I hope that this book will aid the reader to see more clearly the kinds of problems that control the distribution of natural vegetation and the yield of crops in various climates and will show how interdisciplinary research can expedite the solution of some of them.

REFERENCES

1. T. C. Hsiao, *Annu. Rev. Plant Physiol.*, **24**, 519 (1973).
2. J. Levitt, *Responses of Plants to Environmental Stresses,* Academic Press, New York–London, 1972.
3. J. B. St. John and M. N. Christiansen, *Plant Physiol.*, **57**, 257 (1976).
4. R. D. Powell and K. W. Huffman, *Plant Physiol., Suppl.*, **61**, 5 (1978).
5. E. E. J. Akpan and E. W, Bean, *Ann. Bot. (London)*, **41**, 689 (1977).
6. J. F. Thomas and C. D. Raper, Jr., *Tobacco Sci.*, **19**, 37 (1975).
7. J. A. Harris, *The Physico-Chemical Properties of Plant Saps in Relation to Phytogeography*, University of Minnesota Press, Minneapolis, 1934.
8. E. C. Miller, *Plant Physiology*, 2nd ed., McGraw-Hill, New York, 1938.
9. H. D. Barrs, in T. T. Kozlowski, Ed., *Water Deficits and Plant Growth*, Vol. 1, Academic Press, New York–London, 1968, p. 235.
10. B. Slavík, *Methods of Studying Plant-Water Relations*, Academia, Prague, and Springer-Verlag, Berlin–Heidelberg–New York, 1974.
11. M. T. Tyree and H. T. Hammel, *J. Exp. Bot.*, **23**, 267 (1972).
12. M. F. Sanchez-Diaz and P. J. Kramer, *Plant Physiol.*, **48**, 613 (1971).
13. L. H. May and F. L. Milthorpe, *Field Crop Abstr.*, **15**, 171 (1962).
14. C. H. M. Van Bavel and F. J. Verlinden, *Agricultural Drought in North Carolina, N. C. Agric. Exp. St. Tech. Bull.*, **122** (1956).
15. A. J. Wilson and J. A. Sarles, *Agron. J.*, **70**, 231 (1978).
16. L. P. Reitz, *Agric. Meteorol.*, **14**, 3 (1974).
17. J. E. Begg and N. C. Turner, *Adv. Agron.*, **28**, 161 (1976).
18. K. L. Giles, D. Cohen, and M. F. Beardsell, *Plant Physiol.*, **57**, 11 (1976).
19. J. Vieira da Silva, in O. L. Lange, L. Kappen, and E.-D. Schulze, Eds., *Water and Plant Life: Problems and Modern Approaches,* Springer-Verlag, Berlin–Heidelberg–New York, 1976, p. 207.
20. J. L. Harper and J. Ogden, *J. Ecol.*, **58**, 681 (1970).
21. J. B. Passioura, in I. F. Wardlaw and J. B. Passioura, Eds., *Transport and Transfer Processes in Plants*, Academic Press, New York–San Francisco–London, 1976, p. 373.
22. G. C. Williams, *Adaptation and Natural Selection*, Princeton University Press, Princeton, New Jersey, 1966.
23. A. D. Bradshaw, *Adv. Genet.*, **13**, 115 (1965).
24. R. A. Fischer and N. C. Turner, *Annu. Rev. Plant Physiol.*, **29**, 277 (1978).
25. H. A. Mooney, O. Björkman, and G. J. Collatz, *Carnegie Inst. Washington Yearb.*, **76**, 328 (1977).
26. R. O. Slatyer and P. A. Morrow, *Aust. J. Bot.*, **25**, 1 (1977).
27. P. A. Henckel, *Annu. Rev. Plant Physiol.*, **15**, 363 (1964).
28. J. F. Stone, Ed., *Plant Modification for More Efficient Water Use*, Elsevier, Amsterdam–Oxford–New York, 1975.
29. W. H. Pierre, D. Kirkham, J. Pesek, and R. Shaw, Eds., *Plant Environment and Efficient Water Use*, Am. Soc. Agron. and Soil Sci. Soc. Am., Madison, Wisconsin, 1966.

2 | Plant Stress Research and Crop Production: The Challenge Ahead

JOE T. RITCHIE

USDA–SEA–AR, Grassland, Soil and Water Research Laboratory, Temple, Texas

1. WORLD POPULATION AND CROP PRODUCTION

The world is becoming increasingly concerned about agricultural production, particularly food production. Agricultural crop productivity results from complex interactions of plants with the climate and soil. This chapter examines the scope of the problem of producing enough food for a rapidly expanding world population and considers the great need for research on plant water stress and soil water conservation in the years ahead.

Most people are aware of the narrowing gap between production and demand for food in the world. However it is sometimes difficult to grasp the magnitude of the problem and to understand the urgent need for research to develop better plants and management techniques that will keep food production at levels near to the demand for food in spite of the limitations of land and water resources.

Present world food reserves are estimated as adequate to compensate for only a single year of worldwide bad harvests (1). To keep up with population growth, food production will have to increase about 100% by the year 2000. It took all the history of man, except for the past 130 years, for the world population to reach 1 billion. Only 70, 40, and 15 more years were required

21

to reach 2, 3, and 4 billion, respectively. Even with birthrates falling in many parts of the world, the population is expected to continue increasing and will probably reach between 6 and 8 billion by the year 2000. The majority of this increase will take place in the less developed countries, where life expectancy has increased greatly in the past 40 years. This means that by the end of the career of readers who are 40 years old or less, the good earth will need to produce as much as it has in all past generations combined to keep pace with world demand and with present food habits. It is estimated that about 75% of the increased world demand results from population increases and 25% from changes in food habits (2). Because it is likely that changes in food habits will continue to demand more food per person in less developed countries, the call to increase production by a factor of 2 within the next 25 years seems to be a conservative one.

Although some may tire of these statistics in a time when the more developed countries are clearly capable of producing more than the demand in "good years," we need to be continually reminded of the task ahead. Trends in yields of the major crops in the United States have shown a tendency to level off since 1970. The reason for the appearance of this plateau is complicated by rather unusual weather and economic changes during the 1970s as compared to the 15 year period prior to 1970. Since present yield levels are generally well below the genetic potential (as expressed in record yields of some major crops), it follows that if the price of farm products were to increase sufficiently, production could be increased within a few years, provided fertilizer, water, and fuel are available. Figure 2.1 illustrates the trend in grain yields since 1950 as a function of nitrogen fertilizer use in both more developed countries and less developed countries (3). Figure 2.2 shows the record yields of several crops as a function of nitrogen fertilizer use; these record yields are 5 to 10 times greater than the maximum average yield of 2510 kg/ha obtained in the more developed countries in 1973 (Figure 2.1). These statistics give us hope of increasing food production per unit area in the foreseeable future if the necessary inputs for higher production are available, especially in less developed countries.

Total crop production can be increased in two ways: by increasing the production per unit area, and by increasing the area cultivated. It is obvious from Figures 2.1 and 2.2 that the record yields and average yields per unit area for large areas have paralleled the increase in use of nitrogen fertilizer. Hardy (3) estimated that up to half the increased yields can be attributed to increased rates of nitrogen fertilization.

Another important input required to achieve the high yields shown in Figure 2.2 was irrigation. It is difficult to generalize about how much irrigation water was required to reach the high levels because of the complexity of rainfall patterns and stored soil water, but we can safely conclude that unless more land is brought under irrigation, yields cannot be expected to increase to levels implied by the record yields.

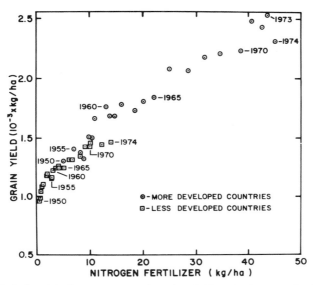

Figure 2.1. Relationship between cereal grain yield and nitrogen fertilizer use for more developed countries and less developed countries over the past 25 years. The nitrogen fertilizer rate was assumed to be half the total amount used for all purposes. Taken from Ref. 3.

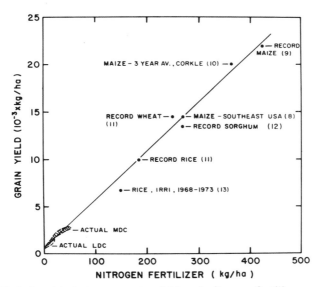

Figure 2.2. Relationship between grain yield and nitrogen fertilizer use for several cereal crops. Note the comparison of these record and near record yields to the actual yields from Figure 2.1 for more developed countries (MDC) and less developed countries (LDC). Sources are noted in the references indicated by the numbers in parentheses.

23

Extra production possibilities from new land being put into cultivation is also encouraging. The estimated 1.5×10^9 ha of cultivated land in the world could be expected to approximately double if needed (4), but new land brought into cultivation in general would be less productive. When comparing the magnitude of possible production increases from placing more land in production or obtaining more production per unit area, evidence suggests that increasing yield per unit area is more promising than increasing the area of cultivated land, unless desalinization of water makes it possible to cultivate arid regions of the world.

2. PLANT WATER STRESS

The major cause of year-to-year variation in crop yields is the fluctuation in weather. The component of the weather subject to the greatest variability is the distribution and amount of precipitation, which in turn causes uncertainty in the soil water content; the degree of uncertainty is moderated by the soil water holding capacity and by the crop water requirement.

In a rough breakdown of specific stress features that characterize soils worldwide, Dudal (4) reported that 28% were affected by drought stress. An additional 24% were affected by shallowness, a condition that causes initial soil water deficits to develop during short periods in which there is no precipitation. This analysis suggests that plant water stress at some stage of crop growth is the primary reason that crop yields fall below their potential and vary so much from year to year.

Although I am pessimistic about developing crops that will produce at high levels with reduced transpiration, research needs to be done to develop cropping systems for greater production without large new inputs of irrigation water. Given increased priority for research into plant response to water deficits, achievement of this objective over the next 25 years or so seems feasible.

Evidence of concern for increased plant water stress research is seen in several important documents recently published in the United States:

1. The National Agricultural Research Extension and Teaching Policy Act of 1977, a part of Public Law 95-113, passed by Congress in September 1977, listed "research on climate, drought, and weather modification as factors in food and agricultural production" as one of 15 critical areas calling for new federal initiatives.

2. "World Food and Nutrition Study—The Potential Contributions of Research," a National Research Council report (5) requested by the President and published in 1977, lists 22 areas in which research should be intensified. Three of those are resistance to environmental stresses, weather and climate, and irrigation and water management.

3. A conference on "Crop Productivity—Research Imperatives," sponsored by Michigan State University and the Charles F. Kettering Foundation in October 1975, evaluated research imperatives in six important areas (6). The expanded research imperatives related to environmental stresses listed the three highest priority items as the manipulation of crops and their environments to avoid or reduce stress injury and productivity, the exploitation of genetic potential for developing new varieties of crops resistant to environmental stresses, and the elucidation of the basic principles of stress injury and resistance and evaluation of the scope and nature of stress damage to crops.

4. A National Research Council study by the Committee of Climate and Weather Fluctuations and Agricultural Production published "Climate and Food" (1) in 1976, a report recommending increased research on (a) the fundamental responses of crops to weather and climate, (b) breeding crops with higher water use efficiencies and greater drought tolerance, (c) techniques for screening plant characteristics associated with tolerance to drought and temperature stresses, and other vagaries in the weather, and for speeding plant breeding progress in general, and (d) defining quantitative relationships of plants to weather, other environmental factors, and their interactions.

5. In response to a presidential request in 1975, the USDA held a workshop on "Research to Meet U.S. and World Food Needs." They placed climatically induced stress research in the top 10% of the problems that need additional research to enable more stable productivity.

6. A study requested by Congress in the 1978 Agriculture and Related Agencies Appropriation Bill was carried out by a team of seven federal research scientists. The committee was asked to study the feasibility of a new plant stress and soil water conservation laboratory. The study team's findings, reported to Congress in 1978, were that a new laboratory was essential to bring together multidisciplinary teams and to provide facilities for needed research programs in plant stress and soil conservation. The 1979 Agricultural Appropriation Bill appropriated funds for the purpose of planning a plant stress and water conservation laboratory. Although the proposed laboratory may not survive the funding process, progress thus far serves to emphasize the need for research to improve productivity in climates where water shortages are frequent, but the soil resources are good.

3. THE CHALLENGE AHEAD

My career has been mostly devoted to field studies of crop responses to soil water deficits, and early on I was very fortunate to become acquainted with a plant physiologist. Together we attempted to develop a balanced view of the important role of the plant as an integrator of climate and soil water conditions. Without an understanding of the whole soil-plant-climate system, it was difficult to make the kind of progress needed. My general philosophical approach to plant stress research is outlined below; it is the result of trying to maintain a balance between the use of plant physiology to understand plant responses to the environment and the use of plant breeding to make beneficial changes in the response of plants to water deficits.

Figure 2.3 presents an approach to developing plant and management systems that will minimize risks resulting from plant water deficits and maintain acceptable levels of crop production. The hierarchy of the approach suggests the need to develop and optimize a system of crop produc-

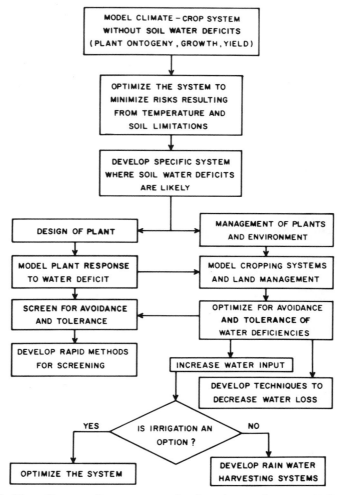

Figure 2.3. Flow diagram of a strategy to develop plant and management systems to optimize crop production with limited soil water.

tion that considers first the major constraints, except water deficits. This provides a better understanding of the potential productivity of genetic materials available for a given climate and soil in the absence of water deficits. It will be difficult to maximize the benefits from research on plant water stress without a workable model of climate-crop interactions in the absence of stress. Stresses caused by nutrient deficiencies and plant competition all interact with plant water stress, and their individual relationship to yield must be understood before the sometimes dominating influence of water stress is added.

The development of an optimum production system in which there are constraints from plant water deficits has two equally important parts. One is to design through breeding and selection plants that are best suited to the environment, and the other is to manage plants to suit the environment. These two areas are closely linked and must be combined to achieve an optimum production strategy. For example, we know that if no water inputs are expected during grain filling, stored soil water can be saved during vegetative growth for use during grain filling by limiting the leaf area of plants, thereby reducing transpiration rates. The same result may be obtained by using a short season plant designed to fill grain before the soil water is fully depleted. General understanding and cooperation among researchers in several disciplines will be required to provide the optimum production strategies for various regions of the world.

The flow chart (Figure 2.3) lists the need for a model of plant response to water deficits under "Design of Plant." Such a model is really necessary before we can understand what traits to screen for in a breeding program, or before a specific cropping system can be developed to optimize avoidance of water deficits. The varying responses of different plant processes to water deficits must be described quantitatively to be of greatest benefit in helping plant breeders screen for well-defined characteristics, and this requires the use of plant models of water stress. Once the specific traits for avoidance or tolerance of water stress have been defined in models, rapid screening techniques are required to find suitable genetic materials that contain those traits from among thousands of possible lines. To be rapid enough for practical testing of enough material, screening techniques should be amenable to assembly-line operation.

In the flow chart module "Management of Plants and Environment" it is important to develop a model of cropping systems and land management where water deficits are major constraints. The goal is to optimize for avoidance or tolerance of water deficiencies through increasing water input or decreasing water loss.

To achieve the goal of avoiding water deficits, two approaches are needed: one is to increase the supply of usable water, the other is to reduce the demand for water. A report prepared by a group of specialists for the National Academy of Sciences in 1974 (7) summarized the following major opportunities to save water: (a) improvement of existing water-conducting systems, (b) improvement of water management on the farm, (c) design of systems that deliver the right amount of water when needed, (d) conjunctive use of groundwater and surface water for optimal use of the total water resources, and (e) reduction of irrigation amounts through use of modern scheduling techniques.

To equip a new generation of research scientists to cope with the major task ahead, our educational institutions must strengthen interdisciplinary programs in climate-plant-soil relationships. Under traditional educational systems, this goal will be difficult to achieve because of the strong emphasis on specialization within a discipline for advanced degrees. Griffith University in Brisbane, Australia, is a good example of an educational institution having a school that is attempting to operate an integrated team with a specific mission—in this case, the improvement of the environment. This school should provide a model for the type of interdisciplinary program needed to meet the challenge of increasing production through improving plants and management techniques.

4. CONCLUSIONS

Evidence suggests that the world is not close to universal famine despite a rapidly growing population and improved diets. However in the not too distant future, this may not be true. One answer to meeting the need for increased production is to increase the scientific and technological innovations to derive maximum benefit of our natural resources. We must continue the search to unravel the mysteries of the plant and its response to the environment through both basic and applied research. Because the loss in world crop yields from water stress is large compared with that from other factors, and because of recent international interest in this subject, now is an excellent time for scientists to study plant stresses and to ensure adequate food, feed, and fiber supplies for future generations.

A specific strategy to guide research to meet future production demands requires close linkage among scientists of several disciplines, especially plant breeding, plant physiology, climatology, and soil and crop management. Multidisciplinary teams will be needed to meet the challenge of producing optimum crop production systems that avoid or tolerate plant water stress.

ACKNOWLEDGMENTS

Contribution from the USDA, Science and Education Administration, Agricultural Research, in cooperation with the Texas Agricultural Experiment Station, Texas A&M University. The help of W. R. Jordan and A. Blum in developing the ideas presented in this chapter is acknowledged.

REFERENCES

1. National Academy of Sciences–National Research Council, *Climate and Food—Climate Fluctuation and U.S. Agricultural Production*, National Academy of Sciences, Washington, DC, 1976.
2. J. Mayer, in A. W. A. Brown, T. C. Byerly, M. Gibbs, and A. Sam Pietro, Eds., *Crop Productivity—Research Imperatives*, Mich. Agric. Exp. Stn., East Lansing, and Charles F. Kettering Foundation, Yellow Springs, Ohio, 1975, p. 97.
3. R. W. F. Hardy, in *Proceedings of 24th Annual Meeting of Agricultural Research Institute,* National Academy of Sciences, Washington, DC, 1975, p. 115.
4. R. Dudal, in M. J. Wright, Ed., *Plant Adaptation to Mineral Stress in Problem Soils*, Cornell Univ. Agric. Exp. Stn., Cornell University, Ithaca, New York, 1976, p. 3.
5. National Academy of Sciences–National Research Council, *World Food and Nutrition Study—The Potential Contributions of Research*, National Academy of Sciences, Washington, DC, 1977.
6. A. W. A. Brown, T. C. Byerly, M. Gibbs, and A. San Pietro, Eds., *Crop Productivity—Research Imperatives*, Mich. Agric. Exp. Stn., East Lansing, and Charles F. Kettering Foundation, Yellow Springs, Ohio, 1975.
7. National Academy of Sciences, *More Water for Arid Lands—Promising Technologies and Research Opportunities*, National Academy of Sciences, Washington, DC, 1974.
8. D. L. Wright, F. M. Rhoads, and R. L. Stanley, Jr., *Crops Soils*, **31**, 10 (1978).
9. Unpublished leaflet distributed by Dekalb Seed Company, Dekalb, Illinois, 1977.
10. R. Ross, Ed., *Irrig. Age Mag.*, May–June 1978, p. 22.
11. L. T. Evans, in L. T. Evans, Ed., *Crop Physiology; Some Case Histories*, Cambridge University Press, London, 1975, p. 6.
12. Private communication with B. A. Stewart, USDA Southwestern Great Plains Research Center, Bushland, Texas, 1978.
13. N. C. Brady, in A. W. A. Brown, T. C. Byerly, M. Gibbs, and A. San Pietro, Eds., *Crop Productivity—Research Imperatives*, Mich. Agric. Exp. Stn., East Lansing, and Charles F. Kettering Foundation, Yellow Springs, Ohio, 1975, p. 62.

MORPHOLOGICAL ADAPTATIONS TO WATER STRESS | II

Just as there is a diversity of plant types inhabiting dry habitats, there is a diversity of mechanisms of adaptation to these habitats. Chapter 1 proposed that the adaptations by which plants survive regions subjected to drought can be classified into two major categories: namely, drought escape and drought tolerance, and that the adaptations contributing to drought tolerance be subdivided further into those that postpone dehydration and those that tolerate dehydration. Alternatively, the mechanisms of adaptation can be divided into morphological adaptations and physiological adaptations (1). This is the classification used in the remainder of this book.

In a survey in 1960 Oppenheimer (2) listed almost 20 characters that were considered to confer some adaptive advantage in xerophytes. The majority of these were structural and anatomical characters (i.e., morphological adaptations), and although a few of the characters were associated with the roots and stems, the majority involved leaf structure and anatomy. Several of the characters, such as smaller or fewer leaves, increased cutinization of the epidermis, reduction in the stomatal pore area, and radiation shedding are features that reduce water loss, whereas others such as early lignification, water storage, and accumulation of mucilage probably help the plant to tolerate dehydration.

Part II considers morphological adaptations of leaves and roots to water stress in agricultural and natural ecosystems. We deal elsewhere with the role of radiation shedding by leaf hairs on both water and temperature stress (Chapter 19) and with some morphological adaptations of the apex (Chapter 13).

REFERENCES

1. J. E. Begg and N. C. Turner, *Adv. Agron.*, **28**, 161 (1976).
2. H. R. Oppenheimer, in *Plant-Water Relationships in Arid and Semi-Arid Conditions: Reviews of Research*, UNESCO, Paris, 1960, p. 105.

3 | Morphological Adaptations of Leaves to Water Stress

JOHN E. BEGG

Division of Plant Industry, CSIRO, Canberra, Australian Capital Territory, Australia

1. INTRODUCTION

It has been argued that "nothing in biology has meaning except in the light of evolution" (1, 2). What then has the evolutionary record to teach us about adaptation to stress? What has been the succession of successful mechanisms of adaptation that have evolved through natural selection, enabling plants so effectively to colonize such a diverse range of terrestrial environments?

It is now generally accepted that life originated in the form of anaerobic bacteria some 3.5 billion years ago. These earliest life forms were the simple aquatic, unicellular, nonnucleate cells of the prokaryotes. During the course of the next 1.5 billion years these primitive prokaryotes became well adapted to the initially anaerobic and finally aerobic atmospheric conditions. It was during this period that the anaerobic and photosynthetic bacteria, and the blue-green algae, evolved. This was a grand period of biochemical evolution, during which the metabolic pathways associated with fermentation, photosynthesis, respiration, photorespiration, and nitrogen fixation evolved. So complete and diverse was the evolution of these bioenergetic processes among the primitive aquatic prokaryotes that during the evolution of higher plants over a subsequent 1.5 billion years the basic biochemistry of the cells underwent relatively little further change (3). It has been suggested that 93% of all enzymes present in the eukaryotes or higher plants are also present in the primitive prokaryotes (4).

33

In contrast to this period of evolution of great biochemical diversity and little morphological change, the next 1.5 billion years was marked by dramatic variations in plant morphology that developed as plants became adapted to the stress conditions of a wide range of terrestrial environments.

The early colonizers of terrestrial environments were the poikilohydric plants such as the algae, fungi, lichens, and mosses. However the result of their lack of an effective cuticle and any stomatal control mechanism was that the water status of their thalli matched the ambient moisture conditions. Thus their metabolic activity was restricted to moist sites and seasons, their biomass remained small, and their effectiveness in colonizing terrestrial habitats was limited (5). Some idea of the limitations facing plants in equilibrium with ambient moisture can be gauged from the realization that assuming a threshold water potential for effective biological activity of approximately -50 bars, metabolic activity will be restricted to the daylight hours when the relative humidity exceeds 96%.

Thus although poikilohydric plants achieved only limited success in colonizing the land, they gained a foothold and provided the basis for the development of the homoiohydric higher plants. The specialized structures for water uptake, water transport, and control over water loss of homoiohydric plants enabled them to maintain favorable water relations for continued metabolic activity within their tissues even under dry atmospheric conditions when, for example, ambient water potentials drop below -1000 bars (i.e., a relative humidity of approximately 48%). Keeton (6) outlined some of the problems facing plants making the transition from water to land and pointed out that only the vascular plants have effectively solved these problems. Through the evolution of a wide range of morphological adaptations associated with the uptake and conservation of water, vascular plants have successfully colonized all but the most inhospitable terrestrial environments.

The persistence with relatively little change during the evolution of higher plants of the basic metabolic pathways, which evolved more than 2 billion years ago, points to the possibility of a dearth of genetic diversity in the biochemistry of higher plants in the context of adaptation to water deficits and offers little hope for rapid improvement through conventional breeding techniques. By contrast, the successful colonization of terrestrial environments by higher plants, largely as a result of the evolution of a wide range of morphological adaptations, indicates that this may provide a more fruitful ground for adapting crops to water stress environments through selection and breeding.

Thus we need a better understanding of the various static and dynamic features of plant morphology that may enable plants to become better adapted to stress. As Passioura (7) stated, "It is the control of leaf area and morphology which is often the most powerful means a mesophytic plant has for influencing its fate when subject to long term water stress in the field." Because of the enormous range of morphological diversity evident in higher plants, this chapter is restricted to aspects of leaf morphology associated with their growth and development, orientation, and senescence. Leaf anatomy in relation to water use efficiency, and leaf morphology in the context of reflectance during water and temperature stress, are covered in Chapters 4 and 19.

2. MORPHOLOGICAL ADAPTATIONS DURING GROWTH AND DEVELOPMENT

It is now generally accepted that a reduction in cell growth is one of the earliest discernible effects of water deficits and that other less sensitive plant processes are affected in sequence as more severe deficits develop (8, 9). There are, however, conflicting reports about the relative sensitivity of the two immediate components of growth, cell division and cell enlargement. Data for palisade cells in tobacco (10) and for epidermal cells in soybean (11) indicate that cell enlargement is more sensitive to water deficits than is cell division. Evidence of short-term compensatory growth during recovery from water deficits in maize (12), *Nitella* (13), rye (14), and soybean (15) also suggests that cell enlargement may be more sensitive to water deficits than is cell division. However under certain conditions cell division and cell elongation may be equally sensitive to water deficits (17). For more detailed discussion of this subject, see the review by Begg and Turner (16).

The level of stress that results in a reduction in growth varies with the conditions under which the plants are grown. For maize in growth chambers, leaf extension declined rapidly at leaf water potentials below -2 bars and ceased at potentials of -7 to -9 bars (12, 18, 19). However field data have shown that a decline in leaf water potential to -8 or -9 bars had little effect on the rate of leaf extension in maize (20), and a decline in leaf water potential to -13 bars did not affect leaf extension in sorghum or soybean (17, 21, 22). The data of McCree and Davis (17) and Watts (20) also indicate that leaf expansion continues day and night at the same rate, despite a diurnal change in leaf water potential from -1 to about -8 bars. This apparent insensitivity of field-grown plants arises partly from attempts to relate extension to leaf water potential rather than to the turgor potential of the leaf, thus not accounting for active and/or passive changes in osmotic potential. Other factors include the likelihood of large gradients in leaf water potential between the point of measurement and the site of cell enlargement at the base of actively transpiring leaves in the field (23). Also Hsiao et al. (9), Green et al. (13), Green and Cummins (14), and Green (24) have indicated that gross extensibility and threshold turgor are not constant and can change with time, thereby permitting a resumption of growth at reduced turgor and also checking the transitory rapid growth response following rewatering. Thus the growth rate over the longer term is stabilized during small diurnal changes in turgor potential, provided they are above the minimum threshold value.

One of the most important consequences of the sensitivity of cell growth to small water deficits is a marked reduction in leaf area. Since leaf growth is generally more sensitive to water stress than stomatal conductance and carbon dioxide assimilation, crop growth can be reduced by water deficits too small to cause a reduction in stomatal aperture and photosynthesis (12, 18, 25, 26). A reduction in leaf area will reduce crop growth, particularly during establishment when there is incomplete light interception. An important consequence of this reduction in leaf area is the associated reduction in water loss, particularly at leaf area indices less than 3 (27–29), thus providing a mechanism for reducing the rate of water use and delaying the onset of more severe stress.

In addition to reducing the rate of water use, there is evidence for increased access to soil water through an increase in the root/shoot ratio (30–33). In some cases stress appears to enhance root growth not only relative to shoot growth, but also absolutely (26, 34). Hsiao and Acevedo (26) presented evidence for this effect in maize, and it might benefit root crops such as sugar beet. The effect is likely to occur at levels of water deficit sufficient to significantly reduce shoot growth but not carbon dioxide assimilation. The increase in available assimilates, resulting from the reduction in shoot growth, permits osmotic adjustment and additional root growth.

Water stress also reduces leaf area by accelerating the rate of senescence of the physiologically older leaves (25, 35–38) and through its effect on leaf shedding; see Turner (39) for discussion. In determinate crops such as wheat, where the leaf area is fixed at flowering, yield under dryland conditions has been shown to be inversely related to the rate of leaf senescence after flowering, which in turn was related to plant water stress (35). However Hall et al. (40) have argued that leaf senescence in cereals could confer an adaptive advantage if it is accompanied by a substantial reduction in transpiration, since the older lower leaves, which senesce first, are supplying relatively little carbohydrate to the developing grain. Furthermore, Evans et al. (41) proposed that awns (i.e., modified leaves) make a major contribution of photosynthate to developing grains, and that this source of carbohydrate is more stable under drought than is the supply from flag leaves. And in the context of perennials, as shown by Mooney and Dunn (42), enhanced leaf senescence is an important adaptive mechanism in drought tolerant deciduous shrubs.

A disadvantage of the morphological adaptations that occur during the growth and development of the crop is that they are largely irreversible, particularly in determinate crops that have no scope for compensation through an increase in the number of leaves. Thus the growth or yield potential that has been lost cannot be fully recovered if there is a return to more favorable conditions.

3. MORPHOLOGICAL ADAPTATIONS FOLLOWING FULL LEAF AREA DEVELOPMENT

Once leaf area development is complete, one of the main mechanisms for adapting to stress is through changes in leaf angle. This can be an effective mechanism for reducing the radiation load on water-stressed leaves when less water is available to dissipate energy as latent heat, and a change in leaf orientation may prevent overheating or firing. Important features of the mechanisms responsible for changing leaf angle or orientation are reversibility, rapidity of recovery on the relief of stress, and the possibility that yield will not be seriously reduced.

The passive wilting response in sunflower and cotton leaves is an effective mechanism for shedding radiation and reducing the rate of development of severe water deficits. This is illustrated in Figure 3.1, which shows the incident radiation on wilted sunflower leaves oriented N, S, E, and W, compared with the radiation incident on an unstressed diaphotonastic leaf and a horizontal surface, on a cloudless day in the southern hemisphere. In contrast to these passive leaf movements, active leaf movements orienting the blade or lamina parallel to the incident radiation (parahelionasty) when under stress have been reported in several leguminous species (43–45). Townsville stylo (*Stylosanthes humilis* H.B.K.) is of particular interest

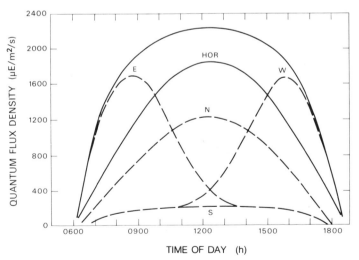

Figure 3.1. Diurnal changes in the quantum flux density (400–700 nm) incident upon wilted, vertically hanging sunflower leaves (dashed curves) oriented north (N), south (S), east (E), or west (W), and on an unstressed sunflower leaf (solid curve) exhibiting diaphotonastic leaf movements that orient it normal to direct radiation. The quantum flux density on a horizontal plane (HOR) is shown for comparison. Data for a cloudless day (March 19) at Canberra, Australia (lat. 35°S). From Ref. 16.

because its leaves exhibit both diaphotonastic and parahelionastic movements depending on the degree of water stress (44). In the absence of stress, the leaflets are oriented at right angles to the direction of the incident radiation (diaphotonasty) and follow the course of the sun in this position, giving a cos α value, that is, the angle that the direct solar beam makes with the normal to the leaf, of ~1.0 (curve A, Figure 3.2). Under severe water stress, the leaflets are oriented parallel to the incident radiation (parahelionasty) and follow the course of the sun in this position, giving a cos α value of ~0.2 (curve B). The mechanism involves light-dependent turgor changes in the pulvini at the base of each leaflet: the turgor changes are mediated by phytochrome and involve the movement of potassium through cell membranes. Thus the mechanism is basically similar to stomatal opening and closing in response to turgor changes in the guard cells. Townsville stylo is therefore adapted to growth under favorable conditions, particularly when growing in competition with taller species, and for shedding radiation and survival under stress conditions. A similar response has been reported for the desert lupin *Lupinus arizonicus* (45).

An interesting contrast to the active leaf movements displayed by Townsville stylo is the more static leaf orientation of many perennials, particularly the desert perennials that are exposed to stress almost permanently. A common example is the vertical display of the leaves of many desert species. A more advanced version of this is the opposite and decussate arrangement of the vertical, sessile leaves of the Sonoran Desert shrub jojoba [*Simmondsia chinensis* (Link) Schneid]. This leaf arrangement and orientation achieve a considerable reduction in the interception of radiation,

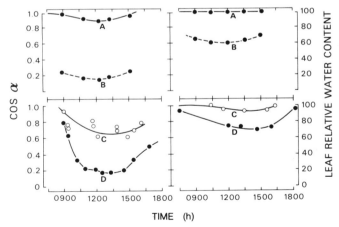

Figure 3.2. *Left*: diurnal variations in cos α (the angle that the direct solar beam makes with the normal to the leaf); *right*: leaf relative water content. Curves A and C are for plants in water culture, A in the glasshouse and C in drier air outside. Curve B is for a potted plant in dry soil, and curve D is for a plant in the field at the beginning of the dry season. The minimum leaf water potentials for curves A, B, and D were -2, -28, -8, and -26 bars, respectively. From Ref. 44.

particularly during the midday period of peak intensity. In terms of the interception of direct radiation, the effective leaf area is approximately 10 to 15% at solar noon, and reaches a maximum of only 25 to 30% of the total leaf area at sunset and sunrise. Thus maximum exposure of the leaf area to direct radiation occurs when the evaporative demand is low and water use efficiency high (46).

In grasses, leaf rolling is a common response to stress and results in a marked reduction in effective leaf area and a more vertical leaf orientation. This is frequently cited in the literature as an adaptive mechanism, which in addition to reducing the effective leaf area, and the energy load on the plant, may markedly reduce transpiration by, for example, 50 to 70% (47), as well as increasing water use efficiency (48). The rolling response results from a loss of water from the bulliform cells in the upper epidermis of the leaf (49, 50) and can be very sensitive to diurnal changes in leaf water deficits. It thus enables the plant to respond rapidly to periods of high evaporative demand, to avoid some of the radiation load, then to recover during periods of low evaporative demand, when the water use efficiency is higher.

Previous work on leaf rolling has been based on subjective assessment using visual scoring systems of the leaf symptoms (e.g., Ref. 51). Since the visual ratings of different workers cannot be compared or related adequately to other quantitative measurements of plant water deficits, a quantitative index of leaf rolling has been devised. Leaf rolling was characterized by the ratio of the projected width of the leaf to its maximum width and termed the rolling index (RI). Thus a leaf that has rolled to the extent that the margins are touching has reduced its effective or projected leaf area by approximately 68% and has an RI of 0.32. A diurnal study of RI in sorghum (*Sorghum bicolor* cv. 100 M) clearly showed that it is very sensitive to both onset and recovery from stress (Figure 3.3). The relationship between RI and leaf water potential (Figure 3.4) shows a marked reduction in the index

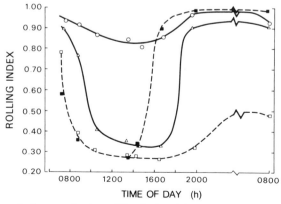

Figure 3.3. Diurnal changes in the rolling index (the ratio of the projected leaf width to maximum leaf width) for sorghum that was well watered (○), dry (△), or very dry (□, ■). Half the very dry plot was irrigated at 1315 h (■).

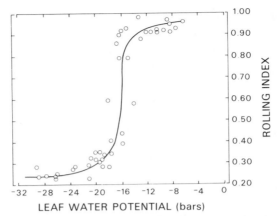

Figure 3.4. Relationship between the rolling index and leaf water potential for sorghum leaves with osmotic potentials ranging from −12 to −13 bars at dawn, to −20 to −22 bars during the afternoon.

at leaf water potentials of less than −14 bars. The osmotic potential of this tissue ranged from −12 to −13 bars at dawn, to −20 to −22 bars during the afternoon, indicating that RI declined rapidly as the turgor potential approached zero. Supporting data on stomatal conductance indicated that leaf rolling preceded the closure of abaxial stomata. The sensitivity and ease of measurement of RI makes it a useful selection criteria, if it can be related to the ability of a cultivar to achieve under stress a higher yield than a less responsive selection.

Thus, more attention should be paid to the various short- and long-term morphological responses such as leaf area development, duration, and orientation, since they may be among the most effective means a mesophytic plant has for adapting to water stress in the field.

REFERENCES

1. T. Dobzhansky, *Am. Biol. Teach.* **March,** 125 (1973).
2. J. L. Harper, in J. R. Wilson, Ed., *Plant Relations in Pastures*, CSIRO, Melbourne, Australia, 1978, p. 3.
3. E. Broda, *The Evolution of the Bioenergetic Processes*, 1st ed., Pergamon Press, New York, 1975.
4. J. H. Troughton, in J. P. Cooper, Ed., *Photosynthesis and Productivity in Different Environments*, I.B.P. 3, Cambridge University Press, London, 1975, p. 357.
5. O. L. Lange, L. Kappen, and E.-D. Schulze, Eds., *Water and Plant Life: Problems and Modern Approaches,* Springer-Verlag, Berlin–Heidelberg–New York, 1976.
6. W. T. Keeton, *Biological Sciences*, Norton, New York, 1972.

7. J. B. Passioura, in I. F. Wardlaw and J. B. Passioura, Eds., *Transport and Transfer Processes in Plants*, Academic Press, New York–San Francisco–London, 1976, p. 373.
8. T. C. Hsiao, *Annu. Rev. Plant Physiol.*, **24**, 519 (1973).
9. T. C. Hsiao, E. Acevedo, E. Fereres, and D. W. Henderson, *Philos. Trans. R. Soc. London, Ser. B.*, **273**, 479 (1976).
10. B. F. Clough and F. L. Milthorpe, *Aust. J. Plant Physiol.*, **2**, 291 (1975).
11. J. A. Bunce, *J. Exp. Bot.*, **28**, 156 (1977).
12. E. Acevedo, T. C. Hsiao, and D. W. Henderson, *Plant Physiol.*, **48**, 631 (1971).
13. P. B. Green, R. O. Erickson, and J. Buggy, *Plant Physiol.*, **47**, 423 (1971).
14. P. B. Green and W. R. Cummins, *Plant Physiol.*, **54**, 863 (1974).
15. W. Wenkert, E. R. Lemon, and T. R. Sinclair, *Agron. J.*, **70**, 761 (1978).
16. J. E. Begg and N. C. Turner, *Adv. Agron.*, **28**, 161 (1976).
17. K. J. McCree and S. D. Davis, *Crop Sci.*, **14**, 751 (1974).
18. J. S. Boyer, *Plant Physiol.*, **46**, 233 (1970).
19. T. C. Hsiao, E. Acevedo, and D. W. Henderson, *Science*, **168**, 590 (1970).
20. W. R. Watts, *J. Exp. Bot.*, **25**, 1085 (1974).
21. A. C. P. Chu and J. P. Kerr, *N.Z. J. Agric. Res.*, **20**, 467 (1977).
22. N. C. Turner and G. J. Burch, in I. D. Teare, Ed., *Crop-water Relations*, Interscience, New York–London, 1980 (in press).
23. S. J. Yang and E. de Jong, *Can. J. Plant Sci.*, **51**, 333 (1971).
24. P. B. Green, *Plant Physiol.*, **43**, 1169 (1968).
25. R. A. Fischer and R. M. Hagan, *Exp. Agric.*, **1**, 161 (1965).
26. T. C. Hsiao and E. Acevedo, *Agric. Meteorol.*, **14**, 59 (1974).
27. R. A. Fischer and G. D. Kohn, *Aust. J. Agric. Res.*, **17**, 255 (1966).
28. J. T. Ritchie and E. Burnett, *Agron. J.*, **63**, 56 (1971).
29. J. T. Ritchie, *Agric. Meteorol.*, **14**, 183 (1974).
30. R. W. Pearson, in W. H. Pierre, D. Kirkham, J. Pesek, and R. Shaw, Eds., *Plant Environment and Efficient Water Use*, Am. Soc. Agron. and Soil Sci. Soc. Am., Madison, Wisconsin, 1966, p. 95.
31. R. L. Davidson, *Ann. Bot. (London)*, **33**, 571 (1969).
32. A. H. El Nadi, R. Brouwer, and J. T. Locher, *Neth. J. Agric. Sci.*, **17**, 133 (1969).
33. G. J. Hoffman, S. L. Rawlins, M. J. Garber, and E. M. Cullen, *Agron. J.*, **63**, 822 (1971).
34. O. L. Bennett and B. D. Doss, *Agron. J.*, **52**, 204 (1960).
35. R. A. Fischer and G. D. Kohn, *Aust. J. Agric. Res.*, **17**, 281 (1966).
36. R. A. Fischer, in R. O. Slatyer, Ed., *Plant Response to Climatic Factors. Proceedings of the Uppsala Symposium, 1970*, UNESCO, Paris, 1973, p. 233.
37. R. O. Slatyer, in R. O. Slatyer, Ed., *Plant Response to Climatic Factors. Proceedings of the Uppsala Symposium, 1970*, UNESCO, Paris, 1973, p. 177.
38. M. M. Ludlow, in R. Marcelle, Ed., *Environmental and Biological Control of Photosynthesis*, Junk, The Hague, 1975, p. 123.
39. N. C. Turner, in H. Mussell and R. C. Staples, Eds., *Stress Physiology in Crop Plants*, Interscience, New York–London, 1979, p. 343.
40. A. E. Hall, K. W. Foster, and J. G. Waines, in A. E. Hall, G. H. Cannell, and H. W. Lawton, Eds., *Agriculture in Semi-Arid Environments*, Springer-Verlag, Berlin–Heidelberg–New York, 1979, p. 148.

41. L. T. Evans, J. Bingham, P. Jackson, and J. Sutherland, *Ann. Appl. Biol.*, **70,** 67 (1972).
42. H. A. Mooney and E. L. Dunn, *Am. Nat.*, **104,** 447 (1970).
43. S. Dubetz, *Can. J. Bot.*, **47,** 1640 (1969).
44. J. E. Begg and B. W. R. Torssell, in R. L. Bieleski, A. R. Ferguson, and M. M. Creswell, Eds., *Mechanisms of Regulation of Plant Growth*, Bulletin 12, Royal Society of New Zealand, Wellington, 1974, p. 277.
45. C. M. Wainwright, *Am. J. Bot.*, **64,** 1032 (1977).
46. H. M. Rawson, N. C. Turner, and J. E. Begg, *Aust. J. Plant Physiol.*, **5,** 195 (1978).
47. H. R. Oppenheimer, in *Plant-Water Relationships in Arid and Semi-Arid Conditions: Reviews of Research*, UNESCO, Paris, 1960, p. 105.
48. G. G. Johns, *Aust. J. Plant Physiol.*, **5,** 113 (1978).
49. A. Fahn, *Plant Anatomy,* 2nd ed., Pergamon Press, Oxford, 1974, pp. 172–247.
50. K. Esau, *Anatomy of Seed Plants*, 2nd ed., Wiley, New York, 1977, pp. 360–364.
51. J. C. O'Toole and T. B. Moya, *Crop Sci.*, **18,** 873 (1978).

4 | Leaf Anatomy and Water Use Efficiency

PARK S. NOBEL

Department of Biology and Division of Environmental Biology of the Laboratory of Nuclear Medicine and Radiation Biology, University of California, Los Angeles, California

1. INTRODUCTION

Environmental parameters can affect leaf anatomy by causing changes in the number of layers of mesophyll cells and/or in the cellular dimensions. This results in different amounts of internal leaf area being available for the absorption of carbon dioxide (CO_2) per unit leaf surface area. Through effects on net CO_2 exchange, changes in leaf anatomy thus affect both photosynthesis and water use efficiency (the ratio of photosynthesis to transpiration).

The environmental factor having the most influence on leaf anatomy is the illumination level during leaf development (1, 2). However important effects of growth temperature and water relations on leaf anatomy occur, as discussed below. Photosynthesis, and therefore water use efficiency, can also be influenced by changes in the properties of mesophyll cells themselves. To separate effects of mesophyll cell numbers and geometry from changes within the cells, the mesophyll surface area per unit leaf area (A^{mes}/A)* is

*Abbreviations: A^{mes}/A—mesophyll surface area per unit leaf area. $g^{cell}_{CO_2}$—CO_2 cellular con-

43

used to handle anatomy (3–8). If stomatal responses and factors within mesophyll cells remain unchanged, increases in A^{mes}/A (which lead to a greater region for inward CO_2 diffusion) should result in higher photosynthetic rates and higher water use efficiencies. This chapter develops a theory for integrating environmental effects on the anatomical parameter A^{mes}/A with the effects on CO_2 diffusional conductances. Differences in illumination, temperature, and water relations during leaf development are then discussed in terms of their effect on A^{mes}/A, conductances, and water use efficiency. Thus emphasis is on the consequences for photosynthesis and transpiration of environmentally induced anatomical variation occurring during leaf development.

2. THEORY—WATER USE EFFICIENCY

The transpiration rate (J_{wv}) of a leaf can be related to the leaf water vapor conductance (g_{wv}) and a drop in water vapor concentration:

$$J_{wv} = g_{wv} \cdot \Delta c_{wv} = g_{wv} \cdot (c_{wv}^* - c_{wv}^{ext}) \tag{4.1}$$

where c_{wv}^* is the saturation water vapor concentration at the leaf temperature and c_{wv}^{ext} is the water vapor concentration in the ambient air. Similarly, the net rate of CO_2 uptake (J_{CO_2}) equals a total CO_2 conductance (g_{CO_2}) times the drop in CO_2 concentration from the ambient air ($c_{CO_2}^{ext}$) to some intrachloroplast location ($c_{CO_2}^{int}$):

$$J_{CO_2} = g_{CO_2} \cdot \Delta c_{CO_2} = g_{CO_2} (c_{CO_2}^{ext} - c_{CO_2}^{int}) = g_{CO_2}^{met} (c_{CO_2}^{ias} - c_{CO_2}^{int}) \tag{4.2}$$

where $g_{CO_2}^{mes}$ is the mesophyll conductance for CO_2 across which the CO_2 concentration drops from the value in the intercellular air spaces of the leaf ($c_{CO_2}^{ias}$) to the value within the chloroplasts. In turn, $c_{CO_2}^{ias}$ equals the ambient CO_2 concentrations minus the drop across the stomates ($\Delta c_{CO_2}^{st}$):

$$c_{CO_2}^{ias} = c_{CO_2}^{ext} - \Delta c_{CO_2}^{st} = c_{CO_2}^{ext} - \frac{J_{CO_2}}{g_{CO_2}^{st}}$$
$$= c_{CO_2}^{ext} - 1.56 \frac{J_{CO_2}}{g_{wv}} \tag{4.3}$$

where 1.56 is the ratio of the diffusion coefficient of CO_2 to that of water vapor.

The water use efficiency (WUE) of leaves can be equated to the rate of net CO_2 uptake divided by the transpiration rate (J_{CO_2}/J_{wv}). Using Equation

---

ductance expressed on a mesophyll surface area basis. $g_{CO_2}^{mes}$—CO_2 mesophyll conductance. g_{wv}—water vapor conductance. J_{CO_2}—net rate of CO_2 uptake (photosynthetic rate). J_{wv}—net rate of water vapor loss (transpiration rate).   PAR—photosynthetically active radiation. WUE—water use efficiency (J_{CO_2}/J_{wv}).

4.1 for J_{wv} and substituting the form of $c_{CO_2}^{ias}$ involving g_{wv} (Equation 4.3) into the last equality in Equation 4.2, suitable manipulations lead to the following convenient form:

$$\text{WUE} = \frac{J_{CO_2}}{J_{wv}} = \frac{g_{CO_2}^{mes}\,(c_{CO_2}^{ext} - c_{CO_2}^{int})}{(g_{wv} + 1.56\,g_{CO_2}^{mes})\,(c_{wv}^{*} - c_{wv}^{ext})} \tag{4.4}$$

The contributions of cellular properties to $g_{CO_2}^{mes}$ can be separated from anatomical ones by using a CO_2 cellular conductance ($g_{CO_2}^{cell}$, expressed on the basis of mesophyll cell wall area) and the ratio of mesophyll cell wall area to leaf area (A^{mes}/A):

$$g_{CO_2}^{mes} = g_{CO_2}^{cell}\,\frac{A^{mes}}{A} \tag{4.5}$$

The cellular conductance incorporates all intracellular factors, including diffusion across the cell wall and membranes, photochemistry, and biochemistry. Substituting this expression into Equation 4.4 leads to the following relation:

$$\text{WUE} = \frac{g_{CO_2}^{cell}\,(A^{mes}/A)\,(c_{CO_2}^{ext} - c_{CO_2}^{int})}{[g_{wv} + 1.56\,g_{CO_2}^{cell}\,(A^{mes}/A)](c_{wv}^{*} - c_{wv}^{ext})} \tag{4.6}$$

Equation 4.6 specifically recognizes effects on WUE of the atmosphere ($c_{CO_2}^{ext}$, c_{wv}^{ext}), leaf temperature (c_{wv}^{*}, $g_{CO_2}^{cell}$), stomatal opening (g_{wv}), cellular properties ($g_{CO_2}^{cell}$), and leaf anatomy (A^{mes}/A). The quantity $c_{CO_2}^{ext} - c_{CO_2}^{int}$ represents the overall drop in CO_2 concentration; if $c_{CO_2}^{int}$ is assumed to be zero, as it generally is in analysis, this drop becomes simply $c_{CO_2}^{ext}$. For a leaf at 28°C in ambient air at 25°C and 1 bar total pressure with 340 μl/l CO_2 (0.603 g/m³) and 30% relative humidity (6.92 g/m³), $c_{CO_2}^{ext}/(c_{wv}^{*} - c_{wv}^{ext})$ is 29.4 g CO_2/kg H_2O.

3. MEASUREMENT OF LEAF ANATOMY

A^{mes}/A depends on the geometry of the mesophyll cells. For a single layer of spheres in an orthogonal array (i.e., a cubic lattice), A^{mes}/A would be $4\,\pi r^2/(2r)^2$ or π. Hence four layers of spheres would represent an A^{mes}/A of 12.6 (Figure 4.1). If the radius of the spheres were cut in half but the thickness of the array remained the same, A^{mes}/A would double. In a more realistic representation of a leaf with one layer of palisade cells, (overall length 4 times the diameter) and two layers of spherical spongy cells, A^{mes}/A would be 18.8, two-thirds of which comes from the palisade cells (Figure 4.1).

For the results presented here, mesophyll cells were examined in paradermal and transverse sections at a magnification of about 400×. They were modeled as spheres or as cylinders with hemispherical ends (8, 9). Ignoring local surface irregularities leads to an underestimate of A^{mes}/A; ignoring the fraction of the surface area of mesophyll cells touching each other leads to a potentially compensating overestimate. In experiments on *Lolium perenne*

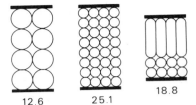

12.6   25.1   18.8

Figure 4.1. Geometric representations of A^{mes}/A. Spheres or cylinders with hemispherical ends in an orthogonal array are illustrated together with the corresponding A^{mes}/A.

where selection was made for smaller mesophyll cell size, photosynthesis was negatively correlated with cell size in the F_1 generation (10), indicating that increases in internal leaf area can increase net CO_2 uptake.

As indicated above, A^{mes}/A can be readily incorporated into gas flux equations. On the other hand, parameters such as leaf thickness and specific leaf weight (dry weight per unit leaf area) are easier to measure and have often been used to express environmental effects on leaf anatomy. Indeed, increases in maximum rates of photosynthesis have been correlated with increased leaf thickness (11, 12), increased mesophyll thickness (13), increased specific leaf weight (12, 14–16), and increased fresh weight per unit area (17) for plants having either the C_3 or C_4 pathway for photosynthesis. However these parameters are not always appropriate as a measure of photosynthetic capacity because of the inclusion of one or more of the following: pubescence, cuticle, epidermal cells, leaf ridges, protruding veins, and vascular tissue in general. Also, fluctuating water content can affect fresh weight per unit area, and specific leaf weight changes diurnally because of fluctuations in starch content; for example, in soybean (*Glycine max*) it can increase 33% during the day (18). Thus A^{mes}/A can be a more useful parameter for handling anatomical effects on photosynthesis, since it considers only the mesophyll region and incorporates both cell sizes and shapes.

Table 4.1 indicates that A^{mes}/A can vary considerably for the same leaf thickness, largely because of species differences in the size of mesophyll cells. The specific leaf weight increased from *Beta vulgaris* to *Plectranthus parviflorus* to *Hyptis emoryi*, but A^{mes}/A was lower in *P. parviflorus* than in *B. vulgaris* (Table 4.1). Within a single species (*Encelia farinosa*), A^{mes}/A increased 50% at the same leaf thickness for droughted plants with small cell dimensions compared to plants under well-watered conditions, smaller mesophyll cell size being a common observation under water stress conditions (19). The extremes of A^{mes}/A so far reported for leaves is represented by the moss *Mnium ciliare* with a leaf one cell thick (20) and a succulent, *Agave deserti* (21), whose leaves can be more than 2 cm thick (Table 4.1). A^{mes}/A was only 2.0 for the moss, since the lateral walls were nearly completely touching, thus only the end walls (abaxial and adaxial sides of the leaf) were readily available for CO_2 entry.

Table 4.1. Relation Between A^{mes}/A, Leaf Thickness, and Specific Leaf Weight in Several Species

Leaf thickness and A^{mes}/A were determined as described in Nobel et al. (8). Dry weight was determined by heating leaves at 90°C for 2 days (5 days for *A. deserti*), after which no further weight decrease occurred.

Species	Thickness (μm)	Specific leaf dry weight (g/m²)	A^{mes}/A
Plectranthus parviflorus	255	23	9
Beta vulgaris	256	17	17
Hyptis emoryi	247	41	24
Encelia farinosa	281	41	26
Encelia farinosa (droughted)	280	49	39
Mnium ciliare	27	3	2
Agave deserti	25,200	2900	176

4. ENVIRONMENTAL INFLUENCES ON LEAF ANATOMY

When the illumination level during development was raised, A^{mes}/A increased fairly linearly with mesophyll thickness (Figure 4.2). The thickest leaves had slightly tighter packing of mesophyll cells, which led to a greater A^{mes}/A for a given leaf thickness. The overall range of changes in A^{mes}/A was greater than eight fold for *P. parviflorus*. As the illumination level (for 12 h days) was increased from 20 to 900 μE/m²/s photosynthetically active radiation (PAR) from 400 to 700 nm, A^{mes}/A increased from 11.3 to 49.9 for *P. parviflorus* (8). The A^{mes}/A increase for the upper two palisade levels represented an increase in length with little change in diameter. For the spongy

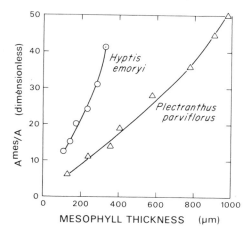

Figure 4.2. Relationship between mesophyll thickness and mesophyll surface area per unit area (A^{mes}/A) in two species.

mesophyll, which contributed about 35% of the total A^{mes}/A, the average diameter increased 26% and the number of cells per unit leaf area increased 2.4 fold as the PAR increased from 20 to 900 $\mu E/m^2/s$.

Field studies on *H. emoryi* indicated that A^{mes}/A did not respond to the maximum PAR, but rather to the total integrated PAR during the day (9). For example, it was 40 for an unshaded leaf and 28 for an intermittently shaded leaf that received about the same maximum PAR. A leaf at the bottom of the bush received appreciable PAR only at the beginning and end of the day and had an A^{mes}/A of 18. Figure 4.3 shows that A^{mes}/A continuously increased as the total daily PAR increased for three different species. Data for *H. emoryi* were obtained by integrating under PAR curves measured in the field (9), whereas *P. parviflorus* (8, 20) and *Fragaria vesca* (22) were studied in the laboratory using square-wave illumination patterns of constant photoperiod.

Although PAR exerted the major influence on leaf anatomy, leading to greater than three fold increases in A^{mes}/A for *H. emoryi* and *P. parviflorus*, other environmental parameters can also influence A^{mes}/A. As the air temperature during leaf development was increased, A^{mes}/A increased about 40% for *P. parviflorus* and *F. vesca* (Figure 4.4*a*). Higher temperatures led to smaller cell sizes, which in turn increased A^{mes}/A; for example, for *F. vesca*, A^{mes}/A increased 27% as the daytime temperature was raised from 10 to 30°C, although leaf thickness decreased 22% (22). Other studies have also

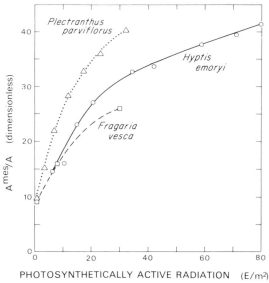

Figure 4.3. Effect of total integrated photosynthetically active radiation on A^{mes}/A in three species. Results for *Fragaria vesca* were obtained by Chabot and Chabot (22); the other curves represent previously unpublished data of Nobel.

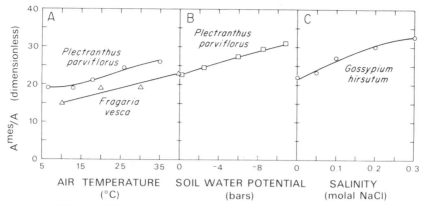

Figure 4.4. Effects of air temperature, soil water potential, and salinity on A^{mes}/A in three species. Data for *Fragaria vesca* were obtained by Chabot and Chabot (22).

indicated that leaf thickness could decrease as growth temperature was increased; thus leaves of *Populus euramericana* decreased 28% in thickness as the growth temperature was raised from 16 to 25°C (23). The specific leaf weight for *Festuca arundinacea* decreased 13% from 10 to 25°C (24); both thickness and specific leaf weight decreased as growth temperatures were raised for *Glycine max* and *Gossypium hirsutum* (cotton) (25). Since A^{mes}/A was not determined in the latter studies, direct comparison with the results in Figure 4.4 is not possible.

When the soil water potential was decreased for *P. parviflorus*, A^{mes}/A increased about 40% (Figure 4.4*b*) because of more layers of smaller mesophyll cells. A 50% increase in A^{mes}/A similarly occurred for *E. farinosa* under water stress conditions (Table 4.1). The mesophyll surface area per unit leaf area was rather unaffected as the soil water potential was reduced from −0.2 to −12 bars for *G. hirsutum* (Table 4.2), although the leaf was 24% thinner for the drier condition. The thinnest leaves and smallest cells also occurred under water stress for *Pisum sativum*, although the thickest leaves occurred at an intermediate water status (26). Reduced cell size commonly accompanies water stress, but changes in A^{mes}/A apparently depend on the species.

When salinity was increased for *G. hirsutum* (Figure 4.4*c*), A^{mes}/A increased 50% because palisade cells were longer. By comparison, increasing salinity for *Atriplex patula* ssp. *hastata* led to cells larger in all dimensions and so there was little change in A^{mes}/A, although the leaves were again thicker at the higher salinities (27). An increase in mesophyll cell size can also be induced by hormone application, for example, gibberellin A_1 led to a 40% increase in cell dimensions for *Poa alpina* (28). Thus leaf anatomical features can vary considerably with environmental conditions.

Table 4.2. Influence of Daily Integrated Photosynthetically Active Radiation (PAR) and Soil Water Potential During Development on Cellular Properties of *Gossypium hirsutum*

Water use efficiency (WUE) was calculated using Equation 4.6 assuming $(c_{CO_2}^{ext} - c_{CO_2}^{int})/(c_{wv}^* - c_{wv}^{ext}) = 29.4$ g CO_2/kg H_2O. Plants were grown and measurements were made as in Longstreth and Nobel (27).

Growth conditions					
Daily PAR (E/m^2)	Soil water potential (bars)	A^{mes}/A	$g_{CO_2}^{cell}$ (cm/s)	g_{wv} (cm/s)	WUE (g CO_2/kg H_2O)
4.8	−0.2	16.7	0.012	0.68	5.9
9.2	−0.2	19.2	0.013	0.98	5.4
18	−0.2	22.3	0.014	0.93	6.5
46	−0.2	28.6	0.016	1.11	7.4
81	−0.2	30.4	0.016	1.12	7.2
17	−0.2	20.8	0.014	1.25	5.0
17	−7	22.2	0.013	0.50	8.9
17	−12	21.8	0.011	0.16	13.2

5. ENVIRONMENTAL INFLUENCES ON WATER VAPOR CONDUCTANCE AND CELL CONDUCTANCE

Equation 4.2 is more appropriate when respiration and photorespiration can be ignored; therefore the gas phase for CO_2 uptake studies described here contained only 1% oxygen (the balance was 340 μl/l CO_2, water vapor equivalent to 40 ± 5% relative humidity, and nitrogen). Measurements were made where J_{CO_2} was linear with $c_{CO_2}^{ias}$ (20). The data are for light saturation and leaf temperatures optimal for net CO_2 uptake.

Figure 4.5 shows that as the PAR during leaf development for *P. parviflorus* was increased, J_{CO_2} and $g_{CO_2}^{mes}$ determined for mature leaves both increased, while g_{wv} and $g_{CO_2}^{cell}$ remained essentially constant. Since $g_{CO_2}^{cell}$ was constant, Equation 4.5 indicates that the four fold changes in $g_{CO_2}^{mes}$ can be attributed to the four fold changes in A^{mes}/A (Figure 4.3).

Contrary to the results for *P. parviflorus* (Figure 4.5), differences in PAR during growth can affect g_{wv} for some species; for example, for *A. patula* a ten fold increase in daily PAR to 53 E/m^2 led to a three fold increase in g_{wv} (29), in qualitative agreement with certain other observations (1, 2). On the other hand, g_{wv} was fairly constant for *Abutilon theophrasti* and *G. hirsutum* from 5 to 41 E/m^2 (30). For *G. hirsutum*, g_{wv} increased from 0.7 to 1.2 cm/s as the PAR during leaf development was raised from 5 to 81 E/m^2 (Table 4.2). The level of PAR during development also influenced $g_{CO_2}^{cell}$ in this case (Table 4.2), in contrast to the results for *P. parviflorus* (Figure 4.5). As for other plants, PAR during development had a major effect on A^{mes}/A for cotton.

Differences in air temperature, soil water potential, or salinity during

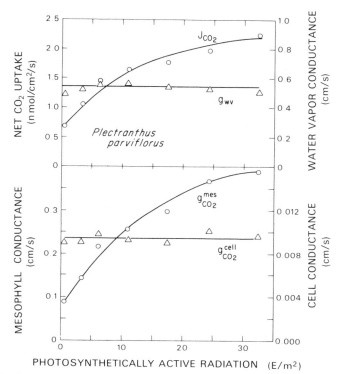

Figure 4.5. Gas exchange for *Plectranthus parviflorus* developing under the indicated total daily photosynthetically active radiation. Data were obtained at light saturation (1600 $\mu E/m^2/s$) and optimal temperature ($29 \pm 1°C$) for plants grown as in Nobel (20), except with a daytime air temperature of $26°C$, a nighttime temperature of $20°C$, and a soil water potential of -0.1 bar.

growth led to different values for $g_{CO_2}^{cell}$ (Figure 4.6). The maximum $g_{CO_2}^{cell}$ for *P. parviflorus* occurred for growth air temperatures near $26°C$ (Figure 4.6*a*), similar to results on *Lolium perenne* and *L. multiflorum* (31). As the soil water potential was lowered from -0.1 to -10.8 bars (Figure 4.6*b*), $g_{CO_2}^{cell}$ decreased nearly three fold (plants did not survive at -12 bars). Figure 4.6*c* indicates that cellular properties were also adversely affected by increasing salinity for *G. hirsutum* (growth ceased at 0.4 molal NaCl added to the usual nutrient solution). Preliminary experiments have indicated that nutritional factors can also influence $g_{CO_2}^{cell}$; for example, for *P. parviflorus* it can be halved by replacing a nutrient solution with distilled water, which leads to little effect on A^{mes}/A (20).

The water vapor conductance determined at light saturation and optimal temperatures for net CO_2 uptake also depended on the environmental conditions during leaf development (Figure 4.7). In particular, g_{wv} of *P. parviflorus* was maximum near a growth air temperature of $26°C$ (Figure

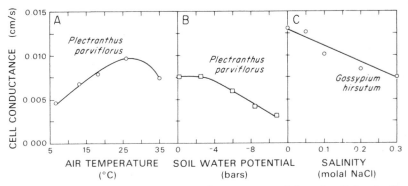

Figure 4.6. Effects of air temperature, soil water potential, and salinity during leaf development on cell conductance in two species. Data for *Plectranthus parviflorus* were obtained as in Figure 4.5, except that the growth temperature was 20°C for the soil water potential studies, and data for *Gossypium hirsutum* are from Longstreth and Nobel (27).

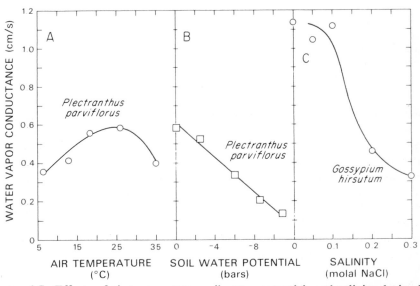

Figure 4.7. Effects of air temperature, soil water potential, and salinity during leaf development on water vapor conductance in two species. Data as for Figure 4.6.

4.7a), similar to the location of maximum $g_{CO_2}^{cell}$. The water vapor conductance decreased about four fold as the soil water potential during leaf development was reduced from -0.1 to -10.8 bars for *P. parviflorus* (Figure 4.7b) and as the salinity was raised from 0.0 to 0.3 molal NaCl for *G. hirsutum* (Figure 4.7c). For *G. hirsutum* lowering the soil water potential from -0.2 to -12 bars led to an 87% decrease in stomatal conductance (Table 4.2). Similar effects of reduced soil water potential and increased salinity on stomatal opening have also been reported previously (32–34).

6. ENVIRONMENTAL INFLUENCES ON WATER USE EFFICIENCY

Figure 4.8 presents the effects on water use efficiency of the four environmental factors considered. For the sake of uniformity $(c_{CO_2}^{ext} = c_{CO_2}^{int})/(c_{wv}^* - c_{wv}^{ext})$ in Equation 4.6 has been set equal to 29.4 g CO_2/kg H_2O. Values of A^{mes}/A, $g_{CO_2}^{cell}$, and g_{wv} are those presented in Figures 4.3 to 4.7. As the PAR

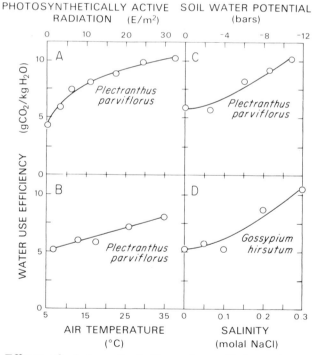

Figure 4.8. Effects of photosynthetically active radiation, air temperature, soil water potential, and salinity during leaf development on water use efficiency in two species.

during leaf development increased, WUE for *P. parviflorus* more than doubled (Figure 4.8*a*) as a consequence of the greatly increased mesophyll surface area per unit leaf area (Figure 4.3). An increase in developmental temperature caused a 50% increase in WUE (Figure 4.8*b*), reflecting g_{wv} and $g_{CO_2}^{cell}$ curves peaking near 26°C, together with a steady increase in A^{mes}/A with temperature. For soil water potential decreases and salinity increases, the two environmental changes directly affecting leaf water relations, the WUE doubled as the stress increased (Figure 4.8*c* and *d*). This primarily reflected a decrease in water vapor conductance as soil water potential decreased for *P. parviflorus* (Figure 4.7*b*) and as salinity increased for *G. hirsutum* (Figure 4.7*c*), although some anatomical changes did occur, leading to 40% variations in A^{mes}/A.

7. CONCLUSIONS

The responses to water stress and to salinity were mainly stomatal as far as the water use efficiency was concerned. The increase in WUE with growth temperature represented a complicated interplay of A^{mes}/A, $g_{CO_2}^{cell}$, and g_{wv}. The increase in WUE for *P. parviflorus* as the growth illumination was raised clearly reflected the increased CO_2 uptake due to the enhanced internal leaf area. So far hormonal and nutrient effects on WUE caused by changed leaf anatomy and the effects of changes in combinations of environmental factors have not been experimentally elucidated. Apparently the most critical environmental factor affecting leaf anatomy is the illumination level present during leaf development, but the effects at various developmental stages and the pigments responsible are not quantitatively understood. Enhanced WUE caused by increased A^{mes}/A may have significance in agronomic practice.

ACKNOWLEDGMENTS

Financial support from Department of Energy Contract EY-76-C-03-0012 and National Science Foundation grant DEB 77-11128 is gratefully acknowledged. Dr. David J. Longstreth provided many useful comments on the manuscript.

REFERENCES

1. H. C. Hanson, *Am. J. Bot.*, **4**, 533 (1917).
2. N. K. Boardman, *Annu. Rev. Plant Physiol.*, **28**, 355 (1977).
3. F. M. Turrell, *Am. J. Bot.*, **23**, 255 (1936).
4. F. M. Turrell, in A. Wexler, Ed., *Humidity and Moisture*, Vol. 2, Reinhold, New York, 1965, p. 39.

5. M. El-Sharkawy and J. Hesketh, *Crop Sci.*, **5**, 517 (1965).
6. A. Laisk, V. Oja, and M. Rahi, *Sov. Plant Physiol.* (*Engl. trans.*), **17**, 31 (1970).
7. A. E. Hall, *Carnegie Inst. Washington Yearb.*, **70**, 530 (1971).
8. P. S. Nobel, L. J. Zaragoza, and W. K. Smith, *Plant Physiol.*, **55**, 1067 (1975).
9. P. S. Nobel, *Plant Physiol.*, **58**, 218 (1976).
10. D. Wilson and J. P. Cooper, *New Phytol.*, **69**, 233 (1970).
11. D. A. Charles-Edwards and L. J. Ludwig, in R. Marcelle, Ed., *Environmental and Biological Control of Photosynthesis*, Junk, The Hague, 1975, p. 37.
12. G. M. Dornhoff and R. Shibles, *Crop Sci.*, **16**, 377 (1976).
13. G. A. Pieters, *Meded. Landbouwhogesch. Wageningen*, **60–17**, 1 (1960).
14. P. Holmgren, *Physiol. Plant.*, **21**, 676 (1968).
15. R. B. Pearce, G. E. Carlson, D. K. Barnes, R. H. Hart, and C. H. Hanson, *Crop Sci.*, **9**, 423 (1969).
16. W. Louwerse and W. V. D. Zweerde, *Photosynthetica*, **11**, 11 (1977).
17. J. H. McClendon, *Am. J. Bot.*, **49**, 320 (1962).
18. I. J. Warrington, M. Peet, D. T. Patterson, J. Bunce, R. M. Haslemore, and H. Hellmers, *Aust. J. Plant Physiol.*, **4**, 371 (1977).
19. J. M. Cutler, D. W. Rains, and R. S. Loomis, *Physiol. Plant.*, **40**, 255 (1977).
20. P. S. Nobel, *Physiol. Plant.*, **40**, 137 (1977).
21. P. S. Nobel, *Plant Physiol.*, **58**, 576 (1976).
22. B. F. Chabot and J. F. Chabot, *Oecologia*, **26**, 363 (1977).
23. G. A. Pieters, *Meded. Landbouwhogesch. Wageningen*, **74–11**, 1 (1974).
24. C. J. Nelson, K. J. Treharne, and J. P. Cooper, *Crop Sci.*, **18**, 217 (1978).
25. E. Van Volkenburg and W. J. Davies, *Crop Sci.*, **17**, 353 (1977).
26. C. E. Manning, D. G. Miller, and I. D. Teare, *J. Am. Soc. Hortic. Sci.*, **102**, 756 (1977).
27. D. J. Longstreth and P. S. Nobel, *Plant Physiol.*, **63**, 700 (1979).
28. B. F. Slade, *Bot. Gaz. Chicago*, **131**, 83 (1970).
29. O. Björkman, J. Troughton, and M. Nobs, in *Basic Mechanisms in Plant Morphogenesis*, Brookhaven Symposia in Biology, No. 25, Brookhaven National Laboratory, Upton, New York, 1974, p. 206.
30. D. T. Patterson, S. O. Duke, and R. E. Hoagland, *Plant Physiol.*, **61**, 402 (1978).
31. D. A. Charles-Edwards, J. Charles-Edwards, and F. I. Sant, *J. Exp. Bot.*, **25**, 715 (1974).
32. R. A. Fischer and N. C. Turner, *Annu. Rev. Plant Physiol.*, **29**, 277 (1978).
33. J. Gale, in A. Poljakoff-Mayber and J. Gale, Eds., *Plants in Saline Environments*, Springer-Verlag, Berlin–Heidelberg–New York, 1975, p. 168.
34. A. E. Hall, E.-D. Schulze, and O. L. Lange, in O. L. Lange, L. Kappen, and E.-D. Schulze, Eds., *Water and Plant Life: Problems and Modern Approaches*, Springer-Verlag, Berlin–Heidelberg–New York, 1976, p. 169.

5 | Adaptation of Roots in Water-Stressed Native Vegetation

J. KUMMEROW

Department of Botany, San Diego State University, San Diego, California

1. INTRODUCTION

The two most critical areas with respect to water flow through the soil-plant-atmosphere continuum are the soil-root and the leaf-air interfaces. It is reasonable to suppose that adaptations to drought have developed in roots, as they have in leaves and stems. However relatively little information is available concerning adaptations in roots because direct observation is difficult. Rhizotrons or costly and labor-intensive excavations are required, and there are questions concerning the validity of the data obtained by either method. The environment in a rhizotron differs from that in nature, and it is very difficult to extract all the roots from a volume of soil large enough to provide statistically significant data. Methods for the study of roots and root systems were described by Schuurman and Goedewaagen (1) and Böhm (2).

This chapter is limited to a discussion of the perennial vegetation of mediterranean and desert ecosystems, which are characterized by extended droughts and high temperatures.

2. ROOT SYSTEMS

2.1. Horizontal and Vertical Root Extension

Since water is the major limiting factor for plant growth in arid areas, it can be expected that root systems develop in a way that assures optimum water

57

absorption. This can be achieved by a fibrous root system with a large surface area, characteristic of many annuals: the plants absorb water as long as it is available and complete their life cycle before drought makes further development impossible. These "drought escapers," which are mostly ephemeral desert plants, are not discussed further.

Another possibility for assuring optimum water absorption is an extensive root system. However a root system extending over several meters requires an investment of carbon and nutrient that is too high to be lost and replaced every year: hence an extensive root system is possibly suitable only for perennial plants. The following discussion focuses on the hypothesis that the extension of root systems of trees and shrubs of deserts and other areas exposed to seasonal drought can be considered to be adaptive. We compare representative data from different sources and discuss the results in the context of the hypothesis formulated above.

Root systems of perennials in arid zones seem to fall into two or three categories. Cannon (3) classified desert root systems into "specialized tap," "specialized lateral," and "generalized" according to the existence of major tap roots, a main lateral root system without a major tap root, and a combined tap root and lateral root system, respectively. The subjectivity of such a classification is obvious, and Ludwig (4) tried to reduce this. He suggested the term "specialized tap root system" if more than 75% of the root biomass was found in the tap root, whereas if the amounts of tap root and lateral root biomass were about equal, he spoke of a "generalized root system." Specht and Rayson (5) also classified the observed root systems of 91 species, but their classification was again subjective. Such a classification may be satisfactory for a specific study, but it is not workable on a worldwide scale. Since edaphic conditions and excavation methods vary considerably, comparing results from root excavations by different researchers and in different areas is difficult, but an attempt was made at a general comparison of the results of several independent groups (Table 5.1). The data in Table 5.1 are arranged according to a gradient of yearly precipitation, ranging from 250 mm in Israel to 650 mm in Arizona. Only a small fraction of the plant species that have been reported in the literature are represented in the table, but the material seems to be broad enough to justify some conclusions. In all cases the crown radius was found to be substantially smaller than the root extension. However rooting depths are highly variable and apparently are more dependent on soil conditions than on the plant's genetic makeup. The existence of a tap root is by no means a general phenomenon. In some cases tap roots might have developed if the plants had been found on deeper soils. For example, *Adenostoma fasciculatum* excavated in southern California was described by Hellmers et al. (9) as having a tap root that reached a depth of 8 m, and Kummerow et al. (7) found a shrub of the same species on a shallow soil only 60 cm deep without any indication of the existence of a tap root. Evenari's statement that "the root systems of Israelian desert plants do not show special peculiarities and the observed

Table 5.1. Horizontal and Vertical Root Extensions of Trees and Shrubs from Several Mediterranean and Desert Ecosystems

The values of horizontal root extensions are means. The species are arranged according to increasing annual precipitation (numbers in parentheses indicate the number of years in which rainfall was measured in the specific research area).

Species	Horizontal extension radius (m)	Vertical extension (m)	Tap root present	R:S biomass ratio	Crown radius	Geographical area	Precipitation (mm/yr)	Ref.
Haplophyllum tuberculatum	0.4	0.3	—	0.5	0.1	Israel	250	6
Retama retam	2.2	1.2	+	0.2	0.5	Israel		6
Sueda asphaltica	1.1	0.1	—	—	0.5	Israel		6
Zygophyllum dumosum	4.0	1.6	—	0.8	0.7	Israel		6
Banksia ornata	7	3	+	—	1.3	Southern Australia	441(1)	5
Casuarina mülleriana	3	2.5	+	—	1.3	Southern Australia		5
Leucopogon costatus	0.3	0.3	—	—	0.2	Southern Australia		5
Xanthorroea australis	2.7	3	+	—	0.7	Southern Australia		5
Adenostoma fasciculatum	3	0.6	—	0.6	1.0	Southern California	476(7)	7
Arctostaphylos pungens	4.5	0.6	—	0.4	1.5	Southern California		7
Ceanothus greggii	3.5	0.6	—	0.2	1.5	Southern California		7
Heteromeles arbutifolia	0.6	0.3	—	0.7	0.4	Southern California		7
Colliguaya odorifera	3.5	0.6	—	0.6	1.0	Central Chile	543(7)	8
Cryptocarya alba	>3	>0.6	—	1.4	1.5	Central Chile		8
Lithraea caustica	>4	>0.6	—	2.9	1.8	Central Chile		8
Satureja gilliesii	1.0	0.2	—	0.3	0.2	Central Chile		8
Trichocereus chiloensis	0.5	0.4	+	0.5	0.1	Central Chile		8
Adenostoma fasciculatum	4	8	—	—	0.9	Southern California	550	9
Arctostaphylos glauca	10	2.8	—	—	1.2	Southern California		9
Ceanothus greggii	4	1.5	—	—	0.5	Southern California		9
Quercus dumosa	3.3	9	+	—	0.8	Southern California		9
Quercus turbinella	6.9	6.4	+	1.9	2.4	Central Arizona	653(18)	10

extensions lie within the same limits as those recognized for other habitats''
(6) probably can be generalized to plants from other deserts.

Thus the hypothesis that root extension is an adaptation to dry soil
conditions is not supported by facts. Root systems show a high degree of
morphological plasticity that enables them to cope with the most variable
soil and soil moisture conditions. The data confirm the generally accepted
view that desert plants frequently have more shallow and widespread root
systems, although this is not an exclusive feature. In less extreme dry areas,
deep-rooting species become more abundant. However there are certainly
other morphological characters, more important than extension and
branching of root systems, that make xerophytes adapted to arid conditions.

2.2. Root/Shoot Biomass Ratio

The previous section considered the extension of the root system alone. If
the aboveground plant parts are not taken into account, however, the dimensions of root systems have limited significance. Table 5.1 also gives the
widths of plant crowns. A conventional way to describe the varying degrees
of carbon allocation to the above and below ground plant parts is the
root/shoot biomass ratio (R:S ratio). Since it is usually difficult to completely extract the whole root system from the soil, many R:S ratios may be
underestimates of the true values. Furthermore it must be remembered that
roots and shoots show considerable seasonal fluctuation of their biomass.
Thus the published data may reflect the R:S ratios at different seasonal
stages of development, and values should be compared with caution.

In 1973 a review by Barbour (11) led to the conclusion that perennial
plants in arid areas rarely show R:S ratios above 1. However Rodin et al.
(12) found R:S ratios of 7 to 8 in the Syrian *Artemisieta*. Rodin and
Bazilevich (13) reported similarly high values from the saline steppes of the
USSR, and Fernandez and Caldwell (14) obtained analogous results from
Atriplex- and *Ceratoides*-dominated cold deserts in the Great Basin of Utah.

It now appears that R:S ratios do not offer any clue regarding the
adaptation of roots to water stress because although R:S ratios are frequently considerably below unity, ratios well above 1 have also been found
in both warm and cold deserts. Again it can be concluded that factors other
than water stress are decisive for carbon and nutrient partitioning between
root and shoot systems in arid areas.

2.3. Seasonal Changes of the Fine Root Biomass

The methods used for root excavation in the analysis above precluded
measurement of the physiologically important fine roots (diameter < 1 mm).

Lyr and Hoffmann (15) addressed the question of fine root longevity and
found that rootlets can live through one growing season and in some species

up to 2 yr. These, however, are extreme values and in the context of the present discussion it seems more relevant to ask for average fine root longevities than for extreme values. Under temperate climatic conditions, Bode (16) found that 90% of walnut (*Juglans regia*) rootlets died at the beginning of winter. On the other hand, Childers and White (17) reported that under favorable growth conditions fine roots of apple trees died after only 1 to 2 weeks, an observation supported by Rogers and Booth (18), who studied rootlets through vertical glass windows. These observations made Head (19) conclude that rootlets were ephemeral. It seems legitimate therefore to ask whether the ephemeral fine roots of perennial plants growing under seasonally arid conditions are shed simultaneously like leaves of deciduous trees, or whether a large portion of the fine root fraction remains alive, to make optimal use of the sporadic rainfall characteristic of many desert areas.

Development of the elegant $^{14}C/^{12}C$ dilution technique allowed Caldwell and Camp (20) to assess the turnover coefficients of root systems in a uniform, cold desert plant community dominated by *Ceratoides lanata* and *Atriplex confertifolia* in northern Utah. These perennial shrubs have very large root systems and are characterized by R:S ratios of 7 to 11. Their productivity below ground was calculated as 3 times that of the aboveground plant parts. The root system consists mainly of fine roots with an average diameter less than 0.5 mm. The annual rainfall in the area is erratic, but averages 230 mm. The soil is also saline, giving a total soil water potential of -35 bars (21). Air temperatures frequently reach 40°C in the summer and -30°C in winter. The results of the study indicated a turnover coefficient of 0.24 for the fine roots of *Atriplex confertifolia*. This means that from a root biomass of 1823 g/m^2 in spring, about one-fourth had been renewed at the end of the growing season in September. Direct observations suggested that by midsummer 17% of the *Atriplex* root system consisted of dead roots. Phenological observations of fine roots growing against oblique window panes at the same location showed a gradual shift of growth activity from the upper soil layers in spring to the 1 m layer in late November. Thus the data suggest a more or less continuous fine root turnover during the year, there being no evidence of a period of intensive root shedding. Observations of root darkening suggested that suberization of roots took place 1 to 2 weeks after rapid root extension in spring and after an even shorter time in August. Many of these roots may stay alive for a considerable time (14). Irrigation and fertilizer experiments by Hodgkinson et al. (22) in the same plant community tend to support this conclusion. Even in summer with soil water potentials of -30 bars, turgid fine roots were found in the soil. Thus there was no evidence for strong seasonal root shedding in these desert shrubs.

Seasonal changes in the fine root density in the chaparral of southern California were studied by Kummerow et al. (23). The investigation concentrated on the root systems of *Adenostoma fasciculatum, Arctostaphylos glauca, Ceanothus greggii,* and *Rhus ovata.* The technique consisted of soil core sampling and extraction of roots from these cores (23). Table 5.2 shows that during the summer the fraction of dead roots in the samples increased significantly. Consequently it appears that fine roots were shed during the summer, but not in as short a period as 3 to 4 weeks. It is fortunate that the low soil moisture contents during the summer (0.05 g water/g soil) in the rooting zone seemed to have inhibited the decomposition processes, thus allowing an assessment of the dead biomass fraction. The values given in Table 5.2 are minimal, since only well-structured dead roots were considered, and no effort was made to separate dead root fragments from the organic debris in the soil samples. This study suggests that in contrast to *Atriplex confertifolia,* a cold desert shrub with relatively deep-reaching fine root systems, the chaparral shrubs on shallow soils apparently have a long and intensive root-shedding period during the summer.

2.4. Ratio of Root Area to Leaf Area

Section 2.2 demonstrated that the R:S ratios do not seem to reflect any specific adaptation of trees and shrubs to their arid environment. It is also clear (Section 2.3) that the physiologically important fine roots represent a very important fraction of the total root biomass, 93% in the case of *A. confertifolia* in a cold desert ecosystem (20) or 50% in the more shallow-rooting shrubs in the southern California chaparral (23). However the relevant factor is not the root/shoot biomass ratios, but the ratio of the absorbing root surface to the photosynthetically active leaf area of the plant.

Table 5.2. Seasonal Biomass Increment Coefficients of the Fine Roots of Four Chaparral Shrub Species

Roots extracted from the 10–20 cm depth of soils shaded by the shrub canopies. Each value is the average of 36 soil core extractions of 125 cm³ each. Winter and summer values are from January to March and from June to August, respectively.

| | Fine roots (g dry weight/dm³) | | | | | |
| | Living | | | Dead | | |
Species	Winter	Summer	Increment coefficient	Winter	Summer	Increment coefficient
Adenostoma fasciculatum	0.21	1.43	6.8	0.44	2.08	4.7
Arctostaphylos glauca	0.34	1.57	4.6	0.25	1.98	7.9
Ceanothus greggii	0.54	1.31	2.4	0.58	1.40	2.4
Rhus ovata	0.37	0.79	2.1	0.59	2.36	4.0

The technical difficulties of making a reasonably accurate estimate of the fine root surface area are substantial: Kummerow and Brinkman (24) undertook a root box experiment to estimate the root surface area and leaf areas of young chaparral shrubs. Redwood boxes, 75 cm high, 60 cm wide, and 20 cm deep, filled with a fertilized mixture of sand and peat moss, were planted in December with 1.5 yr old *Adenostoma fasciculatum, Arctostaphylos pungens, A. glauca,* and *Yucca whipplei.* Six boxes were placed in a relatively shaded area and watered once a week. Three boxes were fully exposed to the sun and watered once every 2 weeks: these plants showed signs of water stress from time to time. Four boxes were placed into previously excavated ditches in the chaparral at 1000 m elevation. These boxes received irrigation only once a month and showed symptoms of severe water stress during the summer. In fact, only two plants of the original eight of this group survived. The following August the stems and leaves were removed and the total leaf area was measured. By using the pinboard technique (1) a sample of the root system was chosen and analyzed. For each shrub, the following features were measured: the dry weight of the total root system and the fine root fraction (diameter < 1.0 mm), the mean fine root diameter, the length of 1 g of fine roots and the ratio of suberized to unsuberized fine roots. Thus the fine root surface per shrub could be estimated and the ratio of root surface to leaf area could be established (Table 5.3). The results show that about half the fine roots were unsuberized at harvest time. This surprisingly high fraction of unsuberized roots may have been obtained because these were young and partially irrigated shrubs—this conclusion is supported insofar as the two plants that were kept under severe water stress had only 13 and 21% of their fine roots unsuberized. The fine root surface was calculated from the total fine root fraction (suberized and unsuberized), since Kramer and Bullock (25), Chung and Kramer (26), and Queen (27) had shown that in *Liriodendron tulipifera, Pinus taeda,* and *Vitis vinifera* suberized fine roots absorb substantial amounts of water: the frequent cracks observed in the root bark of chaparral shrubs (Figure 5.1) make it likely that the suberized roots also absorb water in these species.

Based on this experience and a different sampling technique, Kummerow et al. (23) estimated fine root area indices (i.e., the fine root surface area in square meters per square meter of ground area) and related these to the respective leaf area indices (Table 5.4). The root area index (RAI) and leaf area index (LAI) have about the same numerical values, and their ratios fluctuate between 0.7 for *Adenostoma fasciculatum* and 1.6 for *Rhus ovata.* The fine root density of a 36 yr old Douglas fir stand under the temperate climatic conditions of Oxford (England) was analyzed by Reynolds (28). The values are included in Table 5.4 for comparison. The RAI/LAI ratio is close to unity and very similar to the ratios found for chaparral shrubs.

Undoubtedly much more field work is needed to test whether these results are indicative of the general situation. The data obtained seem to suggest that low soil water potentials and high soil temperatures trigger abscission of

Table 5.3. Root Growth and Ratios of Root Surface to Leaf Area and Root to Shoot Biomass of Four Southern California Chaparral Species

The 1.5 yr old seedling plants were cultivated over 7 months in redwood boxes and harvested in midsummer after cessation of the growing period. The soil was a fertilized mixture of sand and peat moss.

Water regime	Fine roots (g)	Length of 1 g fine roots (m)	Fine root diameter (mm)	Total fine root surface (m²)	Total leaf area (m²)	Ratio of root surface to leaf area	Ratio of root to shoot biomass
Arctostaphylos pungens							
High	691	8.6	0.17	3.2	0.54	5.9	0.89
High	317	10.5	0.17	1.8	0.38	4.7	0.61
Medium	491	11.2	0.17	2.9	0.40	7.2	0.88
Low	543	9.5	0.20	3.2	0.22	14.5	1.73
Arctostaphylos glauca							
High	785	7.1	0.24	4.2	0.80	5.2	0.66
Medium	599	8.7	0.17	2.8	0.29	9.7	1.63
Yucca whipplei							
High	290	3.5	0.45	1.4	0.33	4.2	0.30
High	242	3.5	0.45	1.3	0.36	3.6	0.23
Low	469	4.8	0.50	3.5	0.48	7.3	0.29
Adenostoma fasciculatum							
High	92	23.9	0.24	1.7	0.17	10	0.71
Medium	242	20.5	0.21	3.2	0.32	10	0.78

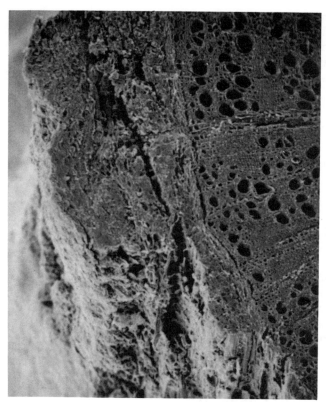

Figure 5.1. Scanning electron micrograph (×100) of a root of *Ceanothus greggii* (diameter 3 mm). Careful cutting and observations on numerous other roots guarantee that the cracks in the bark are natural and not artifacts.

Table 5.4. Length and Diameter of Fine Roots, Root Area Index, and Leaf Area Index of Four Chaparral Shrubs[a] and *Pseudotsuga menziesii*[b]

Roots with diameters > 1.0 mm were not included.

Species	Fine root length (km/m²)	Fine root diameter (mm)	Root area index (m²/m²)	Leaf area index (m²/m²)
Adenostoma fasciculatum	2.20	0.30	2.11	3.09
Arctostaphylos glauca	3.52	0.37	4.10	3.58
Ceanothus greggii	2.11	0.36	2.39	2.46
Rhus ovata	2.82	0.35	3.10	1.95
Pseudotsuga menziesii	7.00	0.50	11.0	10[c]

[a]After Ref. 23.   [b]After Ref. 28.   [c]After Ref. 29.

a substantial fraction of the fine root system (see Section 2.3), and it would be worthwhile to determine whether the period of intensive leaf abscision in the chaparral (30) is related to the fine root shedding below ground. We consider such a feature to be an interesting adaptation of root systems to water stress. Primary maize roots have an average respiration of 5 ml O_2/h/g dry weight (31); that is, in one week they consume energy equivaleht to the dry weight of the roots. Root shedding under water stress may therefore be an energy-conserving mechanism.

2.5. Contractile Roots

Active root shortening is an interesting phenomenon observed in a number of species. Even though the process of root contraction is widespread throughout the plant kingdom—Rimbach (32) counted 450 species from 315 genera and 82 families—contractile roots appear to have special significance for plants in arid areas. Plants with contractile roots form a conspicuous floristic element in the mediterranean areas of the world. Moreover, the finding of many mature bulbs at depths of 20 to 30 cm would seem to justify speculations on the biological significance of contractile roots. It is possible that the downward movement pulls the young bulbs not only out of the reach of predators, but also out of the zone of dangerously high soil temperatures and into soil where moisture could be available longer or water loss more restricted.

3. ADAPTATION OF ROOTS

3.1. Roots as Water-Storing Organs

Water storage often occurs in plants of arid environments. All the basic plant organs (i.e., leaves, stems, and roots) can be found to have a water storage function. A brief summary on this subject was published by Gessner (33). Plants with roots that store water have been studied by Marloth (34) in the Cape [of Good Hope] flora. There are many genera in the Umbelliferae and Geraniaceae especially that show this specific adaptation. Walter (35) found a specimen of *Pachypodium succulentum* (Apocynaceae) with spiny, leafless branches 20 to 30 cm high and a bulblike root that contained 266 g of water (28 g dry weight). Likewise Marloth (34) found a *Pachypodium* with a fresh weight of root of 71 kg and a leafless shoot of 68 g fresh weight. These roots can obviously supply water for a long time: Walter (35) estimated that only about 1% of the fresh weight of *P. succulentum* was lost by transpiration daily. Large water storage roots can also be found in the genera *Cissus* and *Cussonia*. The dry weight of roots in these genera is about 5% of their fresh weight, a value indicating that the roots are mainly water.

In *Oxalis cernua* (South Africa) fingerlike roots are formed during the winter rain period. In the following spring and summer these roots shrink, and the water is used to support the development of summer buds, which remain dormant until the next spring (34). In this species the stored water seems to be used preferentially for reproduction. In other cases, such as that of *Citrullus colocynthus* with storage roots up to 7 cm thick (36), it was found that the accumulated water was used to maintain transpiration and to prevent the leaf from overheating (37).

It can be concluded that water storage in roots is relatively common in arid areas of the world. The phenomenon is not limited to specific taxa of the higher plants, yet it is found in some families more frequently than in others.

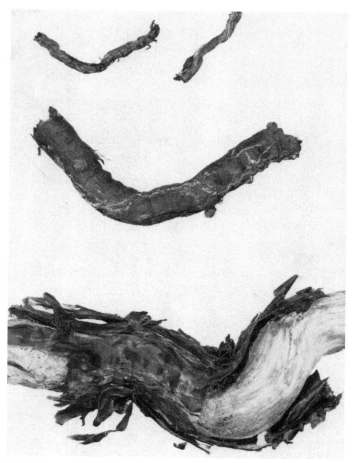

Figure 5.2. Roots of *Fouquieria splendens* (×1). *Below*: dead root showing the exfoliating bark; *above*, pieces of living roots with a bark still intact.

3.2. Root Bark Anatomy

Knowledge of root anatomy in relation to arid environments is very limited. There is not only a lack of data from roots of xerophytes, but also there is little information on the root structure of plants from mesic ecosystems needed for comparative purposes.

According to Henrickson (38), the larger roots of the Fouquieriaceae possess distinct water storage cells in the cortex. The roots of the common ocotillo (*Fouquieria splendens*) are shallow and wide spreading. The main roots are found at a depth of 10 to 20 cm, where soil moisture is below the 5% level during many months and soil temperatures often reach 40°C. No data are available on the water storage capacity of ocotillo roots, but the roots have a leathery, exfoliating bark common to many species in arid regions (Figure 5.2). Cross sections through roots of several species of chaparral shrubs revealed no specific water storage tissue. However the bark of each species possessed a characteristic structure, confirming Oppenheimer's (39)

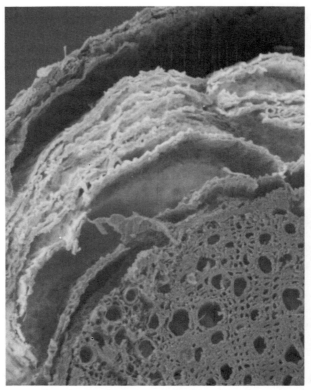

Figure 5.3. Scanning electron micrograph (×150) of large root cross section (diameter 6 to 8 mm) of *Arctostaphylos glauca* showing the characteristic bark exfoliation.

view that root structure has diagnostic value. Scanning electron micrographs of the thicker roots of chaparral shrubs showed the marked exfoliation common to many arid species (Figures 5.3 and 5.4). It is possible that the leathery exfoliating bark may reduce water loss from the roots. The root bark of *Rhus ovata* does not, however, show exfoliation, although it is thick and spongy because of the presence of many large resin ducts (Figure 5.5). Thick bark has been observed on the roots of other desert shrubs; the bark contributes 90% of the dry weight of roots less than 0.5 cm in diameter and more than 50% of roots larger than 0.5 cm in diameter in *Colliguaya odorifera* (40). However it has not been determined whether significant quantities of water are stored in these thick roots.

3.3. Metacutization

The process of metacutization refers to the suberization of cell walls of the outer layer of the root cap and the region where the exodermis begins to

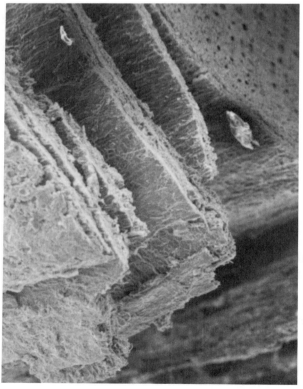

Figure 5.4. Scanning electron micrograph (×100) of large root cross section (diameter 6 to 8 mm) of *Adenostoma fasciculatum* showing the characteristic bark exfoliation.

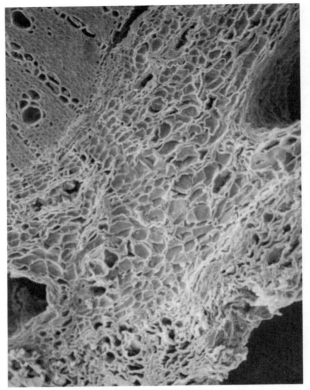

Figure 5.5. Scanning electron micrograph (×150) of the root of *Rhus ovata*. Note that the very wide cortex is characterized by large lactifers and relatively thin-walled parenchyma responsible for the spongy consistency of the cortex.

form (41). The metacutis of *Pinus halepensis* was described by Leshem (42) as a layer of suberized cells that join the endodermis behind the apex, thus isolating the meristem from the root cap and the primary cortex. Fernandez and Caldwell (14) observed metacutization in *Artemisia tridentata*, a common shrub in the cold desert ecosystem in Utah.

Investigations regarding the adaptive advantage of metacutization in arid areas are limited to those communicated by Leshem (42). He showed that the roots of potted pine trees produced an increasing number of dormant tips after cessation of irrigation; but how well this reflects the situation in mediterranean or desert ecosystems is unclear.

Ecologically, metacutization is understood as a protection of the root meristem during periods of dormancy. If the shrub vegetation entered the summer and fall dormancy with a large number of fine roots, respiration at the high soil temperatures prevailing would be detrimental to the energy balance of the plants. But even where fine roots are shed abundantly during drought periods, it is necessary to maintain a framework of somewhat larger roots to initiate the reexploitation of the soil after adequate precipitation.

4. DISCUSSION AND OUTLOOK

Three root structures are generally cited as adaptations to xeric conditions:

1. Root systems that are wide spreading and shallow.
2. Roots that are frequently succulent.
3. Root bark that is thick, sclerified, and frequently exfoliating.

Examples of these are abundant and well documented. The situation becomes more complex when roots from plants in less xeric habitats are analyzed. *Adenostoma fasciculatum* was described as deep rooting by Hellmers et al. (9). However Kummerow et al. (7) found the same species to be extremely shallow rooting. In the first case soil conditions permitted deep rooting, and in the second the shallow soil forced the roots to spread out widely. Yet in both habitats *Adenostoma* is a very successful species. In this example, shallow- or deep-rooting habits are expressions of morphological plasticity rather than adaptations to water stress. Compared with the true desert, mediterranean regions are more mesic, and winter rainfall can be abundant enough to percolate into deeper soils, favoring deeper penetration of root systems. Similar arguments could be formulated regarding the R:S biomass ratios.

There is no evidence that root shedding is an adaptive character. However the summer-deciduous habit of the foliage of quite a number of shrub species is a conspicuous adaptation to the dry summer season of the mediterranean climates (43). It is therefore attractive to speculate that an analogous adaptation below ground may exist, especially when the mutual dependence of root and shoot is taken into account. The data on fine root turnover seem to point to the same conclusion, there being an increase in the death of fine roots during the summer. Studies are needed to decide whether this phenomenon is a facultative adaptation that occurs only under water stress or whether truly constitutive elements are involved. The estimates of ratios of root surface area to leaf area could be a tool for further analysis of the hypothesis.

The adaptive nature of root succulence remains unchallenged. An open question is the degree to which the thick bark, found on roots of perennials in mediterranean and desert areas, contributes to the water balance of the shrubs. Again, experimental work is needed to clarify whether thick bark conveys an adaptive advantage to species under prolonged water stress. The situation regarding metacutization is unclear. Root dormancy is certainly a phenomenon that could have substantial adaptive value in summer dry areas, but so far only limited material is available to assess the ecological importance of this observation.

In conclusion, it appears that roots and root systems have relatively few adaptive structures when compared with leaves. Roots appear to be much more plastic than aboveground plant parts in their dealing with environmental stress. Leaf adaptations appear to be chiefly responsible for the success of a species in a water-stressed environment. However some of these speculations may be the result of our still very limited knowledge regarding the comparative morphology of roots and root systems.

ACKNOWLEDGMENTS

I thank Dr. J. V. Alexander and Ms. Linda Allen for the scanning electron micrographs. The National Science Foundation (DEB-7713944) provided financial support.

REFERENCES

1. J. J. Schuurman and M. A. J. Goedewaagen, *Methods for the Examination of Root Systems and Roots*, 2nd ed., Center for Agricultural Publishing and Documentation, Wageningen, 1971.
2. W. Böhm, *Kali-Briefe*, **14**, 91 (1978).
3. W. A. Cannon, *Ecology*, **30**, 542 (1949).
4. J. A. Ludwig, in J. K. Marshall, Ed., *The Below-Ground Ecosystem: A Synthesis of Plant Associated Processes*, Range Science Department, Science Series No. 26, Colorado State University, Fort Collins, 1977, p. 85.
5. R. L. Specht and P. Rayson, *Aust. J. Bot.*, **5**, 103 (1957).
6. M. Evenari, *J. Linn. Soc. London, Bot.*, **51**, 383 (1938).
7. J. Kummerow, D. Krause, and W. Jow, *Oecologia*, **29**, 163 (1977).
8. A. Hoffmann and J. Kummerow, *Oecologia*, **32**, 57 (1978).
9. H. Hellmers, J. S. Horton, G. Juhren, and J. O'Keefe, *Ecology*, **36**, 667 (1955).
10. E. A. Davis and C. P. Pase, *J. Soil Water Conserv.*, **32**, 174 (1977).
11. M. G. Barbour, *Am. Midl. Nat.*, **89**, 41 (1973).
12. L. E. Rodin, N. I. Bazilevich, and T. M. Miroshnichenko, in L. E. Rodin, Ed., *Ecophysiological Foundation of Ecosystems Productivity in Arid Zones*, Nauka, Leningrad, 1972.
13. L. E. Rodin and N. I. Bazilevich, *Production and Mineral Cycling in Terrestrial Vegetation*, Oliver & Boyd, Edinburgh, 1967.
14. O. A. Fernandez and M. M. Caldwell, *J. Ecol.*, **63**, 703 (1975).
15. H. Lyr and G. Hoffmann, *Int. Rev. For. Res.*, **1**, 181 (1967).
16. H. R. Bode, *Ber. Dtsch. Bot. Ges.*, **72**, 93 (1959).
17. N. F. Childers and D. G. White, *Plant Physiol.*, **17**, 603 (1942).
18. W. S. Rogers and G. A. Booth, *Sci. Hortic.*, **14**, 27 (1959).

19. G. C. Head, in T. T. Kozlowski, Ed., *Shedding of Plant Parts,* Academic Press, New York–San Francisco–London, 1973, p. 237.
20. M. M. Caldwell and L. B. Camp, *Oecologia,* 17, 123 (1974).
21. R. T. Moore and M. M. Caldwell, in R. W. Brown and B. P. Van Haveren, Eds., *Psychrometry in Water Relations Research,* Utah Agricultural Experiment Station, Logan, 1972, p. 155.
22. K. C. Hodgkinson, P. S. Johnson, and B. E. Norton, *Oecologia,* 34, 353 (1978).
23. J. Kummerow, D. Krause, and W. Jow, *Oecologia,* 37, 201 (1978).
24. J. Kummerow and M. Brinkman, *Bot. Soc. Am., Abstr., Misc. Ser.,* 154, 25 (1977).
25. P. J. Kramer and H. C. Bullock, *Am. J. Bot.,* 53, 200 (1966).
26. H. H. Chung and P. J. Kramer, *Can. J. For. Res.,* 5, 229 (1975).
27. W. H. Queen, *Radial Movement of Water and ^{32}P Through Suberized and Unsuberized Roots of Grape,* Ph.D. thesis, Duke University, Durham, North Carolina, 1967.
28. E. R. C. Reynolds, *Plant Soil,* 32, 501 (1970).
29. C. C. Grier and S. W. Running, *Ecology,* 58, 893 (1977).
30. A. Hoffmann, H. A. Mooney, and J. Kummerow, in N. J. W. Thrower and D. E. Bradbury, Eds., *Chile-California Mediterranean Scrub Atlas: A Comparative Analysis,* Dowden, Hutchinson & Ross, Stroudsburg, Pennsylvania, 1977, p. 102.
31. A. R. Grable, *Adv. Agron.,* 18, 57 (1966).
32. A. Rimbach, *Ber. Dtsch. Bot. Ges.,* 47, 22 (1929).
33. F. Gessner, in O. Stocker, Ed., *Handbuch der Pflanzenphysiologie,* Vol. 3, Springer-Verlag, Berlin–Göttingen–Heidelberg, 1956, p. 247.
34. R. Marloth, "Das Kapland, insonderheit das Reich der Kapflora, das Waldgebiet und die Karoo, pflanzengeographisch dargestellt," *Ergebnisse der Deutschen Tiefsee-Expedition 1889/99,* Fischer, Jena, 1908.
35. H. Walter, *Die Vegetation der Erde in Öko-physiologischer Betrachtung,* Vol. 1, 2nd ed., Fischer, Jena, 1964, pp. 439–440.
36. M. Zohary, in *Plant-Water Relationships in Arid and Semi-Arid Conditions: Proceedings of the Madrid Symposium, 1959,* UNESCO, Paris, 1961, p. 199.
37. O. Stocker, *Flora (Jena),* 163, 46 (1974).
38. J. Henrickson, *Aliso,* 7, 97 (1969).
39. H. R. Oppenheimer, *Bull. Res. Counc. Isr., Sect. D,* 6, 18 (1957).
40. G. Avila, S. Araya, F. Riveros, and J. Kummerow, *Oecol. Plant.,* 13, 367 (1978).
41. A. Fahn, *Plant Anatomy,* Pergamon Press, Oxford, 1967, p. 263.
42. B. Leshem, *For. Sci.,* 11, 291 (1965).
43. H. A. Mooney and E. L. Dunn, *Am. Nat.,* 104, 447 (1970).

6 | Modifying Root Systems of Cotton and Soybean to Increase Water Absorption

HOWARD M. TAYLOR

USDA–SEA–AR, Soil & Water Management Unit, Iowa State University, Ames, Iowa

1. INTRODUCTION

The pioneering water uptake models of Philip (1) and Gardner (2) provided valuable background information on water uptake by roots. These investigators proposed that uptake by a single root can be described by a conductivity function (a reciprocal of the resistance to water flow from bulk soil to soil at the root surface), a potential energy gradient function (the difference in potential energy between water located in bulk soil and that located at the root surface divided by the distance between these two locations), and the total root length. Gardner (3) extended this analysis to include water uptake from different soil layers, each with different root lengths and water contents. The assumption by Philip and Gardner that plant root resistance to water flow was negligible was questioned (4, 5), and recent field and greenhouse experiments have confirmed that resistance of plant roots to water flow, indeed, is appreciable over most of the available soil water range (6, 7). Recently Faiz and Weatherley (8) concluded that resistance to water flow across the interface between root and soil also may be appreciable.

Thus appropriate variables to consider in a discussion of modifying water uptake from soil by root systems of cotton and soybean plants are (*a*) transpiration rate (which affects the water potential in the plant xylem elements at crown level), (*b*) axial resistance to water flow along the xylem elements (which affects the water potential in the root and shoot xylem at substantial distances from the crown), (*c*) radial resistance to flow of water from its location in the soil to the soil-root interface, across that interface, and from the outer cells of the root to the lumen of xylem vessels, (*d*) total length of roots on the plant, and (*e*) energy status and quantity of water in soil around these roots.

This chapter discusses the significance of some observed (or postulated) differences in root development among species and among genotypes of a species. The discussion is focused on the rooting characteristics of cotton and soybean that seem to be most easily manipulated in a selection program to delay onset of severe water stress. These factors are root depth, root length density within each soil layer, axial resistance along the root xylem elements, and radial resistance from the outer layers of the root to the xylem lumen.

2. ROOT DEPTH

Reviewers of research on soil-plant-water relations usually conclude explicitly or implicitly that plants need deep, vigorous root systems during drought. This conclusion arises because if all other conditions are equal and if the entire soil profile is at field capacity when the crop is planted, increases in rooting depth increase the total quantity of water available for extraction during the growing season.

Most plant scientists believe that some plants inherently have shallow-rooted systems and others have deep-rooted systems. This idea is only partly valid. Indeed, supposedly deep-rooted plants are often shallow rooted in the field, especially if rooting depth is determined by adverse soil conditions or only a short period of growth. In contrast, even in onions (*Allium sativum L.*), a species considered to be shallow rooted, the rooting depths may exceed 1 m and lateral spreads of roots may exceed 0.5 m if soil conditions permit (9).

Root systems of most annual field crops extend to considerable depths before seed harvest if soil conditions are satisfactory. For example, 13 species of field crops were screened for rooting depth at the Auburn, Alabama, rhizotron (10) during the 1969 to 1972 growing seasons. These species were peanuts (*Arachis hypogaea* L.), oats (*Avena sativa* L.), bermuda grass [*Cynodon dactylon* (L.) Pers. 'Coastal'], soybean [*Glycine max* (L.) Merr.], cotton (*Gossypium hirsutum* L.), barley (*Hordeum vulgare* L.), sweet potato [*Ipomoea batatas* (L.) Lamarck], tomato (*Lycopersicon esculentum* Mill.), millet [*Pennisetum typhoides* (Burm.) Staph and Hubbard], rye (*Secale cereale* L.), grain sorghum [*Sorghum bicolor* (L.) Moench.], wheat (*Triticum aestivum* L.), cowpeas [*Vigna sinensis* (L.) Endl. ex Hassk.], and maize (*Zea mays* L.). Each species was planted in Cahaba loamy fine sand (Typic Hapludults) on about the usual planting date for that species when grown in Alabama. Roots of each species penetrated to the bottom of the rhizotron compartments (1.88 m) by maturity, or for bermuda grass, by the end of the first growing season. The species differed, however, in the rates at which their roots grew downwards under uniform soil conditions during one growing season (6, 11, 12). For example, maize roots grew downwards at an average rate of 7.7 cm/day and tomato roots at a rate of 8.6 cm/day (12), whereas peanut roots grew downwards at a rate of 3.7 cm/day (unpublished data). In addition, cotton roots grew downwards at different rates during different growing seasons (6, 13–15) and in different soils during one growing season (16).

Rates at which rooting depths increase, and the total rooting depths during a growing season, can be changed by removing a soil condition that limits root extension. *The Plant Root and Its Environment* (17) describes about 20 different soil conditions (biological, chemical, and physical) that affect root growth and function, and that coverage probably is incomplete. Removing any one limiting condition from among several affects root depth only slightly, if at all, but removing the *only* limiting soil condition often greatly alters the rate of downward root growth (18). No satisfactory models are available presently to predict root depth from a knowledge of soil environmental conditions that partially limit root extension rates.

Soybean and cotton root elongation rates are reduced as chemical activity of aluminum is increased in the soil solution (19, 20). There are differences among soybean genotypes (and presumably among cotton genotypes as well) in responses of root elongation to aluminum toxicity (21). Cotton roots subjected to aluminum toxicity are still effective in water uptake, even though their elongation rates are slowed (22). A distinct possibility, therefore, exists that cotton and soybean genotypes can be selected to extract substantially more water than normal from deep within soil profiles that contain toxic concentrations of aluminum ions and probably others as well.

Genotypes differ in their root extension rates through soil layers that are favorable for rooting. Taylor et al. (23) examined the taproot extension rates of 29 soybean cultivars under uniform soil conditions. After 27 days, roots of 'Blackhawk,' the slowest cultivar, averaged 83.5 cm, whereas those of 'Sciota,' the fastest cultivar, averaged 117.3 cm, 1.25 cm/day faster than 'Blackhawk.' Only small fractions of the increase were associated with seed weight at planting and with top weight at harvest.

Even where the roots of cotton and soybean cultivars currently grown already extend as deep as the soil profile is wetted, selecting cultivars for faster root extension rates might delay onset of moderate water stresses, because faster extension rates allow more time for root length density to increase at depths where some water remains available. In some circumstances, however, the faster root extension rates can result in more even rooting within the profile, and thus will cause plant water status to shift rather abruptly from a nonstressed to a severely stressed condition when the water supply is exhausted.

3. ROOT LENGTH DENSITY

Scientists, who mathematically describe water uptake by root systems, usually consider that uptake rate is proportional to rooting density if other conditions are equal. Using this assumption, Taylor and Klepper (6) and Herkelrath et al. (24) found that they could satisfactorily distribute water uptake of a root system among various soil layers.

Root length density (RLD, cm roots/cm^3 soil) varies substantially among species. Allmaras et al. (25), who compared RLDs of maize and soybean, found that RLD in the upper 1.3 m of soil averaged 0.22 cm/cm^3 for maize and only 0.08 cm/cm^3 for soybean. Burch et al. (26) found that RLD of sorghum (cv. RS610) ranged from 4 cm/cm^3 at the 5 cm depth to almost zero at the 95 cm depth, whereas that of soybean (cv. Ruse, rainfed) ranged from slightly more than 2 cm/cm^3 to about 0.8 cm/cm^3 at the same soil depths. Taylor and Klepper (6, 11) published RLDs for cotton (cv. Auburn 623b) and for maize (cv. Funk's G4949) grown in Cahaba loamy fine sand. Their data showed that maize RLDs on July 31 ranged from 6.2 to 3.7 cm/cm^3 in a 180 cm profile (11), whereas on the same date cotton RLDs ranged from 1.8 to 1.1 cm/cm^3 in the profile (6). Data from these experiments indicated that monocotyledonous crop plants have greater RLDs than dicotyledonous ones grown under the same conditions. This trend also was evident when the RLD data were examined for the 13 species grown in the rhizotron at Auburn (Section 2). Root length density not only differs among species but also varies with soil depth, as shown for sorghum and soybean (26) in the study reported above. Gerwitz and Page (27) found that root dry weight density decreased exponentially with depth in 71 of 101 case histories. In a substantial number of these case histories RLD also decreased with depth. Development of dry soil layers can seriously distort this density-depth pattern. Sivakumar et al. (28) showed that on July 8 soybean RLDs decreased exponentially with depth to 105 cm in Ida silt loam (Typic Udorthents), but 28 days later many roots had died in the drying upper soil layers. The RLDs had decreased at all depths to 60 cm, but had increased at lower depths. As a result, RLD was greater at the 75 to 90 cm depth than at any depth between 22 and 75 cm. Cotton showed the same trends in RLD changes as the soil profile dried (13).

RLD in any specific layer usually increases with time until the developing seeds create a major sink for photosynthates. At that time the RLD starts to decrease in plants that cease vegetative growth. However in indeterminate soybeans and presumably in other crops that continue vegetative growth during seed development, root development continues in deep soil layers during grain-filling (25, 29).

RLDs for soybean, and presumably for all other crops, differ substantially when the crop is grown in different locations, when it is grown in the same location during different years, or when different cultivars are grown at the same time. Burch et al. (26) compared RLDs for the rainfed 'Ruse' cultivar with those for both rainfed and irrigated 'Bragg' cultivar. RLDs were about 3.0 cm/cm^3 at the 5 cm depth in all three cases. The irrigated 'Bragg' RLD decreased to about 1 cm/cm^3 at 50 cm and to about 0.2 cm/cm^3 at 95 cm. The rainfed 'Ruse' RLD decreased to 0.4 cm/cm^3 at 40 cm and did not further change with depth to 1 m. The rainfed 'Bragg' was intermediate between the other two in the pattern of RLD with depth. Willatt and Taylor (29) measured RLD of 'Wayne' soybeans 5 times during a growing season and found that RLD was always less than 1.0 cm/cm^3 at all depths and times in Ida silt loam (Typic Udorthents). Allmaras et al. (25) examined both determinate and indeterminate isolines of 'Harosoy' soybeans grown in Nicollet sandy clay loam (Aquic Hapludolls) in Minnesota at four dates and found that RLD never exceeded 0.3 cm/cm^3 at any depth. In contrast, Arya (30), also working in Minnesota, found that the soybean cultivar '79.648' had RLDs greater than 5.0 cm/cm^3 at some locations in the profile of a Waukegan loam (Typic Hapludolls).

Part of the RLD differences in the various experiments may have been due to sampling or measuring technique differences and part may have been caused by differences in rooting depth among the various experiments. These two factors cannot account for all the RLD differences, however. Even though the plants had comparable leaf areas and their roots extracted about equal amounts of water, the soybeans of Willatt and Taylor (29) had about 10 times the total root length of those of Allmaras et al. (25).

Evaluation of RLD as a factor in delaying onset of water stress in cotton and soybeans is further complicated by uncertainties about the fraction of total root length that is "effective" in water uptake and about the importance of uniform dispersion of that length within the soil volume. RLD is probably one of the controlling variables in allocating water uptake among various soil layers within a particular experiment, but cotton and soybean plants can compensate for low rooting densities much better than for shallow rooting depths.

4. AXIAL RESISTANCE

Xylem elements conduct water from the soil volume where absorption occurs to the crown at ground surface and, finally, through the shoots to the leaves. These xylem elements vary widely in their number and size within any one root cross section. They also vary in length and in continuity within the total root system. The resistance to water flow within any one xylem element probably varies partly with the fourth power of its radius (Poiseuille equation), but several morphological features of the vessel elements modify the fourth-power function (31, 32). Roots with a few small-diameter xylem elements may have large potential energy decreases along their lengths when transpiration rates are great.

Evidence concerning the magnitude of axial resistance in various plant species is scanty and often circumstantial. The greatest reported axial resistance in healthy plants occurred in the seminal root of blue grama [*Bouteloua gracilis* (H.B.K.) Lag. ex Steud.] seedlings. Wilson et al. (33) estimated that the potential energy decrease was 13 bars when water flowed through a 5 cm segment at 8 cm/s. The potential energy decrease thus was 2.6 bars/cm. Hellkvist et al. (34) found that midday xylem water potential along a root of Sitka spruce [*Picea sitchensis* (Bong.) Carr.] decreased as much as 0.06 bar/cm. Taylor and Klepper (31) used data of Willatt and Taylor (29) to estimate that decrease in water potential along the roots of soybean might have averaged 0.07 bar/cm. Nnyamah et al. (35) found a decrease in water potential of about 0.01 bar/cm in a root of Douglas fir [*Pseudotsuga menziesii* (Mirbr) Franco] at a high transpiration rate. In contrast to these situations, in which axial resistances might have been significant, Taylor and Klepper (6), who conducted an experiment using cotton, found the same relationship for all six soil depths when water uptake rate per unit root length per unit water potential difference between soil and leaf was plotted against soil water content. Their interpretation was that axial resistance was not significant in that experiment.

This unresolved question about magnitude of axial resistance in the various species is very important. Significant axial resistance tends to concentrate water use in the surface layers and to retard its use at deep soil layers. Passioura (36) proposed that a wheat cultivar, selected for small metaxylem vessels in its seminal roots to increase axial resistance, might possess a yield advantage over one with larger vessels when grown under drought conditions. Meyer (37) showed that the diameter of wheat metaxylem vessels is subject to selection in breeding programs.

The consequence of significant axial resistance is important enough to have been incorporated in several recent models (31, 38–41). Probably the plant selection process can be used to alter (either increase or decrease) axial resistance, but further experiments are necessary to determine the circumstances under which plants benefit from this alteration. In addition, further experiments are needed to determine, for an individual root genotype, whether it is more efficient to alter axial resistance by increasing the radii of xylem elements, the total number of elements in a specific root member, or the total number of roots that carry water to the crown from a specific volume of soil.

5. RADIAL RESISTANCE

There is considerable controversy in the literature about the effects of transpiration rates on radial root resistance. Some researchers (42–46) have reported that leaf water potential does not change with flow rate, thus implying that radial root resistance decreases as flow rate increases. Other experiments (47–49) have shown a linear decrease in leaf water potential with an increase in transpiration rate, thus implying that radial root resistance remains constant within the range of experimental conditions. Bunce (49) has shown that part but not all of the controversy exists because leaf water potentials did not reach equilibrium.

Questions about the magnitude and constancy of radial root resistance are extremely important in any program for delaying onset of water stress in cotton and soybean plants. Recent research has shown that major control of resistance often resides within the root tissue as water flows from bulk soil to the root xylem (31). Therefore RLD and radial root resistance are inversely related. As radial resistance is lowered, the RLD required to furnish a specific quantity of water to the shoot is lowered.

Taylor and Klepper (6) studied the resistance to water flow through a soil-root system for cotton. They measured leaf water potential as a function of time and, in addition, soil water contents and root length densities as functions of soil depth and time. They found that the combined resistance (soil to soil, soil to root epidermis, and root epidermis to root xylem) increased linearly as soil water content decreased in a particular layer. The relationship between soil-root hydraulic conductivity (q'_r, the inverse of resistance) and soil water content (θ) was

$$q'_r = 10^{-7} (73.44\theta - 2.459) \tag{6.1}$$

where q'_r is cm^3 H$_2$O/cm root/cm $\psi_l - \psi_s$/day, θ is average soil water content (cm^3 H$_2$O/cm^3 soil) for the time period, and ψ_l and ψ_s are the leaf and soil water potentials, respectively. There seemed to be no depth or time effect on the relationship between the hydraulic conductivity and soil water content, which implies that axial resistance was negligible and that changes in leaf water potential with time did not affect q'_r. We should emphasize, however, that Taylor and Klepper (6) used only one set of plants withdrawing water from one soil column during one drying cycle. Bunce (49) has shown that the quantity of leaf area being supplied with water, the radiation intensity onto the leaves, and the duration of radiation all affected his calculated radial root resistances for cotton.

Several investigators (25, 26, 29, 30, 49–52) have determined radial root resistance (or conductance) of field-grown soybeans. Allmaras et al. (25) found that specific root uptake rate of 'Harosoy' soybeans ranged from 197.5 × 10^{-2} cm^3 H$_2$O/cm root/day in wet soil to almost zero in dry soil. Burch et al. (26) found that the specific uptake rate of rainfed 'Ruse' soybeans ranged from 9.9×10^{-3} cm³/cm/day in wet soil to zero in dry soil. Willatt and Taylor (29) determined that the specific uptake rate of rainfed 'Wayne' soybeans ranged from 20×10^{-2} cm³/day in wet soil to zero in dry soil. These three experiments show the wide range in specific root uptake rates when soybeans are grown in field experiments. The results not only point out the large effect of decreases in soil water content on specific uptake rate, but also show that major differences in uptake rate occur in moist soil even when soil hydraulic resistance is low.

The results of Bunce (49) and Eavis and Taylor (50) help to explain the differences in specific root water uptake rate in wet soil. Bunce (49) removed half the soybean leaf area in one of his experiments. This removal, at a specific radiation load, did not alter leaf water potential, but it reduced transpiration of soybean to less than half the original value. Leaf removal thus halved specific rate of root water uptake. Eavis and Taylor (50) altered soil fertility levels, watering history, and container size, and they determined soybean leaf areas, root lengths, and soil water contents during a drought cycle. The treatments altered the ratios of root length to leaf area four fold. At a soil water potential of about -0.33 bar, the specific root water uptake rate was about 4.2×10^{-3} cm³ H_2O/cm/day with 39 cm roots/cm² leaf area, and it was about 1.6×10^{-3} cm³/cm/day with 140 cm roots/cm² leaf area. When compared at equal soil water contents, there were no differences in transpiration rates or in leaf water potentials among the various treatments. In these two experiments a decrease in the leaf area relative to root length caused a decrease in the transpiration rate, did not alter the leaf water potential, and reduced the specific root water uptake rate. In terms of most of the water uptake models, the reduction in specific root water uptake rate was associated with an increase in the calculated root resistance.

Plants obviously need some roots to absorb water and delay water stress, but it is not at all clear that cotton and soybean plants need as many roots as they produce, even near wilting point. Plant scientists probably cannot delay the onset of severe water stress by selecting cotton and soybean plants with root length densities greater than about 0.2 to 0.3 cm/cm³ within the upper half of the soil profile. The inverse relationship between total root length that is available for uptake and the specific root water uptake rate seems to preclude any significant advantages either from increased RLD or reduced radial resistance (or increased specific root water uptake rate). This negative finding for cotton and soybean cannot be extended to any of the grasses, where altered RLD in the upper half of the root zone may significantly affect axial resistance.

6. SUMMARY AND CONCLUSIONS

Plant scientists have suggested that plants need deep, vigorous root systems in drought situations. Many experiments have been conducted to define the attributes of such a root system. Current water uptake models include root depth, RLD, axial resistance to water flow through the xylem, and radial resistance to water flow from the epidermis to the root xylem as appropriate characteristics for use in efforts to delay the onset of severe water stress through selection and breeding techniques.

This chapter has discussed root depth, RLDs, and axial resistance and radial resistance of cotton and soybean (and other plants as appropriate). It has been suggested that crop plants selected for greater than normal root depth often will not suffer water stress as soon as their normally rooted counterparts. An inverse relationship seems to exist between total root length and specific root water uptake rate. Axial resistance seems to be negligible in cotton but may be significant in soybeans for root depths greater than 1 m.

Programs to increase root depth (soil or crop management or plant selection) show more promise for delaying the onset of water stress in cotton or soybean than do programs to increase RLD or specific root water uptake rate. Information is too scanty to permit speculation on the effects of modifying axial resistance in crops like soybean.

REFERENCES

1. J. R. Philip, "The Physical Principles of Soil Water Movement During the Irrigation Cycle," *3rd Cong. Int. Comm. Irrig. Drainage*, **8,** 125 (1957).
2. W. R. Gardner, *Soil Sci.*, **89,** 63 (1960).
3. W. R. Gardner, *Agron. J.*, **56,** 41 (1964).
4. E. I. Newman, *J. Appl. Ecol.*, **6,** 1 (1969).
5. E. I. Newman, *J. Appl. Ecol.*, **6,** 261 (1969).
6. H. M. Taylor and B. Klepper, *Soil Sci.*, **120,** 57 (1975).
7. D. C. Reicosky and J. T. Ritchie, *Soil Sci. Soc. Am. J.*, **40,** 293 (1976).
8. S. M. A. Faiz and P. E. Weatherley, *New Phytol.*, **81,** 19 (1978).
9. J. E. Weaver and W. E. Bruner, *Root Development of Vegetable Crops*, McGraw-Hill, New York. 1927.
10. H. M. Taylor, *Auburn Univ. Agric. Exp. Stn. Circ.* **171** (1969).
11. H. M. Taylor and B. Klepper, *Agron. J.*, **65,** 965 (1973).
12. H. M. Taylor, M. G. Huck, B. Klepper, and Z. F. Lund, *Agron. J.*, **62,** 807 (1970).
13. B. Klepper, H. M. Taylor, M. G. Huck, and E. L. Fiscus, *Agron. J.*, **65,** 307 (1973).
14. H. M. Taylor and B. Klepper, *Aust. J. Biol. Sci.*, **24,** 853 (1971).
15. H. M. Taylor and B. Klepper, *Agron. J.*, **66,** 584 (1974).
16. V. D. Browning, H. M. Taylor, M. G. Huck, and B. Klepper, *Auburn Univ. Agric. Exp. Stn. Bull.* **467** (1975).
17. E. W. Carson, Ed., *The Plant Root and Its Environment*, University of Virginia Press, Charlottesville. 1974.
18. R. W. Pearson, L. F. Ratliff, and H. M. Taylor, *Agron. J.*, **62,** 243 (1970).

19. J. B. Sartain and E. J. Kamprath, *Agron. J.*, **67**, 507 (1975).
20. F. Adams and Z. F. Lund, *Soil Sci.* **101**, 193 (1966).
21. C. D. Foy, in E. W. Carson, Ed., *The Plant Root and Its Environment*, University of Virginia Press, Charlottesville, 1974, p. 601.
22. J. C. Lance and R. W. Pearson, *Soil Sci. Soc. Am. Proc.*, **33**, 95 (1969).
23. H. M. Taylor, E. Burnett, and G. D. Booth, *Z. Acker. Pflanzenbau*, **146**, 33 (1978).
24. W. N. Herkelrath, E. E. Miller, and W. R. Gardner, *Soil Sci. Soc. Am. J.*, **41**, 1039 (1977).
25. R. R. Allmaras, W. W. Nelson, and W. B. Voorhees, *Soil Sci. Soc. Am. Proc.*, **39**, 771 (1975).
26. G. J. Burch, R. C. G. Smith, and W. K. Mason, *Aust. J. Plant Physiol.*, **5**, 169 (1978).
27. A. Gerwitz and E. R. Page, *J. Appl. Ecol.*, **11**, 773 (1974).
28. M. V. K. Sivakumar, H. M. Taylor, and R. H. Shaw, *Agron. J.*, **69**, 470 (1977).
29. S. T. Willatt and H. M. Taylor, *J. Agric. Sci.*, **90**, 205 (1978).
30. L. M. Arya, *University of Minnesota Water Resources Research Center Bulletin* 60 (1973).
31. H. M. Taylor and B. Klepper, *Adv. Agron.*, **30**, 99 (1978).
32. M. H. Zimmermann, *Phytopathology*, **68**, 253 (1978).
33. A. M. Wilson, D. N. Hyder, and D. D. Briske, *Agron. J.*, **68**, 479 (1976).
34. J. Hellkvist, G. P. Richards, and P. G. Jarvis, *J. Appl. Ecol.*, **11**, 637 (1974).
35. J. U. Nnyamah, T. A. Black, and C. S. Tan, *Soil Sci.*, **126**, 63 (1978).
36. J. B. Passioura, *Aust. J. Agric. Res.*, **23**, 745 (1972).
37. W. S. Meyer, *Seminal Roots of Wheat: Manipulation of Their Geometry to Increase the Availability of Soil Water and to Improve the Efficiency of Water Use,* Ph.D. Thesis, University of Adelaide, Adelaide, Australia, 1976.
38. M. N. Nimah and R. J. Hanks, *Soil Sci. Soc. Am. Proc.*, **37**, 522 (1973).
39. C. H. M. Van Bavel and J. Ahmed, *Ecol. Modell.*, **2**, 189 (1976).
40. D. Hillel, H. Talpaz, and H. van Keulen, *Soil Sci.*, **121**, 242 (1976).
41. J. J. Landsberg and N. D. Fowkes, *Ann. Bot. (London)*, **42**, 493 (1978).
42. H. D. Barrs, in R. O. Slatyer, Ed., *Plant Response to Climatic Factors. Proceedings of the Uppsala Symposium, 1970*, UNESCO, Paris, 1973, p. 249.
43. S. E. Camacho-B., A. E. Hall, and M. R. Kaufmann, *Plant Physiol.*, **54**, 169 (1974).
44. D. W. Lawlor and G. F. J. Milford, *J. Exp. Bot.*, **26**, 657 (1975).
45. R. Stoker and P. E. Weatherley, *New Phytol.*, **70**, 547 (1971).
46. R. Tinklin and P. E. Weatherley, *New Phytol.*, **65**, 509 (1966).
47. B. E. Janes, *Plant Physiol.*, **45**, 95 (1970).
48. J. L. Hailey, E. A. Hiler, W. R. Jordan, and C. H. M. Van Bavel, *Crop Sci.*, **13**, 264 (1973).
49. J. A. Bunce, *J. Exp. Bot.*, **29**, 595 (1978).
50. B. W. Eavis and H. M. Taylor, *Agron. J.*, **71**, 441 (1979).
51. B. E. Michel, Environmental Research Center Report 1674, Georgia Institute of Technology, Atlanta, 1974.
52. D. C. Reicosky, R. J. Millington, A. Klute, and D. B. Peters, *Agron. J.*, **64**, 292 (1972).

PHYSIOLOGICAL ADAPTATIONS TO WATER STRESS | III

As indicated in Part I, there has been considerable interest in the last decade in the physiological responses of plants to water deficits. In this part these responses are discussed not simply as responses to water stress, but in light of their adaptive significance. Thus there have been many reports previously of the effects of water stress on stomatal behavior and photosynthesis, but in Chapters 9 and 10 the significance of adaptation of the stomatal apparatus and photosynthetic apparatus to water stress is discussed. In the preceding decade proline and abscisic acid, too, have been reported to increase in response to water stress in a wide range of species, but the authors of Chapters 11 and 12 discuss the role that these compounds play in adaptation to water stress.

The last line of defense that a plant has in its battle against water stress is a protoplasm that is tolerant of dehydration. Although almost all land plants can withstand severe desiccation at certain stages of their life cycle, as when they exist as seeds or underground rhizomes, the "resurrection plants" provide an example of a class of plants with leaf and stem tissue that can withstand extreme water stress and then quickly revive when water is available again. They are therefore worthy of discussion in a book of this nature on adaptation to stress, and the basis of their tolerance of desiccation is examined in the last chapter in this part.

7 | Turgor Maintenance by Osmotic Adjustment: A Review and Evaluation

NEIL C. TURNER

Division of Plant Industry, CSIRO, Canberra, Australian Capital Territory, Australia

MADELEINE M. JONES

Department of Environmental Biology, Research School of Biological Sciences, Australian National University, Canberra, Australian Capital Territory, Australia

1. INTRODUCTION

Small changes in turgor potential are now regarded as the most likely means whereby a plant transduces changes in water status into changes in metabolism (1–4). Therefore the maintenance of turgor during a change in plant water status should maintain the metabolic processes of the plant and aid in its growth and survival (2, 5–7). In higher plants the turgor potential at a particular water potential depends on the osmotic potential and elasticity of the tissue (7, 8). The osmotic potential and elasticity are, in turn, influenced by other factors such as solute accumulation, cell size, osmotic volume, and cell wall thickness (7–9). As has been shown elsewhere (7, 8), a low osmotic potential or the capacity to accumulate solutes, highly elastic cells, and small cells, aid in maintaining positive turgor as the water content of plant tissue decreases.

Water deficits have been shown to induce a lowering of the osmotic potential in some species and cultivars, thereby contributing to the maintenance of turgor as water potential decreases (2, 5–8, 10–13). The effects of water deficits on elasticity are less clear (13). Although water stress generally reduces cell elasticity (12–14), water deficits have been reported to increase elasticity in expanding leaves (15). Furthermore, water stress generally induces a decrease in cell size (16), which tends to increase the elasticity of the cells (9), and induces an increase in the dry weight/turgid weight ratio of tissue, thereby altering the hydration properties of the cells (17).

This chapter considers the effects of water stress on changes in osmotic potential alone; the effects of elasticity have been covered briefly elsewhere (7, 13). Evidence of osmotic adjustment is provided, and the importance of this on turgor maintenance and physiological and morphological processes is evaluated.

2. TERMINOLOGY

A lowering of osmotic potential in response to water deficits can arise from the concentration of osmotic solutes as water is withdrawn from the vacuole and the cell volume decreases, or additionally from the accumulation of solutes in the cell. If cells act as perfect osmometers, the osmotic potential π will change as a result of the passive concentration of solutes according to the following equation:

$$\pi = \frac{\pi_0 V_0}{V} \qquad (7.1)$$

where V is the osmotic volume and π_0 and V_0 are the osmotic potential and osmotic volume, respectively, at a reference value such as full turgor or zero turgor. Active accumulation of solutes in the cell will result in a lowering of the osmotic potential greater than that predicted by Equation 7.1.

The terms osmoregulation, turgor regulation, water activity regulation, osmotic adjustment, and osmotic adaptation have been used to describe the concentration of solutes, the active accumulation of solutes, or both. The failure to distinguish between the concentration of osmotic solutes and their active accumulation is leading to confusion, and we suggest that adherence to the following convention of usage will reduce confusion in the future:

Osmoregulation or *turgor regulation:* "osmoregulation" has been defined as the regulation of osmotic potential within a cell by addition or removal of solutes from solution until the intracellular osmotic potential is approximately equal to the potential of the medium surrounding the cell (18). This phenomenon is widely used by those studying responses of plants and microorganisms to external salinity. It is not clear whether cell turgor or the solute concentration (i.e., the osmotic potential) is the regulatory factor, hence the use of both "osmoregulation" and "turgor regulation" in the literature (3, 4).

Osmotic adjustment: in higher plants this term refers to the lowering of osmotic potential arising from the net accumulation of solutes in response to water deficits or salinity. It refers to the net solute increase and should be used to distinguish the active accumulation of solutes from the passive concentration of solutes. We recommend that "osmotic adjustment" be used only for the accumulation of solutes in higher plants in response to water deficits and that "osmoregulation" and "turgor regulation" be retained for use in relation to lower plants and microorganisms, or the change in osmotic potential of higher plants in response to salinity.

Turgor maintenance: osmotic adjustment in response to water deficits can give rise to either *partial* or *full turgor maintenance*. The changes in osmotic potential arising from osmotic adjustment and solute concentration are shown diagrammatically in Figure 7.1. Full turgor maintenance is represented by line *a*, partial turgor maintenance by line *b*, and no osmotic adjustment by line *c*. Line *d* gives the case for constant internal osmotic potential; that is, it represents a decrease in cell solutes as water deficits increase.

The degree of osmotic adjustment is measured as the change in osmotic potential π at a particular water potential or water content. For convenience and purposes of comparison it is usual to measure the degree of osmotic adjustment at either full turgor (i.e., when $\psi = 0$ and $P = \pi$, where ψ is the total water potential and P is the turgor potential) or at zero turgor (i.e., when $P = 0$ and $\psi = \pi$). The degree of osmotic adjustment measured at full turgor is somewhat smaller than that measured at zero turgor (8). In this chapter, unless otherwise stated, the reported degree of osmotic adjustment refers to the change in osmotic potential at full turgor.

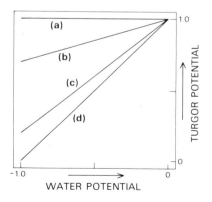

Figure 7.1. Schematic representation of the relationship between turgor potential and total water potential for (*a*) full turgor maintenance, (*b*) partial turgor maintenance, (*c*) no turgor maintenance, and (*d*) constant internal osmotic potential in plant tissue. Arrows show direction of increase in potential.

3. EVIDENCE FOR OSMOTIC ADJUSTMENT

Evidence that plants lower their osmotic potential in response to water deficits has been available for many years, but earlier investigators failed to see the advantages of osmotic adjustment in higher plants (19–21). Recently considerable evidence has accumulated showing that osmotic adjustment takes place in the leaves, hypocotyls, roots, and reproductive organs of several plant species, resulting in full or partial turgor maintenance. An example of turgor maintenance in wheat leaves as the water potential decreases from -1 to -13 bars is shown in Figure 7.2a: one genotype (*Triticum dicoccum*) showed full turgor maintenance, whereas a second genotype (*T. aestivum*) showed no osmotic adjustment over the same range (22, 23). Table 7.1 summarizes the current evidence for full turgor maintenance in leaves, stems, roots, and reproductive organs of higher plants as a result of water deficits.

Consider next an example of partial turgor maintenance in sorghum leaves. Figure 7.2b shows the relationship between turgor potential and water potential for leaves of sorghum plants dried slowly over several weeks compared with those of leaves sampled from well-watered, unstressed plants that were dried quickly over several hours to minimize osmotic adjustment. It is clear that as water potentials decreased below -5 bars, the leaves of the slowly dried plants had higher turgor potentials at similar leaf water potentials compared with the leaves of the rapidly dried plants. In Table 7.2 we summarize data in which osmotic adjustment has been shown to partly maintain turgor: the literature contains numerous examples in which the relationship between turgor potential and leaf water potential ($\Delta P/\Delta \psi_l$) has

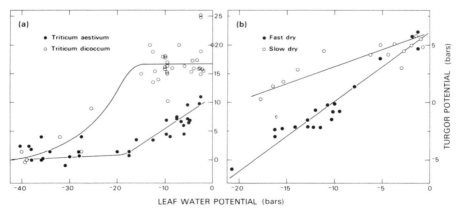

Figure 7.2. Relationship between turgor potential and leaf water potential for (a) two wheat genotypes *Triticum aestivum* spp. *vulgare* (AUS 3850) and *T. dicoccum* (AUS 3582), and (b) sorghum plants dried quickly or slowly. Redrawn from Refs. 22 and 12 and unpublished data of M. M. Jones.

Table 7.1. The Range of Water Potentials Over Which Full Turgor Maintenance Was Observed in Several Organs of Different Species and Cultivars

Organ	Species	Water potential range (bars)	Ref.
Leaves	Sorghum (*Sorghum bicolor*)	−10 to −16[a]	2, 59
	Maize (*Zea mays*)	− 9 to −18[a]	59
	Wheat (*Triticum aestivum* cv. AUS 3850)	0 to −13	22
	Wheat (*Triticum dicoccum*)	0 to −13	23, 23
	Wheat (*Triticum aestivum* cv. Condor)	0 to −10	23
	Wheat (*Triticum aestivum* cv. Heron) (expanding leaves)	−17 to −40	32
	Orchard grass (*Dactylis glomerata*)	−12 to −20	36
	Sugar beet (*Beta vulgaris*)	− 6 to − 8	57
	Apple (*Malus pumila*)	−18 to −26[a]	58
	Silver birch (*Betula verrucosa*)	−12 to −16	25
	Dryas integrifolia	− 7 to −14	60
Branchlet	*Hammada scoparia*	−15 to −30[a]	14
Hypocotyl	Soybean (*Glycine max*)	− 4 to − 9	11
Root	Pea (*Pisum sativum*)	− 3 to − 8[b]	10
Floral apex	Wheat (*Triticum aestivum* cv. Heron)	−18 to −42	32
Spikelet	Wheat (*Triticum* sp.)	− 3 to −20	23

[a]Changes in osmotic potential from water deficits confounded with seasonal changes in osmotic potential.
[b]Soil water potential.

been studied, but these have been omitted unless the $\Delta P/\Delta \psi_l$ for an irrigated crop or well-watered plants can be compared with an unirrigated crop or plants allowed to dry slowly. Recent claims of osmotic adjustment and partial turgor maintenance in two cultivars of sorghum (24) and in oak seedlings (25) are not included in Table 7.2 because of insufficient data on the relationship between turgor potential and leaf water potential in both irrigated and unirrigated plants.

Turgor maintenance by osmotic adjustment has also been shown to occur in response to the daily changes in water status of leaf tissue. Turner (26) observed a daily change in osmotic potential of 6 bars in maize; Hsiao et al. (2) and Wenkert et al. (27) also provide evidence of diurnal changes in osmotic potential greater than that arising from dehydration alone, thereby at least partially maintaining turgor.

Evidence for osmotic adjustment also comes from studies on the effects of water deficits on the concentration of solutes, particularly sugars and free

Table 7.2. Partial Turgor Maintenance by Osmotic Adjustment in Leaves of Plants of Different Species and Cultivars

A difference in slope of the relationship between turgor potential and water potential ($\Delta P/\Delta\psi_l$) between rapidly dried, well-watered plants and plants allowed to dry slowly is used as evidence for osmotic adjustment.

Species	$\Delta P/\Delta\psi_l$		Ref.
	Rapid dry	Slow dry	
Sorghum (*Sorghum bicolor* cv. RS610)	0.65	0.36	12
Sorghum (*Sorghum bicolor* cv. RS610)	0.88	0.48	38
Sorghum (*Sorghum bicolor* cv. RS610)	0.50	0.35	39
Sorghum (*Sorghum bicolor* cv. Shallu)	0.59	0.29	12
Sunflower (*Helianthus annuus* cv. Hysun 30)	0.88	0.46	38
Buffel grass (*Cenchrus cilaris* cv. Biloela)	0.71	0.46	17
Green panic (*Panicum maximum* var. *trichoglume,* cv. Petrie)	0.75	0.47	17
Spear grass (*Heteropogon contortus*)	0.70	0.47	17
Siratro (*Macroptilium atropurpureum* cv. Siratro)	0.74	0.57	17

amino acids (e.g., Refs. 28–31). However few of the studies assessed the contribution of these solutes to the osmotic potential. Increases in the concentration of potassium, sugars, and amino acids accounted for 60 to 100% of the osmotic adjustment observed in the apex and expanding leaves of wheat (32), whereas increases in chloride and carboxylic acids, in addition to potassium, sugars, and free amino acids, were needed to account for the osmotic adjustment in fully expanded sorghum leaves (13). The observation that the osmoles of solute per unit leaf area increased when water deficits developed in sorghum is supporting evidence for osmotic adjustment in leaves of this species (24). Unfortunately, however, changes in the solute concentration reported in cotton in response to leaf water deficits were not accompanied by measurements of the leaf osmotic potential (33).

Not all species or cultivars show evidence of osmotic adjustment. In many instances the lack of osmotic adjustment is attributable to too rapid a rate of drying of the plant (Section 4.1). On the other hand, no evidence of osmotic adjustment was found in two cultivars of soybean (*Glycine max*) or *Agropyron dasystachyum* subjected to slowly developing water deficits in the field (34, 35). Likewise, Gavande and Taylor (36) and Morgan (22, 23) found no evidence of osmotic adjustment in leaves of tomato or three wheat genotypes, respectively, although grown and allowed to dry under conditions similar to those used for other cultivars or species showing full turgor maintenance. From the current evidence it is not clear whether the absence of osmotic adjustment in soybean is common to all cultivars. Certainly Sionit and Kramer (37) observed no osmotic adjustment in a third cultivar of soybean at the pod-formation and pod-filling stages of development, although there was evidence of a limited degree of adjustment while the plants were still vegetative. On the other hand, Wenkert et al. (27) found that the osmotic potential at maximum hydration decreased throughout the season in a fourth cultivar, but it is not clear whether this represents osmotic adjustment in response to water deficits or simply the effect of plant age (2).

There is limited evidence that different plant organs may exhibit different degrees of osmotic adjustment. For example, Morgan (23) observed a greater degree of osmotic adjustment in the apex and expanding leaves of wheat than in fully expanded leaves. Further study is necessary to determine whether these differences reveal true differences in the capacity of organs to adjust osmotically or merely reflect differences in the degree and rate of stress in the different organs (Ref. 32, and Chapter 24, this book).

4. FACTORS AFFECTING OSMOTIC ADJUSTMENT

Several factors are now known to influence the degree of osmotic adjustment.

4.1. Rate of Development of Stress

The rate of development of stress has a major effect on the degree of osmotic adjustment. Hsiao et al. (2) observed full turgor maintenance when the water potential of sorghum leaves decreased at the rate of 0.08 bar/day, whereas Jones and Turner (12) and Turner et al. (38) observed only partial turgor maintenance when the water potential of sorghum leaves decreased at 1 bar/day. Furthermore Jones and Rawson (39) found in another study that the same degree of osmotic adjustment of 5 to 6 bars occurred whether the rate of drying was 1.5 or 7 bars/day, but no osmotic adjustment occurred when drying was as rapid as 12 bars/day. Indeed, Figure 7.2*b* and Table 7.2 show that less adjustment occurs with rapid compared to slow rates of stress development.

The rate of development of water deficits is a function of the soil volume occupied by roots, the hydraulic conductivities of the soil and plant, the stomatal conductance, the leaf area, and the atmospheric demand for water. Gavande and Taylor (36) showed that when orchard grass (*Dactylis glomerata*) dried slowly in a humid, cool atmosphere, turgor was maintained as the leaf water potential decreased from −10 to −16 bars, whereas no such turgor maintenance occurred when the same plants dried rapidly (presumably) in a warm, dry atmosphere.

4.2. Degree of Stress

A recent field study reported a 0.64 and a 0.54 bar change in osmotic potential, measured at zero turgor, for each bar decrease in leaf water potential when the midday leaf water potential decreased over a range of −12 to −22 bars in sorghum and sunflower, respectively (38). In a more detailed study in a glasshouse, sorghum plants allowed to dry until the predawn leaf water potential decreased to −4 bars adjusted osmotically by approximately 4 bars, whereas plants allowed to dry to a predawn leaf water potential of −16 bars had an osmotic adjustment of approximately 7 bars (12). These results suggest that at least in fully expanded sorghum leaves, osmotic adjustment was initially able to maintain turgor fully, but that this capacity decreased as water deficits became more severe. The results contrast with those in the expanding leaves and apex of wheat in which full turgor maintenance occurred only when leaf water deficits were severe enough to stop growth (32). The differences between wheat and sorghum may arise from the dependence of osmotic adjustment in fully expanded leaves on current assimilates for its source of solutes and because competition for these becomes more severe as water deficits increase, whereas solute accumulation in the developing apex and expanding leaves may be a survival mechanism dependent on the transfer of assimilates from elsewhere in the plant.

4.3. Environmental Conditions

In addition to the effects that environmental factors have on the rate of drying, temperature and light appear to exert a direct influence on the degree of osmotic adjustment. Recently Johnson (40) showed that leaf turgor potential at a particular leaf water potential was higher in six range grasses when grown at 10/5°C day/night temperatures than when grown at 15/10°C in controlled environment chambers or at 25/15°C in a glasshouse. This is consistent with the observation that solute accumulation is greater at low temperatures, particularly in temperate grasses, because photosynthesis is less affected than is growth (41, 42).

With respect to the effect of light, Turner and Long (43) found that osmotic adjustment was only 3 bars when the quantum flux density at the leaf level was 650 $\mu E/m^2/s$, but was about 6 bars at a higher light level of 1300 $\mu E/m^2/s$.

4.4. Differences in Cultivars

Morgan (22, 23) observed differences in the presence and absence of turgor maintenance in two wheat genotypes (e.g., Figure 7.2a) and also observed differences in turgor maintenance in cultivars of *Triticum aestivum*. 'AUS 2067' and 'Cappelle Desprez' showed little or no osmotic adjustment in leaves, whereas 'AUS 3850' and 'Condor' showed full turgor maintenance over the water potential range of −1 to −15 bars. However no differences in osmotic adjustment were observed in the leaves of 'Shallu' and 'RS610' sorghum cultivars (Table 7.2) or in two soybean cultivars (12, 34). Varietal differences in osmotic adjustment are dealt with in greater detail in Chapter 24.

5. BENEFITS OF OSMOTIC ADJUSTMENT

Osmotic adjustment is an important mechanism in the drought tolerance of plants (7, 13), and we now turn to a detailed discussion of some of the reasons for this.

5.1. Maintenance of Cell Elongation

Osmotic adjustment enables continuation of cell elongation. Greacen and Oh (10) showed that in soils of similar mechanical resistance, root elongation of wheat seedlings was similar at a range of soil water potentials between -3 and -8 bars as a result of full turgor maintenance by osmotic adjustment in the roots. Hsiao et al. (2) reported a similar degree of osmotic adjustment in maize roots. Meyer and Boyer (11) also showed that growth of the soybean hypocotyl was maintained over a greater water potential range when slow drying allowed turgor maintenance of the hypocotyl compared with the rapid cessation of growth when the plant was rapidly dried and turgor was not maintained (but see Section 6.3). Finally Hsiao et al. (2) suggested that the diurnal changes in osmotic potential and the consequent partial maintenance of turgor are what maintains leaf expansion in the field to a degree not seen in earlier controlled environment studies.

5.2. Maintenance of Stomatal Opening

Osmotic adjustment maintains higher stomatal conductances at lower leaf water potentials in plants that adjust osmotically than in plants that do not adjust osmotically. Turner et al. (38) showed that there was a decrease in the water potential at which a stomatal conductance of 0.17 cm/s was reached as water stress developed, and as the sorghum and sunflower plants adjusted osmotically: a similar result was obtained in Sitka spruce grown under different environmental conditions and with different osmotic potentials (44). Jones and Rawson (39) observed a decrease from -21 to -24 bars in the water potential at which a stomatal conductance of 0.17 cm/s was reached in sorghum plants that adjusted osmotically by 2 and 5 bars, respectively; however with no further increase in osmotic adjustment, a slower rate of stress reduced the leaf water potential for a stomatal conductance of 0.17 cm/s to -28 bars. Chapter 9 provides further evidence that osmotic adjustment plays a role in the adjustment of the stomata to water stress.

5.3. Maintenance of Photosynthesis

Although nonstomatal factors may influence the rate of photosynthesis, a decrease in stomatal conductance will invariably decrease photosynthesis (see, e.g., Ref. 34). Evidence is provided in Chapter 10 that stomata and the internal photosynthetic pathway respond in concert to water deficits in such a way as to maintain the concentration of carbon dioxide within the leaf constant, thereby minimizing photoinhibitory damage to the photosynthetic pathway. Therefore the maintenance of higher conductances at a particular water potential by osmotic adjustment would be expected to maintain higher rates of photosynthesis at a particular leaf water potential. This has been explored recently by Jones and Rawson (39), who showed that sorghum plants allowed to dry slowly and adjust osmotically maintained a higher rate of photosynthesis at low leaf water potentials than sorghum plants in which little adjustment occurred. For example, the rate of net photosynthesis near anthesis decreased to 1 nmol $CO_2/cm^2/s$ at a water potential of -22 bars in sorghum that adjusted minimally, compared to -24 and -29 bars in plants that adjusted osmotically by 5 bars at rates of stress of 7 and 1.5 bars/day, respectively (but see Section 6.3).

5.4. Survival of Dehydration

Barlow et al. (45) showed that the apex of a wheat plant can withstand lower water potentials than an expanded leaf and still recover from the stress: the apex survived a water potential of -60 bars, but the expanded leaf died as the water potential fell below -33 bars. Subsequently Munns et al. (32) demonstrated that the apex and expanding leaves of wheat adjust osmotically and maintain turgor once water stress has stopped leaf expansion, whereas expanded leaves show no turgor maintenance at low leaf water potentials. Whether it is turgor maintenance or protection from rapid water loss by surrounding leaf sheath tissue or other factors that aids survival is not clear, however.

Sugar and proline have been implicated as important contributors to cell survival under dehydration, but it is uncertain whether this is because of their contribution as osmotic solutes to the maintenance of turgor or through the maintenance of the structural and physiological integrity of the cytoplasm (46, 47). The role of proline is discussed in Chapter 12.

5.5. Exploration of Greater Soil Volume for Water

If leaf conductance in osmotically adjusted plants does not decrease as quickly with a decrease in leaf water potential as in nonadjusted plants, transpiration must also not decrease as quickly in plants that have lowered their osmotic potential. Provided the resistances in the plant remain constant, the lowering of water potential in the leaf and in other parts of the pathway in the plant will result in a lowering of the soil water potential. Low osmotic potentials and osmotic adjustment presumably account for the low water potentials of soils reported under some desert species (48). If turgor maintenance allows root growth to continue in drying soils, as reported for wheat and maize (2, 10), this enables a greater volume of soil to be explored or leads to a greater density of roots in a fixed volume of soil. The latter will also decrease the average soil water potential lower than in situations of low root density. Indeed, osmotic adjustment would be of little value to the plant, except for survival (see Section 5.4), if the organism could not maintain transpiration by exploring a greater volume of soil or could not exploit the water in a given soil volume more fully.

6. LIMITATIONS TO OSMOTIC ADJUSTMENT

It is now becoming clear that there are limits to the degree, persistence, and utility of osmotic adjustment.

6.1. Transience of Osmotic Adjustment

The studies by Wilson et al. (17) and Turner et al. (38) showed that osmotic adjustments of from 4 to 7 bars, measured at zero turgor, had disappeared within 10 days of relief of stress in buffel grass, Siratro, sunflower, and sorghum. The recovery of the osmotic potential on rewatering was followed in greater detail in a subsequent study with sorghum (39). Within 3 days an initial osmotic adjustment of 5 bars had decreased to almost 1.5 bars, and within 11 days the osmotic adjustment had completely disappeared. This suggests that if rehydration of the soil is sufficient to prevent development of severe soil water deficits until the solutes contributing to osmotic adjustment have disappeared, there will be no benefit in a subsequent stress cycle. Indeed, when rewatering resulted in the maintenance of high predawn water potentials for 10 days in sorghum, no advantage of a previous stress cycle was observed in osmotic adjustment during a subsequent stress cycle (39). Presumably, however, a partial recharging of the soil profile as a result of light rainfall followed by a decrease in the water potential may result in a greater degree of osmotic adjustment in a subsequent drying cycle. Several studies have reported a change in the response of stomata to water potential, suggestive of a shift in the osmotic potential, as a result of a series of short cycles of stress (49–52). Reports of a decrease in osmotic potential of cotton leaves subjected to a series of cycles of water stress (49) were not confirmed by a subsequent study (53), possibly because in the latter case the plants were fully watered for 3 days after prestressing and before restressing, and this may have been sufficient for loss of all the solutes that had accumulated.

6.2. Osmotic Adjustment Is Finite

The degree of osmotic adjustment is limited. Turgor was not maintained in orchard grass (*Dactylis glomerata*) when the leaf water potential decreased below −20 bars (36). Several wheat genotypes, which had showed osmotic adjustment in the leaves and spikelets when subjected to water deficits, also no longer maintained turgor as the leaf water potentials fell below −20 bars (22, 23; Figure 7.2*a*). Furthermore, in several studies of sorghum, the degree of osmotic adjustment has always been in the range of 5 to 8 bars, even though the conditions under which the studies were undertaken have varied from field to glasshouse to controlled environment cabinets (2, 12, 38, 39, 43).

That the degree of osmotic adjustment must be limited where plants are grown in the restricted soil volume of a container and where there is a limit to the available soil water is obvious (12, 22, 23, 36, 39, 43). However the same degree of adjustment was observed in the field studies with sorghum (2, 38), perhaps indicating that contrary to the observation of Greacen and Oh (10), the ability to adjust osmotically does not fully maintain root growth.

6.3. Osmotic Adjustment Does Not Fully Maintain Physiological and Morphological Processes

Although osmotic adjustment did maintain the hypocotyl growth of soybean (11), turgor maintenance did not prevent a decrease in the growth rate of the hypocotyl as shown in Figure 7.3*a*. Turgor was maintained between 4 and 5 bars as the total water potential decreased from −2 to −8 bars, but the hypocotyl growth rate decreased from 1.5 to 0.15 mm/h. The authors attribute the reduction in growth to an increase in the minimum turgor at which growth occurs as a result of dehydration, although this is contrary to the experience with leaves that showed a decrease in the minimum turgor for growth and/or an increase in the extensibility of the cell as a result of water stress (54).

Similarly, Simmelsgaard (55) observed an increase in stomatal resistance (i.e., a decrease in stomatal conductance) of the adaxial stomata of wheat as leaf water potential decreased, but this was not associated with a decrease in leaf turgor as is shown in Figure 7.3*b*. However this study should be interpreted with caution, since the polyethylene glycol that was used to induce the water deficits may have been absorbed by the roots, causing damage to the stomatal apparatus. Indeed, the study by Simmelsgaard (55) contrasts with those by Jones and Rawson (39) and Ludlow (Chapter 9), who observed a greater degree of stomatal adjustment than osmotic adjustment. As reported previously, the leaf water potential of sorghum at which a conductance of 0.17 cm/s was reached decreased from −21 to −24 to −28

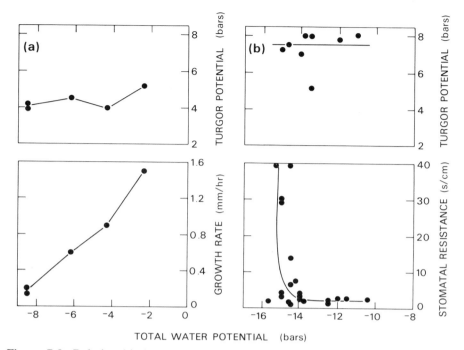

Figure 7.3. Relationship between total water potential and (*a*) turgor potential and the rate of growth of soybean hypocotyls, and (*b*) turgor potential and stomatal resistance of wheat leaves. Redrawn from Refs. 11 and 55.

bars as the rate of drying decreased from 12 to 7 to 1.5 bars/day, respectively, but the osmotic adjustments were only 2, 5, and 5 bars, respectively. In all the studies just cited the change in stomatal conductance was related to the change in turgor and osmotic potentials in the bulk leaf tissue, not to that in the guard cells. The results of Turner et al. (38), Beadle et al. (44), and others (26, 39, 49, 55, 56) suggest only a loose relationship between bulk leaf turgor and turgor of the guard cells.

Because the turgor of the guard cells is not closely coupled with that of the bulk leaf, leaf photosynthesis may not necessarily be maintained in concert with the degree of osmotic adjustment. As pointed out earlier (Section 5.3), when sorghum plants were dried quickly (12 bars/day), osmotic adjustment was 2 bars and the rate of net photosynthesis dropped below 1 nmol CO_2/cm^2/s when the leaf water potential decreased below -22 bars (Figure 7.4). When the plants were dried at 7 and 1.5 bars/day, the osmotic adjustment was about 5 bars in both cases, and the rate of net photosynthesis decreased below 1 nmol CO_2/cm^2/s at values of leaf water potential of -24 and -29 bars, respectively (Figure 7.4).

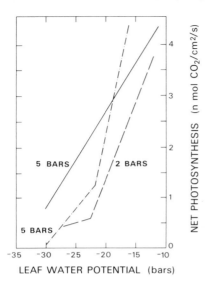

Figure 7.4. Relationship between water potential and the rate of net photosynthesis of sorghum leaves subjected to three rates of stress: —— ——, 12 bars/day; —— · ——, 7 bars/day; and ————, 1.5 bars/day. Bar amounts on the graph indicate the approximate degrees of osmotic adjustment. Redrawn from Ref. 39.

7. CONCLUDING REMARKS

There is now good evidence that osmotic adjustment occurs in some species and cultivars in response to water stress. That such a mechanism can play a role in dehydration tolerance as water deficits develop is unquestioned. However the importance of this role is still open to question. Early promise that osmotic adjustment fully accounted for the lowering of the water potential for stomatal closure has not held up after further experimentation. Furthermore, there is some evidence that in some circumstances increases in the minimum turgor for growth or changes in cell extensibility may override any beneficial effects of osmotic adjustment in the maintenance of extension growth.

A variety of mechanisms enable plants to survive drought, but these vary in cost as far as productivity is concerned. Stomatal control or a reduction in leaf area will almost certainly give a significant reduction in productivity and yield. Osmotic adjustment, however, provides the potential for maintaining photosynthesis and growth of at least some parts of the plant as water deficits increase. Thus the cost of osmotic adjustment ought to be lower, although the solute accumulated in osmotic adjustment cannot be used elsewhere. The range of compounds involved in osmotic adjustment is principally soluble sugars, potassium, organic acids, chloride, and free amino acids. Since these compounds can account for complete osmotic adjustment in some cases, it appears unlikely that new biochemical pathways or syntheses of new compounds are involved during osmotic adjustment; rather, the lowering of the solute potential seems to arise from alteration of existing pathways and translocation patterns in the plant.

The degree and presence or absence of osmotic adjustment varies with species or cultivars. The variation and heritability of osmotic adjustment certainly warrants further investigation. Unfortunately, the range of compounds involved in solute accumulation likely precludes the use of a simple chemical test that would aid in the screening of a wide range of genotypes for osmotic adjustment.

Thus several aspects of osmotic adjustment deserve further study:

1. Its role in maintaining turgor relative to the role of cell size and elasticity.

2. The cost of producing the solutes involved in osmotic adjustment and the role of changed translocation in osmotic adjustment.

3. Its variation among cultivars and genotypes and its effect on their drought tolerance.

ACKNOWLEDGMENTS

We thank Drs. J. R. Wilson, R. E. Munns, and J. M. Morgan for access to manuscripts prior to publication and Professor P. J. Kramer, Dr. J. R. Wilson, and Dr. R. E. Munns for helpful comments on the manuscript.

REFERENCES

1. T. C. Hsiao, *Annu. Rev. Plant Physiol.*, **24**, 519 (1973).
2. T. C. Hsiao, E. Acevedo, E. Fereres, and D. W. Henderson, *Philos. Trans. R. Soc. London, Ser. B*, **273**, 479 (1976).
3. J. Gutknecht and M. A. Bisson, in A. M. Jungreis, T. K. Hodges, A. Kleinzeller, and S. G. Schulze, Eds., *Water Relations in Membrane Transport in Plants and Animals*, Academic Press, New York–San Francisco–London, 1977, p. 3.
4. U. Zimmermann, *Annu. Rev. Plant Physiol.*, **29**, 121 (1978).
5. J. E. Begg and N. C. Turner, *Adv. Agron.*, **28**, 161 (1976).
6. N. C. Turner and J. E. Begg, in J. R. Wilson, Ed., *Plant Relations in Pastures*, CSIRO, Melbourne, Australia, 1978, p. 50.
7. N. C. Turner, in H. Mussell and R. C. Staples, Eds., *Stress Physiology in Crop Plants*, Interscience, New York–London, 1979 p. 343.
8. M. M. Jones, *Physiological Responses of Sorghum and Sunflower to Leaf Water Deficits*, Ph.D. thesis, Australian National University, Canberra, Australia, 1979.
9. E. Steudle, U. Zimmermann, and U. Lüttge, *Plant Physiol.*, **59**, 285 (1977).
10. E. L. Greacen and J. S. Oh, *Nature (London) New Biol.*, **235**, 24 (1972).
11. R. F. Meyer and J. S. Boyer, *Planta*, **108**, 77 (1972).
12. M. M. Jones and N. C. Turner, *Plant Physiol.*, **61**, 122 (1978).

13. M. M. Jones, N. C. Turner, and C. B. Osmond, in L. G. Paleg and D. Aspinall, Eds., *The Physiology and Biochemistry of Drought Resistance*, Academic Press, New York–San Francisco–London, 1980 (in press).

14. L. Kappen, J. J. Oertli, O. L. Lange, E.-D. Schulze, M. Evenari, and U. Buschbom, *Oecologia*, **21**, 175 (1975).

15. J. Elston, A. J. Karamanos, A. H. Kassam, and R. M. Wadsworth, *Philos. Trans. R. Soc. London, Ser. B*, **273**, 581 (1976).

16. S. A. Quarrie and H. G. Jones, *J. Exp. Bot.*, **28**, 192 (1977).

17. J. R. Wilson, M. M. Ludlow, M. J. Fisher, and E.-D. Schulze, *Aust. J. Plant Physiol.*, **7**, 207 (1980).

18. L. Borowitzka, in L. G. Paleg and D. Aspinall, Eds., *The Physiology and Biochemistry of Drought Resistance*, Academic Press, New York–San Francisco–London, 1980 (in press).

19. H. Walter, *Die Hydratur de Pflanze und ihre physiologisch-ökologische Bedeutung (Untersuchungen über den osmotischen Wert)*, Fischer, Jena, 1931.

20. H. Walter and E. Stadelmann, in G. W. Brown, Ed., *Desert Biology*, Vol. 2, Academic Press, New York–San Francisco–London, 1974, p. 213.

21. K. Kreeb, in A. J. Rutter and F. H. Whitehead, Eds., *The Water Relations of Plants*, Blackwell, London, 1963, p. 272.

22. J. M. Morgan, *Nature (London)*, **270**, 234 (1977).

23. J. M. Morgan, *J. Exp. Bot.* **31**, (in press), (1980).

24. D. G. Stout and G. M. Simpson, *Can. J. Plant Sci.*, **58**, 213 (1978).

25. O. Osonubi and W. J. Davies, *Oecologia*, **32**, 323 (1978).

26. N. C. Turner, *Plant Physiol.*, **55**, 932 (1975).

27. W. Wenkert, E. R. Lemon, and T. R. Sinclair, *Ann. Bot. (London)*, **42**, 295 (1978).

28. J. A. Hellebust, *Annu. Rev. Plant Physiol.*, **27**, 485 (1976).

29. N. M. Barnett and A. W. Naylor, *Plant Physiol.*, **41**, 1222 (1966).

30. T. N. Singh, L. G. Paleg, and D. Aspinall, *Aust. J. Biol. Sci.*, **26**, 45 (1973).

31. H.-J. Jäger and H. R. Meyer, *Oecologia*, **30**, 83 (1977).

32. R. E. Munns, C. J. Brady, and E. W. R. Barlow, *Aust. J. Plant Physiol.*, **6**, 379 (1979).

33. J. M. Cutler, D. W. Rains, and R. S. Loomis, *Agron. J.*, **69**, 773 (1977).

34. N. C. Turner, J. E. Begg, H. M. Rawson, S. D. English, and A. B. Hearn, *Aust. J. Plant Physiol.*, **5**, 179 (1978).

35. R. E. Redmann, *Oecologia*, **23**, 283 (1976).

36. S. A. Gavande and S. A. Taylor, *Agron. J.*, **59**, 4 (1967).

37. N. Sionit and P. J. Kramer, *Agron. J.*, **69**, 274 (1977).

38. N. C. Turner, J. E. Begg, and M. L. Tonnet, *Aust. J. Plant Physiol.*, **5**, 597 (1978).

39. M. M. Jones and H. M. Rawson, *Physiol. Plant.*, **45**, 103 (1979).

40. D. A. Johnson, *Crop Sci.* **18**, 945 (1978).

41. J. R. Wilson, *Neth. J. Agric. Sci.*, **23**, 48 (1975).

42. J. R. Wilson, *Neth. J. Agric. Sci.*, **23**, 104 (1975).

43. N. C. Turner and M. J. Long, *Aust. CSIRO Div. Plant Ind. Ann. Rep.*, **1978**, p. 93.

44. C. L. Beadle, N. C. Turner, and P. G. Jarvis, *Physiol. Plant.*, **43**, 160 (1978).

45. E. W. R. Barlow, R. Munns, N. S. Scott, and A. H. Reisner, *J. Exp. Bot.*, **28**, 909 (1977).
46. O. Y. Lee-Stadelmann and E. J. Stadelmann, in O. L. Lange, L. Kappen, and E.-D. Schulze, Eds., *Water and Plant Life: Problems and Modern Approaches*, Springer-Verlag, Berlin–Heidelberg–New York, 1976, p. 268.
47. B. Schobert, *J. Theor. Biol.*, **68**, 17 (1977).
48. I. Noy-Meir, *Annu. Rev. Ecol. Syst.*, **4**, 25 (1973).
49. K. W. Brown, W. R. Jordan, and J. C. Thomas, *Physiol. Plant.*, **37**, 1 (1976).
50. K. J. McCree, *Crop Sci.*, **14**, 273 (1974).
51. J. C. Thomas, K. W. Brown, and W. R. Jordan, *Agron. J.*, **68**, 706 (1976).
52. J. M. Cutler and D. W. Rains, *Crop Sci.*, **17**, 329 (1977).
53. J. M. Cutler and D. W. Rains, *Physiol. Plant.*, **42**, 261 (1978).
54. J. A. Bunce, *J. Exp. Bot.*, **28**, 156 (1977).
55. S. E. Simmelsgaard, *Physiol. Plant.*, **37**, 167 (1976).
56. N. C. Turner, in R. L. Bieleski, A. R. Ferguson, and M. M. Creswell, Eds., *Mechanisms of Regulation of Plant Growth*, Bulletin 12, Royal Society of New Zealand, Wellington, 1974, p. 423.
57. P. V. Biscoe, *J. Exp. Bot.*, **23**, 930 (1972).
58. J. E. Goode and K. H. Higgs, *J. Hortic. Sci.*, **48**, 203 (1973).
59. E. Fereres, E. Acevedo, D. W. Henderson, and T. C. Hsiao, *Physiol. Plant.*, **44**, 261 (1978).
60. A. P. Hartgerink and J. M. Mayo, *Can. J. Bot.*, **54**, 1884 (1976).

8 | Stomatal Response to Water Stress in Conifers

PAUL G. JARVIS

Department of Forestry and Natural Resources, University of Edinburgh, Edinburgh, U.K.

1. INTRODUCTION

Conifers tolerate normal diurnal variations in water potential without apparent ill effects. More severe water stress in the first instance is avoided by stomatal closure. Consequently knowledge of how stomatal conductance responds to environmental variables is needed to form any hypothesis about acclimation or adaptation to water stress. In addition, a proper quantitative appreciation of stomatal responses to environmental variables is required for two very practical purposes. First, at the present state of development any mechanistic models of production processes in canopies include stomatal conductance as a variable that must be known. Second, stomatal conductance is required in all physically based models of water loss by transpiration from canopies. Consequently production ecologists, water relations ecologists, and hydrologists concerned with the carbon and water balances of afforested catchments need a means of predicting stomatal conductance, insofar as is possible, from environmental variables.

This chapter systematically describes the response of the stomata of conifers to environmental variables and draws attention to some anomalies that on the face of it suggest that the mechanism of stomatal action in conifers may be somewhat different from current conceptions. For this purpose I make use of information obtained in laboratory experiments on the following two species: *Picea sitchensis* (Bong.) Carr. or Sitka spruce, which

105

is a native of northwestern North America, and *Pinus sylvestris* L. or Scots pine, which is a native of northern Europe. Despite the considerable differences in morphology and canopy characteristics between pines and spruces, and the very different geographical distribution of these two species, their stomata respond to environmental variables in broadly similar ways, suggesting that it may be reasonable to generalize about stomatal responses in conifers as a result.

Because coverage by wax tubes prevents direct observation of the stomatal pore in conifers (1), I refer mainly to measurements of stomatal conductance either derived from the measurement of water vapor exchange in an assimilation chamber, following conventional theory (2), or made with a null balance diffusion porometer (3). Measurements made in the assimilation chamber have been on parts of potted plants, grown in growth rooms or in a glasshouse, or on detached shoots from the forest canopy. Porometers have been used on potted plants in growth rooms and in a wind tunnel and in forest canopies.

2. CONTROL OF TRANSPIRATION BY CONIFEROUS FOREST

The transpiration of water from foliage depends strongly on the available energy A, the saturation deficit of the air δe, the wind speed u, and, to a lesser extent on the air temperature θ. In addition, transpiration is strongly dependent on the stomatal conductance g_s of the needles and to a lesser extent on structural characteristics of the foliage that influence the boundary layer conductance g_a in addition to the wind speed. These variables are frequently associated in the following equation (4):

$$\lambda E = \frac{sA + c_p \rho \delta e g_a}{s + \gamma(1 + g_a/g_s)} \tag{8.1}$$

The physical parameters c_p, ρ, γ, and λ—the specific heat of air, the density of air, the psychrometric constant, and the latent heat of vaporization of water, respectively—are all weak functions of temperature. The parameter s is the slope of the curve relating the saturated vapor pressure of water e_s to temperature (i.e., $de_s/d\theta$) at the approximate temperature and varies appreciably with temperature (see appendices in Ref. 4).

Equation 8.1 applies to single leaves, plants, or extensive canopies and can be applied successfully over periods of weeks, days, hours, or minutes, provided appropriate values of the variables are available. Application over short periods of time, such as an hour, is usually limited by the availability of hourly meteorological data and hourly measurements or reliable estimates of stomatal conductance.

For the aerodynamically rough canopy of coniferous forest, the term containing the saturation deficit δe is usually much larger than the term containing the available energy A, and the ratio g_a/g_s is also large: typically g_a is 10 to 30 cm/s, whereas g_s is less than 2 cm/s (both expressed per unit ground area) (5). Therefore Equation 8.1 can be simplified and the rate of transpiration approximated by:

$$E = \frac{c_p\rho}{\lambda\gamma}\, \delta e h_s = K \delta e g_s \qquad (8.2)$$

where K is the factor for converting partial pressure of water vapor into mass concentration and has a value of 740×10^{-6} kg/m³/mbar at a temperature of 20°C and atmospheric pressure of 1013 mbar. This approximation is only reasonable (within, say, 10%) for aerodynamically rough vegetation.

In principle the rate of transpiration could vary widely and could reach very high values depending on the values of the variables in Equation 8.1, particularly δe and g_s (Equation 8.2). In practice, however, the range is restricted by the *combinations* of values of the main variables that occur. These combinations depend both on the environmental conditions and on the morphological and physiological properties of the foliage of the trees, and particularly on the response of the stomata to the environmental variables.

Measured mean hourly rates of transpiration from extensive plantations of Sitka spruce in Scotland (Forest of Mearns) and Scots pine in England (Thetford Forest) can reach 0.3 mm/h (200 W/m²), but are more commonly in the range 0.1 to 0.2 mm/h (70 to 140 W/m²), depending on the weather and time of day (5). Daily rates can reach 3 mm/day in the summer, but are more commonly about 2 mm/day in summer and 0.3 mm/day in winter. These rates of transpiration are less than one-third of what would be expected if the stomatal conductances of the canopies had remained constant throughout the day at the measured maximum values of 1 or 2 cm/s for Scots pine and Sitka spruce, respectively.

Why are the measured transpiration rates much lower than might be expected? From Equation 8.2 it is clear that high transpiration rates will result only if g_s remains large, when δe is also large. In practice we find that g_s is large in humid, cool weather when saturation deficits are small, but because of the small deficits, transpiration is not large. On the other hand, in warm, dry weather when saturation deficits are large, the stomatal conductance is small. Since stomatal conductance and saturation deficit are never both large at the same time, transpiration never reaches high rates. Although the rate of transpiration does increase as δe increases (Figure 8.1), it does not increase to the extent that would be expected, because g_s decreases as δe increases.

It is therefore very pertinent to consider how g_s is controlled in general by environmental variables, and in particular by saturation deficit.

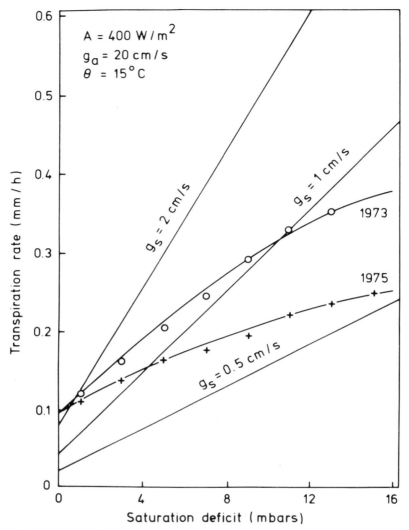

Figure 8.1. Relationship between transpiration rate and saturation deficit calculated from Equation 8.1 at three different values of stomatal conductance g_s. The data points are medians of a large number of mean hourly values within saturation deficit classes for Sitka spruce at Forest of Mearns in June and July 1973 (O) and 1975 (+): A = available energy, g_a = boundary layer conductance, and θ = air temperature. The stand was thinned in 1974.

108

3. RESPONSE OF STOMATAL CONDUCTANCE TO ENVIRONMENTAL VARIABLES

3.1. Response to Photon Flux Density (Light)

As in other plants, the stomata of conifers open and close in response to light. In Scots pine and Sitka spruce the relationship between g_s and light is hyperbolic (6, 7), maximum opening being achieved at moderate photon flux densities with bilateral illumination but rarely with unilateral illumination (6). Figure 8.2 shows marked hysteresis in the response of g_s to increasing and decreasing photon flux. This hysteresis results from the long time constant of stomatal response to a change in photon flux density and largely disappears after 2 h of equilibration. In addition there is a carryover effect from the previous treatment. When the photon flux density is increased or decreased, the value of g_s after an equilibration of 1 h still depends to some extent on the opening or closing trend during the previous hour (6).

The stomata open and close in response to light in air with or without carbon dioxide (CO_2) present (Figure 8.3). Over a wide range of CO_2 concentrations (20 to 8000 $\mu l/l$), g_s in both species seems to be unaffected by CO_2

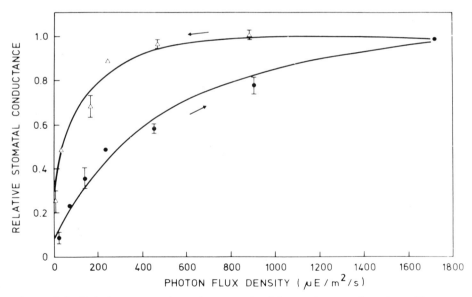

Figure 8.2. Relative stomatal conductance G of Scots pine as a function of increasing (●) and decreasing (△) photon flux density with equilibrium periods of 1 h; fitted curves. Average stomatal conductance for three shoots at $G = 1.0$ was 0.45 cm/s. Other conditions as follows: forest shoots; bilateral illumination; leaf temperature, 10.1°C; leaf-air vapor pressure difference, 5.3 mbars. Two standard errors are shown on representative points. From Ref. 6.

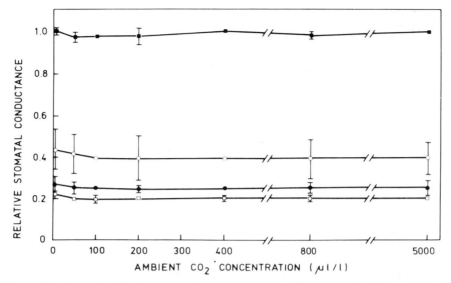

Figure 8.3. Relationship between relative stomatal conductance G of Scots pine and ambient CO_2 concentration at four levels of photon flux density: 1300 (■), 170 (○), 25 (●), 0(□)$\mu E/m^2/s$. Average stomatal conductance of three shoots at $G = 1.0$ was 0.33 cm/s; leaf temperature, 20.6°C; leaf-air vapor pressure difference, 5.8 mbars. Two standard errors are shown on representative points. From Ref. 6.

concentration at both high and low photon flux densities, even though the lower end of the range is below the CO_2 compensation concentration. Conventional hypotheses assume that the response of g_s to light is actually a response to the concentration of CO_2 in the intercellular spaces, which changes as a result of changing rates of photosynthesis (8). Figure 8.3 suggests that this is not the case, thereby raising questions about the mode of action of light. The hypothesis we are now testing is discussed later in relation to the effects of CO_2.

The photon flux density at which saturation occurs is much lower in the field than in the laboratory, sometimes being no more than 200 $\mu E/m^2/s$. This would seem to result partly from the general sluggishness of the stomata and the carryover effect of the previous conditions, and from acclimation. Acclimation to the light environment in which needle development has occurred is evident in the forest canopy. "Shade" needles from the middle and lower parts of the canopy open to smaller values of g_s in response to photon flux density than do the "sun" needles from the top of the canopy (7). Similar acclimation phenomena have been induced by growing potted plants in different photon flux densities in controlled environments (7).

3.2 Response to Temperature

Effects of temperature are frequently confounded by simultaneous changes in saturation deficit so that the response of g_s to temperature alone is poorly known. In conifers the response curve is bell shaped, with a distinct optimum temperature for maximum g_s and low and high temperatures at which g_s is almost zero (Figure 8.4).

Acclimation of g_s to temperature can be induced in controlled environments (9) and without doubt occurs in the field. However it is difficult to characterize in the field because of the comparatively narrow range of temperature usually occurring over a few days in any season (10). Whereas in growth room plants the high temperature at which g_s is zero is quite low (30 to 35°C), in the field it is much higher (45°C), with a consequent broadening of the range of temperature in which g_s is fairly close to maximum.

Interpretation of the response of g_s to temperature is difficult because we have no firm idea about which processes or enzymes in the guard cells are most likely to be determining the response.

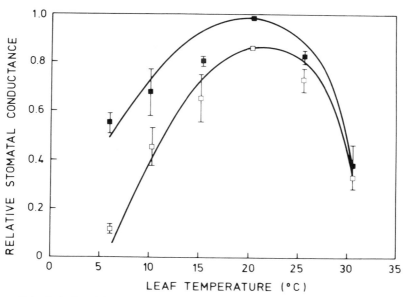

Figure 8.4. Relationship between relative stomatal conductance G and leaf temperature at leaf-air vapor pressure differences of 6.2 (□) and 9.5 mbar (■) in Scots pine; fitted curves. Average stomatal conductance at $G = 1.0$ was 0.44 cm/s; photon flux density, 1260 $\mu E/m^2/s$. All points are means of five measurements with two standard errors shown on representative points. From Ref. 6.

3.3. Response to Carbon Dioxide

Over a wide range of ambient C_a and mean intercellular space C_i concentrations of CO_2, the stomata of several species of conifer do not seem to respond strongly and rapidly to CO_2 in white light (6, 9, 11–13).

In Scots pine (6) and Sitka spruce (R. E. Neilson, personal communication), this lack of response is not restricted to high photon flux densities, as is apparently the case in some other species. Furthermore, Beadle et al. (12) have shown that a response is not induced by water stress, as apparently occurs in some other species (14). Clearly, too, in contrast to some other species (15), g_s in conifers cannot adjust to maintain a more or less constant C_i over a wide range of photon flux density and C_a.

These results suggest that stomatal action in conifers may be independent of photosynthetic carbon fixation. Conventionally it has been supposed that separate photochemical processes, activated by blue and red light, are involved in stomatal opening and that the process activated by red light was CO_2 dependent, involving photosynthetic carbon fixation, whereas the blue light effect was CO_2 independent (8). Recent studies have shown that the stomata of Scots pine are much more sensitive to blue light than to red light and that opening occurs in both blue and red light in the presence or absence of CO_2 (J. I. L. Morison, personal communication). Moreover, opening induced by both blue and red light can be stimulated by prolonged exposure in CO_2-free air, although the degree of opening in CO_2-free air in red light is still less than in air in blue light of the same photon flux density. In addition, removal of oxygen strongly stimulates opening in red light after prolonged exposure. These observations suggest that photosynthesis is involved in stomatal action in conifers, but not perhaps as a result of carbon fixation. Tentatively, we suggest that photosynthesis may at least partly affect stomatal aperture through a step in the catena preceding carbon fixation such as the production of ATP in cyclic photophosphorylation to drive the movement of potassium (K^+) ions in and out of guard cells.

3.4. Response to Leaf Water Potential

Stomatal conductance in conifers is sensitive to leaf water potential over a wide but variable range (see, e.g., Ref. 16), as indicated by Figure 8.5. There is a threshold or critical water potential for closure that in Sitka spruce is similar in growth room and glasshouse plants, and in field-grown trees (17). In Sitka spruce the critical leaf water potential is about −20 bars and in Scots pine it is about −8 bars. Similar threshold values, generally in the range −14 to −25 bars, have been reported for other species of conifer (see, e.g., Refs. 16, 18). In conifers as in other species (Chapter 9), acclimation results in considerable variation in the value of the critical water potential. Beadle et al. (17) give a range of values from −16 to −27 bars for Sitka spruce needles at different times of the year. The critical potential for needles in different

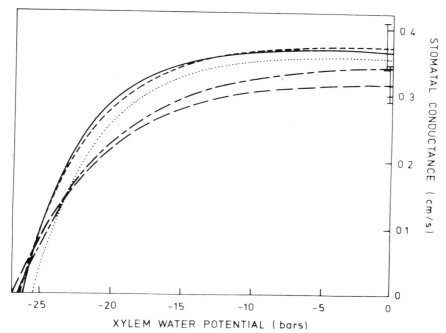

XYLEM WATER POTENTIAL (bars)

Figure 8.5. Relationship between stomatal conductance of Sitka spruce and leaf water potential at ambient CO_2 concentrations of 600(———), 300 (————), 150 (——-——), 75 (————) and 20 (·····) $\mu l/l$. The lines are fitted as two parameter exponentials; needle temperature, 20°C; leaf-air vapor pressure difference, 6 mbars; photon flux density, 1000 $\mu E/m^2/s$. From Ref. 12.

positions in the canopy or in different light environments in growth rooms also falls within this range.

The ecological consequence of the threshold is that the normal diurnal variations in leaf water potential, which result from the diurnal changes in transpiration rate, do not cause significant stomatal closure. Significant closure occurs only when reductions in soil water potential result in much lower leaf water potentials. Hence the reduction in leaf water potential comes into play as a significant variable, feeding back to control water loss, only when soil drought has ensued.

Interpretation of this type of response depends essentially on an assumption regarding the turgor relations of the guard cells and the subsidiary cells. As the water content, turgor potential, and water potential of the guard cells fall over the threshold range of values, stomatal conductance remains the same. From the work of Meidner and Edwards (19), we may conclude that over the threshold range, the difference in turgor potential (TPD) between guard cells and subsidiary cells remains constant despite the fall in turgor potential in each cell. At some stage, however, a further loss of water from the complex results in a larger fall in turgor potential in the guard cells than in the subsidiary cells, and closure ensues. The following assumptions are involved:

1. Stomatal conductance is proportional to the difference in turgor potential between the guard cells and the subsidiary cells (TPD).
2. The relationship between TPD and volume of water in the guard cells has a distinct threshold.
3. Guard cell water relations follow a Höfler diagram similar to that determined for leaf tissue.
4. Maximum stomatal conductance occurs at less than maximum turgor potential; full closure occurs at more than the minimum possible turgor potential.
5. Guard cell water potential is loosely related to bulk mesophyll water potential.
6. Solute accumulation occurs as a result of ion transport and metabolic processes during water stress; some time may be required for appreciable accumulation.

Figure 8.6 shows the changes that could be expected to occur in guard cell water relations and in stomatal conductance on the basis of these assumptions. As water stress develops, guard cell water potential falls along the line $AXYZ$. Stomatal closure begins when the water potential reaches X and the TPD begins to decline, and is complete when the water potential reaches Y. Thus stomatal conductance remains constant over a substantial range of changing water potentials, then declines rapidly over a comparatively narrow range of water potentials. As a result of osmotic adjustment (Chapter 7), shown by the lower osmotic potential lines π_2 and π_3, the threshold water potential and the water potentials over the range in which stomatal closure occurs are lower than would be the case if the osmotic potentials remained on the π_1 line. After water again becomes readily available, recovery probably occurs along the line $ZCBA$. The subsequent extrusion of ions and dispersal of organic solutes would explain in part the often observed lag in recovery of stomatal conductance behind water potential.

3.5. Response to Saturation Deficit

The stomata of conifers close in response to increasing leaf-air vapor pressure difference (LAVD) or saturation deficit of the air (see, e.g., Refs. 20, 21). In the forest canopy stomatal conductance falls linearly in response to increasing saturation deficit over a wide range of deficits in both Sitka spruce (23) and Scots pine (C. L. Beadle et al., unpublished). On the other hand, since g_s of shoots in the assimilation chamber often is insensitive to LAVD until a threshold is reached, the response essentially can be described by a

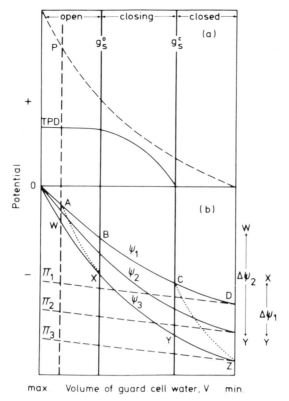

Figure 8.6. A diagram of (*a*) the relationships between turgor potential of the guard cells (*P*) and the turgor potential difference (TPD) between guard cells and subsidiary cells and the volume of water in the guard cells (*V*) and (*b*) the relationships between osmotic potential (π), water potential (ψ), and volume of water (*V*) in the guard cells during stomatal closure and opening. The symbols g_s^o and g_s^c represent stomatal conductance when stomata are open and closed, respectively. As water stress occurs, guard cell water potential falls along the line *AXYZ*, moving from *A* to *X* as guard cell osmotic potential falls from π_1 to π_2 to π_3. At *Y* stomatal closure is complete. During recovery from stress, water potential rises along the line *ZCBA*.

negative exponential (Figure 8.7). The response of the stomata of potted plants in the growth room and wind tunnel is generally intermediate in form, often with no threshold or only a small threshold (Figure 8.8). These observations suggest that the sharpness of the response depends on the base level of water potential in the plant. In an experiment designed to test this suggestion, Ng (6) showed that sensitivity to LAVD in Scots pine was lost completely if shoots were detached from potted plants in the wind tunnel and placed with their base in water. The water potential of the detached shoots

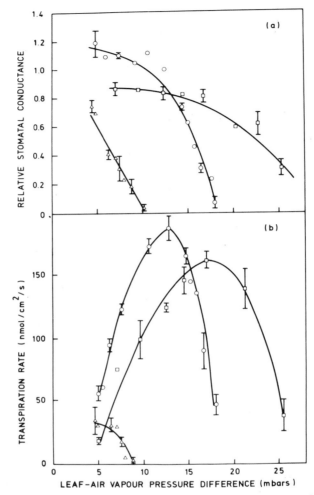

Figure 8.7. Relative stomatal conductance G and transpiration rate of Scots pine as a function of leaf-air vapor pressure difference at leaf temperatures of 25°C (□), 20°C (○), and at 10°C (△); fitted curves. Average stomatal conductance at $G = 1.0$ was 0.43 cm/s; photon flux density, 1320 μE/m²/s. All points are mean of five measurements with two standard errors shown on representative points. From Ref. 6.

rose to about -2 bars, whereas the water potential of the attached shoots, which remained sensitive, was about -7 bars.

In addition to an interaction with water potential, there appears to be an interaction with temperature. The value of LAVD at which g_s extrapolates to zero in Scots pine decreases with decreasing temperature; moreover, at temperatures substantially below the optimum the threshold disappears (Figure 8.7).

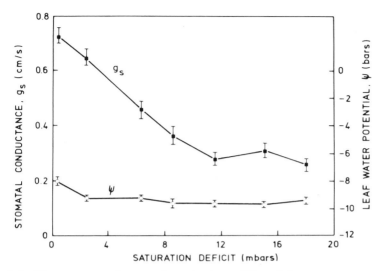

Figure 8.8. The response of stomatal conductance (■) and shoot water potential (○) of Sitka spruce to increasing vapor pressure deficit. All points are means of five measurements with two standard errors shown.

The mechanism by which g_s responds to LAVD in other species has been discussed many times recently (e.g., Ref. 22) and earlier with respect to Sitka spruce (20, 21, 23). With new information from Scots pine, we are now in a better position to assess the mode of action in conifers.

I start with the following assumptions:

1. The leaf does not contain a relative humidity sensor, so that LAVD, or δe, is sensed through its influence on an evaporation rate, as specified in Equation 8.2.

2. Reduction in stomatal aperture results from a larger decline in guard cell turgor potential than in subsidiary cell turgor potential (i.e., a decline in TPD).

Given these assumptions, at least four possible hypotheses could explain how an increase in δe could cause apparent stomatal closure as a result of an increase in an evaporation rate. An increase in δe might have one of the following effects:

1. Cause "incipient drying" of the mesophyll cell walls lining the substomatal cavities rather than decreasing stomatal aperture itself.

2. Increase the rate of evaporation from the mesophyll cells, leading to a reduction in bulk leaf water potential and bulk leaf turgor, which would feed back to cause a fall in TPD.

3. Increase the rate of evaporation from the guard cells themselves, or from the hypodermal cells close to the guard cells, within the leaf, thereby leading to a reduction in the TPD; that is, the so-called peristomatal transpiration hypothesis.

4. Increase the rate of evaporation from the guard cells and subsidiary cells and other closely adjacent epidermal cells external to the stomatal pore, thereby causing a reduction in the TPD.

Hypothesis 1 is unlikely because an improbably large reduction in vapor pressure at the liquid-air interface in the leaf would be needed to account for the apparent reduction in g_s, if g_s had in fact not changed (23). Furthermore, in other species with more suitable stomata a change in the stomatal aperture has been observed with the microscope or with a viscous flow porometer.

Two straightforward observations are crucial to the evaluation of the other three hypotheses. These are the bulk leaf water potential and the transpiration rate, as g_s changes in response to δe.

If the bulk leaf water potential remains more or less constant and reasonably high, as shown in Figure 8.8, hypothesis 2 can be eliminated. The very small change in water potential suggests that there can be only a small hydraulic resistance from the source of water in the stele across the main stores of water in the mesophyll. On the other hand, sensitivity of stomatal control suggests the presence of a large hydraulic resistance between the main stores of water and the guard cell vacuoles. The stele is surrounded by an endodermis, and there are several mesophyll cells between it and the epidermis. This pathway must have a low resistance to flow of water, for otherwise an appreciable fall in water potential would develop in the mesophyll at high transpiration rates. However near the end of the pathway there must be an appreciable resistance to flow, to ensure that the TPD decreases sensitively in response to increasing transpiration rate. This resistance to flow may lie in the hypodermis.

The simultaneous measurement of transpiration rate as g_s falls in response to decreasing δe is crucial to a decision regarding hypotheses 3 and 4.

The rate of transpiration increases approximately linearly with both δe and g_s (Equation 8.2). Hence if g_s decreases linearly in response to increasing δe, the relationship between transpiration rate E and δe is a quadratic of the form:

$$E = Kg_s^1 \delta e \, \frac{1 - \delta e}{\delta e_0} \tag{8.3}$$

where g_s^1 is the stomatal conductance at $\delta e = 0$ and δe_0 is the saturation deficit at $g_s = 0$ (Figure 8.9). Thus the transpiration rate would be expected to reach a maximum and, as δe further increases, g_s and E might decline together. I immediately question whether this is possible or likely. If g_s responds to an increase in the rate of a transpiration flux from a source within the leaf (hypothesis 3), closure of the stomata would be expected to

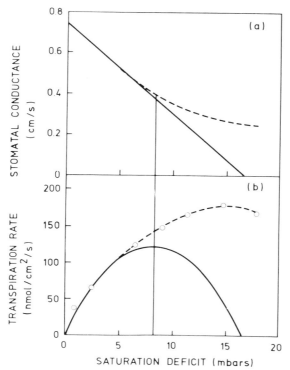

Figure 8.9. The influence of ambient saturation deficit on (*a*) stomatal conductance and (*b*) transpiration rate in Sitka spruce. Assuming the linear solid line in *a*, the solid line in *b* predicts the effect of saturation deficit on transpiration according to Equation 8.3. The dashed lines and open circles show the measured responses (in *a* taken from Figure 8.7).

reduce that transpiration flux, thus preventing further closure of the stomata. A falling rate of transpiration and declining g_s as δe increases would not be possible: the transpiration rate would not pass through a maximum, and g_s would not continue to decline linearly to near zero, but would level off at a value that would result in transpiration at the maximum rate. This appears to be the case with Sitka spruce (Figure 8.9), but not in Scots pine, in which transpiration and g_s continue to fall together as δe increases (Figure 8.7). Thus in Scots pine evaporation from sites external to the stomatal pore must be controlling TPD and g_s (hypothesis 4). Anatomical evidence (24; W. J. Davies, personal communication) indicates that the possible sites of evaporation in pines are the anticlinal walls between the guard cells and the subsidiary cells in the stomatal antechamber. Hypothesis 4, as we suppose it to occur in Scots pine, is shown diagrammatically in Figure 8.10.

Finally I consider why the base level of water potential in the plant should influence the response of g_s to δe, using assumptions 1 to 4 made previously with respect to the effect of water potential (Section 3.4); that is, we apply a

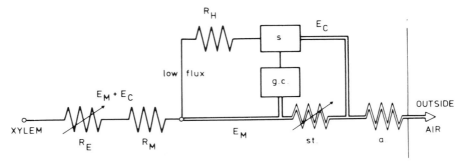

Figure 8.10. A scheme showing liquid flow (——) and vapor flow (═══) from the xylem of a leaf or needle to the outside: s = subsidiary cell; $gc.$ = guard cell. Liquid flow resistance in hypodermis = R_H, in endodermis = R_E, and in mesophyll = R_M. Vapor flow resistance in stomata = $st.$, and in stomatal antechamber and boundary layer = a. Transpiratory flux from mesophyll = E_M and from subsidiary cell = E_C. Based on Raschke (25).

Höfler diagram to the guard cell. In this case a fifth assumption is that stomatal opening is associated with an influx of K^+ and anion production and that closure in response of δe is followed by K^+ efflux and anion dispersal (26).

From Figure 8.10 it is clear that the water potential of the guard cells must depend on both the water potential in the mesophyll and on the flux of water along the extrastomatal pathway. If the mesophyll water potential is high, the guard cell water potential will also be high. Making use of Figure 8.6, we see that rapid stomatal closure will occur as guard cell water potential falls in response to increases in saturation deficit along the line $WXYZ$. Because of the TPD threshold, a large drop in water potential $\Delta\psi_2$ is necessary to move from W to Y and close the stomata, and part of the fall in water potential in moving from W to X will have no effect whatsoever on g_s.

On the other hand, if the guard cell water potential is initially lower at X, a very much smaller reduction in water potential $\Delta\psi_1$ is necessary to move from X to Y and close the stomata. Consequently, an increase in the flow of water through the extrastomatal pathway with a consequent decrease in guard cell water potential, as the result of a small increase in saturation deficit, will have no effect on g_s when the water potential is high (i.e., at W) but will result in an appreciable, nearly linear, reduction in g_s when the initial water potential is low (i.e., at X). This effect is further accentuated by the curvilinear relationship between the water potential and volume of guard cell water. As the saturation deficit and transpiration rate vary, stomatal conductance will also vary reversibly as guard cell water potential moves up and down the line $WXYZ$. However maintenance of a large saturation deficit for a long period may result in K^+ efflux and anion dispersal (26), so that the osmotic potential increases from π_3 to π_1 and subsequent reopening would be along the water potential line $DCBA$ rather than $ZYXW$.

ACKNOWLEDGMENTS

Many of the experimental results drawn on in this chapter were obtained by C. L. Beadle, R. E. Neilson, W. R. Watts, J. I. L. Morison, and P. Ai Peng Ng.

REFERENCES

1. C. E. Jeffree, R. P. C. Johnson, and P. G. Jarvis, *Planta,* **98,** 1 (1971).
2. P. G. Jarvis, in Z. Šesták, J. Čatský, and P. G. Jarvis, Eds., *Plant Photosynthetic Production, Manual of Methods,* Junk, The Hague, 1971, p. 566.
3. M. F. Beardsell, P. G. Jarvis, and B. Davidson, *J. Appl. Ecol.,* **9,** 677 (1972).
4. J. L. Monteith, *Principles of Environmental Physics*, Arnold, London, 1973, p. 241.
5. P. G. Jarvis, G. B. James, and J. J. Landsberg, in J. L. Monteith, Ed., *Vegetation and the Atmosphere,* Vol. 2, Academic Press, London–New York–San Francisco, 1976, p. 171.
6. P. A. P. Ng, *Response of Stomata to Environmental Variables in* Pinus sylvestris *L.,* Ph.D. thesis, University of Edinburgh, Edinburgh, U.K., 1978.
7. J. W. Leverenz, *The Effects of Light Flux Density and Direction on Net Photosynthesis in Sitka Spruce,* Ph.D. thesis, University of Aberdeen, Aberdeen, U.K., 1978.
8. H. Meidner and T. A. Mansfield, *Physiology of Stomata,* McGraw-Hill, London, 1968.
9. R. E. Neilson and P. G. Jarvis, *J. Appl. Ecol.,* **12,** 879 (1975).
10. R. E. Neilson, M. M. Ludlow, and P. G. Jarvis, *J. Appl. Ecol.,* **9,** 721 (1972).
11. M. M. Ludlow and P. G. Jarvis, *J. Appl. Ecol.,* **8,** 925 (1971).
12. C. L. Beadle, P. G. Jarvis, and R. E. Neilson, *Physiol. Plant.,* **45,** 158 (1979).
13. K. E. Fry, *A Study of Transpiration and Photosynthesis in Relation to the Stomatal Resistance and Internal Water Potential in Douglas Fir,* Ph.D. thesis, University of Washington, Seattle.
14. K. Raschke and U. Kuhl, *Planta,* **87,** 36 (1969).
15. I. R. Cowan and G. D. Farquhar, in D. H. Jennings, Ed., *Integration of Activity in the Higher Plant,* Society for Experimental Biology, Symposium No. 31, Cambridge University Press, Cambridge, 1977, p. 471.
16. W. Lopushinsky, *Bot. Gaz. (Chicago),* **130,** 258 (1969).
17. C. L. Beadle, N. C. Turner, and P. G. Jarvis, *Physiol. Plant.,* **43,** 160 (1978).
18. P. E. Waggoner and N. C. Turner, *Conn. Agric. Exp. Stn., New Haven, Bull.* **726** (1971).
19. H. Meidner and M. Edwards, *J. Exp. Bot.,* **26,** 319 (1975).
20. W. R. Watts and R. E. Neilson, *J. Appl. Ecol.,* **15,** 245 (1978).
21. J. Grace, D. C. Malcolm, and I. K. Bradbury, *J. Appl. Ecol.,* **12,** 931 (1975).
22. A. E. Hall, E.-D. Schulze, and O. L. Lange, in O. L. Lange, L. Kappen, and E.-D. Schulze, Eds., *Water and Plant Life: Problems and Modern Approaches,* Springer-Verlag, Berlin–Heidelberg–New York, 1976, p. 169.

23. W. R. Watts, R. E. Neilson, and P. G. Jarvis, *J. Appl. Ecol.*, **13**, 623 (1976).
24. K. Esau, *Anatomy of Seed Plants*, 2nd ed., Wiley, New York–London, 1977, p. 550.
25. K. Raschke, *Plant Physiol.*, **45**, 415 (1970).
26. R. Lösch, *Oecologia*, **29**, 85 (1977).

9 | Adaptive Significance of Stomatal Responses to Water Stress

M. M. LUDLOW

Division of Tropical Crops and Pastures, CSIRO, Brisbane, Queensland, Australia

1. INTRODUCTION

Autotrophic plants are faced with a dilemma. To facilitate carbon dioxide (CO_2) uptake for photosynthesis, they must expose to a comparatively dry atmosphere (-1000 to -2000 bars $\equiv$ 50 to 20% relative humidity at 20°C) chlorophyll-bearing cells that cannot withstand water potential below -30 to -100 bars ($\equiv$ 98 to 92% relative humidity at 20°C), but they must avoid dehydration. Evolution has not produced a membrane that allows CO_2 to pass from a gas phase on one side into aqueous solution on the other, while impeding the transfer of water vapor in the opposite direction (1). To reconcile this dilemma, the assimilating parts of higher plants have evolved an epidermis with a cuticle relatively impermeable to both CO_2 and water vapor, and with turgor-operated valves (i.e., the stomata) (2). The presence of an epidermis not only reduces the rates of CO_2 and water vapor exchange, but it also provides the means of controlling assimilation and transpiration through the size of the stomatal pores. Thus stomata play a pivotal role in controlling the balance between assimilation and transpiration.

123

In the absence of limitations to water uptake by roots and transport within the plant, water stress usually arises from a diminished soil water supply or from high atmospheric demand, both of which result in lowered plant water status. Thus the relationship between stomatal conductance and leaf water potential is a key functional response in understanding how stomata react to stress and how they exercise their regulatory role. In addition, however, stomata of many species respond directly to the dryness of the air (or "humidity") independently of the bulk leaf water potential (3–6). This functional relationship also needs to be understood, particularly for plants experiencing water stress.

A plethora of excellent reviews on the general subject of stomatal control of water loss have been published (1–4, 7–13). I confine my discussion to (a) the functional relationships between stomatal conductance, and leaf water potential and humidity of the air, (b) the extent to which these relationships change or adjust as plants experience water stress, (c) how they adjust, and (d) the consequences of these changes for growth and survival of autotrophic higher plants.

2. RESPONSES OF STOMATAL CONDUCTANCE TO LEAF WATER POTENTIAL AND TO HUMIDITY

2.1. Definitions and Approach

I have expressed the degree of stomatal opening as *conductance* rather than *resistance* because fluxes of CO_2 and water vapor are proportional to conductance, and because the former term has more biological meaning (1, 3, 4). No distinction is made between stomatal conductance g_s of different leaf surfaces, but leaf conductance g_l is used to describe fluxes from ad- and abaxial surfaces in series, which are in parallel with the cuticular conductances g_c. Leaf water status is given by water potential ψ_l and its osmotic π_l and pressure or turgor P_l components, assuming that matric potential is negligible. Stomatal conductance is determined by the turgor balance between guard and subsidiary cells, but P_l and more particularly the turgor potential of guard cells P_g are difficult to measure and rarely are reported. Because of this, I am forced to use ψ_l even though it is an imperfect measure of the water status of the stomatal apparatus. However the errors involved may not be too great because there appears to be only a loose relationship between P_g and P_l (14, 15), and g_s seems more related to ψ_l than to P_l (1, 16).

It is not known whether stomata respond to the dryness of air expressed as relative humidity or saturation deficit (δe = saturation vapor pressure minus vapor pressure at dry bulb temperature), or to leaf-air vapor pressure difference (LAVD). Both δe and LAVD are used here. They are numerically equal if leaf and air temperatures are the same.

2.2. Response to Leaf Water Potential and to Saturation Deficit

There is usually a range of ψ_l where g_s is unaffected, but below which it declines in a linear or curvilinear manner until it approaches zero or cuticular conductance (12) (Figure 9.1). Sometimes the plateau is absent and g_s declines from high values of ψ_l. The water potential at which g_s begins to decline (i.e., the critical or threshold water potential: Chapter 8) and that when zero conductance is approached ($\psi_l^{g \to 0}$) can be used to describe the relationship; the latter is particularly useful because it is a robust characteristic and is probably most closely related to loss of stomatal turgor. Stomata respond to ψ_l in a classical feedback manner (2); when ψ_l falls below a critical value, stomata begin to close, thereby reducing water loss and allowing ψ_l to recover. If ψ_l then increases, g_s usually increases again, and with moderately stressed plants this leads to series of partial closures and reopenings during the day (17).

Stomata respond to δe both directly and indirectly via lowered ψ_l, but it is often difficult to determine which is operating. The indirect response is analogous to the ψ_l response and is not considered separately in this chapter. For the direct response, g_s declines in a linear or curvilinear manner as δe increases (3–6; Figure 9.2), and the relationship is best characterized by its slope. In some species such as kenaf (*Hibiscus cannabinus*), there is a range of δe over which g_s is insensitive before it begins to decline as δe increases (17).

Stomata respond directly to humidity in a feed-forward manner; local water deficits develop in the stomatal apparatus which induce stomatal closure before deficits occur in the remainder of the leaf (4). This response acts as an early warning system that prevents deleterious deficits from

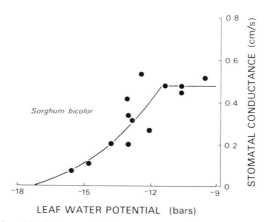

Figure 9.1. Relationship between leaf water potential and stomatal conductance of the abaxial surface of *Sorghum bicolor* leaves grown in the field and measured at quantum fluxes greater than 800 $\mu E/m^2/s$. From Ref. 17.

developing in the bulk of the leaf and reduces the possibility of permanent injury or protracted impaired function. The exact mechanism of the response is not known, but those that have been proposed are discussed in Chapter 8. Whatever the mechanism, it is implicit that there is a degree of hydraulic isolation between the stomatal apparatus and the rest of the leaf that enables gradients of water status to develop.

3. ADJUSTMENT OF STOMATAL RESPONSE DURING WATER STRESS

I define stomatal adjustment to water stress as the *process* by which stomatal responses are modified to suit the new situation of water stress. The word "adaptation" is used in a more general sense like "stress."

3.1. Adjustment to Humidity

Changes may occur in plants exposed to water stress to make stomata either more or less responsive to δe. The latter is by far the most common type of response and is exhibited by kenaf (17), avocado (18), engleman spruce (19), three C_4 grasses, and siratro (*Macroptilium atropurpureum*) (20, 21). Stomata of leaves under water stress respond less in both relative and absolute terms to δe than do stomata of unstressed leaves (Figure 9.2). I

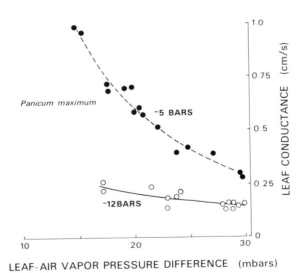

Figure 9.2. Relationship between leaf-air vapor pressure difference and leaf conductance of *Panicum maximum* leaves in the field at leaf water potentials of −5 and −12 bars. Measured at 2200 $\mu E/m^2/s$ and 30°C. From Ref. 21.

have interpreted this on the basis of limiting factors: maximum g_l of stressed leaves is limited by water potential or associated factors—for example, high levels of abscisic acid (ABA); therefore the stomata are less responsive to δe. Furthermore, species with high conductances respond more to δe than those with low conductances (4). I conclude therefore that the reduced sensitivity of stressed stomata to δe is not adjustment to stress, rather, it is a *passive* reaction to the limitation imposed by water stress.

On the other hand, when the stomata on plants under water stress became more sensitive to δe as in grape (*Vitis vinifera*, Ref. 22), apricot (*Prunus armeniaca*, Ref. 23), and two desert shrubs (*Hammada scoparia* and *Zygophyllum dumosum*, Ref. 24), this suggests a form of adjustment. For example, Schulze et al. (23, 24) found for apricot in the Negev Desert that the slope of the $g_l/\delta e$ relationship was steeper in unirrigated than in irrigated plants (Figure 9.3) and that the slope of the relationship between $1/g_l$ and δe became steeper as soil water declined and as ψ_l decreased from spring to summer.

Exposure of plants to water stress does not always alter the $g_l/\delta e$ relationship. For example, *Atriplex hastata* (6) and *Citrus jambhiri* (25) show no change in humidity response.

3.2. Adjustment to Leaf Water Potential

Evidence for the existence of stomatal adjustment to leaf water potential comes from two sources. First, when potted plants are subjected to a series of drying and rewetting cycles, there is a shift in the g_s/ψ_l response of species with the capacity to adjust such that the sensitivity of stomata to water

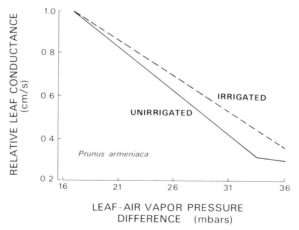

Figure 9.3. Influence of leaf-air vapor pressure difference on the relative leaf conductance of apricot (*Prunus armeniaca*) from irrigated at unirrigated trees in the Negev Desert. From Ref. 24.

deficits decreases with the number of cycles (26–30). In other words, there is a decrease in $\psi_l^{q \to 0}$ and the threshold ψ_l. Similarly when plants were grown in solutions with osmotic potentials of -0.7, -4, or -7 bars, the corresponding $\psi_l^{q \to 0}$ were -13.7, -15.0, and -19.2 bars (31).

Second, the threshold potential for stomatal closure and $\psi_l^{q \to 0}$ is lower for plants grown under field conditions than for those grown under controlled conditions. This was first shown by Jordan and Ritchie (32) for cotton and subsequently was confirmed for a number of species in various situations (7, 8, 33, 34). The reduced sensitivity of field-grown plants to water deficits (Figure 9.4) may arise partly from the greater evaporative demand there because the sensitivity of previously well-watered plants to imposed water stress decreases as evaporative demand increases from controlled environments, to glasshouses, to the field (21, 34), but not if the evaporative demand is similar in controlled environments and outdoors (35). On the other hand, in some instances increased evaporative demand both under controlled conditions and in the field does not induce stomatal adjustment (21, 26), and field-grown plants are more sensitive than those grown under controlled conditions (27, 36). Thus, at best, the greater evaporative demand of the field can only account for a small part of the difference between the field and controlled environments (cf. C.E.R. with Week 1 and Field in Figure 9.4). I believe that the main cause of the difference is the rate at which leaf water potential declines. Plants are invariably grown in small pots in controlled environment experiments. When watering ceases, it takes only *days* to reach

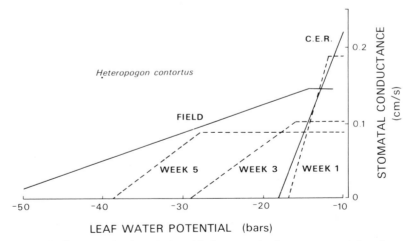

Figure 9.4. Differences in the relationship between leaf water potential and stomatal conductance of *Heteropogon contortus* leaves during single cycles lasting several days for plants grown in small pots in controlled environment rooms (C.E.R.) and lasting 5 weeks for plants grown in the field, and changes in the relationship for tillers (dashed lines) that have been cut, rehydrated, and dehydrated in weeks 1, 3, and 5 of a soil drying cycle in the field. Adapted from Ref. 21.

the same water potentials that occur after *weeks* of drying in the field because plants there have access to large volumes of soil. The additional time allows plants that have the capacity to adjust to do so, thus partially offsetting the effects of the developing water deficits (7, 21, 37). For example, we found that the g_s/ψ_l responses of three C_4 grasses (*Pancium maximum, Cenchrus ciliaris,* and *Heteropogon contortus*) changed during single, continuous drying cycles in the field (Figure 9.4, Table 9.1), but there was no change in well-watered controls (21). The g_s/ψ_l responses were obtained at various stages of the soil drying cycle by cutting stressed tillers, rehydrating them, and allowing them to dry to various levels of ψ_l before measurement. This stomatal adjustment has been confirmed under controlled conditions where the same species growing in large containers were subjected to single, continuous drying cycles of approximately the same duration as the field studies. In this instance stressed plants were rehydrated; then rapid drying was induced by root pruning (21). After rewatering stressed grasses in the field, the degree of stomatal adjustment fell, but even after 10 days it had not returned to the level of the continuously watered plants (21); for example, the difference between the $\psi_l^{\to 0}$ value for well-watered and stressed leaves fell from -13, -13, and -17 bars during stress to -5, -5, and -4 bars after rewatering for *Panicum maximum, Cenchrus ciliaris,* and *Heteropogon contortus,* respectively.

Variation in Stomatal Adjustment Among Species. Table 9.1 shows stomatal adjustment as measured by the difference in $\psi_l^{\to 0}$ between plants grown under controlled conditions and in the field, or the change in $\psi_l^{\to 0}$ when plants are subjected to water stress as multiple drying and rewetting cycles or as a single drying cycle in the field. Both C_3 and C_4 species show stomatal adjustment, but in general the degree of adjustment is least in highly bred and selected agricultural crop species. From the limited data available, the capacity for stomatal adjustment seems to be associated more with the environment to which plants have adapted than with any taxonomic, morphological, or other grouping (33). The greater capacity of sorghum to adjust compared with maize, the ranking of the three C_4 pasture grasses, and the greater capacity of the subtropical crops (soybean and cotton) than the temperate grain crops, are consistent with the ecological distribution of these species. So too is the limited capacity for stomatal adjustment of kenaf, which is thought to have evolved on the edges of swamps in southern Africa. Siratro, a species that depends mainly on dehydration avoidance for survival and has only a limited capacity to adjust to stress physiologically (21, 38), also shows no stomatal adjustment after seven drying cycles in small pots (39) under controlled conditions and only a 2 bar shift in $\psi_l^{\to 0}$ at the end of two soil drying cycles in the field (38).

A similar ranking of species is obtained if one compares the threshold water potential for stomatal closure or ψ_l at which g_s is half the maximum for plants grown under controlled conditions and in the field (8, 38).

Table 9.1. Stomatal Adjustment as Measured by the Difference Between the Maximum and Minimum Values of the Leaf Water Potential at which Stomatal Conductance Approaches Zero ($\psi_l^{g\to0}$)

Variation in $\psi_l^{g\to0}$ arises from comparing stressed with unstressed plants, and those from controlled environments with those from the field. Values given are the highest recorded for each species, those in parentheses are mean values

Species	$\psi_l^{g\to0}$ (bars)			Ref.
	Maximum	Minimum	Difference	
C_4 Plants				
Zea mays	−11	−18	7	58, 59
Sorghum bicolor	−15	−28	13	26, 41
Panicum maximum	−17	−30	13(11)	21
Cenchrus ciliaris	−18	−37	19(15)	21
Heteropogon contortus	−17	−53	36(34)	21
C_3 Plants				
Agricultural				
Hibiscus cannabinus	−21	−21	0	17
Macroptilium atropurpureum	−13	−15	2	21
Vicia faba	− 6	<−10	> 4	34
Triticum aestivum	−14	−19	5	31
Helianthus annuus	−17	−27	10	41
Glycine max	−16	−29	13(11)	34
Gossypium hirsutum	−30	−43	13(8)	28, 29, 32
Semiarid Shrubs				
Eucalyptus socialis	−25	−43	18	30
Larrea divaricata	−40	−58	18	60

4. MECHANISMS OF ADJUSTMENT

4.1. To Water Deficits

I am not aware of evidence that stomatal adjustment results from changes in the membrane characteristics of guard and subsidiary cells. Furthermore, the massive literature on plant growth substances sheds little light on the mechanism of stomatal adjustment. Although Davies (40) has suggested that stomatal adjustment results from a decrease in the threshold water potential for initiation of ABA synthesis, there is no experimental support for this hypothesis.

The most plausible explanation is that osmotic adjustment within the leaf (Chapter 7), and particularly in the guard and subsidiary cells of plants undergoing stress, assists the maintenance of turgor (or turgor difference) by partially offsetting the decline in water potential and allowing partial stomatal opening (7). This explanation assumes that the relationship between g_s and P_l (or P of guard and subsidiary cells) is unaffected by stress.

Evidence to support osmotic adjustment as the main mechanism of stomatal adjustment depends on an association between the two processes when plants grown in controlled environments and the field are compared, and when plants experience water stress either in multiple drying and rewetting cycles or in single drying cycles in the field (7, 8, 10, 21, 28, 29, 31, 34, 40). For example, Brown et al. (28) measured a 14 bar stomatal adjustment in cotton that was accompanied by osmotic adjustment of 7 bars in guard cells and 9 bars in the whole leaf. Furthermore, the decrease of $\psi_l^{g\to 0}$ was the same as the osmotic adjustment in leaves of rainfed and irrigated crops of sorghum and sunflower and of Sitka spruce (*Picea sitchensis*) (10, 12, 41). We also observed a strong correlation between $\psi_l^{g\to 0}$ and osmotic potential at zero turgor π_t of bulk leaf tissue in three C_4 grasses and a C_3 legume subjected in controlled environments to single drying cycles in small and large pots, and to a single drying cycle in the field (21, 38; Figure 9.5). At the more extreme levels of stress the osmotic adjustment failed to match the degree of stomatal adjustment. This could have arisen because (*a*) errors were made in estimating $\psi_l^{g\to 0}$ and π_t in stressed leaves, (*b*) the guard cells adjusted more osmotically than bulk leaf tissue (21) [although the data of Brown et al. (28) do not support this suggestion, differences in π between guard cells and other leaf cells are known to exist (42, 43)], (*c*) other water relations characteristics of stomata such as the water capacitances of guard cells (44) changed (21), or (*d*) the relationship between g_s and P_l (or P_g) may have become steeper, or the threshold value may have decreased such that a higher g_s was achieved at lower values of P_l (or P_g) (45).

Thus I believe that there is strong correlative and mechanistic evidence that osmotic adjustment is the main process responsible for stomatal adjustment.

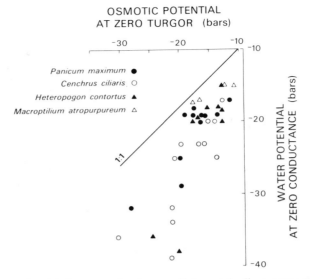

Figure 9.5. Relationship between the degree of stomatal adjustment as shown by the leaf water potential at which stomatal conductance approaches zero and the degree of osmotic adjustment as shown by the osmotic potential at zero turgor for *Panicum maximum, Cenchrus ciliaris, Heteropogon contortus,* and *Macroptilium atropurpureum* subjected to single drying cycles in small and large pots in controlled environments and in the field. From Refs. 21 and 38.

4.2. To Humidity

The active response of stomata that makes them more sensitive to δe as water stress increases could be considered to be adjustment because it reduces water loss and plant deficits, and potentially it increases water use efficiency. In the absence of any studies on the mechanisms of this response I am forced to speculate as follows: (*a*) the relationship between g_l and P_g could change such that g_l becomes more sensitive to P_g as water stress increases; (*b*) an increase in inhibitors such as ABA or farnesol could sensitize stomata to δe [although this seems unlikely because Farquhar (46) found that ABA reduced the response of stomata to δe]; or (*c*) leaves produced during stress may have larger substomatal cavities, which would render them more responsive to δe (47) (see Chapter 8). More work on the adjustment of stomatal responses to δe is obviously needed.

5. ECOPHYSIOLOGICAL SIGNIFICANCE OF ADAPTIVE RESPONSES

When assessing adaptive responses, it is pertinent to remember that there can be conflict between the requirements for survival that have been important in the evolution of natural vegetation and the requirements for production necessary in our agricultural plants (9, 10). It is axiomatic that survival is an important component of production, but the reverse does not always apply. Thus in many agricultural plants that have been intensively bred and selected for production, survival strategies are less likely to be evident than in the plants that have been bred little or not at all. Distinction also needs to be made between perennial plants, which must survive seasonal and temporal water shortages in the vegetative state, and those from harsher environments that survive the most extreme periods as seeds (annuals) or as dormant buds or bulbs (perennials).

5.1. Responses to Humidity

Many species that grow and survive in arid and semiarid areas exhibit a direct humidity response (9, 20, 21, 48). Since the closure resulting from the feed-forward response to humidity is rapid, the stomata can respond to sudden increases in evaporative demand, and it is readily reversible (13). Moreover, it has few long-term effects on growth compared to the feedback response, which depends on a decline of leaf water potential and can be associated with aftereffects such as impaired stomatal opening. The feed-forward response seems to be a desirable adaptive characteristic to prevent deleterious deficits from developing for plants that grow in areas characterized by periods of high evaporative demand. This is important for species lacking dehydration tolerance, for unhardened plants before they have had the opportunity to adapt, and at critical developmental stages such as floral initiation and anthesis. [I use the terms ''dehydration avoidance'' and ''dehydration tolerance'' instead of ''drought avoidance'' and ''drought tolerance'' as defined by Levitt (Ref. 49 and Chapter 28), and instead of Kramer's terms (Chapter 2). These two terms are components of ''drought resistance'' where ''drought'' is a significant period without rain.]

The feed-forward response to humidity is required to maximize daily water use efficiency (WUE: units of CO_2 fixed per unit of water lost) in Cowan and Farquhar's (11) optimization theory. The WUE is increased daily if stomata close in response to high δe in the middle of the day because most gas exchange is restricted to the early morning and late afternoon when δe is lower and when instantaneous WUE is higher (9, 50). Even in *Opuntia inermis*, which has high WUE because of its crassulacean acid metabolism characteristics, the stomata respond to δe at night (51), presumably to maintain the high WUE.

The increased sensitivity of stomata to δe as leaf water status declines is a significant adaptive feature. The survival value of such a response in reducing water loss and plant water deficits is clear in arid areas such as the Negev Desert, where plants with high dehydration tolerance grow during spring and into hot, dry summers on diminishing soil water supplies. Moreover, it allows WUE to remain relatively constant for this period (9, 52). However it is not an obligatory characteristic because other plants that grow and survive there do not exhibit this response (23).

In areas of intermittent water shortage, stomatal response to δe and adjustment of this response may be of no particular advantage to species with a high degree of dehydration tolerance. Indeed it may reduce production as well as reducing water loss, but whether it increases WUE, or only maintains it, depends on whether the species is C_3 or C_4 and on the environmental conditions (9). On the other hand, I see few disadvantages in these responses for species that lack dehydration tolerance characteristics growing in these situations. Such responses would seem to be highly desirable for both native and agricultural plants growing in semiarid areas, especially when seed yield can be influenced by water deficits at critical developmental stages.

5.2. Responses to Water Potential

The ecological significance of stomatal adjustment to water deficits can be illustrated by comparing the C_4 pasture grasses, which show considerable adjustment, with the legume siratro, which scarcely adjusts (21, 38). Water loss in the grasses is reduced, but not prevented, by partial closure as stress develops. Since the stomata adjust osmotically and remain partially open, water loss continues and leaf water potentials of the grasses can reach values of about −120 bars. Thus it is evident that plants exhibiting stomatal adjustment cannot survive without moderate to high dehydration tolerance characteristics. However partial stomatal opening in the grasses does allow some CO_2 fixation, especially during the early morning and late afternoon when δe is lower and WUE is higher. Stomatal adjustment can be seen to have distinct advantages if the additional carbon is used (*a*) to produce roots that can explore new soil volume and extract more water; this may explain the higher root/shoot ratio and greater root growth of stressed plants (4, 53); (*b*) to produce seeds in annuals or replenish reserves in perennials that survive as dormant buds or bulbs; (*c*) to assist maintenance of existing roots and shoots; and (*d*) in osmotic adjustment that assists maintenance of turgor for continued metabolic activity, and for root (54) and shoot growth (55). By comparison with the grasses, the stomata of siratro respond strongly to both δe and ψ_l (20), closing at about −15 bars in the field (21), and little stomatal adjustment occurs. The greatly reduced water loss resulting from stomatal closure, together with paraheliotropic leaf movements and leaf shedding, keeps water potentials above −20 bars (20, 56). However photosynthesis and growth are inhibited: this is a classical dehydration avoidance strategy that is essential for survival of this species, which has low dehydration tolerance.

It is not surprising, therefore, that these perennial grasses and legumes differ in their ability to survive droughts (56). For example, in 1968–1969 all *Cenchrus ciliaris,* half the *Pancicum maximum* var. *trichoglume,* a third of the siratro, and none of the *Medicago sativa* plants survived in an experiment at Narayen Research Station. Both legumes have low dehydration tolerance and only a limited capacity for stomatal adjustment, but *Medicago sativa* has poor stomatal control of water loss compared with siratro. On the other hand, *Cenchrus ciliaris* exhibits more stomatal adjustment than *Panicum maximum* (Table 9.1), and it probably has higher dehydration tolerance.

In a broader context, the relative merits of stomatal adjustment versus stomatal closure with no adjustment depend on many factors, such as the dehydration tolerance of the species; the length, frequency, and severity of dry spells; whether production or survival is being considered; the modes of reproduction and survival; whether plants are growing in pure or mixed communities; and whether the plant is a perennial or an annual (33). In dehydration tolerant species native to arid and semiarid regions where survival is more important than production, stomatal adjustment appears to be beneficial and has few disadvantages. Indeed the greatest degree of stomatal adjustment occurs in shrubs and the C_4 pasture grasses that are native to semiarid parts of Africa and Australia (Table 9.1). High dehydration tolerance is usually associated with slow rates of growth and development, and most plants with good dehydration avoidance lack tolerance. Therefore Sullivan (57) advocates selection of agricultural plants that are intermediate in both avoidance and tolerance. In crop species with moderate dehydration tolerance such as sorghum, stomatal adjustment is probably a desirable characteristic for semiarid areas. The higher dehydration tolerance of sorghum compared with maize (10, 33) may explain why sorghum has a greater degree of stomatal adjustment. Moreover, the greater tolerance and adjustment probably contribute to its higher grain yield in semiarid areas.

However in annual crops and, to a lesser extent, in annual native species with poor dehydration tolerance, where production or survival depends on seed set and grain yield, stomatal adjustment would seem to be disadvantageous because it may increase the likelihood of severe deficits occurring at critical stages of development such as floral initiation and anthesis. For these plants stomatal closure, which keeps internal water status high, would seem to be less hazardous. Indeed, stomatal adjustment is inherently more hazardous, in general, for annuals than for perennials. The relative advantage of stomatal closure over stomatal adjustment for annuals depends on their susceptibility to deficits and on the probability, timing, and severity of deficits. Clearly the risk of severe deficits at critical phases of development is traded off against the loss of carbon when stomata are closed, which reduces the ability to adjust osmotically, obtain more soil water, maintain the existing plant, and fill grains or seeds. If lethal deficits are not reached by the time stress is relieved by rainfall or irrigation, the surviving plant is able to continue to grow and to produce seed or grain yield. On the other hand, if stomatal closure is not sufficient to prevent the development of lethal deficits, because of high cuticular loss even when stomata are closed such as in maize (10) or because the dry spell is protracted, the leaves and ultimately the plants will die. Thus stomatal closure of plants that lack dehydration tolerance probably works well for temporary water shortages. However it is of little benefit in drier environments or, as in the case of siratro compared with the C_4 pasture grasses (56), in drier-than-average seasons in wetter environments.

6. CONCLUSIONS

Much has been learned about stomatal adjustment to water potential since Jordan and Ritchie's (32) data were published. There is good evidence that osmotic adjustment is the main mechanism involved. However there is a need for data on a wider range of species. On the other hand, the mechanism by which stomata respond to humidity and adapt to atmospheric stress is poorly understood. Moreover, variation among species in these responses and in adaptive changes has scarcely been studied.

Because of the speculative nature of my remarks, it is obvious that the physiological and ecological significance of adaptive responses to both leaf water deficits and humidity is poorly understood. Yet an understanding of these matters is required to define desirable physiological characteristics that can be incorporated in plants useful to man, and to assist in the management of natural plant communities. Without this understanding, these responses and adaptive changes are destined to remain nothing more than physiological curiosities.

ACKNOWLEDGMENTS

Most of the unpublished data quoted on pasture grasses and legumes arises from cooperative work with the following people: M. J. Fisher, J. R. Wilson, E.-D. Schulze, R. Davis, and S. Radford. I thank John Wilson and Myles Fisher for their comments on the manuscript, and the Chief, Division of Tropical Crops and Pastures, CSIRO, for allowing me to attend the workshop.

REFERENCES

1. I. R. Cowan, *Adv. Bot. Res.*, **4**, 117 (1978).
2. K. Raschke, *Annu. Rev. Plant Physiol.*, **26**, 309 (1975).
3. F. J. Burrows and F. L. Milthorpe, in T. T. Kozlowski, Ed., *Water Deficits in Plant Growth*, Vol. 4, Academic Press, New York–San Francisco–London, 1976, p. 103.
4. A. E. Hall, E.-D. Schulze, and O. L. Lange, in O. L. Lange, L. Kappen, and E.-D. Schulze, Eds., *Water and Plant Life: Problems and Modern Approaches*, Springer-Verlag, Berlin–Heidelberg–New York, 1976, p. 169.
5. D. W. Sheriff and P. E. Kaye, *Ann. Bot. (London)*, **41**, 653 (1977).
6. D. W. Sheriff and P. E. Kaye, *Z. Pflanzenphysiol.*, **83**, 463 (1977).

7. J. E. Begg and N. C. Turner, *Adv. Agron.*, **28**, 161 (1976).
8. N. C. Turner and J. E. Begg, in J. R. Wilson, Ed., *Plant Relations in Pastures*, CSIRO, Melbourne, Australia, 1978, p. 50.
9. R. A. Fischer and N. C. Turner, *Annu. Rev. Plant Physiol.*, **29**, 277 (1978).
10. N. C. Turner, in H. Mussell and R. C. Staples, Eds., *Stress Physiology in Crop Plants*, Interscience, New York–London, 1979, p. 343.
11. I. R. Cowan and G. D. Farquhar, in D. H. Jennings, Ed., *Integration of Activity in the Higher Plant*, Society for Experimental Biology, Symposium No. 31, Cambridge University Press, Cambridge, 1977, p. 471.
12. N. C. Turner, in R. L. Bieleski, A. R. Ferguson, and M. M. Creswell, Eds., *Mechanisms of Regulation of Plant Growth*, Bulletin 12, Royal Society of New Zealand, Wellington, 1974, p. 423.
13. H. G. Jones, in H. Mussell and R. C. Staples, Eds., *Stress Physiology in Crop Plants*, Interscience, New York–London, 1979, p. 407.
14. N. C. Turner, *Plant Physiol.*, **55**, 932 (1975).
15. C. L. Beadle, N. C. Turner, and P. G. Jarvis, *Physiol. Plant.*, **43**, 160 (1978).
16. P. V. Biscoe, Y. Cohen, and J. S. Wallace, *Philos. Trans. R. Soc. London, Ser. B*, **273**, 565 (1976).
17. R. C. Muchow, M. J. Fisher, M. M. Ludlow, and R. J. K. Myers, *Aust. J. Plant Physiol.*, **7**, (in press) (1980).
18. R. E. Sterne, M. R. Kaufmann, and G. A. Zentmyer, *Physiol. Plant.*, **41**, 1 (1977).
19. M. R. Kaufmann, *Plant Physiol.*, **57**, 898 (1976).
20. M. M. Ludlow and K. Ibaraki, *Ann. Bot. (London)*, **43**, 639 (1979).
21. M. M. Ludlow et al., unpublished data.
22. H. During, *Vitis*, **15**, 82 (1976).
23. E.-D. Schulze, O. L. Lange, L. Kappen, M. Evenari, and U. Buschbom, *Oecologia*, **18**, 219 (1975).
24. E.-D. Schulze, O. L. Lange, U. Buschbom, L. Kappen, and M. Evenari, *Planta*, **108**, 259 (1972).
25. M. R. Kaufmann and Y. Levy, *Physiol. Plant.*, **38**, 105 (1976).
26. K. J. McCree, *Crop Sci.*, **14**, 273 (1974).
27. J. C. Thomas, K. W. Brown, and W. R. Jordan, *Agron. J.*, **68**, 706 (1976).
28. K. W. Brown, W. R. Jordan, and J. C. Thomas, *Physiol. Plant.*, **37**, 1 (1976).
29. J. M. Culter and D. W. Rains, *Crop Sci.*, **17**, 329 (1977).
30. J. Collatz, P. J. Ferrar, and R. O. Slatyer, *Oecologia*, **23**, 95 (1976).
31. S. E. Simmelsgaard, *Physiol. Plant.*, **37**, 167 (1976).
32. W. R. Jordan and J. T. Ritchie, *Plant Physiol.*, **48**, 783 (1971).
33. M. M. Ludlow, in O. L. Lange, L. Kappen, and E.-D. Schulze, Eds., *Water and Plant Life: Problems and Modern Approaches*, Springer-Verlag, Berlin–Heidelberg–New York, 1976, p. 364.
34. W. J. Davies, *Crop Sci.*, **17**, 735 (1977).
35. M. M. Ludlow and T. T. Ng, *Aust. J. Plant Physiol.*, **3**, 401 (1976).
36. E. T. Kanemasu and C. B. Tanner, *Plant Physiol.*, **44**, 1547 (1969).
37. M. M. Jones and N. C. Turner, *Plant Physiol.*, **61**, 122 (1978).
38. J. R. Wilson, M. M. Ludlow, M. J. Fisher, and E.-D. Schulze, *Aust. J. Plant Physiol.*, **7**, 207 (1980).

39. M. J. Fisher, D. A. Charles-Edwards, and M. M. Ludlow, *Aust. J. Plant Physiol.*, **7**, (in press) 1980.

40. W. J. Davies, *J. Exp. Bot.*, **29**, 175 (1978).

41. N. C. Turner, J. E. Begg, and M. L. Tonnet, *Aust. J. Plant Physiol.*, **5**, 597 (1978).

42. G. R. Squire and T. A. Mansfield, *New Phytol.*, **71**, 1033 (1976).

43. R. A. Fischer, *J. Exp. Bot.*, **24**, 387 (1973).

44. I. R. Cowan, *Planta*, **106**, 185 (1972).

45. M. M. Jones and H. M. Rawson, *Physiol. Plant.*, **45**, 103 (1979).

46. G. D. Farquhar, personal communication.

47. D. W. Sheriff, *J. Exp. Bot.*, **28**, 1399 (1977).

48. D. A. Johnson and M. M. Caldwell, *Physiol. Plant.*, **36**, 271 (1976).

49. J. Levitt, *Responses of Plants to Environmental Stresses*, Academic Press, New York–London, 1972.

50. H. M. Rawson, N. C. Turner, and J. E. Begg, *Aust. J. Plant Physiol.*, **5**, 195 (1978).

51. C. B. Osmond, M. M. Ludlow, R. Davis, I. R. Cowan, K. Winter, and S. Powles, *Oecologia*, **41**, 65 (1979).

52. E.-D. Schulze, O. L. Lange, M. Evenari, L. Kappen, and U. Buschbom, *Oecologia*, **19**, 303 (1975).

53. A. E. Hall, K. W. Foster, and J. G. Waines, in A. E. Hall, G. H. Cannell, and H. W. Lawton, Eds., *Agriculture in Semi-Arid Environments*, Springer-Verlag, Berlin–Heidelberg–New York, 1979, p. 148.

54. E. L. Greacen and J. S. Oh, *Nature (London), New Biol.*, **235**, 24 (1972).

55. T. C. Hsiao, E. Acevedo, E. Fereres, and D. W. Henderson, *Philos. Trans. R. Soc. London, Ser. B*, **273**, 479 (1976).

56. D. C. I. Peake, G. B. Stirk, and E. F. Henzell, *Aust. J. Exp. Agric. Anim. Husb.*, **15**, 645 (1975).

57. C. Y. Sullivan, in N. G. P. Rao and L. R. House, Eds., *Sorghum in Seventies*, Oxford and IBH, New Delhi–Bombay–Calcutta, 1972, p. 247.

58. C. L. Beadle, K. R. Stevenson, H. H. Neuman, G. W. Thurtell, and K. M. King, *Can. J. Plant Sci.*, **53**, 537 (1973).

59. N. C. Turner, *Plant Physiol.*, **53**, 360 (1974).

60. H. A. Mooney, O. Björkman, and G. J. Collatz, *Carnegie Inst. Washington Yearb.*, **76**, 328 (1977).

10 | Adaptive Significance of Carbon Dioxide Cycling During Photosynthesis in Water-Stressed Plants

C. B. OSMOND, K. WINTER, AND S. B. POWLES

Department of Environmental Biology, Research School of Biological Sciences, Australian National University, Canberra, Australian Capital Territory, Australia

1. INTRODUCTION

It is generally considered to be useful to distinguish two components of the response of photosynthesis to plant water stress, whether resulting from lowered leaf water potential or from high leaf-air vapor pressure deficits: namely, those due to stomatal factors and those due to nonstomatal factors. From different experiments in an extensive literature it has been concluded that inhibition of photosynthesis in water-stressed plants may be largely ascribed to stomatal control of carbon dioxide (CO_2) uptake, partly ascribed to stomatal and nonstomatal responses, or almost entirely ascribed to nonstomatal responses. In a comparison of the effect of low leaf water

139

potential on photosynthesis of 12 woody species native to habitats ranging from streamside to desert, the nonstomatal component of CO_2 fixation (mesophyll conductance) began to decline at the same water potential as stomatal conductance (1). In this chapter we wish to link, rather than separate, these two components.

It seems to us that the periods of potentially greatest water stress in land plants correspond to periods of peak irradiance on a daily or seasonal basis. Stomatal closure to prevent further water loss is a frequently observed response in water-stressed plants, and stomatal closure at peak irradiance might deprive illuminated leaves of external CO_2. We know that nonstomatal components of photosynthesis are inhibited when leaves of plants are deprived of CO_2 in the light. This response, termed *photoinhibition*, is probably due to the photodestruction of the photosynthetic apparatus when photochemical energy cannot be dissipated in an orderly manner during normal CO_2 fixation. We speculate that illumination of water-stressed leaves in which stomata have closed may itself have important consequences for nonstomatal components of photosynthesis. In exploring this speculation we have been obliged to consider the comparative physiology and biochemistry of CO_2 recycling in plants in the light and to consider plausible roles for photorespiration: we discuss plants exhibiting C_3, C_4, and crassulacean acid metabolism (CAM) pathways of photosynthesis.

2. DAILY COURSE OF CARBON DIOXIDE EXCHANGE FOLLOWING PROLONGED WATER STRESS

If it is true that "stomata function so that the total loss of water during the day is a minimum for the total amount of carbon taken up" (the basic assumption underlying the concept of optimal stomatal behavior), daily patterns of CO_2 exchange in water efficient plants in highly evaporative habitats will show pronounced midday depression because of stomatal closure (2). In extreme cases theory predicts that stomata of these plants will respond so that leaves will be isolated from external CO_2 while exposed to peak irradiance for several hours each day. Many examples are known in which stomatal conductance and net photosynthesis of C_3, C_4, and CAM plants decline at peak irradiance. Figure 10.1 shows the daily patterns of CO_2 exchange that result from midday depression of stomatal conductance for *Glycine max*, a C_3 crop species grown under water-limited conditions (3), for *Hammada scoparia*, a wild C_4 species growing in the Negev Desert (4), and for *Mesembryanthemum crystallinum*, a C_3 plant that can be induced to engage in CAM in response to water stress applied by addition of sodium chloride to culture solutions (5).

Low stomatal conductance and negligible CO_2 uptake in the light are the norm for CAM plants in arid habitats (6). It is evidently also common for leaves of C_3 evergreen shrubs in arid habitats to remain functional, but

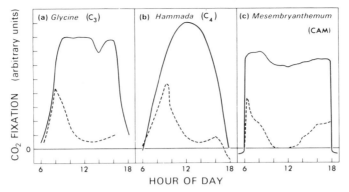

Figure 10.1. Daily course of net CO_2 fixation in C_3, C_4, and CAM plants at high leaf water potential (i.e., control, solid curves) and low leaf water potential (i.e., stressed, dashed curves). *Glycine max* (data redrawn from Ref. 3): dawn water potential; control, -2 bars; stressed, -11 bars. *Hammada scoparia* (data redrawn from Ref. 4): dawn water potential; control, -14 bars; stressed, -42 bars. *Mesembryanthemum crystallinum* (data redrawn from Ref. 5): control, culture solution; stress, culture solution $+400$ mM sodium chloride.

without net CO_2 exchange, during periods of water stress. The stomatal conductance of leaves of some evergreen shrubs in the California chaparral remains very low throughout the summer (7), and leaves of shrubs in the Chilean matorral may remain at CO_2 compensation for 5 months during summer (8). Only in C_3 plants and CAM plants are we aware of examples in which no net CO_2 fixation occurs at peak irradiance, although the C_4 plant *Hammada scoparia* approaches this condition under extreme stress in the Negev Desert (Figure 10.1), as does *Atriplex confertifolia* in the Great Basin (9).

Thus far, particular attention has been given to stomatal mechanisms, such as a direct response to leaf-air water vapor deficits, which may account for the midday depression of stomatal conductance. In terms of the optimization of CO_2 assimilation and water use, we must also consider the behavior of the assimilatory machinery when deprived of external CO_2 at this time. The ability of plants to internally recycle CO_2 by the mechanisms discussed below is an important component of the optimal behavior.

3. CHANGES IN INTERCELLULAR CARBON DIOXIDE CONCENTRATION AND PHOTOSYNTHESIS DURING WATER STRESS

When some plants are rapidly stressed stomata close very effectively and the depression of photosynthesis can be almost totally ascribed to isolation from external CO_2 supply. Lawlor (10) reviewed experiments with C_3 and C_4 plants that were stressed by transfer to nutrient solutions containing up to -10 bars polyethylene glycol. After 6 h in the light, CO_2 fixation and stomatal conductance were measured, and leaves were exposed to $^{14}CO_2$. In C_3 and C_4 plants the incorporation of ^{14}C into glycine and serine and the pool sizes of these compounds increased markedly, and in the C_3 plants the CO_2 compensation point increased. These observations are all consistent with rapid closure of the stomata, which resulted in a rapid decline of intercellular CO_2 concentration and increase in leaf temperature.

This decline of intercellular concentration may be demonstrated in another way. A close approximation to intercellular CO_2 concentration (C_i, $\mu l/l$) is given by the equation:

$$C_i = C_a - \frac{A}{g_s} \tag{10.1}$$

relating ambient CO_2 concentration (C_a, $\mu l/l$), assimilation (A, ng $CO_2/cm^2/s$) and stomatal conductance (g_s, cm/s). An exact derivation of this relationship, in molar units, is given by Cowan and Farquhar (2). It follows from this equation that C_i remains constant when the ratio A/g_s is constant, that is, when the plot of A against g_s is linear. Figure 10.2a shows that the relationship between assimilation and stomatal conductance in *Zea mays* is distinctly nonlinear following the rapid imposition of water stress (1 to 10 bars/h) in the experiments of Lawlor and Fock (11).

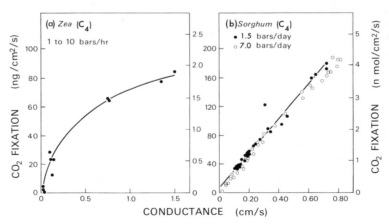

Figure 10.2. Relationship between rate of CO_2 fixation and stomatal conductance for (a) *Zea mays* plants stressed at 1 to 10 bars/h and (b) *Sorghum bicolor* plants stressed at either 1.5 bars/day or 7.0 bars/day. *Z. mays* data taken from Ref. 11; *S. bicolor* data provided by M. M. Jones. Rates of CO_2 fixation in nmol/cm²/s given for comparison with other figures.

When water stress is applied slowly, photosynthetic capacity declines, and it can be shown that stomatal and nonstomatal properties of photosynthesis decline together. When *Sorghum bicolor* is stressed slowly by droughting plants grown in large soil volumes, osmotic adjustment takes place. In these circumstances photosynthetic capacity and stomatal conductance evidently decline in a coordinated way and C_i remains constant (Figure 10.2*b*). The same holds true for slowly stressed seedlings of *Eucalyptus socialis* (C_3), as can be shown by reworking the data of Collatz et al. (12).

These studies suggest that in all but the most rapidly applied cases of water stress, intercellular CO_2 concentration remains constant as stress is applied, as a consequence of concerted adjustments to stomatal conductance and the capacity for CO_2 fixation (13). Controlled changes in stomatal conductance and capacity for CO_2 fixation that maintain constant intercellular CO_2 concentration are evident in other studies. Linear relationships between A and g_s are found when *Atriplex triangularis* or *Eucalyptus pauciflora* is grown at, or subjected to, different irradiances (14, 15) when *Zea mays* (C_4) is grown at different nitrogen levels (S. C. Wong, unpublished), and during senescence of *Phaseolus vulgaris* (C_3) (16).

The coordinated adjustments that take place in the photosynthetic apparatus during slowly applied water stress are not yet understood. The primary carboxylation enzyme RuP$_2$ carboxylase has sometimes been assayed as an indicator of changes in nonstomatal components of photosynthesis in C_3 plants. For example, Jones (17) and O'Toole et al. (18) noted that decreased nonstomatal conductance in slowly stressed *Gossypium hirsutum* and *Phaseolus vulgaris* was accompanied by decreased RuP$_2$ carboxylase activity. It is likely, however, that adjustments will involve more than this and will resemble the adjustment to irradiance in *Atriplex patula*, which entailed a "balanced change in the capacity of a number of component processes—biophysical, biochemical, and photochemical—so that no single leaf factor exclusively limits the overall rate of photosynthesis" (14). The product of these adjustments is the regulation of intercellular CO_2 concentration at a level that is presumably optimal for carbon acquisition in terms of water loss at the prevailing leaf temperature and leaf-air vapor pressure deficit.

If stress is applied rapidly or if stress becomes so severe that plants are illuminated for prolonged periods when stomata are tightly closed, intercellular CO_2 concentration presumably declines to the CO_2 compensation point. The intercellular CO_2 concentrations that prevail in leaves of different species at the CO_2 compensation point are largely a product of the photosynthetic pathway and the prevailing temperature. The biochemical bases for the different CO_2 recycling systems of higher photosynthesis and their significance in water stress are considered in the next sections.

4. CARBON DIOXIDE RECYCLING IN C_3 PLANTS IN RELATION TO WATER STRESS

Photosynthetic carbon metabolism in leaves of C_3 plants takes place through the photosynthetic carbon reduction (PCR) cycle, which is intimately linked to the photorespiratory carbon oxidation (PCO) cycle (19). These cycles are initiated and linked by the activity of RuP_2 carboxylase–oxygenase. The proportion of carbon flowing through the reduction or oxidation cycles is determined by the prevailing CO_2 and oxygen (O_2) concentrations and by the kinetic properties of RuP_2 carboxylase–oxygenase. Photorespiration, the production of CO_2 in the PCO cycle, is thus an inevitable consequence of the oxygenase activity of RuP_2 carboxylase and of the presence of O_2 in the terrestrial environment (19). The linked PCR–PCO cycles and the kinetic properties of RuP_2 carboxylase–oxygenase provide an adequate basis for the modeling of C_3 photosynthesis (13, 20, 21).

These reactions also account for the maintenance in C_3 plants of a CO_2 compensation point in air of about 60 $\mu l/l$ of CO_2 at 25°C. In the present atmosphere it is difficult to imagine situations in which the CO_2 concentration in leaves of illuminated C_3 plants would decline below that of the CO_2 compensation point. In extreme stress, when stomata close completely in the light, the integrated PCR–PCO cycles function to maintain the CO_2 compensation point in illuminated leaves. There has been much speculation about the role of the photorespiratory oxidation cycle, and we may legitimately inquire what the consequences would be if C_3 plants for any reason were unable to attain the CO_2 compensation point while illuminated.

Isolated intact chloroplasts or mesophyll cells of C_3 plants illuminated in the absence of CO_2 show a rapid loss of photosynthetic capacity (22). When leaves or leaf fragments of C_3 plants are illuminated at full sunlight in CO_2-free nitrogen (N_2) and 1% v/v O_2 (conditions that prevent photorespiration), rapid photoinhibition is observed (23, 24). Figure 10.3 shows that the capacity for CO_2 fixation is inhibited about 75% and that the apparent quantum yield is depressed to about the same extent. Photoinhibition with these characteristics has been observed in six C_3 plants, and at full sunlight the process has a half-time of approximately 30 min. As shown in Figure 10.3, the photoinhibition is partly reversed by 24 h in normal air and was fully reversed in 48 h. Consistent with these observations, the fluorescence emission spectrum of chloroplast thylakoids in liquid N_2 indicated changes in photochemical properties, namely, a substantial decrease in photosystem 2 fluorescence relative to photosystem 1 fluorescence following the photoinhibitory treatment (Figure 10.3).

The photoinhibition that results from the illumination of leaves of C_3 plants that cannot photorespire may be prevented by providing a lower CO_2 concentration or a higher O_2 concentration, which permits internal CO_2 production via photorespiration (23, 24). Figure 10.4 shows that when *Phaseolus vulgaris* leaflets are illuminated at high irradiance, but at a CO_2 concen-

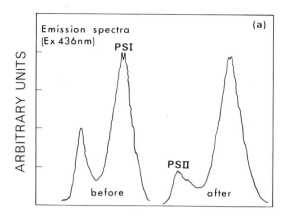

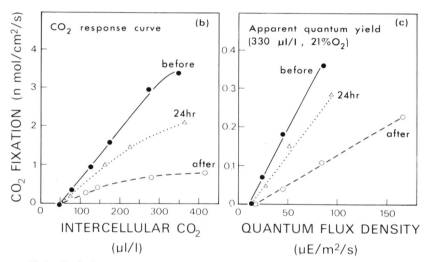

Figure 10.3. Emission spectra of isolated chloroplasts (*a*) and the relationship between CO_2 fixation and intercellular CO_2 concentration (*b*), or incident quantum flux density (*c*) of *Phaseolus vulgaris* (C_3) and 21% (v/v) O_2. Observations were made before, immediately after, and 24 h after exposure to a quantum flux density of 2000 $\mu E/m^2/s$ for 3 h at 30°C in an atmosphere containing 1% (v/v) O_2 at zero CO_2; PS I = photosystem 1 and PS II = photosystem 2.

tration equal to CO_2 compensation in the presence of 21% v/v O_2, no photoinhibition at all is observed. We suggest that the internal recycling of CO_2 as a result of the integrated PCR–PCO cycle metabolism plays a major role in the protection against photoinhibition in leaves of C_3 plants during the midday depression of stomatal conductance and photosynthetic rate (Figure 10.1). Presumably it is responsible for the stability of the photosynthetic

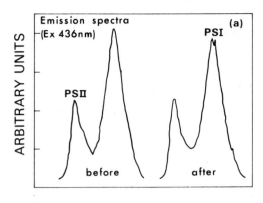

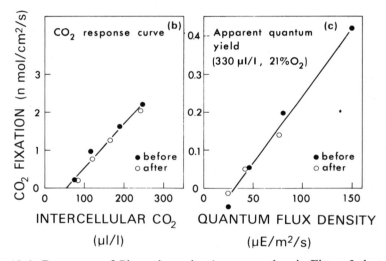

Figure 10.4. Responses of *Phaseolus vulgaris* measured as in Figure 3, but during the exposure to a quantum flux density of 2000 $\mu E/m^2/s$ for 3 h at 30°C the atmosphere contained 21% (v/v) O_2 and 70 $\mu l/l$ CO_2.

apparatus in evergreen sclerophylls that persist at CO_2 compensation point throughout summer. It is more difficult to assess the significance of this internal recycling of CO_2 in the rapid stress that may be encountered in seedling establishment or imposed in laboratory experiments. In laboratory experiments the CO_2 and water exchange patterns of plants during the imposition of stress are rarely followed. From what has been said, if stomata close in the light during imposition of stress, it is important to establish how tightly they are closed, and for how long.

When some C_3 plants are rapidly stressed, nonstomatal components of photosynthesis are substantially inhibited in a way remarkably similar to the photoinhibition of *P. vulgaris* leaves shown in Figure 10.3. Mohanty and Boyer (25) stressed sunflower leaves by depriving plants of water for 3 days. They noted that the quantum yield for CO_2 fixation was depressed to 25% of the control when plant water potential declined from -3.5 to -15.3 bars. Mooney et al. (26) withheld water from *Larrea divaricata* until leaf water potential declined from -10 to -36 bars, and observed that both the light-saturated and light-limited rates of CO_2 fixation were inhibited (Figure 10.5). These authors state that similar responses were obtained when stress was applied within a few minutes by cooling the roots. Rapid application of water stress to the C_4 species *Atriplex hymenelytra* and *Tidestromia oblongifolia* produced similar results (H. A. Mooney, unpublished).

Although these responses are remarkably similar to those observed following photoinhibition (Figure 10.3), it is difficult to imagine how cells of C_3 plant leaves could be exposed to zero CO_2 following stomatal closure in normal air. It is nevertheless possible that the responses in Figures 10.3 and 10.5 may have a common mechanistic basis and may result from photoinhibitory processes. The implication of Figures 10.3 and 10.4 is that at the CO_2 compensation point in leaves of C_3 plants, carbon flux through the integrated PCR–PCO cycles is adequate to permit the orderly dissipation of photochemical energy (24, 27). The PCO cycle itself has a high demand for photochemical energy and is capable of lowering the cellular energy state (28). Rapidly applied water stress may interfere with the utilization of photochemical energy even in the presence of CO_2, resulting in the overenergization of the photochemical apparatus and subsequent photoinhibition. Other explanations are also possible, but it is probable that the illumination of rapidly water-stressed leaves may be responsible for the changes in their photosynthetic properties.

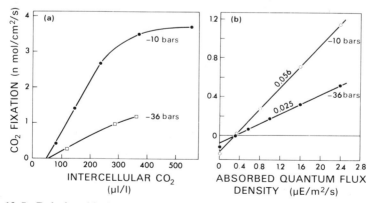

Figure 10.5. Relationship between the rate of net CO_2 fixation and (*a*) intercellular CO_2 concentration and (*b*) absorbed quantum flux density for leaves of *Larrea divaricata* at leaf water potentials of -10 bars and -36 bars. Redrawn from Ref. 26.

The integrated metabolism of the PCR–PCO cycles means that in C_3 plants in air, CO_2 fixation is about 40% less efficient than it might otherwise be (20). This may be the price C_3 plants pay for the option to recycle intercellular CO_2 and to dissipate photochemical energy, should they be deprived of external CO_2 in the light following stomatal closure in response to water stress. In case it ever proves possible to engineer the kinetic properties of RuP_2 carboxylase–oxygenase to improve the productivity of C_3 plants, we should consider whether such plants might become more sensitive to photoinhibition following stomatal closure in the light.

5. CARBON DIOXIDE RECYCLING IN C_4 PLANTS IN RELATION TO WATER STRESS

Preliminary studies with *Zea mays* show that illumination of leaves of this C_4 plant in CO_2-free air containing 1% v/v O_2 results in photoinhibition similar to that observed in leaves of C_3 plants (S. B. Powles, unpublished). If the capacity for internal recycling of CO_2 is important in protecting the photochemical apparatus against CO_2 deprivation in the light and subsequent photoinhibition, we might predict that mesophyll cells of C_4 plants would be particularly sensitive. These cells lack RuP_2 carboxylase–oxygenase and many of the enzymes of the PCR and PCO cycles (29), and hence are unlikely to be capable of the internal CO_2 generation described for mesophyll cells of C_3 plants above. On the other hand, the bundle sheath cells of C_4 plants contain all the enzymes of the PCR and PCO cycles, although the latter are sometimes present at lower activities.

Although mesophyll cells may be incapable of CO_2 recycling, processes other than CO_2 recycling may provide a sink for the orderly dissipation of photochemical energy in mesophyll cells. In NADP malic enzyme-type C_4 plants, which are deficient in photosystem 2 in the bundle sheath, half the carbon reduction requirements of the PCR–PCO cycles may have to be met by metabolite exchange and reduction in mesophyll cells (29). This speculation suggests that metabolites exchanged from the bundle sheath could represent a significant sink for the dissipation of photochemical energy in mesophyll cells when intercellular CO_2 concentration is near zero. The mesophyll chloroplasts of all C_4 plant groups contain the enzymes capable of converting 3-PGA to triose phosphate, but the function of these enzymes during net CO_2 fixation is obscure except in the case of the NADP malic enzyme type. They may represent a general mechanism for NADPH utilization in aspartate-forming C_4 plants and also a substitute for the incapacity of mesophyll cells to recycle CO_2 when external CO_2 supply is interrupted following stomatal closure.

Further experiments are needed to determine whether these processes are an effective substitute for CO_2 recycling in mesophyll cells of water-stressed C_4 plants. Certainly when C_4 plants are rapidly stressed, changes comparable to those in Figure 10.5 are observed (H. A. Mooney, unpublished). The mesophyll cells seem to be particularly sensitive to water stress. Following rapidly applied stress in Z. *mays,* Alberte et al. (30) noted a substantial loss of chlorophyll, all from the light harvesting chlorophyll *a/b*–protein complex, and most of this complex was lost from mesophyll chloroplasts. Giles et al. (31) noted that mesophyll chloroplast ultrastructure was disorganized rather more readily than that of bundle sheath chloroplasts in stressed Z. *mays.* These responses could represent advanced stages of the photoinhibitory responses shown in Figures 10.3 and 10.5.

In C_3 plants, as in C_3 plants, there appears to be a clear distinction between the controlled adjustment of photosynthesis to slowly applied water stress, in which intercellular CO_2 concentration appears to remain constant (Figure 10.2*b*), and rapidly applied stress, when it may not (Figure 10.2*a*). The proposed CO_2 and metabolite recycling systems of C_4 plants do not appear to be particularly effective in preventing damage to mesophyll chloroplasts during rapidly imposed stress. Again, illumination of rapidly water-stressed leaves may be responsible for the changes in their photosynthetic properties.

6. CARBON DIOXIDE RECYCLING IN CAM PLANTS IN RELATION TO WATER STRESS

In CAM plants the prolonged exposure of the photosynthetic apparatus to peak irradiance without access to external CO_2 is a common phenomenon. Depending on plant water status, stomata close shortly after dawn and may remain shut for the whole light period. The stomatal conductance of stem and leaf succulents in the light ranges from 0.01 to 0.001 cm/s during the hottest, most arid months (32, 33). CAM plants accumulate malic acid in the dark period, either by fixation of external CO_2 and the refixation of respiratory carbon at low stress, or by refixation of respiratory carbon alone when stress is more severe. This acid is decarboxylated in the early light period and the internally generated CO_2 is refixed by the PCR–PCO cycles, using the photochemical energy absorbed. It is probable that uncommonly high CO_2 concentrations prevail in the tissues at this time. The CAM plants studied to date show conventional C_3 photosynthesis if stomata open following deacidification (6). Thus we can reasonably project that when stomata remain tightly closed throughout the day, dark fixed malic acid is the source of internally generated CO_2 until the acid is fully consumed. The integrated PCR–PCO cycles then serve to generate CO_2 by way of photorespiration for the remainder of the light period.

Emphasis on the internal recycling of CO_2 in obligate CAM plants as means of protecting the photosynthetic apparatus against photoinhibition when stomata close in the light may be further justified by experiments with the inducible CAM plant *Mesembryanthemum crystallinum* (Aizoaceae) (5).

This succulent species is a conventional C_3 plant that can engage in CAM following water stress. Figure 10.1 shows that when *M. crystallinum* is stressed, stomata close during the period of peak irradiance. This closure differs from that in C_3 and C_4 plants because it is specifically associated with the commencement of deacidification in the light. Cause and effect in this response have yet to be separated, but it is abundantly clear that water stress resulting in stomatal closure in the light is associated with the induction of a more effective means of maintaining intercellular CO_2 concentration when stomata close.

7. ADAPTIVE SIGNIFICANCE OF ALTERNATIVE PHOTOSYNTHETIC PATHWAYS IN RELATION TO WATER STRESS

One neo-Darwinian view of adaptation (34) states that demonstration of adaptation hinges on evidence that an organism is relatively more fit to its environment than are its potential competitors. In Harper's terms (34), relative fitness (the number of descendants left by an individual relative to its fellows) can be determined by any process that confers a competitive advantage in a particular environmental context. Can one hope to assess the adaptive significance of alternative pathways of photosynthesis in these terms? This is unlikely as long as we view photosynthetic adaptation in a limited framework that seeks to match photosynthetic rate to environmental factors. This chapter draws attention away from these performance aspects of photosynthetic metabolism and attempts to refocus metabolic processes of C_3, C_4, and CAM plants in terms that are relevant to stability under water stress.

Reflecting on Harper's definition of adaptation, we might agree that photosynthetic adaptation could be proved if in the life of a plant a change from one photosynthetic pathway to another occurred in response to changes in plant and environment such that the acquired pathway better fitted the plant to the changing environment. Occasional reports that C_4 plants acquire properties of C_3 photosynthesis (or vice versa) during development, or as a result of changed environmental conditions, must be dismissed in the absence of substantive evidence. The most illustrative example of photosynthetic adaptation in these terms is the shift from C_3 photosynthesis to CAM during the life cycle of the annual *M. crystallinum* (35, 36). Figure 10.6 depicts the seasonal progression of leaf water content, nocturnal change in malate content, and δ ^{13}C value in members of a natural population at the shore of the Mediterranean Sea in Israel during the transition from wet to dry season in 1977. Germination took place after heavy rainfall in December, and plants displayed C_3 photosynthesis for the first 3 months of their life cycle as indicated by δ ^{13}C values of about $-26^0/_{00}$ and the absence of day/night fluctuations in malate content of the leaves. During this time adequate soil water was available and leaf water content was high.

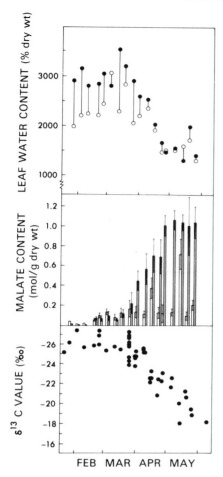

Figure 10.6. Seasonal changes in water content, malate content, and δ ^{13}C values of expanded leaves of *Mesembryanthemum crystallinum* in Israel at the shore of the Mediterranean Sea. Dusk (O, open histogram) and dawn (●, shaded histogram) values of water content and malate content are shown. Redrawn from Ref. 35.

Plants produced large leaves and exhibited high relative growth rates. With onset of the dry season, in April and May, leaf water content declined and pronounced nocturnal changes in malate content developed, indicating the onset of CAM. Values of δ ^{13}C progressively became less negative, providing additional evidence that plants engaged more and more in net CO_2 fixation in the dark.

The initial operation of C_3 photosynthesis, combined with the development of a large leaf area, allows substantial carbon gain and results in high early growth rates, which may be the major explanation of the capacity of *M. crystallinum* to monopolize its site with high ground cover. This phenomenon fits well in the classical concept of competition. The change to CAM permits this annual plant to extend its life cycle 3 to 4 months into the dry season, because the adoption of the CAM photosynthetic pathway lowers the water requirement of the plant as a result of stomatal closure in the light; CO_2 fixation in the dark permits continued assimilation, albeit at lower level compared to C_3 photosynthesis. The change from C_3 to CAM also provides for enhanced internal CO_2 generation and the protection of the photosynthetic apparatus when stomata are closed in the light. This prolonged CO_2 assimilation at low water cost is not very effective in terms of vegetative growth, but is highly effective in terms of reproduction, enabling an extremely large number of fruits to be formed and permitting a high seed yield at the end of the life cycle.

8. CONCLUSIONS

This chapter has attempted to integrate the stomatal and nonstomatal components of photosynthetic responses to water stress in C_3, C_4, and CAM plants. Our intuitive reason for doing this was the correspondence of periods of greatest water stress to periods of peak irradiance and periods when stomata close. Under these conditions, recycling of CO_2 or metabolite recycling could maintain carbon metabolism behind closed stomata and provide a sink for the dissipation of photochemical energy. We have emphasized that in C_3 plants photorespiration via the integrated PCR–PCO cycles can serve this function, that in C_4 plants metabolite recycling between mesophyll and bundle sheath cells may do likewise, and that in CAM plants CO_2 recycling from dark-fixed CO_2 and via photorespiration enables these plants to endure long periods of stomatal closure in the light (37).

We have supported these viewpoints with evidence that when C_3 plants are deprived of CO_2 and cannot photorespire, they are rapidly photoinhibited, and with evidence that in some C_3 plants water stress leading to stomatal closure is accompanied by the induction of CAM. This is, to our knowledge, the only documented case of a change in photosynthetic pathway in response to water stress. The induction of CAM can be viewed as an adaptation in the strict sense, although the competitive advantages thought to be conferred have yet to be proved.

Of perhaps greater significance for future research is the emerging evidence that during slowly applied water stress, plants adjust both stomatal and nonstomatal components of photosynthesis in such a way as to maintain constant intercellular CO_2 concentration. With similar evidence from other environmental responses it suggests that CO_2 homeostasis may be an important feature of photosynthetic function. Whether plants are often exposed to rapid water stress comparable to that applied in many laboratory experiments is doubtful. Whether illumination during rapidly applied water stress is itself responsible, via photoinhibition, for the observed changes in nonstomatal components of photosynthesis, remains to be tested.

ACKNOWLEDGMENTS

This chapter is the product of much provocative discussion with and sympathetic advice from our colleagues, notably I. R. Cowan, G. D. Farquhar, S. C. Wong, M. M. Jones, and M. M. Ludlow. In such circumstances we make no claim to the originality of the ideas presented, but accept responsibility for their presentation in this format.

REFERENCES

1. J. A. Bunce, *Plant Physiol.*, **59**, 348 (1977).
2. I. R. Cowan and G. D. Farquhar, in D. H. Jennings, Ed., *Integration of Activity in the Higher Plant*. Society for Experimental Biology, Symposium No. 31, Cambridge University Press, Cambridge, 1977, p. 471.
3. H. M. Rawson, N. C. Turner, and J. E. Begg, *Aust. J. Plant Physiol.*, **5**, 195 (1978).
4. O. L. Lange, E.-D. Schulze, L. Kappen, U. Buschbom, and M. Evenari, in D. M. Gates and R. B. Schmerl, Eds., *Perspectives of Biophysical Ecology*, Springer-Verlag, Berlin–Heidelberg–New York, 1975, p. 121.
5. K. Winter and U. Lüttge, in O. L. Lange, L. Kappen, and E.-D. Schulze, Eds., *Water and Plant Life: Problems and Modern Approaches*, Springer-Verlag, Berlin–Heidelberg–New York, 1976, p. 323.
6. C. B. Osmond, *Annu. Rev. Plant Physiol.*, **29**, 379 (1978).
7. P. A. Morrow and H. A. Mooney, *Oecologia*, **15**, 205 (1974).
8. E. L. Dunn, F. M. Shropshire, L. C. Song, and H. A. Mooney, in O. L. Lange, L. Kappen, and E.-D. Schulze, Eds., *Water and Plant Life: Problems and Modern Approaches*, Springer-Verlag, Berlin–Heidelberg–New York, 1976, p. 492.
9. M. M. Caldwell, R. S. White, R. T. Moore, and L. B. Camp, *Oecologia*, **29**, 275 (1977).
10. D. W. Lawlor, in H. Mussell and R. C. Staples, Eds., *Stress Physiology in Crop Plants*. Interscience, New York–London, 1979, p. 303.
11. D. W. Lawlor and H. Fock, *J. Exp. Bot.*, **29**, 579 (1978).
12. J. Collatz, P. J. Ferrar, and R. O. Slatyer, *Oecologia*, **23**, 95 (1976).
13. G. D. Farquhar, in R. Marcelle, H. Clijsters, and M. van Pouke, Eds., *Photosynthesis and Plant Development*, Junk, The Hague, 1979, p. 321.
14. O. Björkman, N. K. Boardman, J. M. Anderson, S. W. Thorne, D. J. Goodchild, and N. A. Pyliotis, *Carnegie Inst. Washington Yearb.*, **71**, 115 (1972).
15. S. C. Wong, I. R. Cowan, and G. D. Farquhar, *Plant Physiol.*, **62**, 670 (1978).
16. S. D. Davis and K. J. McCree, *Crop Sci.*, **18**, 280 (1978).
17. H. G. Jones, *New Phytol.*, **72**, 1095 (1973).
18. J. C. O'Toole, R. K. Crookston, K. J. Treharne, and J. L. Ozbun, *Plant Physiol.*, **57**, 465 (1976).
19. G. H. Lorimer, K. C. Woo, J. A. Berry, and C. B. Osmond, in D. O. Hall, J. Coombs, and T. W. Goodwin, Eds., *Photosynthesis 77*, Biochemical Society, London, 1978, p. 311.

20. W. A. Laing, W. L. Ogren, and R. H. Hageman, *Plant Physiol.*, **54**, 678 (1974).
21. M. Peisker, *Kulturpflanze*, **24**, 221 (1976).
22. G. H. Krause, M. Kirk, U. Heber, and C. B. Osmond, *Planta*, **142**, 229 (1978).
23. G. Cornic, *Can. J. Bot.*, **56**, 2128 (1978).
24. S. B. Powles and C. B. Osmond, *Aust. J. Plant Physiol.*, **5**, 619 (1978).
25. P. Mohanty and J. S. Boyer, *Plant Physiol.*, **57**, 704 (1976).
26. H. A. Mooney, O. Björkman, and G. J. Collatz, *Carnegie Inst. Washington Yearb.*, **76**, 328 (1977).
27. C. B. Osmond and O. Björkman, *Carnegie Inst. Washington Yearb.*, **71**, 141 (1972).
28. G. H. Krause, G. H. Lorimer, U. Heber, and M. Kirk, in D. O. Hall, J. Coombs, and T. W. Goodwin, Eds., *Photosynthesis 77*, Biochemical Society, London, 1978, p. 249.
29. M. D. Hatch and C. B. Osmond, in C. R. Stocking and U. Heber, Eds., *Transport in Plants*, Vol. III, *Intracellular Interactions and Transport Processes, Encyclopedia of Plant Physiology*, New Ser., vol. 3, Springer-Verlag, Berlin–Heidelberg–New York, 1976, p. 144.
30. R. S. Alberte, J. P. Thornber, and E. L. Fiscus, *Plant Physiol.*, **59**, 351 (1977).
31. K. L. Giles, M. F. Beardsell, and D. Cohen, *Plant Physiol.*, **54**, 208 (1974).
32. P. S. Nobel, *Plant Physiol.*, **58**, 576 (1976).
33. P. S. Nobel, *Oecologia*, **27**, 117 (1977).
34. J. L. Harper, *The Population Biology of Plants*, Academic Press, New York–San Francisco–London, 1977.
35. K. Winter, U. Lüttge, E. Winter, and J. H. Troughton, *Oecologia*, **34**, 225 (1978).
36. A. J. Bloom and J. H. Troughton, *Oecologia*, **38**, 35 (1979).
37. C. B. Osmond, in R. H. Burris and C. C. Black, Eds., *CO₂ Metabolism and Plant Productivity*, University Park Press, Baltimore, 1976, p. 217.

11 | Role of Abscisic Acid and Other Hormones in Adaptation to Water Stress

D. ASPINALL

Department of Plant Physiology, Waite Agricultural Research Institute, University of Adelaide, Adelaide, South Australia, Australia

1. INTRODUCTION

The evolution of land plants has involved adaptation to an environment of incipient water deficits, and evolutionary progress has generated physiological mechanisms that minimize tissue dehydration, mechanisms that allow metabolism despite a degree of water loss, and mechanisms that preadapt the plant to future stress. Few land plants avoid periods of restricted water supply during their life span and unpredictable water stress, of long or short duration, is a normal feature in the growth of most crops. In these circumstances mechanisms that produce adaptation to restrictions in the current water supply are important. Such adaptive responses are evoked by the onset of water stress, rather than being present as permanent features, and any metabolic or structural response of the plant to water deficit can be examined as a possible adaptive response. Changes in endogenous hormone

155

content, in particular, may represent such a response, and variations in the concentrations of all the major plant hormones with stress have been recorded. There is sufficient evidence to support realistic speculation regarding an adaptive role for abscisic acid (ABA), ethylene, and the cytokinins, and there has been much research concerning the role of ABA in the stomatal regulation of transpiration. This response represents the most evident role of any hormone in drought resistance.

2. ACCUMULATION OF ABSCISIC ACID AS A CONSEQUENCE OF WATER DEFICITS

2.1. The Stimulus for Accumulation

The demonstration that water loss from the detached wheat leaf induces a rapid rise in endogenous ABA content (1) was followed by reports of similar responses in numerous other species (2). The existence of a water potential threshold for the initiation of accumulation has been widely accepted (3), but this requires scrutiny, since the threshold potential has always been calculated from data obtained from plants that were drying progressively. It is likely that the rate of accumulation varies with tissue water potential and is present, although slow, at potentials above the claimed threshold. Indeed, where detached leaves were held for a prolonged period at a range of water potentials, no clear threshold response was found (4). Nevertheless, endogenous ABA levels respond rapidly to a substantial fall in water potential, and accumulation can occur within minutes (5). This rapidity of response allows the consideration of ABA for a role in the dynamic response of the plant to a constantly changing environment.

Although research accounts have generally reported on changes in water potential alone, it has been widely assumed that turgor, rather than total water potential, is the controlling component of leaf water status. There is no direct evidence for this, although Beardsell and Cohen (6) inferred from their data that accumulation commenced at zero turgor, and Loveys et al. (7) found that isolated grape pericarp cells accumulated ABA when plasmolysed. Control by turgor would account for the observation that field-grown cotton plants accumulate less ABA than their greenhouse-grown counterparts despite a lower leaf water potential (8), and would accord with the concept of ABA accumulation controlling transpiration through stomatal regulation.

2.2. Consequences of Prolonged Water Deficits

Continuation of a period of water deficit results in the eventual establishment of a stable elevated concentration of ABA within the plant (9). Synthesis of the hormone continues, but further metabolism accelerates so that the concentration of ABA remains constant or increases slowly (Figure 11.1). Conjugates and metabolites of ABA accumulate with prolonged stress, but little is known of their significance in the stressed plant. The evidence suggests, however, that free ABA is the major, if not the sole, compound of significance and it is apparent that the plant possesses a mechanism to limit the accumulation in prolonged periods of stress. Indeed, when drought continues for a period of weeks, ABA levels ultimately may fall, to approach those in unstressed plants (11).

2.3. Distribution Within the Plant

ABA accumulates in all organs of the intact plant during stress, probably as a result of transport from restricted sites of synthesis (primarily the leaves): ABA has been identified in both phloem (12) and xylem sap (13), although ABA may also be synthesized *in situ* in several organs. The possibility of

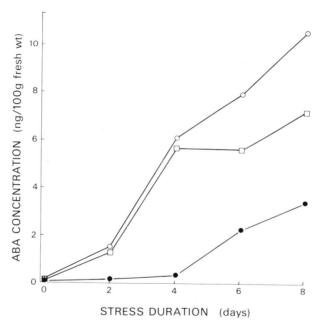

STRESS DURATION   (days)

Figure 11.1. The accumulation of free (□), "bound" (●), and total (○) abscisic acid (ABA) in the third leaf of maize (*Zea mays* cv. Iochief) subjected to water stress. Plants were grown in soil with adequate water until 24 days from germination; water was withheld thereafter, and leaf water potential fell from −4 bars in the well-watered plants to −18 bars following 8 days of stress. The concentrations of free ABA and of ABA released by hydrolysis of the extract ("bound" ABA) were measured by electron capture–gas-liquid chromatography, following extraction in acetone-water. From Ref. 10.

responses to the elevated concentration of ABA in each tissue cannot be ignored.

In the young stressed maize plant, for example, ABA accumulates most rapidly in the developing terminal shoot meristem, but also reaches high concentrations in the axillary meristems (Table 11.1). High concentrations of ABA in inhibited axillary buds (14) and rapid accumulation of ABA in terminal meristems during water stress, noted in other plants (15), have been postulated to follow translocation from the leaves (16).

ABA is not distributed equally among tissues within individual organs in the unstressed leaf. ABA is present primarily in the mesophyll chloroplasts; but additional hormone, accumulated in response to water deficits, appears in the cytoplasm and is presumably transported to the stomata (17). Exogenous ABA applied to epidermal strips is found almost entirely in the stomatal complexes, and little accumulates in other epidermal cells (18). Such reports inject a note of caution into interpretations of relationships between gross ABA content and stomatal response that should be extended to the responses postulated for other organs.

Table 11.1. Leaf Water Potentials and Concentration of Free Abscisic Acid in the Leaf and Developing Inflorescences of Young Maize (*Zea mays*) Plants on Two Dates After Watering Was Withheld (Adapted from Ref. 10)

Maize plants (cv. Iochief) were grown in the glasshouse in individual containers in soil on a high water supply until 24 days from germination (tassel initiation), when water was withheld from half the plants. In this variety of maize the uppermost axillary inflorescence (node 7) always forms a mature cob; those at nodes 5 and 6 initiate floral organs but do not develop into mature cobs on the well-watered plants, although they may do so following water stress; and those at nodes 3 and 4 also initiate flowers but develop no further.

	Day 2		Day 8	
Variable	Watered	Stressed	Watered	Stressed
Leaf water potential (bars)	−4	−8	−5	−18
Free ABA (ng/g fresh weight)				
Leaf 3	40	140	100	700
Developing tassel	40	550	70	810
Developing axillary inflorescence, node 7	80	180	140	630
As above, nodes 5 and 6	80	410	340	1150
As above, nodes 3 and 4	70	160	340	730

2.4. Reduction in Abscisic Acid Concentration
Following Stress Relief

When leaf water potential increases to the prestress level following stress relief, the ABA content of the tissue returns to the prestress concentration (Figure 11.2). This loss of ABA may be slow compared to the initial increase in concentration on the onset of stress. Wright and Hiron (5), for example, found that stressed Brussels sprout plants had an ABA content double that of the control plants 2 days after rewatering, and a similar response has been reported for vines (20). Indeed, Hiron and Wright (21) have suggested that slow recovery may be an adaptation to intermittent stress, since plants subjected to a short daily period of increased transpiration load showed a progressive increase in the basal ABA concentration and a progressive reduction in the wilting response. This phenomenon, if of general occur-

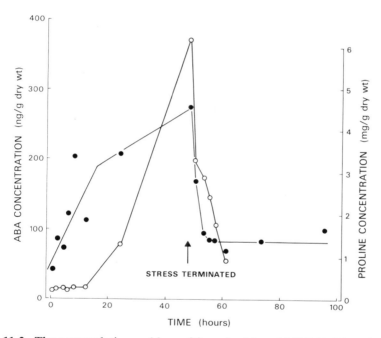

Figure 11.2. The accumulation and loss of free abscisic acid (ABA) (●) and proline (○) in the leaves of young barley (*Hordeum vulgare* cv. Prior) plants subjected to water stress, then relieved of stress. The plants were grown in aerated water culture for 10 days. Polyethyleneglycol (PEG, MW 6000) was added to the nutrient solution at zero time to reduce the osmotic potential of the nutrient solution to −10 bars. First leaves were harvested at intervals and analyzed for proline (19) and free ABA (as in Figure 11.1). Plants were placed in fresh nutrient solution without PEG following 48 h of stress.

rence, could be important in plant response to the diurnal transpiration load and stress encountered in semiarid environments.

3. CONSEQUENCES OF THE ACCUMULATION OF ABSCISIC ACID

3.1. Stomatal Closure

Following the description of the closure of stomata by exogenous ABA (22), it was established that ABA both initiates the closure of open stomata and inhibits the opening of closed stomata (23). These responses to exogenous ABA strongly suggest a role for the endogenous hormone in stomatal aperture control in the fluctuating natural environment. The most compelling evidence for this role has derived from examination of the responses of *flacca* and other tomato mutants (24). These mutants lack the ability to accumulate ABA, the stomata remain open despite turgor loss, and the plants wilt during the daylight hours. Stomatal closure occurs if the plants are supplied with ABA, and turgor is maintained if the hormone is supplied continuously.

Despite some problems in correlating endogenous concentration with stomatal response in normal plants (6, 9), which can be resolved by a consideration of the compartmentation of ABA within the leaf (17), most features of the closure of stomata in mesophytic species in response to water deficits are explicable in terms of the action of endogenous ABA (25). The initial response to a fall in water potential may be a release of preexisting ABA from the mesophyll chloroplasts (17), but this is followed by rapid synthesis. The timing of this response and the concentration required for stomatal closure in stressed plants are closely comparable to the timing and concentration in plants supplied with ABA. The response has been examined with comparatively few mesophytes, however, and others may rely on other mechanisms; hygrophytic species, for example, have been reported to close stomata when stressed, but not to accumulate ABA (26).

Stomata generally open promptly when a supply of ABA through the xylem transpiration stream is terminated, despite the continued presence of ABA in the leaf (27). Such a response has been interpreted as providing prima facie evidence for compartmentation of the hormone. On the other hand, stomata may remain partially closed following water deficits, despite a fall in ABA content to prestress levels and restoration of turgor (9). This frequently described aftereffect of stress (28) may be unrelated to ABA control, but either case counsels caution in interpreting both the response to bulk tissue ABA and the significance of that parameter.

The few reports of the effect of previous stress on the response of stomata to ABA are contradictory. Kriedemann et al. (29) found that a previous episode of water deficit or application of ABA delayed the response of stomata to a subsequent application, but Davies (30) reported that previous stress rendered stomata more sensitive to applied ABA, although less sensitive to stress. Resolution of this conflict is important as diurnal episodes of water deficit are the normal feature of plant growth in the field and the majority of experiments on ABA control of stomatal response have been carried out in artificial circumstances.

In some species ABA may reduce water loss from the leaves in ways other than through the control of stomatal aperture. In the yellow lupin ABA accumulates during water stress, but the stomata are insensitive to all but excessive concentrations of the hormone (31). Nevertheless the apparent diffusive resistance to water loss from the leaf is increased (32). It is suggested that in lupin the reduction in water loss is due to an effect of ABA at the mesophyll cell wall that restricts evaporation at that site. Such a control on transpiration has been suggested previously (33), but others have found no evidence for the effect (34).

3.2. Root Responses

Application of ABA influences both ion uptake and transport, and the permeability of root systems to water. The exudation of water from a detopped root system is stimulated by the application of ABA (35, 36) because of either an increased symplastic permeability to water (35) or an increased transport of ions into the xylem (37). Water exudation from the detopped system is dependent on an osmotic gradient, but applied ABA has a similar effect on flow under an applied hydrostatic gradient (36), suggesting that the response is relevant to the intact transpiring plant. Variations in root resistance consequent on stress-induced ABA accumulation have not been investigated, but three wilty mutants of tomato, with lower than normal ABA content, have a higher than normal root resistance that is decreased by the application of ABA (38).

ABA may also regulate ion uptake, but in separate studies, exogenous ABA has been found to inhibit ion transport through the root (though not the uptake of ions into cortical cells) (39), to inhibit ion accumulation in maize roots (40), or to stimulate ion flux into the xylem with a consequent increase in ion exudation from the root system (37). The response appears to depend on the ion status of the root system, the length of the experiment, and possibly the influence of other hormones, particularly cytokinins. Cram and Pitman (41) suggested that ABA accumulated during water stress may regulate ion transport to the shoot through an inhibition of ion movement into the xylem, hence preventing the accumulation of extracellular ions in the leaf to damaging concentrations. This conceivably could be important when transpiration is relatively unimpeded, but it is unlikely to be as significant during more severe stress when stomata close as a result of ABA accumulation and other causes and transpiration, with consequent passive ion flux to the shoot, is reduced.

3.3. Growth Responses

ABA is generally regarded as a growth inhibitor (42) and the application of ABA to intact plants may inhibit shoot elongation and induce rosette formation (43). There is little correlation between endogenous ABA concentration and elongation, however, (44) and the inhibition of elongation by water stress is as likely to be due to turgor loss as to ABA accumulation. Prolonged exposure to ABA reduced leaf growth in wheat and produced morphogenetic effects similar to those of water stress (45). Although not measured in this experiment, root growth is inhibited by exogenous ABA (46), often more readily than shoot growth (47). This difference in the apparent sensitivity of shoots and roots to applied ABA differs diametrically from the comparative effect of water stress on these organs.

High concentrations of ABA accumulate in terminal and axillary shoot meristems during water stress, and the growth of such meristems is rapidly inhibited during stress (48). It is tempting to correlate this response, together with the marked survival of meristems during prolonged stress, to the accumulated ABA. King and Evans (15) demonstrated that the inhibition of flowering of *Lolium temulentum* by water stress was due to an increase in apical ABA, and ABA treatment of wheat plants reduced the number of spikelets per ear (45), but there is little evidence that apical growth is directly inhibited by ABA. Indeed, the application of ABA to young corn plants evoked all the responses in the shoot meristems produced by a brief period of water stress except the marked inhibition of the growth of the meristematic terminal inflorescence (10).

The response of the terminal and axillary inflorescences of maize (Table 11.2) to water stress and applied ABA appears to be a special example of a

Table 11.2. The Growth of the Terminal and Axillary Inflorescences of Maize (*Zea mays* cv. Iochief) Following Exposure to an Episode of Water Stress or to Applied Abscisic Acid During the Initiation and Early Growth of the Tassel (10)

Plants were subjected to a water deficit, by withholding water, between days 25 and 35 from germination (tassel initiation) or were supplied with ABA (156 μg per plant) on day 25. All plants were harvested and the inflorescences were measured on day 56.

Treatment	Tassel length (cm)	Length (mm) of axillary inflorescences at nodes				
		3	4	5	6	7
Watered	27	28	32	81	145	248
Water deficit	13	32	43	175	209	360
Watered + ABA	25	52	50	145	182	263
LSD ($p = 0.05$)	4	8	12	32	25	n.s.

general effect on apical dominance. In the maize plant, water stress at the time of terminal inflorescence initiation causes a marked increase in the endogenous ABA content of the developing tassel, an inhibition of its growth and, subsequent to stress relief, a promotion of the growth of the axillary inflorescences (10, 49, 50). Each of these responses, except for the inhibition of tassel growth, can be evoked by applied ABA, and the growth of the axillary inflorescences is controlled by the developing tassel. It would appear that the elevation of ABA concentration in the developing tassel during the period of water stress diminishes the capacity of this organ to inhibit the growth of the axillary inflorescences. A very similar response has been found with vegetative pea seedlings (51), and these examples are reminiscent of the effect of a period of water stress on vegetative cereal plants where rewatering is followed by prolific tillering. Lateral bud growth may be inhibited by the application of excessive ABA, and the high endogenous concentrations in inhibited axillary buds have suggested a direct role for the hormone in apical dominance (14). With water stress, however, an elevation of plant ABA content followed by a decrease appears to be associated with a marked reduction in apical dominance.

ABA accumulation may also be involved in the premature cessation of growth and accelerated water loss of cereal grains subjected to water stress (52). The endogenous ABA content of wheat (52) and barley (53) grains increases from anthesis to reach a maximum immediately before the onset of water loss and ripening. Applied ABA hastens maturation and drying of the grain but has no effect on the growth rate (52), and leaf water deficits are followed by a rapid increase in the ABA content of all organs of the barley ear (54). ABA synthesized in the leaves in response to a water deficit appears to be translocated to the grains, where it leads to early cessation of growth and rapid ripening.

3.4. Other Responses

The rapid accumulation of proline in many plants is discussed in Chapter 12. Proline accumulation can be induced in barley by ABA in both excised leaves and intact plants (55). Proline accumulation in response to water stress may also be a response to accumulated ABA, and the kinetics of accumulation of the two compounds permits this hypothesis in relation to barley in that ABA accumulation precedes a detectable change in proline level (Figure 11.2). On rewatering, both the proline level and that of ABA decline rapidly and apparently simultaneously. The response to ABA is not universal, however, since both sunflower (56) and tobacco (57) accumulate proline readily during stress, but do not respond to ABA. Thus either ABA is a necessary intermediary between water stress and proline accumulation in some plants but not others, or the proline response of nonstressed plants to ABA is not related to the control system during stress.

4. GENOTYPIC DIFFERENCES IN ABSCISIC ACID ACCUMULATION

The major differences in ABA production in the tomato mutants *flacca*, *sitiens*, and *notabilis* (38) have established that ABA accumulation is essential to the normal functioning of the plant. The extent of accumulation may also be genetically determined and may have significance in the drought resistance of commercial crop cultivars (58). This suggestion deserves close consideration because it could form the basis of selection in plant breeding programs aimed at improving drought resistance. Differences in the accumulation in response to water deficits in maize and sorghum cultivars were demonstrated by Larqué-Saavedra and Wain (58) and have been found in wheat (59), pearl millet (60), and barley (57). These differences among cultivars range from two- to threefold, but care is required to establish that they reflect real differences in the biochemical response of the plant and do not merely reflect differences in water potential among the cultivars (59). The ABA content of apparently unstressed plants also varies among cultivars, and factors such as leaf age and cultural conditions can have a considerable influence (58, 60). For these reasons quantitative differences among cultivars have been established most clearly by using excised leaves subjected to a known water deficit.

The observed differences among cultivars have been correlated with the known drought resistance of those cultivars in the field, although the information is meager. Quarrie (59) has shown that the cultivar differences in ABA accumulation by detached leaves correlate closely with accumulation by stressed plants, but it remains to be demonstrated that the differences are also apparent in the field environment. For maize and sorghum, Larqué-Saavedra and Wain (58) found that high ABA accumulation was associated with drought resistance, whereas for wheat, Quarrie (59) reported that greater accumulation was correlated with lower resistance. In each case the relationship is based on very few individual cultivars, and drought resistance was assessed from field experience. Although it is possible to account for this discrepancy by postulating fundamental differences among the species in their drought resistance mechanisms, it would seem wiser to suspend judgment until more complete data are available.

In the research published to date, only variation in the accumulation in response to a particular degree of stress has been examined. It is conceivable, however, that variation in the relationship between water potential and accumulation, the level maintained under prolonged stress, translocation to other organs, sensitivity to accumulated ABA, and the rapidity of metabolism of ABA once stress is relieved could be of equal significance. There are major difficulties in comparing cultivars for ABA accumulation in the intact plant, however, which will complicate the measurement of many of these factors. Most significant, it is virtually impossible to produce identical water deficit episodes in different cultivars; indeed inherent differences in ABA accumulation would themselves be expected to lead to differences in stomatal response and consequent leaf water potential. On purely theoretical grounds, therefore, it is perhaps naive to expect to be able to select cultivars for differences in ABA accumulation potential by subjecting intact plants to the same drought treatments. This view is supported by the observation that of a group of eight wheat cultivars subjected to stress by withholding water, the only one to show a markedly different relationship between ABA content and leaf water potential also differed significantly from the others in the rate at which leaf water potential decreased with continuing stress (59). In these circumstances it is logical to proceed by first establishing that cultivar differences in ABA accumulation in response to given leaf water potential regimes exist. At present this appears to be possible only by using excised systems. The second step would be to establish that the demonstrated cultivar differences are heritable, and finally to relate the response of the different cultivars (or hybrids) in drought situations to their known ABA response systems.

5. CYTOKININS

The premise that cytokinin is involved in the response to water deficits rests on the observation that cytokinin is transported from roots to shoots in many plants (61) and that the concentration in sap exuded from a detopped plant is reduced by previous stress (62). These form the basis of the hypothesis advanced by Itai et al. (62) that stress-induced responses in the shoot, including leaf senescence and protein synthesis inhibition, are the consequence of a reduced supply of cytokinin acting as a "signal" of root stress. Such a "signal" would seem to be redundant during water stress, since shoot water potential falls more rapidly than root potential during stress.

Nevertheless the notion that cytokinin-regulated responses are involved in stress-induced effects and are possible candidates for a role in stress resistance requires consideration. Experiments with excised leaves have shown that exogenous cytokinin will induce stomatal opening (63, 64). This response appears to be restricted to the Poaceae (65), and it has been suggested that it is an artifact consequent on the effects of kinetin on mesophyll senescence in the excised leaves (66). However isolated epidermal strips from a grass leaf also respond, and apparently their reaction is too rapid to involve tissue senescence (67). The increased transpiration or wilting of intact plants, including some dicotyledonous species, following kinetin application, can also be interpreted as a promotion or maintenance of stomatal opening (68). Serious further consideration of the hypothesis that stomatal aperture is regulated by cytokinin levels as well as by ABA during stress is limited by the lack of reliable data on the concentrations of cytokinin in the plant, and particularly the leaf, during stress.

The experimental evidence for a connection between reduced cytokinin export from the roots and other plant responses is tenuous, although several possibilities have been suggested, notably leaf senescence and inhibition of protein synthesis. Proline accumulation in excised systems is responsive to applied cytokinin, the accumulation in response to either water stress or applied ABA being reduced by cytokinin applied simultaneously (Figure 11.3). Such a response is in accord with the concept of a stress-induced reduction in cytokinin content contributing to proline accumulation in the intact plant, but attempts to reverse or reduce proline accumulation in stressed intact radish and barley plants with applied cytokinins have been frustrated by significant effects of the treatment on leaf water potential (57).

6. ETHYLENE

The internal concentration of ethylene and its production by plant tissues is enhanced by water stress (69). It is possible that the response is indirect, since applied ABA will also induce ethylene production (70); but in either event the increase in endogenous ethylene appears to exert a controlling influence on leaf abscission (71) and the abscission of young fruit (72). In cotton leaf petioles ethylene production increased rapidly as leaf water

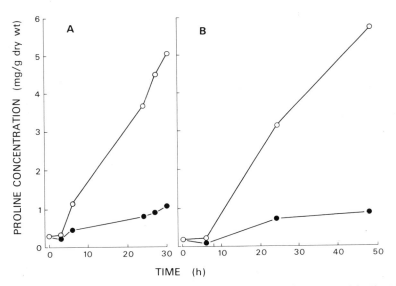

Figure 11.3. The effect of kinetin on proline accumulation in detached barley (*Hordeum vulgare* cv. Prior) leaves in response to (*a*) water stress or (*b*) abscisic acid (ABA). Sections excised from the first leaves on 10 day old barley seedlings were floated on various solutions in Petri dishes and assayed for proline at intervals: solutions were (*a*) either −20 bars osmotic potential polyethylene glycol (PEG) solution (O), or −20 bars PEG with kinetin (5 μg/ml) (●); (*b*) either ABA, 5 μg/ml (O), or ABA, 5 μg/ml with kinetin 5 μg/ml (●).

potential fell, particularly in the older leaves; the ethylene concentration fell when the water stress was relieved, but leaf abscission followed some 6 h after rewatering (73). A similar situation occurred in *Vicia faba,* where again the older leaves were the more likely to be shed (74).

7. HORMONES AND ADAPTATION TO WATER STRESS

7.1. Control of Water Status

Endogenous ABA concentration and intracellular carbon dioxide (CO_2) concentration appear to be major factors in the mechanism regulating stomatal aperture, hence transpiration and leaf turgor in mesophytes (75). Indeed the existence of this control is the major reason for the absence of damaging water deficits in plants growing in normal nonstress land environments, and mutant plants that lack part of the mechanism are permanently wilted. Can we extend this view to classify the mechanism as an adaptation to water stress in more extreme environments?

Stomatal regulation through ABA prevents tissue damage through excess loss of turgor and improves the water use efficiency of plants, allowing more growth on a given supply of water (75). Mizrahi et al. (76) observed longer survival and greater total growth in ABA-treated cereal plants than in controls (plants that were deprived of water until death ensued). Since transpiration during stress, particularly severe stress, is unlikely to be limited by additional ABA (77), and ABA is likely to exert the largest effect during periods of marginal water deficit when stomata are not completely closed, the most logical explanation of the observation by Mizrahi et al. is that the applied ABA limits water loss before stress ensues, thereby sparing water for subsequent growth. This effect of applied ABA, and presumably an analogous effect of a high water potential threshold for the accumulation of endogenous ABA, resembles the strategy of water sparing through a high intrinsic root resistance to water movement (78), which may fit annual crops to exploit a stored-water environment and enable ephemeral species to complete seed production before succumbing to water stress.

Mesophytes respond to a fall in leaf water potential by accumulating ABA in concentrations considerably above the nonstress level. In contrast, the few arid zone plants that have been examined are characterized by a generally lower initial concentration and a much less spectacular increase on stress (20). It has been suggested that this difference in response reflects the necessity for desert species to respond rapidly to a transient amelioration of the environment, but this ability would also seem to be an advantage in mesophytes. The apparently more profligate accumulation of ABA in mesophytes need not inhibit prompt stomatal opening on rewatering (27), but many species show a delay in reopening. Although this aftereffect of stress has been said to confer an advantage, it is difficult to agree that this is clearly so. It is significant that the only arid zone plants investigated appear to be aridoactive species (79) that tend to grow, if only slowly, during periods of stress. A low potential for ABA accumulation may be characteristic of such a growth habit, and it would be instructive to examine aridopassive species, particularly desert ephemerals, to ascertain whether a low potential for ABA accumulation is characteristic of all arid zone plants.

Considering the known sensitivity of stomata to very low concentrations of applied ABA (18), the massive accumulations observed on stress prolonged beyond a few hours seem excessive. Only a small fraction of the accumulated ABA appears to be concerned in stomatal control, and the more modest accumulation in desert species may well suffice to satisfy this role. Do we need to seek an adaptive role for this apparently surplus ABA? The evidence from cultivars differing in ABA accumulation does not clarify this point, and we need to consider the likely consequences of ABA accumulation on processes other than stomatal regulation of transpiration. Although there is little information on the concentrations to be found in roots during stress, it is possible that they are sufficient to induce the effects on hydraulic conductivity described by Glinka (36). In a system where root resistance to water movement is a major component of the total resistance in the soil-plant system, any reduction in this component would tend to increase the flux of water and decrease the potential difference between bulk soil and leaf. Since any increase in plant water potential would ameliorate the effect of stress on plant metabolism, this could improve plant performance during stress. If the increase in leaf water potential were sufficient to lead to transient stomatal opening and an increase in transpiration, however, the response would be counterproductive in a prolonged period of stress.

7.2. Other Responses

It is difficult to accept that each of the several responses to applied hormones that have been shown to mimic the effects of water deficit are necessarily adaptive, although this property has been suggested for many of them. Uncertainty also remains about the actual role of the endogenous hormone in many of the responses, and this needs to be established before the adaptive nature of the response can be examined. Most effects have been postulated from the response of turgid plants to supplied hormone, particularly where the response resembles a stress response, but it is possible that tissue at a lower water potential during stress responds differently to a given concentration of hormone. If this is so, data obtained from turgid plants would be irrelevant to the elucidation of the response to stress.

The strongest case, both for hormone control and for an adaptive role, can be made for leaf shedding in response to stress-induced ethylene production. Such a response has an obvious adaptive effect on subsequent water loss, but an equally obvious penalty in the loss of photosynthetic tissue and the materials in the shed leaves. It is a response tailored to plant survival rather than to subsequent production.

The accumulation of proline and the inhibition of stem elongation during water stress are arguably adaptive responses, but the case for hormonal control in either is not strong. Responses such as the effects of ABA on shoot meristems, which can lead to an inhibition of flowering, loss of apical dominance, or reduction in the size of the flowering spike, have no obvious adaptive value or even appear to damage the success of the plant.

8. CONCLUSION

The role of ABA accumulation in the stomatal control of transpiration is the best documented example of hormone-mediated adaptation to changes in the environment. The evidence suggests strongly that this mechanism operates in mesophytes as a water deficit sensor and control system to minimize the reduction in tissue water potential inherent in a transpiring plant exposed to a fluctuating environment. Although this system of control reduces the effects of relatively short-term and minor water deficits that otherwise might be expected to reduce growth, there are as yet few data to support the view that ABA accumulation has a major role in adaptation to severe or prolonged water stress. Mesophytes accumulate high concentrations of ABA under such circumstances, and despite the presence of mechanisms to metabolize the accumulated ABA, these concentrations may well be detrimental to growth and metabolism. This restricted role for ABA is compatible with the natural evolution and human selection of the species that have so far been examined. Selection over millennia, whether natural or by man, will have been for reproductive success in environments where moderate diurnal changes in tissue water potential are normal, but prolonged drought is a comparative rarity. Deliberate selection for grain yield during drought is a relatively recent innovation in cereal breeding, for instance, and even now has to be coupled with selection for success in the nondrought environment. In these circumstances, mechanisms that adapt the plant to the normal environment would have been strongly selected, and the resulting set of adaptations, including hormone-mediated adaptations, might reasonably be expected to fit the plant to such environments even at the cost of some reduction in adaptation to the extreme stress environment.

An examination of the hormonal responses of species adapted to arid environments would be instructive, but it would not necessarily elucidate the role of hormones in the mesophytes. Such an examination would be particularly valuable if evaluated against a background knowledge of physiological adaptation of the plant to its environment. Our lack of understanding of the nature or extent of the reputed drought resistance of the plants concerned renders inconclusive the few accounts of relationships between hormone content and drought resistance now available.

REFERENCES

1. S. T. C. Wright, *Planta*, **86**, 10 (1969).
2. B. V. Milborrow, *Annu. Rev. Plant Physiol.*, **25**, 259 (1974).
3. T. J. Zabadal, *Plant Physiol.*, **53**, 125 (1974).
4. S. T. C. Wright, *Planta*, **134**, 183 (1977).
5. S. T. C. Wright and R. W. P. Hiron, in D. J. Carr, Ed., *Plant Growth Substances 1970*, Springer-Verlag, Berlin–Heidelberg–New York, 1972, p. 291.
6. M. F. Beardsell and D. Cohen, *Plant Physiol.*, **56**, 207 (1975).
7. B. R. Loveys, C. J. Brien, and P. E. Kriedemann, *Physiol. Plant.*, **33**, 166 (1975).
8. B. L. McMichael and B. W. Hanny, *Agron. J.*, **69**, 979 (1977).
9. M. F. Beardsell and D. Cohen, in R. L. Bieleski, A. R. Ferguson, and M. M. Cresswell, Eds., *Mechanisms of Regulation of Plant Growth*, Bulletin 12, Royal Society of New Zealand, Wellington, 1974, p. 411.
10. H. B. Damptey, B. G. Coombe, and D. Aspinall, *Ann. Bot. (London)*, **42**, 1447 (1978).
11. O. S. Rasmussen, *Physiol. Plant.*, **36**, 208 (1976).
12. M. R. Bowen and G. V. Hoad, *Planta*, **81**, 64 (1968).
13. G. V. Hoad, *Planta*, **124**, 25 (1975).
14. D. J. Tucker and T. A. Mansfield, *J. Exp. Bot.*, **24**, 731 (1973).
15. R. W. King and L. T. Evans, *Aust. J. Plant Physiol.*, **4**, 225 (1977).
16. G. V. Hoad, *Planta*, **113**, 367 (1973).
17. B. R. Loveys, *Physiol. Plant.*, **40**, 6 (1977).
18. C. Itai, J. D. B. Weyers, J. R. Hillman, H. Meidner, and C. Willmer, *Nature (London)*, **271**, 652 (1978).
19. T. N. Singh, L. G. Paleg, and D. Aspinall, *Aust. J. Biol. Sci.*, **26**, 45 (1973).
20. P. E. Kriedemann and B. R. Loveys, in R. L. Bieleski, A. R. Ferguson, and M. M. Cresswell, Eds., *Mechanisms of Regulation of Plant Growth*, Bulletin 12, Royal Society of New Zealand, Wellington, 1974, p. 461.
21. R. W. P. Hiron and S. T. C. Wright, *J. Exp. Bot.*, **24**, 769 (1973).
22. C. J. Mittelheuser and R. F. M. Van Steveninck, *Nature (London)*, **221**, 281 (1969).
23. R. J. Jones and T. A. Mansfield, *J. Exp. Bot.*, **21**, 714 (1970).
24. M. Tal, D. Imber, and I. Gardi, *J. Exp. Bot.*, **25**, 51 (1974).
25. B. R. Loveys and P. E. Kriedemann, in D. N. Sen, Ed., *Structure, Function and Ecology of Stomata*, Bishen Singh Mahendra Pal Singh, Dehradun, India, 1979, p. 77.
26. K. Dörffling, J. Streich, W. Kruse, and B. Muxfeldt, *Z. Pflanzenphysiol.*, **81**, 43 (1977).
27. W. R. Cummins, *Planta*, **114**, 159 (1973).
28. R. A. Fischer, *Plant Physiol. (Suppl.)*, **42**, 18 (1967).
29. P. E. Kriedemann, B. R. Loveys, G. L. Fuller, and A. C. Leopold, *Plant Physiol.*, **49**, 842 (1972).
30. W. J. Davies, *J. Exp. Bot.*, **29**, 175 (1978).
31. J. E. Lancaster, J. D. Mann, and N. G. Porter, *J. Exp. Bot.*, **28**, 184 (1977).
32. J. E. Lancaster and J. D. Mann, *J. Exp. Bot.*, **28**, 1373 (1977).
33. P. G. Jarvis and R. O. Slatyer, *Planta*, **90**, 303 (1970).

34. G. D. Farquhar and K. Raschke, *Plant Physiol.*, **61**, 1000 (1978).
35. Z. Glinka, *Plant Physiol.*, **51**, 217 (1973).
36. Z. Glinka, *Plant Physiol.*, **59**, 933 (1977).
37. J. L. Karmoker and R. F. M. Van Steveninck, *Planta*, **141**, 37 (1978).
38. M. Tal and Y. Nevo, *Biochem. Genet.*, **8**, 291 (1973).
39. M. G. Pitman, U. Lüttge, A. Lauchli, and E. Ball, *J. Exp. Bot.*, **25**, 147 (1974).
40. D. L. Shaner, S. M. Mertz, Jr., and C. J. Arntzen, *Planta*, **122**, 79 (1975).
41. W. J. Cram and M. G. Pitman, *Aust. J. Biol. Sci.*, **25**, 1125 (1972).
42. B. V. Milborrow, *Planta*, **70**, 155 (1966).
43. A. Rikin, M. Waldman, A. E. Richmond, and A. Dovrat, *J. Exp Bot.*, **26**, 175 (1975).
44. J. A. D. Zeevaart, *Plant Physiol.*, **53**, 644 (1974).
45. S. A. Quarrie and H. G. Jones, *J. Exp. Bot.*, **28**, 192 (1977).
46. R. J. Newton, *Physiol. Plant.*, **30**, 108 (1974).
47. G. D. Weston, *HortScience*, **11**, 22 (1976).
48. I. Husain and D. Aspinall, *Ann. Bot. (London)*, **34**, 393 (1970).
49. H. B. Damptey and D. Aspinall, *Ann. Bot. (London)*, **40**, 23 (1976).
50. H. B. Damptey, D. Aspinall, and B. G. Coombe, *Ann. Bot. (London)*, **42**, 849 (1978).
51. D. M. Bellandi and K. Dörffling, *Physiol. Plant.*, **32**, 369 (1974).
52. R. W. King, *Planta*, **132**, 43 (1976).
53. H. Goldbach and G. Michael, *Crop Sci.*, **16**, 797 (1976).
54. H. Goldbach and E. Goldbach, *J. Exp. Bot.*, **28**, 1342 (1977).
55. D. Aspinall, T. N. Singh, and L. G. Paleg, *Aust. J. Biol. Sci.*, **26**, 319 (1973).
56. R. L. Wample and J. D. Bewley, *Can. J. Bot.*, **53**, 2893 (1975).
57. D. Aspinall, unpublished observations.
58. A. Larqué-Saavedra and R. L. Wain, *Ann. Appl. Biol.*, **83**, 291 (1976).
59. S. A. Quarrie, in *Opportunities for Chemical Plant Growth Regulation: Proc. Symp. 4–5th January 1978, Univ. Reading.* British Crop Protection Council, Monograph No. 21, 1978, p. 55.
60. S. A. Quarrie, personal communication.
61. I. E. Henson and P. F. Wareing, *J. Exp. Bot.*, **27**, 1268 (1976).
62. C. Itai, A. Richmond, and Y. Vaadia, *Isr. J. Bot.*, **17**, 187 (1968).
63. A. Livnè and Y. Vaadia, *Physiol. Plant.*, **18**, 658 (1965).
64. H. Meidner, *J. Exp. Bot.*, **18**, 556 (1967).
65. H. H. Luke and T. E. Freeman, *Nature (London)*, **217**, 873 (1968).
66. K. Raschke, *Annu. Rev. Plant Physiol.*, **26**, 309 (1975).
67. L. D. Incoll and G. C. Whitelam, *Planta*, **137**, 243 (1977).
68. M. B. Kirkham, W. R. Gardner, and G. C. Gerloff, *Plant Physiol.*, **53**, 241 (1974).
69. S. Ben-Yehoshua and B. Aloni, *Plant Physiol.*, **53**, 863 (1974).
70. W. C. Cooper, G. K. Rasmussen, B. J. Rogers, P. C. Reece, and W. H. Henry, *Plant Physiol.*, **43**, 1560 (1968).
71. W. R. Jordan, P. W. Morgan, and T. L. Davenport, *Plant Physiol.*, **50**, 756 (1972).
72. J. A. Lipe and P. W. Morgan, *Plant Physiol.*, **51**, 949 (1973).
73. B. L. McMichael, W. R. Jordan, and R. D. Powell, *Plant Physiol.*, **49**, 658 (1972).

74. A. S. El-Beltagy and M. A. Hall, *New Phytol.*, **73**, 47 (1974).
75. K. Raschke, in E. Marrè and O. Ciferri, Eds., *Regulation of Cell Membrane Activities in Plants*, Elsevier/North Holland Biomedical Press, Amsterdam, 1977, p. 173.
76. Y. Mizrahi, S. G. Scherings, S. Malis Arad, and A. E. Richmond, *Physiol. Plant.*, **31**, 44 (1974).
77. M. Talha and P. Larsen, *Physiol. Plant.*, **37**, 104 (1976).
78. J. B. Passioura, *Aust. J. Plant Physiol.*, **3**, 559 (1976).
79. R. A. Fischer and N. C. Turner, *Annu. Rev. Plant Physiol.*, **29**, 277 (1978).

12 | Proline Accumulation as a Metabolic Response to Water Stress

CECIL R. STEWART

Department of Botany and Plant Pathology, Iowa State University, Ames, Iowa

ANDREW D. HANSON

MSU–DOE Plant Research Laboratory, Michigan State University, East Lansing, Michigan

1. INTRODUCTION

The accumulation of free proline in water-stressed leaves, described first by Kemble and MacPherson in ryegrass (1), has been subsequently observed in many species including barley (2, 3), wheat (4), rice (5), sorghum (6), field beans (7), soybeans (8), cotton (9), and tobacco (10). Singh et al. (2) reported that barley genotypes that yielded well in drought prone environments showed higher proline accumulation during water stress at the seedling stage than did drought susceptible genotypes such as Proctor. Accordingly, it was suggested that selection for high proline-accumulating potential could be used as a tool in breeding cereals for drought resistance. This correlation between field performance and seedling tests, and further results (3), encouraged speculation that proline accumulation was adaptive, with the capacity to accumulate high levels of free proline related to improved survival of extreme water stress, and to rapid resumption of growth following stress relief. Such speculation receives support from comparative biology:

173

for example, when growing in the saline conditions to which they are adapted, some halophytes accumulate much free proline—up to 123 μmol/g fresh weight in *Triglochin maritima* (11).

Considerable work has been done to characterize the proline accumulation response physiologically and biochemically, for the most part using detached leaves. This has led to a number of suggestions on ways in which the capacity for proline accumulation could confer adaptation to drought (Table 12.1).

This chapter first reviews the metabolic changes that cause proline to accumulate in excised leaves and asks whether these might repesent adaptive responses. Then we present data on intact plants that bear directly on the adaptive significance of proline accumulation and its proposed application as a metabolic measure of drought resistance in cereal breeding.

2. METABOLIC CAUSES OF PROLINE ACCUMULATION DURING WATER STRESS

The accumulation of a single essential metabolite due to dehydration of the tissue would appear to be a rather simple effect. However it results from more than one metabolic change. Since the pathways and metabolic steps in proline synthesis and oxidation have not been well characterized, additional

Table 12.1. Some Adaptive Roles Proposed for Proline Accumulation in Water-Stressed Leaves

Possible adaptive significance	Refs.
Water Relations	
1. Osmotic substance (especially a cytoplasmic osmoticum) during water stress	10,47
2. "Desiccation protectant"	3, 48–51
Nitrogen Metabolism	
3. Storage of nitrogen (in a readily available, nontoxic form) for utilization on rewatering	3, 14, 52, 53
4. Translocated form of reduced N during water stress	54, 55
5. Detoxification of NH_3 produced during water stress	1, 6
Energy Metabolism	
6. Source of reducing power and/or carbon skeletons for use on rewatering	3, 6, 14, 51–53
7. Sink for surplus reducing power during water stress	56

studies on proline metabolism have been and continue to be necessary to understand the specific metabolic causes of proline accumulation. Proline metabolism in plants has been recently reviewed as a background for the effects of drought stress on proline metabolism (12), and the reader is referred to that review for background information. Proline accumulation is caused by stimulated synthesis, inhibition of oxidation, and impaired protein synthesis.

2.1. Loss of Feedback Inhibition

It was clear from the beginning (1, 13) that the amount of proline that accumulated exceeded the loss of proline from protein of excised leaves. Therefore proline accumulation required a continued synthesis and possibly a stimulation of synthesis. Barnett and Naylor (14) and Morris et al. (15) observed increased incorporation of ^{14}C-glutamate into proline due to wilting. By measuring the total radioactivity in proline (protein + nonprotein proline) and the specific radioactivity when leaves were fed ^{14}C-glutamate, Boggess et al. (16) showed that wilting caused a stimulation of proline synthesis from glutamate. Furthermore, Stewart (17) showed that the time course of proline accumulation in excised barley leaves was almost identical to the time course of the ^{14}C-glutamate to proline conversion. This comparison suggests that stimulation is a major factor in proline accumulation. Since wilting does not cause stimulation in the conversion of ornithine to proline until after proline has started to accumulate, Boggess et al. (16) concluded that the stimulatory effect was on the first enzyme in proline synthesis.

One can reason from the foregoing that proline synthesis in a wilted leaf must not be subject to feedback inhibition because the rate of synthesis increases while the level of proline increases. Boggess et al. (10) showed that proline does inhibit its own synthesis in turgid barley and tobacco leaves, but not in wilted leaves, thus experimentally demonstrating that wilting causes a loss in the sensitivity of proline synthesis to inhibition by proline.

It is presently difficult to say whether this effect of stress could have adaptive significance. If, for example, stimulation of proline synthesis is a response to the presence of a positive effector molecule or to the removal of a negative effector such as that found in glycolysis, which is responsible for the Warburg effect, one might suggest that enhanced proline synthesis reflects operation of a positive regulatory mechanism. If, on the other hand, stimulation of proline synthesis represents a breakdown in a normal metabolic regulatory mechanism, one might consider it to be a disintegrative effect.

2.2. Inhibition of Proline Oxidation

Proline is oxidized via Δ^1-pyrroline-5-carboxylate to glutamate, which is converted to Krebs cycle intermediates, and the carbon is ultimately released as carbon dioxide (7, 18, 19). This process occurs in mitochondria (19), and the rate is dependent on the level of proline in the turgid cell (7, 20). The latter observation means that as proline accumulates, it is increasingly susceptible to utilization by oxidation. Thus the marked accumulation of proline during stress suggests that stress inhibits proline oxidation. This effect of stress has been observed experimentally in barley (20, 21), but calculations of the rate of proline oxidation in turgid leaves indicate that although this effect contributes to proline accumulation, it cannot be solely responsible for it.

Proline oxidation appears to be a mitochondrial process (19). In earlier experiments Boggess et al. (10) showed that the intermediates in proline oxidation are not in equilibrium with the same intermediates of proline synthesis in turgid barley leaves. These results are interpreted to mean that the proline oxidation intermediates are contained within the mitochondria and are therefore not accessible to the enzymes of proline biosynthesis. Recently Stewart and Boggess (21) observed that in wilted barley leaves, intermediates of proline oxidation are reconverted to proline. This result is shown in Reference 21, Figure 3. In that experiment [3,4-^{3}H]-proline was extensively oxidized by barley in the first hour of wilting before the time that wilting inhibits oxidation. Subsequently there was reconversion of the oxidized products to proline in the wilted leaves but not in the turgid leaves. Thus drought stress inhibits proline oxidation by mitochondria and apparently allows oxidized products of proline to leak from the mitochondria.

It is possible that both the inhibition of proline oxidation and the loss of compartmentalization of metabolites result from the same effect of stress on mitochondria. Swollen mitochondria with less distinct cristae have been observed in water-stressed tissue (22, 23). Nir et al. (24) noted that the cytochemical detection of cytochrome oxidase was still possible, but their results indicate the structure of the cristae was altered. Mitochondria from stressed cotton leaves show acid lipase activity, which could be partially responsible for the effects on mitochondrial structure (22). The oxidative capacities of mitochondria from stressed tissue are also diminished compared to those from turgid tissue (24–26). All these observations could result from the effect of water stress on mitochondrial membranes. Certainly metabolite leakage could result from altered membrane structure. The oxidation of proline is probably a membrane-bound process that would be affected by changes in the membrane, but a direct effect of stress on the proline-oxidizing enzyme separate from the effect on the membrane cannot be ruled out.

The acid lipase activity that is present in mitochondria from stressed tissue is an example of a hydrolytic enzyme whose activity is higher in stressed tissue. This higher activity is also observed in other organelles (e.g., chloroplasts). Several hydrolytic enzyme activities have been observed to increase during water stress, including phosphatase(s), lipase(s), ribonuclease(s), and protease(s) (see Ref. 27 for tabulation). These enzymes are usually associated with the disintegration of cell structure and function. If the inhibition of proline oxidation by stress does result from an alteration of membrane structure by enzymic hydrolysis, proline accumulation would be a consequence of disintegrative-type processes in the cell.

2.3. Impaired Protein Synthesis

In normal turgid tissue, protein synthesis is the main metabolic fate of proline because low rates of proline oxidation are observed when proline levels are low. Protein synthesis is severely inhibited by water stress as measured by incorporation of proline into protein (20, 28). Probably the incorporation of all amino acids into protein is inhibited by water stress. As in the case of proline oxidation, inhibition of protein synthesis alone cannot be responsible for proline accumulation, but it does contribute (see Ref. 12 for calculations of rates and the relative contribution to proline accumulation).

The decrease in proline utilization for protein synthesis is undoubtedly a manifestation of the reduction in polyribosome levels caused by water stress in plant tissues, such as corn seedlings (29, 30), and in slowly dehydrated moss (31). Since the time course of the decline in polysomes does not coincide with the time course of an increase in ribonuclease, Hsiao (32) and Dhindsa and Bewley (33) conclude that nucleic acid hydrolysis by RNase is not the cause of the loss of polysomes. Instead, the loss appears to be due to the inability of stressed cells to form an initiation complex and to effect the formation of peptide bonds. The loss of polysomes in stressed tissue is due to ribosome runoff from mRNA and the failure to reinitiate new mRNA-ribosome complexes. Polysomes re-form rapidly when plants are rewatered both in seedlings (29) and in moss (34). The moss *Tortula ruralis* becomes dormant on desiccation and resumes growth on rehydration. In that case the loss of polysomes would appear to be associated with the onset of dormancy rather than with a complete disintegration of the cellular protein-synthesizing mechanism. The similarity between polysome loss in water-stressed seedlings and slowly dried moss suggests that polysome loss in plants under mild stress may also be accompanied by a temporary suspension of protein synthesis rather than by a permanent loss of the capability to synthesize protein.

One might expect the ability of a plant to temporarily suspend protein synthesis under stress to be an adaptive capability. A complete or more permanent loss in protein-synthesizing capability under stress might be associated with drought susceptibility. Either mechanism of impaired protein synthesis would result in decreased utilization of proline for protein synthesis, and the extent to which proline accumulates as a result would not distinguish between the two mechanisms.

2.4. Theoretical Considerations Regarding the Adaptive Nature of Proline Accumulation Based on an Analysis of the Causal Biochemical Mechanisms

Three main changes cause proline to accumulate under stress:

1. Stimulated synthesis due to loss of feedback inhibition.
2. Inhibited oxidation, probably due to effects on mitochondria.
3. Impaired protein synthesis.

To predict a priori whether these changes are adaptive (i.e., able to confer some degree of drought resistance) one might try to characterize them either as "normal" reactions to a change in the environment or as "abnormal" reactions, precipitated by environmental insult, that are associated with the disruption or disintegration of normal cellular processes. Only "normal" reactions would be expected to confer drought resistance; "abnormal" reactions would be associated with drought susceptibility.

A lack of data on the molecular mechanisms involved in stimulated synthesis precludes its characterization as a "normal" positive effect or an "abnormal" consequence of the disintegration of a metabolic control mechanism. Such data would be a valuable contribution to our understanding of the metabolic effects of dehydration.

As for the inhibition of proline oxidation, one might conclude that this response is an example of an "abnormal" disruptive effect that would therefore be associated with drought susceptibility. This conclusion assumes that the inhibition is due to effects of dehydration on mitochondrial structure and function outlined above. However there is no evidence to support this explanation, and further experiments should be undertaken to obtain such evidence. It is also possible that the alteration in mitochondrial function and structure is not associated with irreversible subcellular disintegration. Since it is apparently reversible, it could resemble changes that are associated with the onset of dormancy.

As indicated above, the impairment of protein synthesis could be either a temporary suspension or a disruption of the protein-synthesizing mechanism. Since the impairment is reversible and presumably represents ribosomal runoff, the temporary suspension alternative is suggested. This effect would be similar to what happens as a tissue approaches dormancy. Does this type of response represent an adaptive advantage for a vegetative leaf? It would certainly be associated with slowed metabolism under stress. Perhaps whether this is advantageous or unfavorable depends on how readily normal protein synthesis is resumed on relief of stress.

The level of free proline in stressed tissue is determined by a combination of the effects above, and also by the rate of proline export via the phloem (Section 3.2). Thus free proline level may not be a valid indication of any one of the effects that contribute to it, regardless of whether they are adaptive.

3. IS PROLINE ACCUMULATION AN ADAPTIVE RESPONSE TO WATER STRESS?

Much of the interest in proline accumulation by the leaves of water-stressed crops stems from the suggestion that in barley it is an adaptive metabolic response, for which genotypic variation is available (2, 3). This suggestion raises several issues:

1. What does "adaptation to water stress" mean in the context of crop production in dry regions?
2. What are the heritable characters that combine to make up a crop genotype that is adapted to drought prone environments?
3. How do we know whether a metabolic character is adaptive?
4. Does experimental evidence support either the individual adaptive roles for proline listed in Table 12.1, or the general proposition that proline accumulation is adaptive?

The first three sets of issues are broad and relevant to all attempts to identify physiological adaptations to water stress, and to exploit them in plant-breeding programs. We therefore treat proline accumulation in barley as an illustrative example of a more general case: *the application of metabolic traits in breeding for drought resistance.*

3.1. Adaptation of Crops to Drought Prone Environments

In discussing crop adaptation, both crop and environment must be specified. Spring barley, grown as a winter crop in the cereal bèlts of southern Australia, is used as an example because the suggestion that proline accumulation in barley was adaptive (2, 3) rested in part on the field data of Finlay and Wilkinson (35) from this region. Finlay and Wilkinson tested 277 barley cultivars at up to three sites for three seasons, with a range of 122 to 625 mm of rain during the growing season. Grain yield data for three of these cultivars were used for Figure 12.1, in which the linear regressions of grain yield on seasonal precipitation represent three contrasting responses to low moisture availability. For discussion purposes, Figure 12.1 also includes a theoretical curve relating precipitation to total dry matter yield (calculated from Equation 8 in Ref. 36).

The cultivar 'Atlas' can be taken as a genotype *generally adapted* to the environments tested because it had a high yield when rainfall was adequate, yet still yielded moderately well in low rainfall conditions. Although 'Provost' performed quite well in favorable moisture conditions, its grain yield suffered drastically as moisture availability decreased; 'Provost' is therefore *not generally adapted* to the dry environments in question. The variety

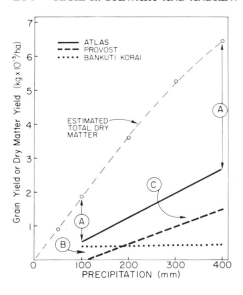

Figure 12.1. Responses to precipitation of three contrasting barley cultivars, plotted schematically as linear regressions of grain yield on seasonal precipitation. Original data were from Finlay and Wilkinson (35) for two sites (Clinton and Minlaton) in South Australia between 1958 and 1960; plot area was taken as 2.81 m². For discussion purposes, an estimate of total dry matter production (calculated from Equation 8 in Ref. 36) is included.

'Bankuti Korai' yielded as well as 'Atlas' in dry conditions, but it did not respond to increasing moisture availability; it, too, is *not generally adapted.*

Three points that bear on crop adaptation to dry environments emerge from Figure 12.1. First, the total dry matter curve is really an upper boundary for growth that is set principally by physical laws to which all vegetation is subject. It was probably similar for all three cultivars, and *cannot be significantly raised* by genetic or cultural means. There are also constraints that probably limit the amount of total plant dry matter in the grains to a maximum of 30 to 40%. Since the grain yield of 'Atlas' probably approached this percentage in all environments, it can be argued that 'Atlas' is already very well adapted, and that the distances *A* between all points on its grain yield curve and the total dry matter curve are therefore essentially irreducible. In short, it is unreasonable to expect much improvement on 'Atlas,' viewed from the standpoint of grain yield alone.

Second, the causes of the shortfall in grain yield of 'Provost' under very low moisture conditions *B* are not necessarily relevant to breeding for drought resistance because the maximum achievable yields in such environments are too low (< 500 kg/ha) for cereal agriculture. Similar considerations apply to the means by which 'Bankuti Korai' achieves some grain production in very dry conditions, because this cultivar was unable to exploit the more favorable environments in which barley is usually grown.

The third point concerns the divergence between the performance of 'Atlas' and 'Provost' in the 250–400 mm rainfall range *C*. Here yields are certainly low (about 1500 to 2500 kg/ha), but are relevant to agricultural production in semiarid regions. It is the adaptive mechanisms operating in this range that are of greatest practical interest to plant breeding. There is no reason to believe that such mechanisms are the same as those operating at *B*.

What are the types of heritable character expressed in region *C* that confer adaptation? They may be split into characters that are expressed at three more or less distinct levels (Table 12.2). The first two levels, are used currently in breeding crops for dry environments because they can be fairly readily detected. Physiological adaptations—including metabolic adaptations—almost certainly exist. They cannot yet be exploited in plant breeding because in general, we do not know which among the many observed metabolic responses to water stress are *adaptive,* which are *deleterious,* and which are simply *incidental consequences* of stress. If such knowledge were available, and if useful genetic variation in adaptive responses could be found, it could be applied in at least two ways to plant breeding. First, metabolic tests could be used to complement current methods for choosing parents for crosses and perhaps for selecting progeny; second, good sources of an adaptive metabolic trait could be identified among otherwise poorly adapted genotypes (46).

Table 12.2. Levels at which Characters Conferring Adaptation to Drought Prone Conditions Are Expressed, with Examples Drawn from Cereals

Level of adaptive character	Examples	Refs.
1. Life cycle or phenological	*Early maturity* in winter wheats of the Great Plains and spring wheats of the Canadian prairies	37, 38
	Short day requirement for flowering, resulting in *late maturity,* in upland rices of areas with a bimodal rainfall pattern	39
2. Morphological	Rapidly growing, extensive root systems (especially deeply penetrating seminal roots) in spring wheats growing on stored water	40, 41
	Awns in wheat	42
3. Physiological/metabolic	Diurnal and seasonal osmotic adjustment in unirrigated maize, sorghum, and wheat	43–45

3.2. Experimental Testing of the Adaptive Significance of Proline Accumulation

The problem of experimental testing of adaptive significance can be approached in various ways. One way is to take a single genotype, to manipulate either the rate of endogenous proline synthesis or the proline content, and to correlate these changes with plant performance during and after stress. Several experiments of this type, mostly seeking to test the significance of proline in tissue-water relations (roles 1 and 2, Table 12.1) have been reported. The results of experiments in which wheat and *Carex* plants were treated with proline (48, 50) are equivocal (see Ref. 46 for a fuller discussion). Similarly, experiments with barley subjected to cycles of stress and relief (3) are compromised by problems with the water potential measurements involved (57, 58). In summary, this apparently straightforward approach has not yielded any definitive results.

A second approach to testing adaptive value also makes use of a single genotype. A "balance sheet" is drawn up for proline accumulation that sets it in the context of other metabolic events. The "balance sheet" can then be used to answer the following questions, for various possible adaptive roles: is there enough proline in the right place, at the right time, for it to have adaptive significance? An obvious application of this approach is in evaluating the contribution of accumulated proline to the osmotic potential in stressed leaves. Calculations from the literature show that the amount of proline accumulated in leaves of crops seldom exceeds 50 μmol/g fresh weight, a level that cannot contribute much to the osmotic potential unless proline is compartmented in a small fraction of the total cell volume.

The so-called balance sheet approach has also been applied with barley seedlings to examine possible roles in nitrogen (N) and energy metabolism (roles 3 to 7 in Table 12.1) (55, 56). Proline did not begin to accumulate rapidly until the second and subsequent days of stress (55), at which time much of the accumulated proline was found to be sequestered in the leaf tips that were irreversibly wilted or "fired" (57). The start of the fast phase of proline accumulation coincided with cessation of N export from the leaves (55). Thus little proline accumulated during early stress when the leaf was still actively exporting N and capable of full recovery from wilting. Though it is probable that much of the proline synthesized at this time was exported, it was not a major species in which N moved out of the stressed leaf for it accounted for only about 13% of N movement (55). Also, although proline came eventually to dominate the free amino acid pool, it represented a small quantity of N relative both to the amount of N translocated from the wilted leaf and to the amount of N that was retained in the leaf. Moreover, since much of the proline was in drought-killed (fired) tissue, it was clearly unavailable for use on relief of stress. These results make a major adaptive role for proline as an N transport form or an N reserve seem unlikely.

If proline synthesis during water stress were to replace or to supplement amide (glutamine and asparagine) formation as a disposal pathway for free ammonia (NH_3), differences between stressed and unstressed leaves in their metabolism of radioactive NH_3 ($^{13}NH_3$) might be expected. In stressed leaves, rapid labeling of proline should occur if free NH_3 is detoxified by immediate conversion (via small intermediate pools of glutamine and/or glutamate) to the imino group of proline. Unstressed and stressed barley leaves both contained very low levels (≤ 0.3 μmol/leaf) of free ammonia (55). When fed tracer $^{13}NH_3$ (< 0.1 μmol/leaf) both as a gas and as a droplet of $^{13}NH_4^+$ solution, ^{13}N label was assimilated within 10 min by illuminated leaves of both unstressed and stressed plants (Tully and Hanson, unpublished results). After incubation times of up to 30 min, however, the major ^{13}N-labeled products were always glutamine and glutamate; proline was essentially unlabeled in both unstressed and stressed leaves. These data make it seem unlikely that proline synthesis is directly involved in ammonia detoxification in water-stressed leaves.

From measurements of the rate and extent of proline accumulation during stress, the reducing power demand for proline synthesis and the reductant storage capacity of proline can be estimated. When this is done, the electron flux to proline synthesis (or from proline oxidation) appears as a very small item in the energy budget of an illuminated leaf. Even assuming that the capacity for photosynthetic electron transport is diminished by 90% in a stressed barley leaf, the amount of reducing power stored in the proline accumulated over 3 days could still be generated by the chloroplasts in less than an hour (56). An adaptive value of proline accumulation in energy metabolism therefore seems implausible.

A third experimental approach to evaluating adaptive significance involves exploiting genotypic differences in rate of proline accumulation, by correlating such differences with genotype performance during and/or after stress. This approach has also been applied to barley (2, 3, 57, 58). Hanson et al. (57, 58) were able to confirm in part the original reports of Singh et al. (2–4) that barley cultivars differ in rate of proline accumulation under water stress; they found, however, that the fastest rates of proline accumulation occurred in the genotypes whose water status declined fastest during stress, and which showed the most severe leaf firing. This was the case when water stress was imposed by flooding the roots of perlite-grown plants with polyethylene glycol (-19 bars), or by withholding water from soil-grown plants. In 'Proctor' (a drought susceptible cultivar similar to 'Provost,' Figure 12.1) leaf water potential (ψ_l) fell more rapidly and proline accumulated faster than in 'Excelsior' (a drought-adapted cultivar similar to 'Atlas,' Figure 12.1) (57). When an empirical plot is made of proline accumulation against ψ_l (Fig. 12.2), it can be seen that there are no differences among barley genotypes with respect to the "threshold" ψ_l at which proline accumulation begins or to the response of proline accumulation to ψ_l values below the "threshold." It therefore seems that cultivar differences in proline accumulation rate can be explained solely by dif-

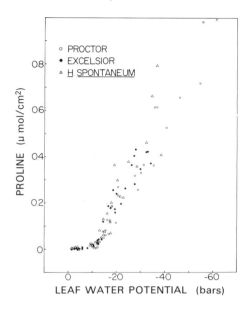

Figure 12.2. The relationship between proline content and leaf water potential (ψ_l) at the midblade position for about 30 individual second leaves of each of three barley genotypes subjected to stress induced by polyethylene glycol. Data were obtained from time course experiments in which the accumulation of proline and the decline of ψ_l were followed in 'Proctor,' 'Excelsior', and a *Hordeum spontaneum* race collected at a dry site. The ranking order for both the rate of proline accumulation and the rate of decline of ψ_l in these experiments was 'Proctor' $>$ *H. spontaneum* $\geqslant$ 'Excelsior.' Adapted from Ref. 58.

ferences in the rate at which water status declines during stress. It can be argued that selecting for high proline-accumulating potential would tend to produce genotypes whose water status would fall more rapidly during stress, thus would be more severely injured by drought.

This possibility was tested using the F_2 to F_4 progeny from the cross 'Proctor' by 'Excelsior.' By selecting the F_2 plants with the highest and lowest proline accumulation during stress (Figure 12.3), and by testing their F_3 and F_4 progenies (Table 12.3), it was shown that the proline accumulation trait was heritable. During these experiments it was noted that F_4 plants of lines selected for high proline-accumulating potential tended to be less vigorous even under nonstressed conditions than F_4 plants of lines selected for low proline. It appeared that selection for high proline-accumulating potential produced generally weak plants. Because the trait for proline accumulation was heritable, the F_3 generation must have been segregating for proline-accumulating potential, as well as for other characters in which the parental cultivars differed. Such a segregating population provides better experimental material for assessing the value of proline accumulation during stress than the contrasting parent cultivars; differences in parental characters other that proline-accumulating potential are effectively "canceled out" in the segregating population since they are, in general, expected to segregate independently of the proline trait. If proline accumulation is adaptive, a *negative* correlation would be expected between the amount of proline accumulated by individual F_3 plants and the drought injury, measured as leaf firing, sustained by those individuals. Figure 12.4 shows that a highly significant *positive* correlation was in fact obtained between proline ac-

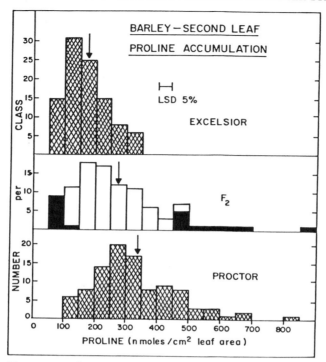

Figure 12.3. Frequency distribution diagrams for proline accumulation in 100 F_2 plants from a 'Proctor' by 'Excelsior' cross, and in 100 plants of each of the parent cultivars. Proline contents of second leaves were measured after 3 days of stress, induced by polyethylene glycol, and expressed on the basis of calculated leaf area (= length × width × 0.75). Vertical arrows show population means; the horizontal bar is the LSD ($p = 0.05$) between population means determined from a two-way analysis of variance. Solid sections in the F_2 distribution histograms identify the high and low proline accumulators advanced to the F_3. From Ref. 58.

cumulation and the extent of leaf firing. This is difficut to reconcile with any adaptive role for proline accumulation during water stress.

3.3. Conclusion

There is little doubt that adaptive metabolic responses to water stress exist; although proline accumulation is probably not one of them. Osmotic adjustment with organic solutes (Table 12.2) can be taken as an example (Chapter 7). In more general terms, some of the best documented instances of adaptive metabolic responses in plants subjected to environmental stresses involve responses to anaerobiosis or to flooding (see, e.g., Refs. 59, 60). It is no coincidence that several adaptive metabolic strategies have been recognized for anaerobic environments; it has happened because both oxygen deficits per se, and their effects on soil chemistry, are relatively simple stresses that

Table 12.3. Inheritance of the Capacity for Proline Accumulation During Water Stress

The proline contents of blades of the second leaf were measured after 3 days of stress induced by polyethylene glycol. All genotypes contained very low levels of proline ($\leqslant$ 5 nmol/cm^2) before stress imposition. In the F_3 generation the progeny of 20 F_2 plants selected for high proline accumulation and of 20 F_2 plants selected for low proline accumulation were screened. In the F_4 generation progeny of 4 F_3 plants reselected for high proline accumulation and 4 F_3 plants reselected for low proline accumulation were tested. Checks of both parental cultivars were included in testing both generations.

Generation	Genotypes	Number of plants tested	Mean proline content (nmol/cm^2 leaf area)
F_3	'Excelsior'	40	126a[a]
	F_3 lines (F_2 plants selected for low proline)	80[b]	185b
	F_3 lines (F_2 plants selected for high proline)	80[b]	263c
	'Proctor'	40	321d
F_4	'Excelsior'	18	322a
	F_4 lines (F_3 plants reselected for low proline)	36[c]	385a
	F_4 lines (F_3 plants reselected for high proline)	36[c]	575b
	'Proctor'	18	619b

[a] Means within an experiment with letters in common not significantly different by LSD ($p = 0.05$).
[b] Four plants of each line.
[c] Nine plants of each line.

are understood reasonably well in terms of the metabolic problems that they pose to any aerobic organism. Water stress is far more complex physically than is anoxia, and we presently lack an adequate theoretical framework on its biophysical, biochemical, and physiological consequences. Because we do not really know which events have the most deleterious impacts for productivity and survival, we cannot yet predict how the damaging effects of water stress might best be alleviated.

ACKNOWLEDGMENTS

The research of Andrew D. Hanson reported in this chapter was supported by the U.S. Department of Energy under Contract No. EY-76-C-02-1338.

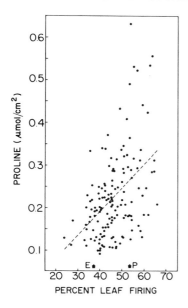

Figure 12.4. The relationship between proline accumulation and leaf firing in 146 plants of the F_3 generation from a 'Proctor' by 'Excelsior' cross. Approximately equal numbers of F_3 progeny of F_2 low accumulators and F_2 high accumulators (see Figure 12.3) were included. Proline accumulation in the second leaves was measured after 3 days of stress. Leaf firing [percent leaf firing = (length of leaf fired/total length) × 100] of the third leaf was measured after 4 days of stress followed by a 14 day recovery period. Broken line shows the linear regression, for which $r = +0.450$, significant at the 0.1% level. Vertical arrows indicate mean leaf firing for 'Excelsior' (E; 38 plants) and 'Proctor' (P; 35 plants) scored in the same experiment. Adapted from Ref. 58.

The contribution of Dr. Samuel F. Boggess to the discussion of the metabolic causes of proline accumulation via numerous discussions with one of us (C.R.S.) is sincerely appreciated.

REFERENCES

1. A. R. Kemble and H. T. MacPherson, *Biochem. J.*, **58**, 46 (1954).
2. T. N. Singh, D. Aspinall, and L. G. Paleg, *Nature (London) New Biol.*, **236**, 188 (1972).
3. T. N. Singh, L. G. Paleg, and D. Aspinall, *Aust. J. Biol. Sci.*, **26**, 65 (1973).
4. T. N. Singh, D. Aspinall, and L. G. Paleg, *Aust. J. Biol. Sci.*, **26**, 77 (1973).
5. International Rice Research Institute, *Annual Report for 1975*, International Rice Research Institute, Los Baños, Philippines, 1976.
6. A. Blum and A. Ebercon, *Crop Sci.*, **16**, 428 (1976).
7. C. R. Stewart, *Plant Physiol.*, **50**, 551 (1972).
8. R. P. Waldren and I. D. Teare, *Plant Soil*, **40**, 689 (1974).
9. B. L. McMichael and C. D. Elmore, *Crop Sci.*, **17**, 905 (1977).
10. S. F. Boggess, D. Aspinall, and L. G. Paleg, *Aust. J. Plant Physiol.*, **3**, 513 (1976).
11. G. R. Stewart and J. A. Lee, *Planta*, **120**, 279 (1974).
12. C. R. Stewart, in L. G. Paleg and D. Aspinall, Eds., *The Physiology and Biochemistry of Drought Resistance*, Academic Press, New York–San Francisco–London, 1980 (in press).

13. J. F. Thompson, C. R. Stewart, and C. J. Morris, *Plant Physiol.*, **41**, 1578 (1966).
14. N. M. Barnett and A. W. Naylor, *Plant Physiol.*, **41**, 1222 (1966).
15. C. J. Morris, J. F. Thompson, and C. M. Johnson, *Plant Physiol.*, **44**, 1023 (1969).
16. S. F. Boggess, C. R. Stewart, D. Aspinall, and L. G. Paleg, *Plant Physiol.*, **58**, 398 (1976).
17. C. R. Stewart, *Plant Physiol.*, **61**, 775 (1978).
18. C. R. Stewart and E. Y. Lai, *Plant Sci. Lett.*, **3**, 173 (1974).
19. S. F. Boggess, D. E. Koeppe, and C. R. Stewart, *Plant Physiol.*, **62**, 22 (1978).
20. C. R. Stewart, S. F. Boggess, D. Aspinall, and L. G. Paleg, *Plant Physiol.*, **59**, 930 (1977).
21. C. R. Stewart and S. F. Boggess, *Plant Physiol.*, **61**, 654 (1978).
22. J. Vieira da Silva, A. W. Naylor, and P. J. Kramer, *Proc. Nat. Acad. Sci. (US)*, **71**, 3243 (1974).
23. E. B. Tucker, J. W. Costerton, and J. D. Bewley, *Can. J. Bot.*, **53**, 94 (1975).
24. I. Nir, A. Poljakoff-Mayber, and S. Klein, *Plant Physiol.*, **45**, 173 (1970).
25. R. J. Miller, D. T. Bell, and D. E. Koeppe, *Plant Physiol.*, **48**, 229 (1971).
26. D. T. Bell, D. E. Koeppe, and R. J. Miller, *Plant Physiol.*, **48**, 413 (1971).
27. G. W. Todd, in T. T. Kozlowski, Ed., *Water Deficits and Plant Growth*, Vol. 3, Academic Press, New York–San Francisco–London, 1972, p. 177.
28. C. R. Stewart, *Plant Physiol.*, **51**, 508 (1972).
29. T. C. Hsiao, *Plant Physiol.*, **46**, 281 (1970).
30. C. A. Morilla, J. S. Boyer, and R. H. Hageman, *Plant Physiol.*, **51**, 817 (1973).
31. J. D. Bewley, *J. Exp. Bot.*, **23**, 692 (1972).
32. T. C. Hsiao, *Annu. Rev. Plant Physiol.*, **24**, 519 (1973).
33. R. S. Dhindsa and J. D. Bewley, *J. Exp. Bot.*, **27**, 513 (1976).
34. J. D. Bewley, *Can. J. Bot.*, **51**, 203 (1973).
35. K. W. Finlay and G. N. Wilkinson, *Aust. J. Agric. Res.*, **14**, 742 (1963).
36. H. Lieth, in O. L. Lange, L. Kappen, and E.-D. Schulze, Eds., *Water and Plant Life: Problems and Modern Approaches*, Springer-Verlag, Berlin–Heidelberg–New York, 1976, p. 392.
37. L. P. Reitz, *Agric. Meteorol.*, **14**, 3 (1974).
38. E. A. Hurd, in K. L. Larson and J. D. Eastin, Ed., *Drought Injury and Resistance in Crops*, Special Publication No. 2, Crop Science Society of America, Madison, Wisconsin, 1971, p. 77.
39. International Rice Research Institute. *Annual Report for 1974*, International Rice Research Institute, Los Baños, Phillippines, 1975.
40. E. A. Hurd, *Agric. Meteorol.*, **14**, 39 (1974).
41. J. B. Passioura, *Aust. J. Agric. Res.*, **32**, 745 (1972).
42. I. Arnon, *Crop Production in Dry Regions*, Vol. 2, Leonard Hill, London, 1972, pp. 13–14.
43. T. C. Hsiao, E. Acevedo, E. Fereres, and D. W. Henderson, *Philos. Trans. R. Soc. London, Ser. B.*, **273**, 479 (1976).
44. M. M. Jones and N. C. Turner, *Plant Physiol.*, **61**, 122 (1978).
45. J. M. Morgan, *Nature (London)*, **270**, 234 (1977).
46. A. D. Hanson and C. E. Nelsen, in P. S. Carlson, Ed., *The Biology of Crop Improvement*, Academic Press, New York–San Francisco–London, 1980, p. 77.

47. V. Rajagopal, V. Balasubramanian, and S. K. Sinha, *Physiol. Plant.*, **40**, 69 (1977).
48. L. A. Tyankova, *C. R. Acad. Bulg. Sci.*, **19**, 847 (1966).
49. C. Hubac, *C. R. Acad. Sci., Ser. D.*, **264**, 1286 (1967).
50. C. Hubac and D. Guerrier, *Oecol. Plant.*, **7**, 147 (1972).
51. G. Pálfi, M. Bitó, R. Nehéz, and R. Sebestyen, *Acta Biol. (Szeged.)*, **20**, 95 (1974).
52. C. R. Stewart, C. J. Morris, and J. F. Thompson, *Plant Physiol.*, **41**, 1585 (1966).
53. D. G. Routley, *Crop Sci.*, **6**, 358 (1966).
54. T. N. Singh, D. Aspinall, L. G. Paleg, and S. F. Boggess, *Aust. J. Biol. Sci.*, **26**, 57 (1973).
55. R. E. Tully, A. D. Hanson, and C. E. Nelsen, *Plant Physiol.*, **63**, 518 (1979).
56. A. D. Hanson and C. E. Nelsen, *Plant Physiol.*, **62**, 305 (1978).
57. A. D. Hanson, C. E. Nelsen, and E. H. Everson, *Crop Sci.*, **17**, 720 (1977).
58. A. D. Hanson, C. E. Nelsen, A. R. Pedersen, and E. H. Everson, *Crop Sci.*, **19**, 489 (1979).
59. R. M. M. Crawford, *Trans. Bot. Soc. Edinburgh* **41**, 309 (1971).
60. C. D. John and H. Greenway, *Aust. J. Plant Physiol.*, **3**, 325 (1976).

13 | Drought Responses of Apical Meristems

E.W.R. BARLOW

School of Biological Sciences, Macquarie University, Sydney, New South Wales, Australia

R.E. MUNNS

Department of Agronomy, Institute of Agriculture, University of Western Australia, Perth, Western Australia, Australia

C.J. BRADY

Plant Physiology Unit, Division of Food Research, CSIRO, Sydney, New South Wales, Australia

1. INTRODUCTION

The apical meristem plays a central role in determining the development of the shoot; hence its response to drought is of vital importance to the survival and productivity of the plant. The response of the apical meristem to drought is somewhat anomalous, for although it displays extreme sensitivity to water stress during growth (1, 2), it is able to survive severe water stress better than most plant tissues (2). Similarly, the meristematic tissue of the wheat embryo can survive desiccation more readily than can mature and differentiated tissue (3). It is probable that the enclosure of the apical meristem within the older leaf sheaths is an important facet of the ability of the Gramineae to adapt to desiccating environments.

191

Despite the general observation of drought survival by apical meristems, there is little knowledge of the physiology of drought response in apices, nor of the mechanisms underlying their drought tolerance. This chapter considers in some detail the adaptive response of the apical meristem to drought, making special reference to the apical meristem of the wheat plant.

2. INITIAL DROUGHT RESPONSE OF THE APEX

2.1. Growth and Differentiation

The differentiation of leaf and spikelet primordia in both wheat and barley is severely retarded by mild water stress (1, 2). Hussain and Aspinall (1) reported that the differentiation of leaf primordia in barley was sensitive to small decreases in soil water potential. In their study plant growth (dry weight increase) was somewhat less sensitive to drought than was differentiation of primordia, and leaf elongation was not measured. Furthermore Hussain and Aspinall (1) reported that some growth and differentiation of lateral spikelets continued after the initiation of terminal spikelet primordia was inhibited by drought. Since the appearance of stamen initials in the most advanced spikelet coincides with the cessation of spikelet primordia formation (4, 5), continued growth and differentiation of lateral spikelets during drought may result in a lower final spikelet number. In these studies the water potential of the barley apex (ψ_a) determined by a comparatively insensitive sucrose equilibration method, remained constant, while the relative turgidity of the exposed leaves decreased to 75%.

In an experiment with wheat in which the ψ_a during drought was measured with a small thermocouple psychrometer, Barlow et al. (2) found that spikelet initiation, apex elongation, and leaf elongation were severely inhibited when the ψ_a declined by 5 bars (Figure 13.1). In this experiment the water potential of the apex and the exposed leaf 4 (ψ_l) declined simultaneously during drought, indicating that in spite of its enclosed position and lack of functional xylem, the apex is in some hydraulic equilibrium with the plant roots and possibly the older exposed leaves. The decrease in ψ_a during drought was confirmed in subsequent experiments (6), in which the osmotic potential of the apex (π_a) was also measured (Figure 13.2). The water relations of an enclosed elongating leaf (leaf 7) and a mature exposed leaf (leaf 5) are included in Figure 13.2 to provide a comparison of ψ within the stressed wheat plant.

The initial decline in ψ_a was associated with a large decline in turgor potential (P_a) and a smaller decline in π_a (Figure 13.2), and was very similar to the initial water stress response of leaf tissue (7). Thus the growth response of this actively dividing tissue to water stress (Figure 13.1) appears to be a turgor-mediated response, which is very similar to that of tissues growing predominantly by cell elongation (8). This may be due to the

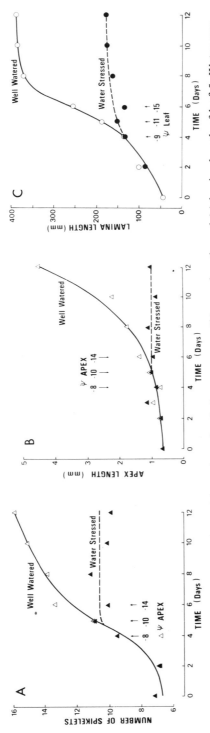

Figure 13.1. The effect of drought on (a) number of spikelets per apex, (b) apex length, and (c) lamina length of leaf 5. Water was withheld from the wheat plants on day 0. Apex; ▲, △; leaf 5; ●, ○:well-watered plants, open symbols; stressed plants, closed symbols. The water potential (bars) of leaf 4 and the apex on days 4, 5, and 6 are shown in a, b, c. The coefficient of variation of the growth measurements was 18% (range 10–25%). After Ref. 2.

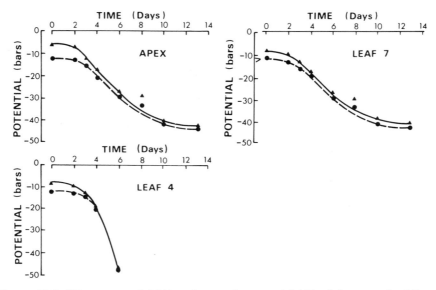

Figure 13.2. Water potential (▲) and osmotic potential (●) of the apex, leaf 7, and leaf 4 in water-stressed wheat plants. Water was withheld from the plants on day 0. The coefficient of variation of the water and osmotic potential measurements was 14% (range 5–18%). After Ref. 6.

requirement of meristematic cells to expand to a certain size before further division (9). The water relations data presented in Figure 13.2 demonstrates that P_a declined from $4·5 ± 0·5$ to $1·5 ± 0·5$ bars in association with the cessation of apex elongation. A P_a of $1·5 ± 0·5$ bars was maintained for the remainder of the drought period, and the apex did not recommence elongation. The maintenance of a slight positive P_a in the absence of growth indicates either that the apex does have a threshold turgor for growth, as has been previously reported for tissues growing primarily by cell expansion (8, 10, 11), or that some metabolic factor is limiting growth. Furthermore the maintenance of a small subthreshold P_a during stress without elongation would indicate that there was no change in the cell wall elasticity within the apex.

Positive P_a may also be important in the survival of the apex at low ψ. In the stress experiment illustrated in Figure 13.1, where ψ_a decreased, but a small positive P_a was maintained, the wheat plants were rewatered on days 6, 8, 10, 12, and 18 of the stress period and the number of spikelets at ear emergence was noted (Figure 13.3). Although the apex was able to survive ψ_a as low as −60 bars, water stress did influence the final spikelet number. When water stress was sufficient to affect spikelet differentiation ($\psi_a = -15$ bars), the final spikelet number was significantly reduced from 16 to 14.5. Reduction in ψ_a to −45 bars over another 6 day period resulted in no

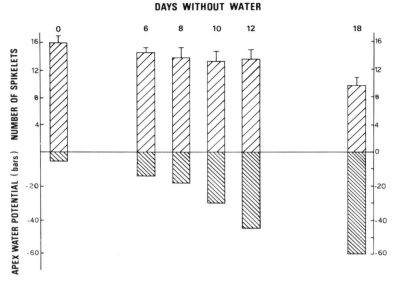

Figure 13.3. The effect of the duration of water stress on the water potential of the apex and the final spikelet number in wheat. The pots were rewatered 6, 8, 10, 12, and 18 days after the initial withholding of water when the water potential of the apex was as shown, and spikelet numbers were counted after ear emergence. Bars indicate standard errors.

significant further reductions in the spikelet number (Figure 13.3). However when ψ_a reached -60 bars after an 18 day stress, the final spikelet number again suffered a significant reduction from 13.7 to 9.7. These plants did not differentiate any new spikelets on rewatering (Figures 13.1 and 13.3).

Reductions in spikelet number may not be entirely due to lowered ψ_a per se, but to continued lateral spikelet differentiation during stress. Hussain and Aspinall (1) have reported continued lateral spikelet differentiation during stress in barley, leading to spikelet maturation and rapid termination of terminal spikelet formation after rewatering (4). In our experiment we were not able to differentiate between these possibilities because the severely stressed plants were also stressed for the longest period. Whatever the mechanism may be, however, it is clear that severe water stress during the spikelet differentiation phase of floral development can severely reduce the final spikelet number.

2.2. Protein Synthesis and Respiration

Protein synthesis is an integral part of growth and differentiation, and actively dividing tissues such as the apex have a large ribosome population. The extent of ribosome organization into polyribosomes is an index of the utilization of the potential for protein synthesis in a tissue. Consequently the occurrence of ribosomes as polyribosomes in the apex during a drought is a measure of the rate of biosynthetic activity in the apical cells at the time of sampling (12).

In wheat, when the ψ_a had decreased by 5 bars, the polyribosome content of the apex decreased synchronously with apex growth and also with leaf growth (Figures 13.1 and 13.4) indicating that the drought responses of all three processes were tightly integrated or similarly sensitive. The polyribosome content of the apex decreased with increasing water stress until polyribosomes were only 10% of the total ribosomes at ψ_a of -25 bars. As the ψ_a fell below -25 bars the polyribosome content stabilized at a level of about 10% and this level was maintained to ψ_a as low as -40 bars (Figure 13.4).

Apex respiration was also depressed by water stress, and the response was analogous to the response of protein synthesis (Figure 13.5). Initially apex respiration decreased rapidly with a small decrease in ψ_a. However further large reductions in ψ_a did not result in concomitantly large reductions in respiration, and at ψ_a of -48 bars apex respiration was 25% of that in nonstressed samples of similar size. The respiration rate maintained by the apex was of the same magnitude as that of excised root tips (13, 14) and appeared to represent an *in vivo* rate, since the tissue exhibited a negligible wounding response, the respiration was sensitive to cyanide and malonate, and the oxygen electrode measurements agreed well with Warburg res-

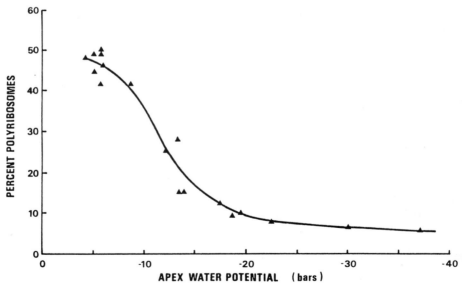

Figure 13.4. The effect of drought on polyribosome content of wheat apices. The wheat plants were droughted by withholding water. Redrawn from Ref. 2.

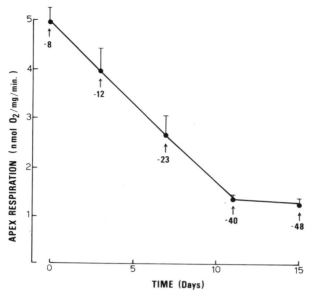

Figure 13.5. *In vitro* respiration rate of wheat apices after water was withheld from plants growing in 15 cm pots on day 0. Numbers on the graph are the apex water potential (bars) at each sampling time. Respiration rates were measured with an oxygen electrode, within 12 min of detachment, in a dilute Ca^{2+} solution. Rehydration effects were insignificant when respiration was measured quickly. Bars indicate standard errors. Experiments of Barlow and Pheloung (unpublished).

pirometer measurements. The initial sensitivity of the apex respiration to water stress was similar to that reported by Flowers and Hanson (15) and Koeppe et al. (16) for coleoptile and seedling tissues, respectively, except that apex respiration did not drop below 20% of the initial nonstress rate at ψ_a of -50 bars. However the rate of imposition of water stress was considerably slower in our wheat apex studies.

The synchronism of these metabolic responses with the growth retardation during stress suggests a strong correlation between growth requirements and metabolic activity with some feedback inhibition a possibility. Since low molecular weight carbohydrates and free amino acids accumulate in the stressed apex (6), it is unlikely that decreased metabolic activity results from substrate limitations.

3. SURVIVAL MECHANISMS

The ability of the wheat apex to survive ψ_a as low as -60 bars and the continued respiration of the apex under these severe stress conditions suggest the existence of powerful survival mechanisms in the apex. This section examines the survival of the apex in terms of its ability to conserve water, accumulate solutes for osmotic adjustment (Chapter 7), and act as a strong sink for continued translocation of respiratory substrates.

3.1. Water Conservation by the Apex During Stress

In the study illustrated in Figure 13.2, the water contents of the apex, an expanding leaf enclosed in older leaf sheaths (leaf 7), and an exposed, fully expanded leaf (leaf 4) were also followed through a 13 day drought period. The water potentials of all three tissues declined simultaneously until day 6, when the exposed leaf turned yellow and subsequently died. The apex and leaf 7 maintained similar water potentials throughout (Figure 13.2). Elongation of the apex and leaf 7 was reduced on day 2 and stopped on day 3 ($\psi_a = -12$ bars). As mentioned in Section 2.1, this decline in elongation was associated with a decrease in turgor potential. The turgor potential of 1 to 1.5 bars was maintained in the apex and in leaf 7 for the remainder of the period, despite a continued decrease in osmotic potential (Figure 13.2).

Initially there were large differences in the percentage water content among the three tissues, and all tissues showed a decrease in water content during the early part of the drought (Figure 13.6). With continued drought the exposed leaf 4 continued to lose water rapidly and desiccated by day 8. In contrast the apex and leaf 7 showed little change in percentage water content after day 4. As the dry weight of the apex continued to increase after

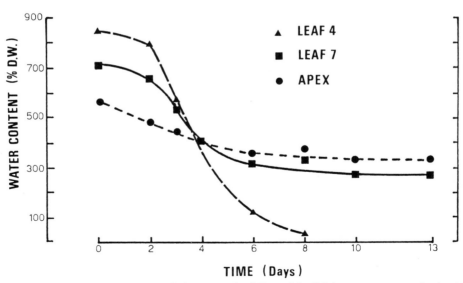

Figure 13.6. Water content of the apex, leaf 7, and leaf 4 in water-stressed wheat plants. Water was withheld from the plants on day 0. The coefficient of variation of water content measurements was 7% (range 2–9%). After Ref. 6.

day 4, it was possible that the decreased apex water content represented a gain in dry matter relative to water. This proposal was confirmed in a separate experiment that entailed the measurement of the water content of individual apices with an electronic balance (Figure 13.7). In this experiment, where higher relative humidities resulted in a more gradual imposition of stress, the apex stopped growth between day 6 and 8 as the ψ_a decreased from -11 to -18 bars, but continued to gain dry matter until day 20. During this period the water per apex remained constant and the apex water content on a dry weight basis fell from 420 to 245%.

The steady increase in apex dry matter after elongation has been inhibited by water stress is primarily due to the accumulation of solutes in the cell. This is illustrated in Table 13.1 for an experiment in which apex elongation occurred before but not after day 4. The ethanol-insoluble residue of the apex remained constant after the fourth day of stress, whereas total dry matter increased by 57% over the next 7 days. Much of the decreases in the

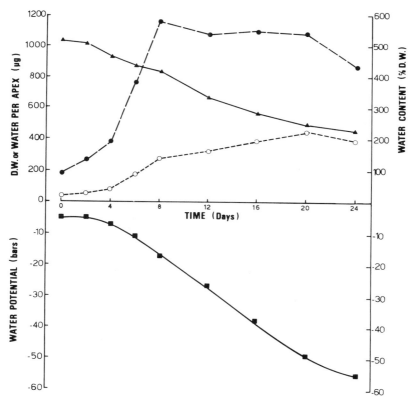

Figure 13.7. The water (●) and dry matter (○) per apex, and the water content (▲) and water potential (■) of the wheat apex during a stress period. Water was withheld on day 0.

Table 13.1. Apex Water Potential (ψ_a) and Total and Ethanol-Insoluble Dry Matter in the Wheat Apex During an 11 day Stress Period

Values of apex dry matter are means ± one standard error.

Days without water	ψ_a (bars)	Apex dry matter (μg/apex)	
		Total	Ethanol Insoluble
0	−8.5	646 ± 54	591 ± 64
4	−11.1	1403 ± 198	1014 ± 158
7	−28.2	1521 ± 73	929 ± 57
11	−39.3	2207 ± 99	1152 ± 109

water content of the apex shown in Figures 13.6 and 13.7 resulted from an accumulation of ethanol-soluble substances in the apex, and water content on an ethanol-insoluble dry matter and per cell basis was essentially constant. The consistent amount of water per cell in the apex does not mean that there is no water loss from the apex during the latter stress period, because there is a continued translocation into the apex during stress (Section 3.3) and production of respiratory water may reach significant proportions. The magnitude of translocation and production of respiratory water is considered in Section 3.3.

Rapid evaporative water losses from the apex during severe stress periods seems to be prevented by the thick waxy cuticle of the encircling older leaf sheaths. During severe stress the exposed leaves wilt and desiccate. The sheaths of these leaves then shrink around the enclosed leaves and apex to provide an effective barrier to rapid water loss. Furthermore, the conservation of water in the apex indicates that during the wilting of the exposed leaves, little water moves from the apex to the wilted leaves at a time when some xylem continuity may remain in the leaf sheaths. Both Anderson and Kerr (17) and Wilson (18) have reported the conservation of water in young growing tissues at the expense of other tissues during stress. It is possible that significant movement of water from the apex to the wilting leaves is prevented by decreases in π_a until xylem continuity is lost.

3.2. Solute Accumulation and Osmotic Adjustment

The increase in the ethanol-soluble materials in the apex (Table 13.1) was investigated in more detail by Munns et al. (6). The accumulation of soluble carbohydrates and free amino acids in the apex, an enclosed leaf 7, and an exposed leaf 4, was monitored over a 13 day stress period. The water relations of the tissues are presented in Figures 13.2 and 13.6.

The accumulation of organic solutes in the enclosed tissues, apex and leaf 7, was very similar except that the maximum solute accumulation occurred on day 8 in leaf 7 and on day 10 in the apex (Figure 13.8). Soluble sugars increased rapidly from the beginning of the stress period in both the apex and the enclosed leaf 7. On day 2 of the stress period the soluble sugar concentration in both enclosed tissues had doubled as the water potential decreased by 1 to 1.5 bars. A rapid accumulation of soluble sugars continued until day 6 in leaf 7 and day 8 in the apex.

Although the free amino acid content did not begin to increase until day 3, it exceeded the accumulation of soluble sugars in both enclosed tissues by day 8. In both the apex and leaf 7 the major components of the increase in free amino acids were proline and asparagine (6).

Solute accumulation in the mature, exposed leaf 4 was much lower than in the enclosed tissues, but followed the same pattern of an immediate increase

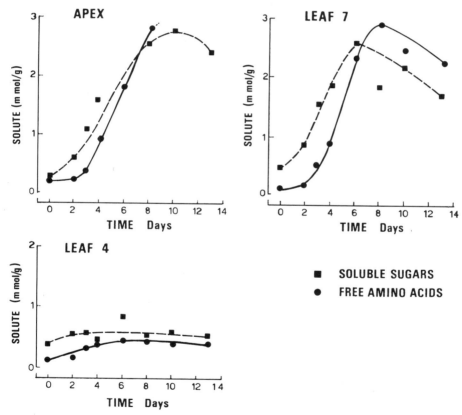

Figure 13.8. Free amino acid and soluble sugar contents, (mmol/g of ethanol-insoluble dry matter) of the apex, leaf 7, and leaf 4 in water-stressed wheat plants. The coefficient of variation of the measurements was 18% (range 12–28%). Redrawn from Ref. 6.

in soluble sugars at the onset of stress, and a subsequent rise in free amino acids (Figure 13.8).

The potassium content of the apex remained constant at 4.4 ± 0.9% of the dry matter throughout this stress period, but because of the increased dry matter content of the apex, this percentage represents a net increase in the amount of potassium in the apex. In a separate experiment the potassium was constant at 3.9 ± 0.2% of the total dry weight, but on an ethanol-insoluble basis this represented an increase from 4.1 to 7.5%. Chloride and organic acids were at a much lower concentration than potassium and did not change significantly during stress.

The calculated contribution of solute accumulations to osmotic adjustment during drought is presented in Figure 13.9. In the first 3 to 4 days of drought the accumulation of soluble sugars made the most important contribution to the lowering of the osmotic potential in the enclosed tissues. After day 3, when growth ceased, free amino acids and potassium became increasingly important in the lowering of the osmotic potential, with free

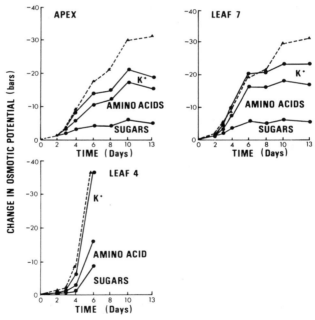

Figure 13.9. Decrease in the osmotic potential of the apex, leaf 7, and leaf 4 of water-stressed wheat plants with time. Osmotic potentials were either measured (▲ — — — ▲) or calculated (● — ●) from the solute contents of potassium (K^+), amino acids, and sugars (Figure 13.8) and water contents (Figure 13.6). All values are expressed as changes from day 0 values and are superimposed on each other. Redrawn from Ref. 6.

amino acids contributing approximately half the solutes for osmotic adjustment from day 6 onward.

Potassium appears to become a significant component of osmotic adjustment in the enclosed tissues during drought, since the initial levels were very high. The increase of less than twofold in the apex, and in leaf 7 made a large contribution to the calculated osmotic potential. This increase in potassium was in proportion to the increase in dry matter during the drought period, indicating that it was probably translocated "passively" in the phloem along with the carbohydrates and nitrogen compounds.

The high levels of potassium in the slightly vacuolated apical tissues raises some interesting questions of compartmentation and the availability of balancing anions. If no compartmentation is assumed, the mean cytoplasmic potassium concentration of stressed and nonstressed apices would be approximately 330 and 200 mM, respectively. The optimal potassium concentration for *in vitro* protein synthesis by wheat polyribosomes is approximately 70 to 130 mM (19), and *in vitro* potassium levels above 200 mM reduce protein synthesis by as much as 80%. The low concentrations of potential balancing anions such as chloride and organic acids appear to preclude the location of large quantities of potassium in fine vacuolar vesicles in the cell. Alternatively the potassium may be balanced by negative charges on proteins, nucleic acids, and amino acids.

The net effect of solute accumulation in the apex and the resultant osmotic adjustment is the maintenance of a positive P_a during the 13 day stress period (Figure 13.2).

3.3. Translocation to the Apex During Stress

The translocation of solutes to the apex continues during stress: translocated organic substances are either oxidized to carbon dioxide or accumulated mainly as sugars and free amino acids. Respiratory substrates do not appear to come from breakdown of large polymers in the apex because the apex contains few fructans and little starch. Furthermore the relative constancy of the ethanol-insoluble residue during stress (Table 13.1) indicates that large molecules, including proteins, are stable.

Table 13.2 presents an estimate of carbon translocation into the apex during water stress. This carbon budget assumes that soluble carbohydrate accumulation results from translocation only and that glucose is the substrate for respiration. The carbon debt for respiration and carbohydrate accumulation is such that we must conclude that a high rate of translocation of carbohydrate into the apex is maintained under severe water stress. Wardlaw (20) has reported translocation to continue during water stress. On this evidence, the accumulation of soluble carbohydrate in the apex, which makes a substantial contribution to osmotic adjustment is less than 35% of the daily carbohydrate translocation.

The continued translocation into and respiration of the apex during water

Table 13.2. Apex Water Potential (ψ_a) and Simplified Soluble Carbohydrate Budget of the Water Stressed Apex

Apex respiration, soluble carbohydrate (anthrone), and ψ_a were measured in triplicate 0, 3, 7, and 11 days after withholding water from wheat plants grown in 15 cm pots in a growth chamber.

Period (days of stress)	ψ_a (bars)	Soluble carbohydrate (μg glucose/apex/day)		
		Respired	Accumulated	Net import
0–3	−10.1	240	16	256
3–7	−17.3	211	76	287
7–11	−31.5	174	65	239

stress has implications in the water balance of the apex. If carbohydrate is translocated into the apex as a 20% sucrose solution, and stoichiometric amounts of water are formed in respiration, the figures in Table 13.2 indicate a net water gain of approximately 1500 μg/day. Given the constancy of apex water relative to dry weight during stress, a 1 mg (dry weight) apex appears to turn over as much as 50% of its water per day under these conditions.

The accumulation of free amino acids may indicate that translocation of nitrogen into the apex continues in the absence of net protein synthesis and is largely independent of continuing growth. Similarly the declining relative importance of soluble sugars during prolonged drought may reflect their use as a substrate for continued respiration in the apex. The continued accumulation of soluble nitrogen compounds in the apex during stress also underlines the need to form storage compounds such as proline (6) that act as compatible solutes within the cell.

4. SUMMARY: IS THE APEX UNIQUE?

The drought tolerance of the apical meristem appears to depend on at least two factors. First, the tissue is protected from rapid evaporative water loss by its location in the mature leaf sheaths. The apex is not connected to the stem by functional xylem vessels because it is not a rapidly transpiring organ at this stage of development. Consequently the water content of the apex changes little during stress. The consequence of this protection is that the apex has time to proceed with the orderly accumulation of solutes for osmotic adjustment. Second, the apex is a strong sink within the plant and remains so throughout severe stress, allowing the continued import of carbohydrates, nitrogenous compounds, and potassium. This is illustrated in Table 13.2, where the net movement of soluble carbohydrate into the wheat apex was relatively constant over an 11 day stress period as ψ_a decreased from −8 to −41 bars. The direction of nutrients to the apex during stress has been commented on. Although solute accumulation leading to osmotic adjustment has not been quantified, there are numerous reports of osmoregulatory responses of meristematic and developing tissue. Meyer and Boyer (21) elegantly demonstrated that soybean hypocotyls are capable of making osmotic adjustments in response to stress and that attachment to the cotyledon is essential for the response. Greacen and Oh (22) have shown that root tips are capable of making osmotic adjustments in response to increased pressure. Consequently it would appear that osmotic adaptation may be a common feature of young, relatively unexpanded tissues.

These characteristics suggest that the privileged position of the apex as an organ may be responsible for a plant's tolerance to drought, rather than the unique qualities of meristematic cells. This contention is supported by the similarity in the response of the apex and leaf 7 to stress (Figures 13.7 and 13.8). Although there are large differences in vacuolation between the tissues, their enclosed nature leads to a similar drought response.

REFERENCES

1. I. Husain and D. Aspinall, *Ann. Bot. (London)*, **34**, 393 (1970).
2. E. W. R. Barlow, R. Munns, N. S. Scott, and A. H. Reisner, *J. Exp. Bot.*, **28**, 909 (1977).
3. F. L. Milthorpe, *Ann. Bot. (London)*, **14**, 79 (1950).
4. P. B. Nicholls and L. H. May, *Aust. J. Biol. Sci.*, **16**, 561 (1963).
5. L. G. Paleg and D. Aspinall, *Aust. J. Biol. Sci.*, **19**, 719 (1966).
6. R. Munns, C. J. Brady, and E. W. R. Barlow, *Aust. J. Plant Physiol.*, **6**, 379 (1979).
7. T. C. Hsaio, *Annu. Rev. Plant Physiol.*, **244**, 519 (1973).
8. T. C. Hsaio, E. Acevedo, E. Fereres, and D. W. Henderson, *Philos. Trans. R. Soc. London, Ser. B.*, **273**, 279 (1976).
9. D. Doley and L. Leyton, *New Phytol.*, **67**, 579 (1968).
10. P. B. Green, R. O. Erickson, and J. Buggy, *Plant Physiol.*, **47**, 423 (1971).
11. J. S. Boyer, *Plant Physiol.*, **43**, 1056 (1968).
12. A. Tissieres, in M. Nomura, A. Tissieres, and P. Lengyel, Eds., *Ribosomes,* Cold Spring Harbour Laboratory, Cold Spring Harbour, New York, 1974, p. 3.
13. P. G. Jarvis and M. S. Jarvis, in B. Slavík, Ed., *Water Stress in Plants,* Junk, The Hague, 1965, p. 167.
14. H. Greenway, *Plant Physiol.*, **46**, 254 (1970).
15. J. J. Flowers and J. B. Hanson, *Plant Physiol.*, **44**, 939 (1969).
16. D. E. Koeppe, R. J. Miller, and D. T. Bell, *Agron. J.*, **65**, 566 (1973).
17. D. B. Anderson and T. Kerr, *Plant Physiol.*, **18**, 261 (1943).
18. C. C. Wilson, *Plant Physiol.*, **23**, 156 (1948).
19. L. A. Weber, E. D. Hickey, P. A. Maroney, and C. Baglioni, *J. Biol. Chem.*, **252**, 4007 (1977).
20. I. F. Wardlaw, *Aust. J. Biol. Sci.*, **20**, 25 (1967).
21. R. F. Meyer and J. S. Boyer, *Planta*, **108**, 77 (1972).
22. E. L. Greacen and J. S. Oh, *Nature (London) New Biol.*, **235**, 24 (1972).

14 | Protoplasmic Tolerance of Extreme Water Stress

D. F. GAFF

Department of Botany, Monash University, Melbourne, Victoria, Australia

1. INTRODUCTION

The ability of plants to survive periods of severe water stress is an indirect but necessary component of productivity in drought prone areas. The overall ability of a plant to survive water stress, that is, its drought resistance, may incorporate (*a*) any one or more of several quite discrete drought avoidance mechanisms (which restrict water loss, increase water supply, or complete life cycles before the return of dry conditions), and (*b*) beyond the competence of these, the capacity of the protoplasm itself to survive drying, that is, its drought tolerance. The terminology used in this chapter follows Levitt (1), not that proposed in Chapter 1. Tolerance is the ultimate resort for survival, and probably is the least understood mechanism of drought resistance.

In this chapter the drought tolerance values are expressed as the lowest water potential survived by 50% of the tissue. Values of tolerance and water potential (ψ) are generally given on a relative humidity (RH)* scale (at 28°C),

Abbreviations: ABA—abscisic acid. ATP—adenosine triphosphate. ATPase—adenosine triphosphatase. DNA—deoxyribosenucleic acid. mRNA—messenger ribonucleic acid. PAGE—polyacrylamide gel electrophoresis. ppm—parts per million. RH—relative humidity. RNA—ribonucleic acid. RNase—ribonuclease. RuBPC—ribulose-1,5-bisphosphate carboxylase. RWC—relative water content. SH—sulfhydryl. SS—disulfide. tRNA—transfer ribonucleic acid.

since in most cases they were determined on this basis and the standard equation for converting RH to ψ in energy or pressure units is inappropriate for very low water potentials (2). The approximately equivalent water potentials in percent relative humidity and bars are given in Figure 14.1, Section 4.1. "Desiccation tolerance" describes extreme drought tolerance, approaching a value of 0% RH.

2. TOLERANCE LIMITS

Leaves of normal agricultural or horticultural plants survive relative water contents (RWC) of 26 to 77% (3–5). The performance of normal plants, however, is far less impressive when viewed from the standpoint of leaf water potentials ψ_{leaf} that the plants can survive: the survival limits cover only a small part of the range of 0 to 100% RH available. For example, Iljin (6) reported tolerance limits of approximately 85 to 99% RH ($\sim$ −220 to −14 bars, respectively). Unfortunately Iljin (6) used a viability test that has since been criticized (7), and certainly the value of −14 bars seems suspiciously oversensitive. Nevertheless, later data obtained with generally accepted methods have established a similar range of tolerance in different species (8). The narrowness of this water potential range compared to the variations seen in ambient relative humidity exemplifies the effectiveness of drought avoidance mechanisms in normal vascular plants. Abnormal plants, however, do exist; leaves of some vascular "resurrection plants" can survive dehydration to the point of air dryness, that is, $\psi_{leaf} \simeq 30$ to 50% RH (2, 9). The drought tolerance of the resurrection plants is extremely high, ranging from 30% RH ($\sim$ −1600 bars) in the less tolerant species to 0% RH, the theoretical optimum tolerance value, in the most tolerant species. The phenomenon is more widespread than is generally appreciated. The foliage of at least 83 angiosperms can survive air dryness to these levels (2, 10, 11, and unpublished data). It is well known that viable seeds usually have very low water contents; if Table 14.1 is representative of flowering plants in general, embryos of most species survive water potentials similar to those tolerated by the leaves of resurrection plants. This is clearly the case with pollen (angiosperm and gymnosperm), which with the notable exception of the grasses, has drought tolerance limits mainly between 0 and 50% RH (13).

In the absence of xylem, cells of nonvascular plants have little capacity to draw on water that is not in direct contact with the tissue; terrestrial species inevitably come under intense selection pressure for improved drought tolerance. It therefore is not surprising that the ability of vegetative protoplasm to survive air-drying is more frequent among the lower plants than in the vascular plants. Abel (14) reports almost one-third of 51 moss species had drought tolerance limits of 0 to 20% RH when freshly taken from wet to xeric habitats. All 14 species of epiphytic mosses studied by Hosokawa and Kubota (15) in Japanese beech forests survived equilibration to 0 to 30% RH.

Table 14.1. Germination Percentages and Water Potentials (ψ) of Commercial Seeds, Determined by Vapor Exchange Techniques (12) at 28°C

Wherever possible seeds were cut open to facilitate vapor exchange. Seed for drought tolerance limits (in parentheses) were equilibrated for 2 months, and their viability was tested with tetrazolium chloride.

Species	Family	Tissue	Germin-ation (%)	ψ (% RH)
Acacia aneura	Fabaceae	Halved seed	100	23
Streptocarpus sp.	Gesneriaceae	Whole seed	100	30
Lychnis chalcedonica	Caryophyllaceae	Whole seed	86	32
Lepidium sativum	Cruciferae	Halved seed	100	35
Cucumis sativus	Cucurbitaceae	Excised embryo	50	37
Allium cepa	Amaryllidaceae	Halved seed	63	38
Geum sp.	Rosaceae	Halved seed	50	41
Pinus pinea	Pinaceae	Excised embryo	100	42.7(5–15)
Pinus pinea	Pinaceae	Endosperm	—	42.4
Portulacca grandiflora	Portulacaceae	Whole seed	92	43
Oryza sativa	Poaceae	Excised embryo	87	43
Zea mays	Poaceae	Excised embryo	67	45
Triticum vulgare	Poaceae	Excised embryo	53	47
Ricinus communis	Euphorbiaceae	Excised embryo	70	47
Matthiola bicornis	Cruciferae	Whole seed	45	48
Brassica oleracea	Cruciferae	Halved seed	60	55
Persea gratissima[a]	Lauraceae	Cotyledon slices	—	99 (>33)
Cocos nucifera[a]	Palmae	Embryo slices	—	99

[a] Fresh from ripe fruit.

A similar proportion of "resurrectable" species has been reported among the liverworts (16, 17), and desiccation tolerance is probably universal among lichens (18).

3. TOLERANCE IMPROVEMENT

Drought tolerance can be increased by "hardening" under the action of moderate levels of water stress or of low temperatures, which produce frost hardening (1).

Drought tolerance improved by 2 to 6% RH units in three out of four crop species subjected to 2 to 6 weeks with low soil water content (8). In pea plants drought tolerances improved by about 2% RH as the root medium dried from water potentials of -0.8 to < -50 bars from the second to fourth week after planting (19). However the timing was critical; tolerance did not improve if drying of the soil was delayed a further 6 to 14 days. Not all crop plants improve tolerance limits under stress. For example, in soybean "hardening off" is due solely to better drought avoidance (20), that is, "pseudohardening" (1).

Many resurrection plant leaves, when dried detached, do not display the extraordinary desiccation tolerance they show when dried on intact plants (Table 14.2). Detached leaves of *Borya nitida* undergo a process of tolerance induction if they are in the 90–98% RH water potential range (~ -27 to $-150.$ bars) for 2 days as they dry. Subsequently they survive water potentials of $\sim$0% RH, an improvement of $\sim$90% RH units (21). Similar improvements in tolerance occur in liverworts; tolerance values improved by 4 to 60% RH units in some species under the action of mild air-drying, especially after long dry periods in the field (22, 23). The apparent tolerance improvement (5 to 80% RH units) reported in mosses (14) probably incorporates intensification of avoidance mechanisms as well as those of tolerance; Abel (14) allowed only 24 h for equilibration, whereas $\sim$7 days appears to be necessary for full equilibration of mosses (24; C. J. Moore, personal communication).

**Table 14.2. Drought Tolerance
Estimates of Freshly Detached Fully
Hydrated Leaves (i.e., the minimum
equilibration RH survived at 28°C)**

Species	Minimum RH survived (%)
Tripogon loliiformis	98
Eragrostis nindensis	94–96
Oropetium capense	94
Sporobolus stapfianus	92–94
Coleochloa setifera	90
Boea hygroscopica	90
Xerophyta viscosa	90
Borya nitida	85–90
Xerophyta villosa	66
Platycerium alicorne	52
Talbotia elegans	0–52
Myrothamnus flabellifolia	
Field grown	11
Laboratory	0–11

4. INJURY: IS IT MULTICAUSAL?

The distribution of injury in relation to water potential and time strongly suggests that injury stems from a number of processes with contrasting characteristics. These are shown in Figure 14.1. The various categories of response are considered in the order that they appear to operate as stress increases.

4.1. Moderately Rapid, High Water Potential Injury

Injury A in Figure 14.1 is moderately rapid, high water potential injury: it decreases as water potential decreases. Tissue samples such as pea leaves (19) maintained in 100% RH begin to degenerate within a few days. In most cases degeneration probably stems from detachment-induced net hydrolysis of macromolecules (25), consequent on interruption of hormone supply

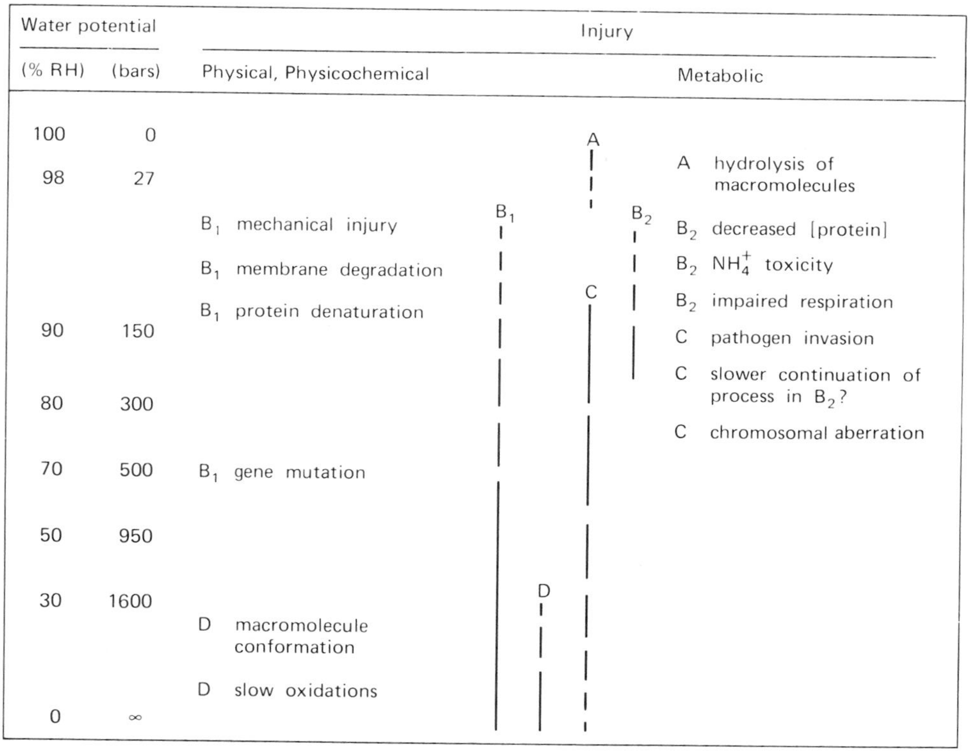

Water potential		Injury	
(% RH) (bars)	Physical, Physicochemical		Metabolic
100 0		A	A hydrolysis of
98 27			macromolecules
	B_1 mechanical injury	B_1 B_2	B_2 decreased [protein]
	B_1 membrane degradation		B_2 NH_4^+ toxicity
	B_1 protein denaturation	C	B_2 impaired respiration
90 150			C pathogen invasion
			C slower continuation of
80 300			process in B_2?
			C chromosomal aberration
70 500	B_1 gene mutation		
50 950			
30 1600		D	
	D macromolecule		
	conformation		
	D slow oxidations		
0 ∞			

Figure 14.1. Types of injury and possible processes affected at the water potentials given. Vertical lines indicate range of water potential over which injury is likely. Length of solid line indicates increasing certainty of injury.

from the rest of the plant. However other factors must be involved also. Free pollen can hardly have such a hormone dependency; yet in a number of species pollen deteriorated after 24 h at 100 to 98% RH, but survived humidities below this (26): the degree of injury decreased as ψ decreased toward 97% RH (-41 bars). In certain pollens, injury A was clearly separated from injury C (at ~85 to 50% RH), described in Section 4.4, by an intermediate zone of ψ where little or no injury occurred.

4.2. Rapid Stress-Parallel Injury

In most vascular plants leaves die at $\psi_{\text{leaf}} \simeq$ to 96% RH (~ -55 to -145 bars), the degree of injury increasing with the intensity of water stress. Injury may be rapid, at least under laboratory conditions and in species with poor drought avoidance in the field. Rapid injury (stress-parallel), denoted as injury B_1 in Figure 14.1, may occur in resurrection plants at any water potential between 0 and 98% RH, depending on the species and its prior treatment. The speed of the effect over such a wide range of ψ_{leaf} is suggestive of major involvement of a physical or physicochemical process triggered at the injurious ψ_{leaf}, though the possibility of a metabolic component cannot be ruled out. The involvement of the following processes has been mooted by various authors.

Mechanical Injury. Iljin (6) suggested that injury to protoplasm is basically mechanical and results from tearing of the protoplasm under the tensions set up by the contraction of the drying protoplasm beyond the point of zero turgor and by the adhesion of the protoplasm to the cell wall. There is reasonable evidence that such "negative turgors" do arise (6, 27, 28). The extent to which they cause injury is debatable, especially since Iljin's methods for testing viability were inadequate (7). Reduced survival in cells repeatedly plasmolysed and deplasmolysed, and the greater damage to cells stressed in air rather than in osmotica, were attributed to mechanical damage (19). The hypothesis has been carried forward into the current literature by Genkel' and Levina (29), who consider that the plasmodesmata of active cells render them liable to mechanical injury, whereas the greater drought tolerance in dormant cells stems from an absence of plasmodesmata.

It is perhaps possible to envisage osmotic adjustment (Chapter 7) as a mode by which, among other things, negative turgor buildup could be postponed as water potential declines. In the resurrection plants, however, it is clear that water content is not maintained effectively by osmotic adjustment as water potential falls (30). Cells of resurrection plants are clearly very shrunken when viewed both under the light microscope and under the scanning electron microscope. The apparent injury seen in electron microscope studies of desiccation sensitive plants is not in the form of massive tears in the protoplasm, but of degeneration of organelles as discrete entities (see next section). Disarray of portions of the plasmalemma are commonly seen in electron micrographs of resurrection cells, but this is probably an artifact due to swelling of the cell wall as it imbibes the fixative, while the protoplasm is maintained close to its original volume due to cross-linking in the fixative. Moreover, plasmodesmata can be clearly seen in resurrection plant cells in a viable air-dry state (27, 31); this contrasts with a report of absence of plasmodesmata in desiccation tolerant mosses and the resurrection plant *Myrothamnus flabellifolia* (29).

Despite their view that mechanical injury was operating in pea cells, Lee-Stadelmann and Stadelmann (19, 32) concluded that intensified tolerance in hardened plants was due to protoplasmic modifications rather than to morphological alterations, since cells that did not alter in size during tolerance-inducing stress nevertheless improved their tolerance, changes in cell permeability coincided with improvement in tolerance, and cells rapidly lost tolerance in darkness.

Degradation of Cell Membranes. Electron microscopy provides evidence of water stress inducing deterioration of cell membrane systems in desiccation sensitive protoplasm. For example, breaks occur in membranes bounding the plastids, mitochondria, and nuclei, in the plasmalemma, and in dictyosome membranes, and the number of cristae falls (33–35). Furthermore, swelling occurs in thylakoids (36) and in the endoplasmic reticulum (37). The possibility of artifacts arising during fixation in media not osmotically adjusted (38) must be borne in mind, however, in assessing these data.

Further evidence of membrane damage stems from observations of leakiness of membranes in tissues at low water contents (39). Soluble components of the cell leaked from liverwort tissue ($\psi \simeq 50$ to 96% RH, ~ -55 to -950 bars) during the first 2 min of rehydration, but were reabsorbed later (40, 41). Some leakage also occurs from viable leaves of resurrection plants rehydrating from air dryness (Table 14.3).

Physical deterioration of membranes might arise from changes in lipid and/or protein structure. Simon (39) considered that reversible leakiness might arise from transitions in phospholipids from a lamellar to a hexagonal conformation at about 20% water content. On the other hand, freeze-fracture faces of membranes of pea roots killed by water stress resemble those of live roots, but not of aldehyde-fixed roots; Buttrose and Swift (35) reasoned from this that membrane damage stemmed from disruption of proteins rather than of the lipid layer. Apparent denaturation of insoluble membrane protein at $\psi \simeq 93\%$ RH (-94 bars) was associated with irreversible injury in water-stressed cabbage leaves (42).

It would seem from the foregoing that leakiness of the plasmalemma need not be fatal. Studies on leaves of some monocotyledonous resurrection plants indicate that loss of many cristae, and of virtually all thylakoids, and breaching of mitochondrial and chloroplast membranes can be survived (27,

Table 14.3. Leakage of Potassium (μg K$^+$/mg dry wt) from Detached Leaf Segments of Resurrection Plants

Air-dry samples taken from plants dried intact. Dead samples were killed with chloroform vapor.

Species and treatment	Time in distilled water (h)			
	1	3	5	24
Xerophyta villosa				
Air dry, viable	1.87	2.49	3.15	5.32
Air dry, dead	4.87	7.11	8.45	10.16
Hydrated, alive	0.48	0.53	0.50	1.07
Hydrated, dead	11.16	9.10	11.10	11.80
Myrothamnus flabellifolia				
Air dry, viable	0.60	0.86	1.04	2.58
Air dry, dead	2.03	2.73	3.47	5.30
Hydrated, alive	0.39	0.59	0.80	1.93
Hydrated, dead	4.69	3.71	4.13	5.51

43). Maintenance of vacuoles and lysosomes as separate compartments theoretically seems more essential, in view of the phenolic and hydrolytic nature of their contents.

Protein Denaturation. Protein denaturation is the crux of Levitt's hypothesis that stress injury is produced by conversion of protein sulfhydryl (SH) to disulfide (SS), which leads to unfolding of the polypeptide chain on reimbibition (1). Little evidence is available on the occurrence or extent of protein denaturation during water stress. The reactive SH level in the insoluble protein fraction increased ~40 to 75% at sublethal ψ_{leaf} in cabbage leaves, whereas a denaturant (SDS) induced increases of ~200% (42); this suggests partial denaturation of the membrane protein. At lethal stress about one-third of the total insoluble protein SH was converted to SS. There was evidence of small amounts of SH/SS conversion in one resurrection plant, *Xerophyta viscosa,* but not in the other two resurrection plants studied to date, nor was there significant SH/SS conversion in a desiccation sensitive plant *Sporobolus pyramidalis* (44). Studies on fraction I protein in *X. viscosa* show a reduction in RuBPC activity in soluble extracts of air-dry plants, whereas fraction I protein levels remained constant in PAGE partitions of the soluble extracts (44). A number of enzymes are known to be inactivated by lyophilization, but this is not true of all enzymes (45).

Gene Mutation. Webb (46) suggested that genetic mutations, which were frequent at 30 to 60% RH, cause rapid injury in bacteria sprayed into air of various humidities. As the relative humidity of the air decreased from 70 to 40%, the number of mutants per 10 million viable cells rose from 2 to 360: the death rate increased as humidity fell over the same range. The bacteria could be protected against injury by the application of various compounds that were rich in —NH_2 or —OH groups, particularly inositol. Since this range of humidity is well below the water potentials at which most angiosperms are killed, only resurrection plants would be at risk. Under field conditions resurrection plants pass through rapid cycles of dehydration and rehydration, perhaps 10 to 20 cycles in a year, yet there are no signs of the aberrant growth that one would expect if mutations were occurring at a near-fatal level. Consequently there must be some means of coping with this source of injury, possibly a repair system similar to that operating in fully imbibed dormant lettuce seed (47).

4.3. Less Rapid Stress-Parallel Injury

Under field conditions the onset of water stress in vascular plants may be slow enough to allow the metabolic repair of physicochemical injury (B_1) if this is not too extensive, but may allow the expression of the incipient disturbances in enzymatic metabolism shown as B_2 in Figure 14.1 and discussed below.

Decreased Protein Content. Protein content, particularly of soluble protein, usually falls to about 40 to 60% of the initial content as water stress becomes intense in drought sensitive plants (1, 48). Protein synthesis may be particularly sensitive to water stress; for example, most polysomes revert to monosomes in 4 h at −5 bars in *Zea mays* roots (49). The relative importance of the various factors contributing to the net protein loss have been studied by Marranville and Paulsen (50); reduced protein synthesis appeared to stem partly from diminished RNA synthesis and partly from a four fold increase in RNase activity at 43% RWC (cf. Ref. 51).

The resurrection plants gave evidence of hydrolysis of the insoluble proteins presumably incorporated in lipoprotein membrane structures, but levels of soluble protein were maintained or in some cases increased. However in detached leaves of the resurrection plant *Xerophyta villosa*, killed by water stress, changes in levels of soluble and insoluble protein were similar to those in viable air-dry leaves; that is, net hydrolysis of proteins was not the basic cause of injury in this instance (48).

Ammonia Toxicity. Accumulation of ammonia was suggested as a potential source of injury by Savitskaya (52). She found that ammonia levels increased by about 40% in barley leaves, reaching levels of about 10 to 25 ppm. However these concentrations are unlikely to be toxic when compared with the 200 ppm of ammonia required for toxic symptoms to appear in cucumber (53). Much higher levels of ammonia (190 to 1100 ppm) were attained in sorghum leaves, but this was following rehydration, not during stress; moreover, injury was not correlated with the maximum ammonia concentration reached (54).

Impaired Respiration. Although respiration frequently increases with moderate water stress (1), rates decline as stress becomes severe (55, 56). It is easy to imagine that shortage of ATP might lead eventually to deterioration of metabolism in general. Gordon et al. (37) reported weakening of electron transport in the early sections of the mitochondrial respiratory chain at about 35% RWC in wheat roots, and uncoupling of electron flow from ATP formation in respiration as the primary source of injury in dehydrating plants was suggested by Zholkevich (57) and Zholkevich and Rogacheva (58). They demonstrated that drastically decreased P/O ratios occurred in severely stressed drought sensitive plants, and that heat release increased in excess of respiratory changes, a situation that was taken to indicate decreased respiratory efficiency. On the other hand, the desiccation tolerant moss *Tortula ruralis* revived from water potentials of ~75% RH (~ −400 bars) even though ATP was virtually absent from stressed plants (59); at least in the short term ATP evidently was not necessary for cell survival. At rates of drying associated with slower recovery, ATP levels of *T. ruralis* were much higher, and the maximum ATP concentrations occurred in air-dry injured leaves of a less tolerant moss *Cratoneuron filicinum*. It seems, therefore, that maintenance of high levels of ATP does not protect the cell.

4.4. Slow Stress-Inverse Injury

Tissues that withstand water potentials well below 50% RH (~ −950 bars) for long periods are injured if held for several days at water potentials of ~85% RH (~ −220 bars): this is listed as injury *C* in Figure 14.1. Injury also occurs at progressively slower rates as water potentials decline. For example, Gaff (2) showed that the majority of 21 species of desiccation tolerant plants had died in 9 days when the RH was 100%, in 46 days when the RH was 76%, and in 79 days when the RH was 52%.

Injury may occur during both dehydration and rehydration in seeds, in resurrection plants, and in pollen. Infections of fungi and bacteria, which can proliferate slowly at ψ approaching 70 to 80% RH (~ −300 to −500 bars) (60, 61), may be implicated in injury, particularly in the upper part of the water potential range, where fungi on leaves are often visible to the naked eye. A similar conclusion has been reached for seed. However since the water potential range of this phase of injury extends well below $\psi \simeq 70\%$ RH, the primary cause probably lies elsewhere. Moreover, although injury was reduced in wheat grains sterilized in hydrochlorite and later shown to be pathogen free, it persisted at a water content of 16 to 18%, where ψ_{seed} was about 80 to 85% RH or −220 to −300 bars (62–64).

The retardation of phase *C* injury as ψ falls is understandable if the injury processes themselves become subject to the progressive slowing evident in metabolic processes such as photosynthesis and respiration at severe stress (56, 65). The general slowing of metabolic processes probably stems from slower diffusion of metabolites and reduced activity of poorly imbibed enzymes (45): for example, self-diffusion of water almost halves as RWC drops from ~80 to ~20% (66). Biochemical changes were observed in rye seed that had lost viability while stored at 15% water content (i.e., $\psi \simeq 75$ to 85% RH, or -300 to -400 bars) (62): spoolable DNA fell to ~30%, and ribosomal RNA to ~45%, whereas on rehydration respiration was only ~12% of the rates in viable seed, and synthesis of protein and RNA failed to recover (67, 68). Acid phosphatase, ATPase, and peroxidase lost some activity in the nonviable seed, particularly after 100 to 5000 years (69). Breaks occurred in the DNA and RNA, but only loss of transferase I activity (involved in aminoacylating tRNA to ribosomes) paralleled loss of viability sufficiently to suggest this as an initial lesion leading to lethal injury (68). A slow accumulation of chromosomal aberration in lettuce seed (Table 14.4) showed stress-inverse rates from 3.6 to 13.5% water contents ($\psi_{seed} = $ ~15% to over 75% RH or ~ -400 to -2600 bars). Presumably such aberrations are involved in this phase C injury.

4.5. Slow, Stress-Parallel Injury in Very Dry Tissues

Injury D in Figure 14.1 (slow, stress-parallel injury to very dry tissues) can be seen in many mosses maintained in air of ~0% RH. The mosses eventually die, although they have survived considerably longer times than those

Table 14.4. Effects of Water Content During Storage on the Germination of Lettuce Seeds and the Rate of Increase in Chromosomal Aberrations in Radicle Tips During Storage (Based on Ref. 70)

RH (%)	Water content during storage (%)	Survival (% germination after 12 months)	Increase in nuclear division aberrations (% per month)
0	3.6	—	0.6
25	5.1	77	1.8
50	7.0	34	3.5
75	9.7	0	6.3
100	13.5	0[b]	27
Fully imbibed, held dormant during storage	—	95	-0.15[a]

[a]That is, a gradual decline from the percentage aberrant in samples at the beginning of the experiment.

[b]% germination after 2 months, not 12 months.

needed for full equilibration. The time of death varies from 2 to 55 weeks depending on the species concerned (18, 71). This bespeaks a very slow phase of degeneration distinguishable from the type *B* injury by the latter's speed of onset and from type *C* by the range of water potential in which it is mainly operative. There is also evidence of slow postequilibration death at very low water potential in seed (72), pollen (26), and leaves of resurrection plants (2). The speed of deterioration decreases as ψ rises toward 30% RH ($\sim$ −1650 bars) in most pollen (26), in mosses with drought tolerances in the 0–40% RH range (15), and in some but not all seed (70, 73, 74).

Again, in view of the stress-parallel pattern of injury and the near cessation of metabolism, it is tempting to speculate that the injury results from effects on physicochemical processes such as the rearrangement of conformation-binding forces in macromolecules. Also gradual oxidative effects may operate in the absence of effective respiration; for example, polar lipids, presumably associated with membranes, were oxidized as cucumber seed aged (75).

The combination of injury arising from *C*- and *D*-type injuries means that desiccation tolerant tissues are often least damaged at values of ψ of about 20 to 40% RH or −1300 to −2200 bars. Maximum longevities are often found in this range.

5. RESURRECTION PLANTS DURING DEHYDRATION AND REVIVAL

Turning to the question of desiccation tolerance from a different point of view, let us examine the processes taking place during a single cycle of dehydration and rehydration in resurrection plants, and attempt to develop a tentative hypothesis on the mechanism of protoplasmic tolerance to drought. Figure 14.2 provides a speculative outline of some of the processes involved.

5.1. Preconditioning

The events before drying may be significant in survival. Leaves of *Borya nitida* survive air-drying as detached organs during the summer months, but perform poorly in winter unless the "donor" plant is kept for a week under high illumination and at temperatures of about 25°C. There seems to be some residual seasonal variation in performance even in plants kept many weeks under constant high light and temperature.

5.2. Variation in Response to Dehydration

It is quite clear that there is considerable variation in response to water stress in the resurrection plants, in terms of the retention or loss of chlorophyll and carotenoids, the degree of retention of organelle fine struc-

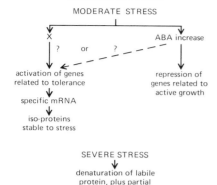

DURING DRYING

MODERATE STRESS

X — ABA increase

? or ?

activation of genes related to tolerance

repression of genes related to active growth

specific mRNA

iso-proteins stable to stress

SEVERE STRESS

denaturation of labile protein, plus partial degradation of membranes

AT AIR DRYNESS

near-cessation of metabolism

DURING REHYDRATION

remnant 'core' metabolism recommences (respiration, minimal protein synthesis)

renovation of full protein synthesis mechanism

renewal of total metabolic system

genetic repair

Figure 14.2. A speculative outline of events related to desiccation tolerance during a drying/revival cycle in an angiosperm resurrection plant.

ture, the hydrolysis of protein, and the direction of change in protein SH/SS levels and proline accumulation (10, 27, 43, 44, 48, 76–79). Rather than indicating a sharp dichotomy of behavior, these species-linked variations form a spectrum of response. At one extreme, in *B. nitida,* loss of fine structure and chlorophyll appears to be associated with time-dependent processes that optimize desiccation tolerance (21, 27); at the other extreme, *Pellaea calomelanos* exhibits almost complete retention of fine structure (80), but not of photosynthetic capacity. The desiccation tolerant moss *T. ruralis* retains considerable metabolic capacity for respiration, photosynthesis, and protein synthesis (59).

Speed of drying may possibly underlie basic differences in tolerance mechanisms. In most higher plants drought avoidance mechanisms extend the drying time sufficiently to allow considerable metabolically mediated adjustment. In lower plants, particularly lichens and terrestrial algae, on the other hand, time for metabolic adjustment is usually very limited. In the latter it is conceivable that tolerance may be a continual intrinsic property of the protoplasm before the onset of drying, that is, protoplasmic components are always stabilized to some degree against the physicochemical sources of injury (B_1), and time for metabolically mediated injury (B_2 and C) is limited. Superficially, the desiccation tolerant *T. ruralis* would appear to be such a case; polysomes are retained in dry plants, and protein synthesis and respiration recover within minutes of rehydration (59). However even in this species protein synthesis recovers more rapidly in plants dried to $\psi \simeq 65$ to 75% RH or -590 to -870 bars over the course of 3 h or more than in those dried over the same range in less than 1 h, thus indicating the ameliorating effect of some time-dependent process.

5.3. Early Events During Drying of Desiccation Tolerant Higher Plants

One of the first responses to water stress is the loss of starch, presumably by hydrolysis to soluble monosaccharides (27, 43, 76, 80–82). No data exist on ammonia production in stressed resurrection plants; if this is a serious potential danger, it is presumably controlled by incorporation into free proline, which increases in most species (79).

Increases in sugar and free amino acids probably result in some osmotic adjustment at mild water stress; since air dryness is eventually attained, however, this can only have a transient effect. It is unlikely for the following reasons that either elevated proline or sugar plays an essential role in desiccation tolerance: (*a*) both remained low in the viable air-dry moss *T. ruralis*, (*b*) sucrose depletion in *T. ruralis* during a drying cycle was only 10 to 20% of the total content, (*c*) both increased in desiccation sensitive plants, with proline increasing considerably more in some nonresurrection angiosperms than in resurrection plants, and (*d*) applied proline had no effect on the tolerance of detached *B. nitida* leaves, and it partially repressed the tolerance of detached *Myrothamnus flabellifolia* leaves (59, 79).

Although ATP is not essential in the air-dry tissue (see earlier), the scheme outlined in Figure 14.2 would require some ATP production for protein synthesis during the earlier phase of drying. The best information available on this phase is in the two mosses *T. ruralis* and *Cratoneuron filicinum*, in which ATP consumption increasingly exceeded production in both as they dried. The level of ATP was slightly higher in the more tolerant moss *T. ruralis* during the first 6 h of drying; it then became slightly lower, reaching approximately zero in both species at $\psi \simeq 75$% RH (59).

5.4. Tolerance Improvement

Section 3 pointed out that drought tolerance in resurrection plants can be increased by stress hardening.

Leaves detached from most species of resurrection plants showed relatively poor drought tolerance when equilibrated directly to various values of ψ without preliminary exposure to moderate stress (Table 14.2). Detached leaves of *Coleochloa setifera* and *Xerophyta villosa* underwent fuller tolerance induction if the donor plant was moderately stressed before the leaves were removed (82, 83). This suggests that materials (possibly hormones?) were needed from the rest of the *X. villosa* plant for full tolerance induction, whereas presumably *B. nitida* and *M. flabellifolia* leaves were able to synthesize these substances independently.

5.5. Trigger Mechanism

An active induction of desiccation tolerance must involve some form of "trigger" mechanism. It seems likely that abscisic acid (ABA) plays a role in this. ABA concentration is known to rise rapidly in many angiosperms at mild stress levels (84). The primary action by which stress stimulates ABA accumulation is, however, a matter of speculation. Loveys et al. (85) suggest that loss of turgor might cause relaxation of a limiting membrane and that metabolism associated with membrane conformation or protein hydration might be involved.

ABA levels increase in *B. nitida* during the early stages of tolerance induction. The concentration reached an apparent maximum at about 20 h (Table 14.5), that is, at the point just before visible chlorophyll loss, where tolerance begins to improve appreciably and thylakoid structure degrades (S. Hetherington, unpublished). Moreover, applied ABA increased the desiccation tolerance of detached leaves of *B. nitida* and *M. flabellifolia* (Gaff, unpublished).

Since plant hormones often influence transcription (86), it is tempting to speculate that ABA is influencing gene activity, repressing genes relating to active growth, and that another component X of the trigger mechanism causes the activation of genes related to tolerance induction. It seems unlikely that ABA itself is activating such genes, since Jacobsen's review (86) contains only examples of ABA decreasing transcription. Apparent

Table 14.5. Abscisic Acid (ABA) Concentration in Detached Leaves of Resurrection Plants at Tolerance-Inducing Levels of Water Stress (in *Borya nitida*) and Under Stresses Injurious to Detached Leaves (B. R. Loveys and D. F. Gaff, unpublished)

Concentrations are expressed as a percentage of the initial concentrations in freshly detached hydrated leaves.

Treatment	ABA
Borya nitida during tolerance induction at 96% RH	
Green leaves at 24 h (maximum concentration) (tolerance >30% RH)	460
Yellow-green leaves at 24 h (50% survive 0% RH)	390
Yellow leaves 48 h (100% survive 0% RH)	370
Borya nitida killed at 11% RH	140
Myrothamnus flabellifolia at 96% RH	
Leaves green, alive at 24 h (maximum concentration)	1610
At 48 h	270
Myrothamnus flabellifolia injured at 0% (~60% killed)	230

conflict of active vegetative growth with drought tolerance has been suggested by Hubac (87) and is supported by the loss of drought tolerance as seedling tissue undergoes auxin-mediated cell expansion (1). Elimination of growth-oriented (labile?) protein synthesis might allow a greater proportion of synthetic capacity to be diverted to stress-stable protein, "oriented" toward survival.

Since ABA also increases to many times the control level in water-stressed nonresurrection plants (88), the essential distinction between plants that are sensitive to desiccation and tolerant of it must lie elsewhere. The postulation of factor X (Figure 14.2) is only one possibility. Senescence also undermines the ability of leaves to tolerate water stress, including those of resurrection plants (2, 21, 32). Given the ability of ABA to stimulate tolerance induction in *B. nitida* and *M. flabellifolia,* tolerance loss in senescent leaves must stem from other factors associated with senescence, possibly ethylene or the unidentified senescence factor (89, 90). Preliminary experiments with other growth substances such as gibberellic acid, zeatin, and auxin have not yet produced any statistically demonstrable influence on tolerance induction in resurrection plants, and endogenous levels of these substances have not yet been studied.

5.6. Protein Synthesis

Figure 14.2 tentatively proposes activation by factor X or ABA of genes for mRNA specific for stress-stable proteins. Considerable protein appears to be synthesized in resurrection angiosperms during drying, since soluble protein levels often increase (48). mRNA activity is retained in dry *X. villosa;* the possibility of changes in specific mRNA is allowed by disproportionately high incorporation of labeled amino acids into some low molecular weight proteins (83). Also new isoenzyme bands of alkaline phosphatase and phenolase appear in drying *X. viscosa* (44).

Full induction of tolerance in *B. nitida* appears to be largely associated with restoration of the electrophoresis pattern of soluble proteins seen in the hydrated control, whereas partial induction and injury at high stress were associated with drastic loss of the lowest and highest molecular weight proteins (44). Although Hsaio (49) reported a high sensitivity of the polysome concentration at water stresses of -5 bars, the drought resistant nonresurrection plant black locust maintained about 80% of its polysomes at $\psi_{leaf} = -25$ bars (91). The resurrection plant *X. villosa* almost doubled its polysome content in leaves during moderate to severe water stress before a final decrease as air-drying ensued (Table 14.6). In electron micrographs of *B. nitida,* polysomes were still abundant at water stress levels that induced desiccation tolerance (S. Hetherington, unpublished). Thus the likelihood that protein synthesis is highly active during tolerance induction renders feasible the suggestion that stress-stable isoenzymes are being produced.

Table 14.6 Polyribosome and Total RNA Content in Leaves of *Xerophyta villosa* Plants Dried Intact to a Viable Air-Dry Condition (83)

RWC (%)	Ribosomes present as polysomes (%)	RNA (μg/g dry weight)
96	~30	396
75–90	~25	—
40–50	~55	—
30	~32	—
20	~1	—
4	0	452

5.7. Stabilization of Protein

Although the content of fraction I protein (RuBPC) of leaves fell by only 10% during air-drying of the resurrection plant *X. viscosa,* enzyme activity declined to 40% of the control (44). On the one hand this indicates denaturation of much of the enzyme, whereas on the other hand the residual activity could be indicative of partial replacement of the original stress-labile enzyme by a stress-stable isoenzyme. The major changes in the fraction I protein amino acid composition were the increases in the content of residues with positively or negatively charged side groups (44).

Levitt (1) suggested that the protein structure is stabilized by native-state SS bonds, but rendered prone to denaturation by SH groups. This is consistent with increases in the proportion of SS to SH in soluble proteins in the resurrection plants *X. viscosa* and *Sporobolus stapfianus* when air-dried (44). Smaller increases in SS/SH occurred in a desiccation sensitive species *Sporobolus pyramidalis.*

5.8. Avoidance of Slow Stress-Inverse Injury

Resurrection plants appear to evade potential injury around $\psi \simeq 85$ to 60% RH (-220 to -700 bars) by passing through this zone too rapidly for injury to occur: first wilting to air dryness usually takes 2 to 3 days (2), whereas injury usually takes 9 to 80 days (Section 4.4). Consequently cuticular transpiration must be moderately high. Leaf cuticles are not particularly thick in resurrection plants (Table 14.7), an adaptation also linked with direct water absorption through the leaf surface following rain (2). Cuticle permeability must be finely tuned to allow the moderately high cuticular water absorption and at the same time an adequate period for tolerance induction during moderate stress.

Table 14.7. Thickness of Leaf Cuticle in Resurrection Plants and Nonresurrection Plants

Measured at the center or side of the epidermal cells, according to where the cuticle was thickest. Standard errors were less than 10% of the mean values given. (From Ref. 92 and C. Katsoris, L. J. Johnston, and D. F. Gaff, unpublished)

Species	Thickness of cuticle (μm)	
	Upper surface	Lower surface
Resurrection Plants		
Myrothamnus flabellifolia	3.1	3.4
Boea hygroscopica	3.7	—
Craterostigma plantagineum	1.4	1.2
Craterostigma wilmsii	2.6	<0.8
Micraira subulifolia	0.1	0.5
Sporobolus stapfianus	1.0	5.5
Coleochloa setifera	0.7	1.9
Xerophyta viscosa	3.5	3.5
Nonresurrection Plants[a]		
Rhamnus glandulosus	2.6	
Rhamnus alaternus	3.3	
Acer syriacum	3.5	
Quercus coccifera	8.0	
Quercus ilex	9.5	
Olea europaea	10.0	
Nerium oleander	13.2	
Olea lancea	13.5	

[a] Surface not specified.

5.9. Low Water Potentials

In air-dry resurrection plants (ψ usually ~30 to 40% RH, -1250 to -1650 bars), general metabolism is probably so slow that metabolic imbalances accumulate only after months of dryness. In most climates, leaves of resurrection plants would be rehydrated at least once in 10 months, and presumably a short period of full rehydration could correct the accrued imbalance (70).

5.10. Rehydration

Plants of the air-dry moss *T. ruralis* can recover full water content in about 30 s, normal respiration rates in about 30 min, protein synthesis in 30 min, and photosynthesis in about 2 h: it recovers so rapidly that much of the metabolic system must be nearly intact (59, 93, 94). The fern *Polypodium polypodioides* recovers from air dryness more slowly, with photosynthesis and respiration recovering in direct proportion to water content (56): lack of water, rather than gross impairment of metabolism, seems to be the main limitation.

In other desiccation tolerant resurrection plants major deficiencies are evident in cell fine structure and function: resurrection monocotyledons particularly show considerable disorganization of chloroplasts, mitochondria, and cytoplasm. Even in the tolerant fern *Pellaea calomelanos,* which retains good chloroplast structure in the dry state, photosynthesis does not begin to recover until after about 8 h of rehydration (Table 14.8). Considerable lags in photosynthetic recovery are evident in all other resurrection plants studied to date.

Clearly then, it is not necessary for the air-dry cell to retain the metabolic system *in toto,* and, in fact, some resurrection plants retain close to a basic minimum of the essential processes. Such an essential "core metabolism" must include æ full complement of genetic information and a reasonable capacity for respiration and protein synthesis. In vascular resurrection plants, respiration increased rapidly in the first 30 to 60 min (Table 14.8), even though water contents were still relatively low. In viable samples of air-dry *T. ruralis* and *Cratoneuron filicinum* respiratory oxygen uptake during rehydration exhibited an initial burst far above control levels. The burst was matched by increased ATP content, and paralleled a 5 to 20% increase in activity of the Embden-Meyerhof-Parnas pathway enzymes. Shifts in the $(1\text{-}^{14}C)/(6\text{-}^{14}C)$ ratio indicated an increased contribution of that pathway immediately on rehydration relative to the pentose phosphate pathway (96).

Nuclei appear intact in air-dry cells of resurrection plants in all species examined. Polysomes are absent in air-dry tissue and often not discernible for 2 to 48 h after rehydration of desiccation tolerant vascular plants: protein synthesis in the first half-day of rehydration is only a small fraction of control rates (48). In *X. villosa* protein synthesis increases gradually, in parallel with increasing RNA synthesis, and by 16 h rehydration it exceeds rates in the controls (48, 83). In the early stage of recovery ^{14}C amino acid incorporation into proteins is disproportionately high in lower molecular weight protein in *X. viscosa* and *X. villosa* (44, 83). Possibly the initial low rates of protein and RNA synthesis represent the renovation of the RNA and protein synthesis system itself, commencing with the remnant capacity in the "core" metabolism. When this system is fully renovated, the total metabolism itself can be renewed. By contrast, *T. ruralis* maintains the major portion of its polysomes in the air-dry state when dried rapidly, and on rehydration resumes rapid protein synthesis at least in part on conserved polysomes (97). Polysomes were not retained in slowly dried *T. ruralis;* rather, they were re-formed during rehydration, allowing protein synthesis to recover more rapidly than in rapidly dried plants, despite inhibition of most RNA synthesis by actinomycin D (98). Evidently even slowly dried *T. ruralis* retains more metabolic capacity than the desiccation tolerant angiosperms.

Table 14.8. Times for Initiation of Processes in Air-Dry Leaves of Desiccation Tolerant Vascular Plants Immersed in Water

Species	Times (h)					Ref.
	Full turgor	First respiration	First photosynthesis	First net photosynthesis	First chlorophyll	
Polypodium polypodioides						
Cut	1.0?	0.1	0.3	0.5	—	56
Uncut	1.7?	0.2	—	0.7	—	56
Pellaea calomelanos	10	0.2	8	25	—	95
Oropetium capense	8	0.2	5	18	—	95
Myrothamnus flabellifolia	12?	0.5	6.5	10	—	10
Xerophyta villosa	9	<3	30	48	22	95

226

6. FUTURE APPLICATIONS

Since the common examples of desiccation tolerance are found among the lower plants, the phenomenon tends to be regarded as a curiosity without practical importance. People vaguely know that seeds survive air-drying, but few think about the implication that angiosperms commonly have this genetic information for desiccation tolerance. The existence of angiosperm resurrection plants clearly demonstrates that this information can be expressed equally well in foliage during the vegetative stage and in the embryonic phase of development. Resurrection grasses and sedges (30) have potential application in semiarid land for the reclamation of dry denuded areas, and as a component of pastures providing green pick after rainfall events that are too small to benefit drought-avoiding perennial species.

Plant breeding has been particularly successful in improving characteristics that can be readily observed. Except for time to flowering, it has been notably unsuccessful in improving drought avoidance characteristics that are not easily perceived. It may prove far easier to breed increased productivity into grasses in which desiccation tolerance is fully developed already than to breed drought avoidance characteristics into plants that are productive in well-watered conditions.

REFERENCES

1. J. Levitt, *Responses of Plants to Environmental Stresses*, Academic Press, New York–London, 1972.
2. D. F. Gaff, *Oecologia*, **31**, 95 (1977).
3. I. Arvidsson, *Oikos*, Suppl. 1 (1951).
4. K. Höfler, H. Migsch, and W. Rottenburg, *Forschungsdienst*, **12**, 50 (1941).
5. H. Oppenheimer, *Ber. Dtsch. Bot. Ges.*, **50a**, 185 (1932).
6. W. S. Iljin, *Protoplasma*, **10**, 379 (1930).
7. H. R. Oppenheimer and B. Jacoby, *Protoplasma*, **57**, 619 (1963).
8. J. Levitt, C. Y. Sullivan, and E. Krull, *Bull. Res. Counc. Isr.*, **8D**, 173 (1960).
9. D. F. Gaff, *Science*, **174**, 1033 (1971).
10. H. Ziegler and G. H. Vieweg, in H. Walter and K. Kreeb, Eds., *Protoplasmatologia*, 2C6, 95 (1970).
11. D. F. Gaff and P. K. Latz, *Aust. J. Bot.*, **26**, 485 (1978).
12. R. O. Slatyer, *Aust. J. Biol. Sci.*, **11**, 349 (1958).
13. T. Visser, *Meded. Landbouwhogesch. Wageningen*, **55**, 1 (1955).
14. W. O. Abel, *S. B. Wien Akad. Wiss., Math. Naturwiss. Kl., Abt. I*, **165**, 619 (1956).
15. T. Hosokawa and H. Kubota, *J. Ecol.*, **45**, 579 (1957).
16. R. Biebl, *Protoplasmatologia*, **12**, 1 (1962).
17. E. Clausen, *Dan. Bot. Ark.*, **15**, 1 (1952).

18. O. L. Lange, *Flora (Jena)*, **140**, 39 (1953).
19. O. Y. Lee-Stadelmann and E. J. Stadelmann, in J. R. Goodin and D. K. Northington, Eds., *Arid Land Plant Resources*, Texas Tech University Press, Lubbock, Texas, 1978, p. 724.
20. J. A. Clark and J. Levitt, *Physiol. Plant.*, **9**, 598 (1956).
21. D. F. Gaff and D. M. Churchill, *Aust. J. Bot.*, **24**, 209 (1976).
22. K. Höfler, *Ber. Dtsch. Bot. Ges.*, **60**, 94 (1943).
23. T. Herzog and K. Höfler, *Hedwigia*, **82**, 1 (1944).
24. K. Mägdefrau and A. Wutz, *Forstwiss. Centralbl.*, **70**, 103 (1951).
25. D. J. Osborne, *Plant Physiol.*, **37**, 595 (1962).
26. S. Pruzsinsky, *Oesterr. Akad. Wiss., Math.-Naturwiss. Kl., Abt. I*, **169**, 43 (1960).
27. D. F. Gaff, S.-Y. Zee, and T. P. O'Brien, *Aust. J. Bot.*, **24**, 225 (1976).
28. B. J. Grieve, *Aust. J. Sci.*, **23**, 375 (1961).
29. P. A. Genkel' and V. V. Levina, *Sov. Plant Physiol. (Engl. transl.)*, **22**, 488 (1975).
30. D. F. Gaff and R. P. Ellis, *Bothalia*, **11**, 305 (1974).
31. E. B. Tucker, J. W. Costerton, and J. D. Bewley, *Can. J. Bot.*, **53**, 94 (1975).
32. O. Y. Lee-Stadelmann and E. J. Stadelmann, in O. L. Lange, L. Kappen, and E.-D. Schulze, Eds., *Water and Plant Life: Problems and Modern Approaches*, Springer-Verlag, Berlin-Heidelberg–New York, 1976, p. 268.
33. E. Schnepf, *Planta*, **57**, 156 (1961).
34. I. Nir, S. Klein, and A. Poljakoff-Mayber, *Aust. J. Biol. Sci.*, **22**, 17 (1969).
35. M. S. Buttrose and J. G. Swift, *Aust. J. Plant Physiol.*, **2**, 225 (1975).
36. E. B. Kurkova, *Sov. Plant Physiol. (Engl. transl.)*, **22**, 981 (1975).
37. L. Kh. Gordon, V. Ya. Alekseeva, A. A. Bichurina, A. I. Golubev, L. A. Kashapova, O. O. Chernysh, and N. N. Gerasimov, *Sov. Plant Physiol. (Engl. transl.)*, **22**, 804 (1975).
38. R. J. Fellows and J. S. Boyer, *Planta*, **132**, 229 (1976).
39. E. W. Simon, *New Phytol.*, **73**, 377 (1974).
40. R. K. Gupta, *Biochem. Physiol. Pflanz.*, **170**, 389 (1976).
41. R. K. Gupta, *Aust. J. Bot.*, **25**, 363 (1977).
42. D. F. Gaff, *Aust. J. Biol. Sci.*, **19**, 291 (1966).
43. N. D. Hallam and D. F. Gaff, *New Phytol.*, **81**, 349 (1978).
44. V. Daniel, *Protein and Sulphydryl Metabolism in Desiccation Tolerant Plants*, Ph.D. thesis, Monash University, Melbourne, Australia, 1976.
45. G. W. Todd, in T. T. Kozlowski, Ed., *Water Deficits and Plant Growth*, Vol. 3, Academic Press, New York–San Francisco–London, 1972, p. 77.
46. S. J. Webb, *Bound Water in Biological Integrity*, Thomas, Springfield, Illinois, 1965.
47. T. A. Villiers and D. J. Edgcumbe, *Seed Sci. Technol.*, **3**, 761 (1975).
48. D. F. Gaff and G. R. McGregor, *Biol. Plant.*, **21**, 92 (1979).
49. T. C. Hsiao, *Plant Physiol.*, **46**, 281 (1970).
50. J. W. Maranville and G. M. Paulsen, *Crop Sci.*, **12**, 660 (1972).
51. B. Kessler, *Rec. Adv. Bot.*, **2**, 1153 (1961).
52. N. N. Savitskaya, *Sov. Plant Physiol. (Engl. transl.)*, **12**, 298 (1965).
53. H. Matsumoto, N. Wakiuchi, and E. Takahashi, *Physiol. Plant.*, **21**, 1210 (1968).

54. A. Blum and A. Ebercon, *Crop Sci.*, **16**, 428 (1976).
55. H. Brix, *Physiol. Plant.*, **15**, 10 (1962).
56. T. S. Stuart, *Planta*, **83**, 185 (1968).
57. V. N. Zholkevich, *Sov. Plant Physiol. (Engl. transl.)*, **8**, 323 (1961).
58. V. N. Zholkevich and A. Ya. Rogacheva, *Sov. Plant Physiol. (Engl. transl.)*, **14**, 424 (1967).
59. J. D. Bewley, P. Halmer, J. E. Kochko, and W. E. Winner, in J. H. Crowe and J. S. Clegg, Eds., *Dry Biological Systems*, Academic Press, New York–San Francisco–London, 1978, p. 185.
60. H. Walter, *Die Hydratur der Pflanze und ihre physiologisch-ökologische Bedeutung (Untersuchungen über den osmotischen Wert)*, Fischer, Jena, 1931.
61. D. M. Griffin, *Proc. Linn. Soc. N.S.W.*, **91**, 84 (1966).
62. E. H. Roberts and D. L. Roberts, in E. H. Roberts, Ed., *Viability of Seeds*, Chapman & Hall, London, 1972, p. 424.
63. B. C. W. Hummel, L. S. Cuendet, C. M. Christensen, and W. F. Geddes, *Cereal Chem.*, **31**, 143 (1954).
64. J. G. Harrison and D. A. Perry, *Ann. Appl. Biol.*, **84**, 57 (1976).
65. S. Koga, A. Echigo, and K. Nunomura, *Biophys. J.*, **6**, 665 (1966).
66. F. G. Miftakhutdinova and N. A. Gusev, *Sov. Plant Physiol. (Engl. transl.)*, **15**, 758 (1969).
67. N. D. Hallam, B. E. Roberts, and D. J. Osborne, *Planta*, **105**, 293 (1972).
68. B. E. Roberts and D. J. Osborne, in W. Heydecker, Ed., *Seed Ecology*, Butterworths, London, 1973, p. 99.
69. N. D. Hallam, in W. Heydecker, Ed., *Seed Ecology*, Butterworths, London, 1973, p. 115.
70. T. A. Villiers, *Plant Physiol.*, **53**, 875 (1974).
71. E. Irmscher, *Jahrb. Wiss. Bot.*, **50**, 387 (1912).
72. A. A. Powell and S. Matthews, *J. Exp. Bot.*, **28**, 225 (1977).
73. L. W. Woodstock, J. Simkin, and E. Schroeder, *Seed Sci. Technol.*, **4**, 301 (1976).
74. H. A. H. Wallace, *Can. J. Bot.*, **38**, 287 (1960).
75. P. Koostra, *Seed Sci. Technol.*, **1**, 417 (1973).
76. N. D. Hallam and D. F. Gaff, *New Phytol.*, **81**, 657 (1978).
77. F. A. M. Wellburn and A. R. Wellburn, *Bot. J. Linn. Soc.*, **72**, 51 (1976).
78. D. F. Gaff and N. D. Hallam, in R. L. Bieleski, A. R. Ferguson, and M. M. Creswell, Eds., *Mechanisms of Regulation of Plant Growth*, Bulletin 12, Royal Society of New Zealand, Wellington, 1974, p. 389.
79. M. J. Tymms and D. F. Gaff, *J. Exp. Bot.*, **30**, 165 (1979).
80. J. A. Owoseye and W. W. Sandford, *J. Ecol.*, **60**, 807 (1972).
81. M. J. Tymms, *Some Aspects of Free Proline in Drought Tolerant Plants*, B.Sc. (Hon.) Report, Botany Department, Monash University, Melbourne, Australia, 1974.
82. M. Bartley, *Aspects of the Ultrastructure and Physiology of the Southern African Sedge* Coleochloa setifera *(Ridley) Gilly under Water Deficit Stress*, B.Sc. (Hon.) Report, Botany Department, Monash University, Melbourne, Australia, 1978.
83. M. J. Tymms, *Protein Synthesis in* Xerophyta villosa, Ph.D. thesis, Botany Department, Monash University, Melbourne, Australia, 1979.

84. T. J. Zabadal, *Plant Physiol.*, **53**, 125 (1974).
85. B. R. Loveys, C. J. Brien, and P. E. Kreidemann, *Physiol. Plant*, **33**, 166 (1975).
86. J. V. Jacobsen, *Annu. Rev. Plant Physiol.*, **28**, 537 (1977).
87. C. Hubac, *Plant-Water Relationships in Arid and Semi-Arid Conditions, Proceedings of the Madrid Symposium*, UNESCO, Paris, 1961, p. 271.
88. S. T. C. Wright and R. W. P. Hiron, *Nature (London)*, **224**, 719 (1969).
89. D. J. Osborne, M. B. Jackson, and B. V. Millborrow, *Nature (London), New Biol.*, **240**, 98 (1972).
90. Y.-P. Chang and W. P. Jacobs, *Am. J. Bot.*, **60**, 10 (1973).
91. J. R. Brandle, T. M. Hinckley, and G. N. Brown, *Physiol. Plant.*, **40**, 1 (1977).
92. H. Kamp, *Jahrb. Wiss Bot.*, **72**, 403 (1930).
93. J. D. Bewley, *Can. J. Bot.*, **51**, 203 (1972).
94. E. A. Gwóźdź and J. D. Bewley, *Plant Physiol.*, **55**, 340 (1975).
95. D. F. Gaff, G. R. McGregor, and T. F. Neales, unpublished.
96. J. D. Bewley and T. A. Thorpe, *Physiol. Plant.*, **32**, 147 (1974).
97. J. D. Bewley, *Plant Physiol.*, **51**, 285 (1973).
98. E. A. Gwóźdź, J. D. Bewley, and E. B. Tucker, *J. Exp. Bot.*, **25**, 599 (1974).

ADAPTATION TO HIGH TEMPERATURE STRESS | IV

Mention of high temperature stress usually brings to mind studies on thermophilic organisms that inhabit hot springs and seed plants occurring in hot deserts. These plants can survive temperatures of 50°C or higher. However we tend to ignore the plants that suffer high temperature stress at temperatures often encountered during the growing season in temperate zones. For example, many plants from alpine habitats are quite unable to survive hot summer weather at low altitudes, and cool season crops such as broccoli, lettuce, and English peas do poorly at midsummer. This is partly because such basic physiological processes as dark respiration and photosynthesis are affected directly by temperature. The maximum rate of photosynthesis occurs at 20 to 30°C in most temperate zone crop plants and as high as 35°C in some tropical plants. However the rate of dark respiration tends to increase with rising temperature up to the point at which injury to the protoplasm begins to occur. As a result, high night temperatures are particularly injurious to some plants because they result in excessive use of photosynthate in respiration, a point made by Lundegardh 50 years ago. This type of chronic high temperature stress usually results in decreased yield, but only rarely in death. Damage also is caused occasionally by short periods of abnormally high temperature, but it often is difficult to separate temperature injury from water stress injury, as shown in Chapter 21 in Part V.

Some years ago injury from high temperature was often attributed to denaturation of proteins, and Soviet workers believed that some injury was caused by ammonia released during protein breakdown. More recently injury has been attributed to inactivation of enzymes and damage to cell membranes. The three chapters in this part report research on the activity of enzymes and changes in membrane integrity with temperature: two chapters focus attention on the enzymes and membranes associated with the photosynthetic apparatus.

Acclimation to temperature is a well-known phenomenon of plants and animals. Research on algae and fungi indicates that successful acclimation to high temperature is correlated with the increase in the amount of saturated fatty acids in cell membranes. Acclimation of the photosynthetic apparatus and the role of enzymes and membranes in this acclimation are other aspects of the research reported in this part.

15 | Response and Adaptation of Photosynthesis to High Temperatures

OLLE BJÖRKMAN, MURRAY R. BADGER, AND PAUL A. ARMOND

Department of Plant Biology, Carnegie Institution of Washington, Stanford, California

1. INTRODUCTION

Death Valley, California, is one of the most arid and hottest environments on the surface of the earth. The mean daily maximum temperature ranges from 18.5°C for the coolest month, to 46.7°C for the hottest month; the range of air temperatures that an evergreen plant might be subjected to during a typical year extends from −3.1 to 50.6°C.

As in many other arid habitats, the hottest period of the year is also the period of most severe drought. This coincidence of the periods of extreme drought and heat makes it difficult to separate the effects of water stress and high temperature stress on plants growing in their natural habitat. In the studies described here, which were designed to evaluate the responses and adaptation of plants to high temperature stress, an attempt was made to eliminate the water stress component.

2. RESPONSE OF GROWTH TO TEMPERATURE

The investigations of plant response to temperature have centered around the study of plant species native to both the cool coastal environment of

233

Bodega Head in northern California, and to the hot environment of the floor of Death Valley. Initial growth experiments established the potential response to temperature of species native to the two types of habitat. These studies included the use of irrigated and nonirrigated transplant gardens on the floor of Death Valley and at Bodega Head (1).

On the basis of their growth and survival in the Death Valley garden, the plants used in the study may be divided into three categories. The first is represented by two coastal species *Atriplex glabriuscula,* which has a C_3 pathway of photosynthesis, and *Atriplex sabulosa,* which has a C_4 pathway of photosynthesis. Like several other cool coastal species tested in the two experimental gardens, these plants were unable to survive the hot summer in Death Valley, even though supplied with ample water. The second category includes evergreen species native to the Death Valley floor: the C_4 species *Atriplex hymenelytra* and *Atriplex lentiformis,* and the C_3 species *Larrea divaricata.* These plants maintained active growth throughout the year, but peak growth rates occurred during spring and fall. The third category has only one representative, the Death Valley native C_4 species *Tidestromia oblongifolia.* This perennial herbaceous shrub restricts its period of active growth to the summer months and is dormant from November until May. In the Bodega Head garden the cool coastal species attained very high rates of growth. The growth rates of the evergreen Death Valley species ranged from slow to moderate, whereas *Tidestromia oblongifolia* was unable to survive in this cool habitat.

It is evident that pronounced differences exist among these plants in their ability to grow and survive under contrasting temperature regimes. The plants that are most productive in one regime evidently are unable to cope with the other extreme. However the evergreen desert species *A. hymenelytra, A. lentiformis,* and *L. divaricata* are able to grow over a very wide range of temperatures; that is, they possess a high acclimation potential. These studies (2) also indicated that the seasonal activity of the plants in their native Death Valley habitat is controlled largely by temperature. Even during the summer the seasonal activities of the various species tested closely followed those of the unirrigated naturally occurring plants.

Controlled growth experiments in which the summer temperature regimes of Death Valley and Bodega Head were simulated, while maintaining a low water vapor deficit, are summarized in Table 15.1. These studies (3) confirmed that the widely different abilities to grow in the two contrasting natural habitats are to a large extent attributable to intrinsic differences in the response of the plants to temperature. Further experiments (3), summarized in Figure 15.1, showed that 45°C day/32°C night temperature regime, which simulates a typical day during summer in Death Valley, is at or very near the optimum for growth in *T. oblongifolia,* but is lethal to the coastal species. Conversely, the 16°C day/11°C night regime, which permits the coastal plants to grow at about 70% of their rate at optimum temperature, is too low for sustained growth in *T. oblongifolia.* The evergreen desert species are

Table 15.1. Total Dry Matter Yields (final weight/initial weight) Obtained in a 22 Day Growth Period Under Two Contrasting Temperature Regimes (From Ref. 2 and O. Björkman, unpublished)

Species	16°C Day/ 11°C Night	45°C Day/ 31°C Night
Atriplex glabriuscula (coastal C_3)	24.4	0.1 (died)
Atriplex sabulosa (coastal C_4)	18.2	0.3 (died)
Larrea divaricata (desert C_3)	5.4	3.2
Tidestromia oblongifolia (desert C_4)	<2.5	88.6

able to tolerate both these extremes, but their growth rates are lower than those of coastal species under the cool regime, and much lower than that of *T. oblongifolia* when growing in the hot regime. Allocation of carbon to the leaves compared with other organs showed a surprisingly similar temperature dependence in all species. This indicates that the contrasting interspecific differences must be related to intrinsic differences in the temperature dependence of primary growth processes.

3. TEMPERATURE DEPENDENCE OF PHOTOSYNTHESIS

To investigate the factors that underlie the contrasting temperature responses of growth, comparative studies of the photosynthetic characteristics were conducted on the cold-adapted C_4 species *Atriplex sabulosa* and the heat-adapted C_4 species *Tidestromia oblongifolia* (4–10). Figure 15.2 shows

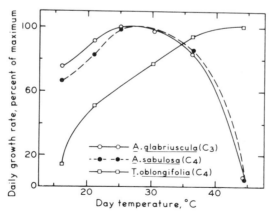

Figure 15.1. Daily growth rate as a function of daytime growth temperature in *Atriplex glabriuscula* (C_3, ○—○), *Atriplex sabulosa* (C_4, ●--●), and *Tidestromia oblongifolia* (C_4, □—□). Growth rates are expressed as daily gain in dry weight, in grams, per gram of total plant dry weight. From Ref. 3.

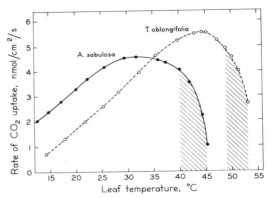

Figure 15.2. Temperature dependence of light-saturated net CO_2 uptake in *Atriplex sabulosa* (grown at 20°C day/15°C night, ●—●) and *Tidestromia oblongifolia* (grown at 45°C day/32°C night, ○—○). Measured quantum flux density was 2000 $\mu E/m^2/s$, CO_2 concentration was 330 $\mu l/l$, and oxygen concentration was 21% v/v. The shaded areas represent the temperature region in which photosynthesis is unstable, showing a time-dependent decline. Unpublished data of O. Björkman.

the temperature dependence of light-saturated net photosynthesis in intact leaves of these two species. *A. sabulosa* was grown at a cool regime and *T. oblongifolia* under a hot regime. It is obvious that the coastal species has a much higher photosynthetic capacity at low temperatures than the desert species. Conversely, the desert species possesses a much superior photosynthetic performance at high temperatures. This high temperature superiority of *T. oblongifolia* is related to a strikingly increased stability of photosynthesis at high temperatures, indicated by the fact that in *A. sabulosa* a time-dependent inactivation of photosynthesis sets in when the leaf temperature exceeds 39 to 40°C, whereas in *T. oblongifolia* a corresponding inactivation is not seen until the temperature exceeds 48 to 49°C. Other experiments show that neither the superior photosynthetic performance of the coastal species at low temperatures nor the superior performance of the desert species at high temperatures could in any part be attributed to stomatal factors. Since photosynthesis was essentially carbon dioxide (CO_2) saturated at all temperatures in these two C_4 species, physical factors determining the diffusion of CO_2 from the leaf intercellular spaces to the sites of CO_2 fixation also cannot be responsible for their contrasting photosynthetic temperature responses. The possibility that the higher photosynthetic capacity of the coastal species might be the result of a generally higher amount of photosynthetic machinery per unit leaf area was not supported by measured values of protein or chlorophyll per unit leaf area (data not shown). A more likely alternative is that the increased capacity at low temperatures is caused by a higher level of one or several rate-limiting, temperature-dependent catalytic steps of photosynthesis. The possible identity of these steps is discussed later in the context of acclimation of photosynthesis to growth temperature.

As indicated by the temperature-response curves in Figure 15.2, the steep decline in photosynthetic rate at high temperature is related to an irreversible inactivation of photosynthesis. This is further illustrated in the upper panel of Figure 15.3, which shows the rate of light-saturated photosynthesis, measured at a constant noninhibitory temperature (30°C), fol-

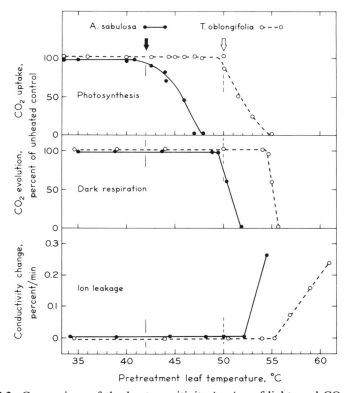

Figure 15.3. Comparison of the heat sensitivity *in vivo* of light- and CO_2-saturated photosynthetic rate, dark respiration, and ion leakage in leaves of *Atriplex sabulosa* (●—●) and *Tidestromia oblongifolia* (O--O). "Control rates" of photosynthesis were measured at a standard noninhibitory temperature (30°C); the attached, illuminated leaves were subsequently treated for 15 min at the temperature indicated on the abscissa, then quickly returned to the standard noninhibitory temperature, and the photosynthetic rate measured again. Data from Ref. 2. The procedure for the respiration measurements was similar to that used for photosynthesis, except that the leaves were kept in the dark. Ion leakage was measured by the increase in conductivity of the medium in which leaf slices were submerged. Unpublished data of O. Björkman. The solid and open arrows and solid and dashed vertical lines denote the temperatures at which time-dependent inactivation of photosynthesis sets in in *A. sabulosa* and *T. oblongifolia*, respectively.

lowing pretreatment of illuminated intact leaves for 15 min at the temperature indicated on the abscissa. The exposure of *A. sabulosa* leaves to temperatures above 40°C results in a progressive, irreversible inactivation of photosynthesis; *T. oblongifolia* tolerates exposure to about 49°C before any irreversible injury can be detected.

High temperature inactivation of photosynthesis could be an indirect result of a general breakdown of cellular components such as loss of semipermeability of the cell membranes. As shown in the lower panels of Figure 15.3, the two species do indeed differ in the sensitivity of other parameters to high temperature inhibition. Both leakage of solutes from the leaves and irreversible inhibition of respiration set in at considerably lower temperatures in *A. sabulosa* than in *T. oblongifolia*. However in each species photosynthesis is completely inactivated before the symptoms of other high temperature injury can be detected. This strongly indicates that the observed thermal inactivation of photosynthesis is caused primarily by a direct effect of temperature on the photosynthetic machinery itself. It follows that the difference in photosynthetic performance between the two species is attributable to differences in the thermal stability of chloroplast components.

Positive experimental evidence for such differences in chloroplast membrane properties is given in Figure 15.4. This figure, which combines the results of several studies (5–7, 9, 10), shows the effect of temperature on the quantum yield of photosynthesis by intact leaves (at rate-limiting light intensities) and the quantum yields of photosystem 1- and photosystem 2-driven electron transport by chloroplasts isolated from leaves pretreated to different temperatures. Also shown in Figure 15.4 is the temperature dependence of chlorophyll fluorescence by detached leaves, measured at an extremely weak excitation light. Inhibition of the quantum yield of photosynthesis in each species occurs at about the same temperatures that cause irreversible inactivation of photosystem 2-driven electron transport. Both the quantum yield and the photosystem 2 activity exhibit heat labilities similar to that of whole leaf CO_2 fixation. Photosystem 1-driven electron transport shows no inactivation in either species, even at temperatures that cause complete inhibition of whole leaf photosynthesis. The fluorescence data show that in each species there is a close correspondence between the temperature at which a sharp rise in the F_0 level of fluorescence occurs and the temperatures at which inhibition of photosynthesis and photosystem 2 activity takes place. All available evidence supports the conclusion that a greatly increased thermal stability of chloroplast membrane reactions, and in particular the integrity of photosystem 2, plays a key role in the superior performance of *T. oblongifolia* at high temperatures.

An enhanced ability to photosynthesize at high temperatures requires that all component reactions be able to maintain their integrity and function at these temperatures. Further studies were therefore directed toward an analysis of the *in vivo* thermal stabilities of a number of enzymes of photo-

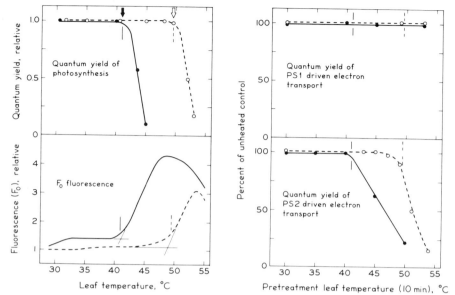

Figure 15.4. Effect of leaf temperature on the quantum yield for CO_2 uptake by intact leaves of *Atriplex sabulosa* (●—●) and *Tidestromia oblongifolia* (○--○) (*top left*), and of pretreating illuminated detached leaves for 10 min at different temperatures on the quantum yields for photosystem 1 (PS1)- and photosystem 2 (PS2)-driven electron transport by chloroplasts isolated from these leaves (*top and bottom right*, respectively). Redrawn from Ref. 7. Also shown is the effect of leaf temperature on the fluorescence yield of detached leaves (*bottom left*). Redrawn from Ref. 10. Solid and open arrows and solid and dashed vertical lines denote the temperatures at which time-dependent inactivation of photosynthesis sets in in *A. sabulosa* and *T. oblongifolia*, respectively.

synthetic carbon metabolism. These enzymes are not bound to the thylakoid membrane but are located in the stroma region of the chloroplast and, in some cases, outside the chloroplast.

Figure 15.5 summarizes the results obtained with some of the 14 enzymes investigated (7, 9). Also shown are the effects of heat pretreating the illuminated leaves for 10 min on the subsequent light-saturated photosynthetic capacity and on the amount of protein extractable from the leaves with an aqueous buffer. It is evident from these results that considerable differences exist between *A. sabulosa* and *T. oblongifolia* in the *in vivo* heat stabilities of a number of stromal and extrachloroplast enzymes. However for several of these enzymes (e.g., RuP_2 carboxylase and PEP carboxylase) the heat stabilities are too high to account for the observed heat inhibition of whole leaf photosynthesis. For certain other enzymes, such as NADP glyceraldehyde-3P-dehydrogenase, Ru5P kinase, and NADP malate dehy-

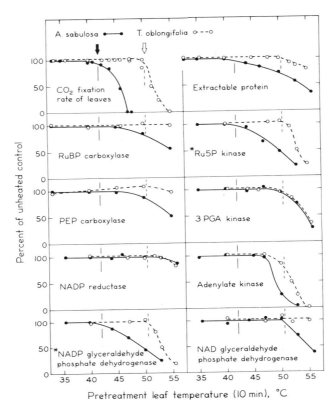

Figure 15.5. Responses of light-saturated photosynthesis, extractable protein, and indicated enzyme activities to pretreatment of illuminated *Atriplex sabulosa* (●—●) and *Tidestromia oblongifolia* (○——○) leaves to different temperatures. From Ref. 9. Solid and open arrows and solid and dashed vertical lines denote the temperatures at which time-dependent inactivation of photosynthesis sets in in *A. sabulosa* and *T. oblongifolia*, respectively.

drogenase (data not shown) in both *A. sabulosa* and *T. oblongifolia*, heat inactivation occurs near the same temperatures that cause inactivation of photosynthesis in each of these species.

These four enzymes may be partly responsible for the observed heat inactivation of photosynthesis. Except for adenylate kinase, however, these enzymes require photochemically generated reducing power for their activation (11, 12). Moreover, full activity can be restored in extracts from heat-treated leaves by addition of an *in vitro* activating system that replaces the *in vivo* photochemically generated reductants (9). The results suggest that the relatively low thermal stability of these "light-activated" enzymes, as observed in the leaf pretreatment experiments, may result from an inactivation of photosystem 2 activity, thus whole-chain electron transport, rather than reflecting a direct thermal inactivation of the enzymes. Nevertheless, it is clear that a substantial loss in the amounts of protein extractable with aqueous buffers occurs when the leaves are pretreated at temperatures just high enough to cause a significant inhibition of photosynthesis. The identity and location in the cell of the enzyme or enzymes responsible for the soluble protein loss remain to be determined.

4. PHOTOSYNTHETIC TEMPERATURE ACCLIMATION

The preceding comparisons of photosynthetic characteristics have been made between plants that are both native to and grown in either a cool or a hot environment. It is well known that the temperature regime under which a given plant is grown may have a pronounced influence on its photosynthetic temperature response characteristics. Examples of such responses to growth under both cool and hot regimes are shown in Figure 15.6. The two coastal species *A. glabriuscula* and *A. sabulosa* are incapable of acclimating to a

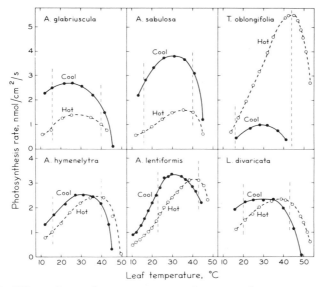

Figure 15.6. Effect of growth temperature on the rate and temperature dependence of light-saturated net CO_2 uptake for a number of C_3 and C_4 species native to habitats with contrasting thermal regimes. The "hot" growth regimes were: 40°C for *Atriplex glabriuscula*, *Atriplex sabulosa*, and *Atriplex hymenelytra*, 43°C for *Atriplex lentiformis*, and 45°C for *Tidestromia oblongifolia* and *Larrea divaricata*. The "cool" growth regimes were: 23°C for *A. lentiformis*, 20°C for *L. divaricata*, and 16°C for the other species. Data for *A. glabriuscula*, *A. sabulosa*, *A. hymenelytra*, and *T. oblongifolia* from Ref. 4 and O. Björkman, unpublished; *A. lentiformis* from Ref. 14; and *L. divaricata* from Ref. 13.

40°C daytime temperature regime. Although an upward shift in the optimum temperature for photosynthesis does take place when the plants are grown under a hot regime, the photosynthetic rates at this temperature do not increase; instead they are reduced to about 50% of the values found in the plants that were grown under a cool regime. The response of *T. oblongifolia* is in sharp contrast with those of the two coastal species. This thermophile completely fails to acclimate to low temperatures; when these plants are kept below 20°C for an extended period, photosynthesis is severely inhibited at any measurement temperature, and growth comes to an almost complete halt. This low photosynthetic activity cannot be attributed to stomatal closure or to other factors affecting the diffusive transport of CO_2.

A very different situation is found with the evergreen Death Valley species *Larrea divaricata* (13) and *Atriplex lentiformis* (14). These plants, which remain active throughout the year in their native habitats, thus experiencing a great seasonal temperature change, are capable of photosynthetic acclimation over a very wide temperature range. Growth under a cool regime results in a higher photosynthetic capacity at low temperatures, whereas growth in a hot regime results in an improved performance at high temperatures.

Another species that is capable of a very wide range of temperature acclimation is the C_3 species *Nerium oleander* (15). This species, a native to the deserts of Southwest Asia and North Africa, is extensively grown in many climatic zones in California, from the cool coast to the floor of Death Valley, where it remains active throughout the year. Figure 15.7 (top panel) shows the temperature dependence of light-saturated photosynthesis in normal air of individuals of the same clone of *N. oleander* grown under contrasting daytime temperature regimes of 20 and 45°C. The 20°C-grown plants have twice the photosynthetic capacity of the 45°C-grown plants at 20°C, whereas at 45°C the reverse is true. Simultaneous measurements of photosynthesis and stomatal conductances show that stomatal factors are not responsible for the differences in photosynthetic rates between the 20°C-grown and 45°C-grown plants, at either low or high measurement temperatures.

The photosynthetic performance determined in normal air is undoubtedly most meaningful from an ecological viewpoint. However, analysis of the effect of growth temperature regime on the intrinsic capacity and the thermal stability of photosynthesis can be made best under conditions of saturating CO_2. Figure 15.7 (bottom panel) shows the photosynthetic temperature response of the plants determined at high and essentially saturating CO_2 concentrations. These results demonstrate that growth at low temperature caused large increases in the intrinsic photosynthetic capacity of the leaves. This higher capacity is evident over a wide temperature range, but is especially pronounced at moderate and low measurement temperatures. Only at temperatures exceeding 40°C do the 45°C-grown plants become superior to the 20°C-grown material. This superiority is attributable to an

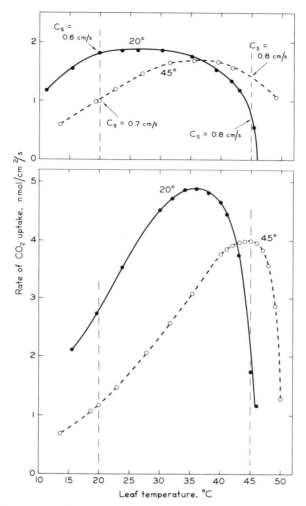

Figure 15.7. Temperature dependence of light-saturated CO_2 uptake for *Nerium oleander*, grown at two contrasting temperature regimes (20°C day/15°C night and 45°C day/32°C night). Measurements were made in normal air (*top*) 330 μl/l CO_2, 21% v/v oxygen, and at high CO_2 concentration (*bottom*) 750 μl/l CO_2, 20% v/v oxygen. Also shown are the stomatal conductances C_s measured at the two daytime growth temperatures. These temperatures are indicated by the broken vertical lines. Data from Ref. 15.

upward shift in the temperature at which severe inhibition of photosynthesis is induced by high temperatures.

It thus appears that temperature acclimation can be considered to involve two distinct and perhaps unrelated changes: low temperature acclimation is attributable to an increased capacity of a limiting, temperature-dependent catalytic step; high temperature acclimation is primarily the result of an increased high temperature stability of one or several chloroplast components. It is noteworthy that even fully expanded leaves are capable of a complete acclimation after transfer of the plants from a cool to hot growth regime, and vice versa, as illustrated in Figure 15.8 (top panel), which shows that less than 2 weeks is required for full acclimation after such a transfer.

As shown above, the superior photosynthetic capacity at low temperatures of 20°C-grown plants compared with 45°C-grown plants is not related to stomatal or other leaf factors that influence the diffusive transport into the chloroplast. This superiority also cannot be attributed to an increased

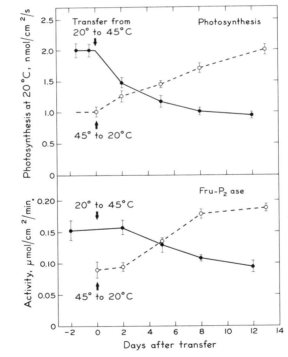

Figure 15.8. Time course of change in light-saturated photosynthetic capacity, measured at 20°C (*top*) and of fructose-1,6,-bisphosphate phosphatase activity (*bottom*) in mature leaves, just prior to and following transfer of 45°C- and 20°C-grown *Nerium oleander* plants to the opposite growth temperature regimes. Unpublished data of M. Badger and O. Björkman.

amount of photosynthetic machinery per unit of leaf area (Table 15.2), and microscopic examination of leaf sections did not reveal any apparent differences in leaf thickness or gross anatomy between leaves from the two growth regimes. Growth at low temperature did not result in a general increase in the level of enzymes of photosynthetic carbon metabolism, or in the capacity for photosynthetic electron transport or photophosphorylation (Table 15.3).

Two Calvin cycle enzymes, RuP_2 carboxylase and $FruP_2$ phosphatase, have been implicated in the literature as potentially limiting steps of photosynthesis under certain conditions. The data in Table 15.3 indicate that the higher RuP_2 carboxylase activity found in the 20°C-grown plants may to some extent contribute to their superior photosynthetic capacity, but the difference in the level of the enzyme is too small to account for the difference in photosynthetic capacity at 20°C between the 20°C-grown and the 45°C-grown plants. $FruP_2$ phosphatase is the only enzyme component examined that is affected by growth temperature to the same extent as photosynthetic capacity. It is also noteworthy that in the transfer experiments shown in Figure 15.8, the time course of increase in the level of this enzyme in mature leaves, following transfer from 45°C to a 20°C growth regime, closely resembles the increase in photosynthetic capacity. Parallel decreases in photosynthetic capacity and $FruP_2$ phosphatase also occurred following transfer of plants from 20 to 45°C, except that the initial decline in photosynthesis was not accompanied by a decrease in enzyme activity. This initial decline is probably a result of a direct thermal inhibition occurring after the abrupt increase in temperature to 45°C. The present data implicate $FruP_2$ phosphatase as a factor that may be predominantly responsible for the higher photosynthetic capacity at low temperature of *N. oleander* grown at low temperature.

As seen previously (Figure 15.7), growth of *N. oleander* at 45°C/32°C resulted in a superior photosynthetic performance above 40°C. We attributed this to an increased high temperature stability of the photosynthetic machinery. Heat treatment experiments with illuminated detached leaves

Table 15.2. Leaf Specific Weights and Chlorophyll and Protein Contents of *Nerium oleander* Leaves, Grown Under Two Contrasting Thermal Regimes (15)

Variable	A: Grown at 20°C	B: Grown at 45°C	A/B
Fresh weight (mg/cm²)	44.2	45.8	0.97
Dry weight (mg/cm²)	12.6	15.4	0.82
Chlorophyll $a + b$ (mg/cm²)	0.076	0.065	1.19
Total protein (mg/cm²)	1.41	1.65	0.85
Soluble protein (mg/cm²)	0.67	0.64	1.05
Insoluble protein (mg/cm²)	0.74	0.94	0.79

Table 15.3. Comparison of Photosynthetic Capacity at 20°C of Intact Leaves, Photosynthetic Electron Transport, and Photophosphorylation Capacities of Isolated Chloroplasts, and Activities of Enzymes of Photosynthetic Carbon Metabolism in Leaf Extracts of *Nerium oleander* Grown Under Two Contrasting Thermal Regimes (15)

Variable	Ratio, $\dfrac{\text{Grown at 20°C}}{\text{Grown at 45°C}}$	
	Fresh weight basis	Chlorophyll basis
Photosynthetic CO_2 uptake	2.40	2.02
Electron transport ($H_2O \rightarrow MV$)	1.37	1.27
Noncyclic photophosphorylation	1.11	1.03
Enzyme activity		
$\quad$ RuP$_2$ carboxylase	1.49	1.32
$\quad$ FruP$_2$ phosphatase	2.46	2.12
$\quad$ RuP kinase	1.17	1.01
$\quad$ 3-phosphoglycerate kinase	1.14	0.88
$\quad$ NADP glyceraldehyde-3P-dehydrogenase	1.02	0.87
$\quad$ FruP$_2$ aldolase	1.17	0.96
$\quad$ Phosphoglucomutase	1.18	0.89
$\quad$ Phosphohexose isomerase	0.86	0.65
$\quad$ NADP malate dehydrogenase	0.81	0.68
$\quad$ Adenylate kinase	1.13	0.95

(Figure 15.9) show that substantial differences exist between the materials grown at low and at high temperatures in the *in vivo* thermal stability of photosystem 2-driven electron transport and photophosphorylation, as well as in several Calvin cycle enzymes. Of these enzymes, those that require light for their activation are especially sensitive to heat inactivation, a situation closely resembling that found in *A. sabulosa* and *T. oblongifolia*. These results again suggest that the high temperature stability of these enzymes may be linked to reducing power production coupled to photosystem 2 activity. These experiments, and the work reported in Chapter 17, suggest that alterations in the heat stability of the thylakoid membranes is a key component in acclimation to high temperatures.

In addition to the role of changes in the thylakoid membrane properties, it appears that an increased high temperature stability of extra thylakoid proteins is also an important component of high temperature acclimation. The loss of soluble leaf protein after heat treatment of the leaves (Figure 15.10) occurs at considerably lower temperatures in the 20°C-grown than in the 45°C-grown plants. This loss begins at about the same temperature at which photosynthesis becomes irreversibly inhibited. The protein loss does not appear to be caused by a disruption of cell membrane integrity,

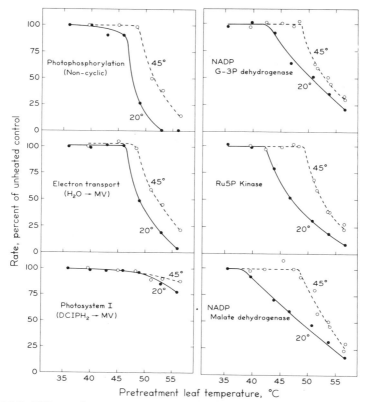

Figure 15.9. Effect of pretreatment of *Nerium oleander* leaves at different leaf temperatures for 10 min in the light on the subsequent activities of electron transport and photophosphorylation of isolated chloroplasts and of three "light-activated" enzymes of photosynthetic carbon metabolism extracted from the leaves. Solid and broken lines and solid and open circles depict the response of 20°C-grown and 45°C-grown plants, respectively. Heat treatments as in Figure 15.5. From Ref. 15.

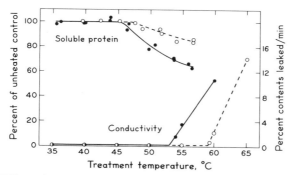

Figure 15.10. Effect of pretreatment leaf temperature on the extractable soluble leaf protein. Also shown is the response of ion leakage to temperature (measured as in Figure 15.3). Solid and broken lines and solid and open circles depict the responses of 20°C-grown and 45°C-grown *Nerium oleander* materials, respectively. Unpublished data of M. Badger and O. Björkman.

247

since ion leakage from the cells does not begin until much higher temperatures. We are currently attempting to determine the identity of the thermolabile protein fraction. Results obtained to date show that the loss of protein is confined to the fraction of relatively low molecular weight ($<100,000$ daltons).

5. CONCLUSIONS

1. Ability to cope with high temperatures is an important factor in determining the success and distribution of plants in hot, arid habitats.

2. Photosynthesis is one of the most heat sensitive aspects of growth.

3. Plants differ greatly in their potential for photosynthetic acclimation to temperature in a manner reflecting the temperature regimes of their native habitats.

4. Superior photosynthetic performance at either high or low temperatures is largely attributed to changes in the intrinsic properties of the photosynthetic apparatus at the chloroplast level. (a) Photosynthesis differences at low temperature are strongly correlated with the capacity of specific rate-limiting components, such as the enzymes RuP_2 carboxylase, and $FruP_2$ phosphatase. (b) At high temperatures the limitations to photosynthesis are imposed primarily by the thermal stability of the chloroplasts. Increased thermal stability evidently involves changes in the properties of the thylakoid membranes, as well as perhaps the soluble enzymes located outside these membranes.

REFERENCES

1. O. Björkman, M. A. Nobs, J. Berry, H. Mooney, F. Nicholson, and B. Catanzaro, *Carnegie Inst. Washington Yearb.*, **72**, 393 (1973).
2. O. Björkman, M. Nobs, H. Mooney, J. Troughton, J. Berry, F. Nicholson, and W. Ward, *Carnegie Inst. Washington Yearb.*, **73**, 748 (1974).
3. O. Björkman, B. Mahall, M. Nobs, W. Ward, F. Nicholson, and H. A. Mooney, *Carnegie Inst. Washington Yearb.*, **73**, 757 (1974).
4. O. Björkman, H. A. Mooney, and J. Ehleringer, *Carnegie Inst. Washington Yearb.*, **74**, 743 (1975).
5. O. Björkman, *Carnegie Inst. Washington Yearb.*, **74**, 748 (1975).
6. J. A. Berry, D. C. Fork, and S. Garrison, *Carnegie Inst. Washington Yearb.*, **74**, 751 (1975).
7. O. Björkman, J. Boynton, and J. Berry, *Carnegie Inst. Washington Yearb.*, **75**, 400 (1976).
8. J. Ehleringer and O. Björkman, *Plant Physiol.*, **59**, 86 (1977).
9. O. Björkman and M. Badger, *Carnegie Inst. Washington Yearb.*, **76**, 346 (1977).
10. U. Schreiber and J. A. Berry, *Planta*, **136**, 233 (1977).

11. L. Anderson, in *Proceedings of the Third International Congress on Photosynthesis,* Elsevier, Amsterdam, 1974, p. 1393.
12. R. A. Wolosiuk and B. B. Buchanan, *Nature (London),* **266,** 565 (1977).
13. H. A. Mooney, O. Björkman, and G. J. Collatz, *Plant Physiol.,* **61,** 406 (1978).
14. R. W. Pearcy, *Plant Physiol.,* **59,** 795 (1977).
15. O. Björkman, M. Badger, and P. A. Armond, *Carnegie Inst. Washington Yearb.,* **77,** 262 (1978).

16 | Adaptation of Kinetic Properties of Enzymes to Temperature Variability

J. A. TEERI

Barnes Laboratory, University of Chicago, Chicago, Illinois

1. INTRODUCTION

There is great variability among plants of different species with regard to the extent to which their metabolic processes can adjust when the organism is exposed to changes in temperature (Chapter 15). The degree of adjustment is subject to both environmental and genetic control. It has been demonstrated in a number of species that the ability of different genotypes to regulate growth over a range of temperatures appears to be correlated with the seasonal and daily temperature ranges of their native habitats (1–3). Figure 16.1 compares the growth at three temperatures of two clones of *Poa pratensis*, one native to the Great Basin and the other from coastal Oregon. The clone from the variable climate of the Great Basin showed less variation in growth over the three temperatures than did the clone from the cooler, more constant climate of coastal Oregon.

A number of mechanisms have been proposed by which an organism may adjust its metabolism in response to environmental change (2, 4–7). Many ectothermic animals possess the ability to adjust their metabolic rates to compensate for changes in habitat temperature. Hazel and Prosser (8) surveyed the temperature dependence of metabolism of poikilothermic animals

251

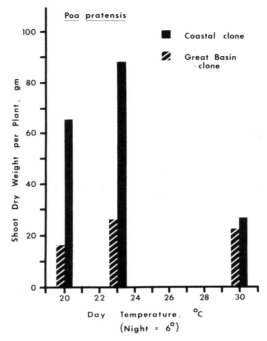

Figure 16.1. The influence of daily temperature range on the growth of two clones of *Poa pratensis* that are native to climates of contrasting daily temperature amplitude. All plants were grown at a night temperature of 6°C. Data from Ref. 1.

and concluded that in many species metabolism exhibits relative insensitivity to temperature over the range of environmental temperatures usually encountered. The evidence for such a thermal homeostasis is that metabolic rate, or growth rate, or both, remain at a single value whether the individual is placed in a cool or a warm environment. Some types of homeostasis require an adaptation period ranging from several hours to many days to become fully adjusted to the new thermal regime. Other forms of adjustment are instantaneous in that the thermal plateau for growth or metabolism is very broad and the Q_{10} of metabolism is near 1.0 over the full range of probable habitat temperatures.

There is increasing evidence (4, 6, 8) that in many cases the ability of an organism to adjust its metabolic rate to changes in temperature is strongly correlated with the degree of temperature dependence of enzyme-catalyzed reactions. The understanding of the relationship between habitat temperature variability and enzyme properties is based primarily on studies of aquatic ectothermic animals (4, 6, 8). Because of the high volumetric heat capacity of water, however, the pattern of tissue temperature variability in aquatic organisms differs considerably from that of terrestrial plants. This chapter summarizes the observations that relate to how closely the catalytic properties of plant enzymes are correlated with the range of likely tissue temperatures encountered during growth.

Data are now available for the temperature dependence of apparent Michaelis constant (K_m) of enzymes for populations of three species of angiosperms. Malate dehydrogenase (MDH, EC 1.1.1.37) has been investigated in *Potentilla glandulosa* (Rosaceae) and *Lathyrus japonicus* (Leguminosae), and glucose-6-phosphate dehydrogenase (G6PDH, EC 1.1.1.49) has been studied in *Arabidopsis thaliana* (Cruciferae). In all three studies the approach has been to compare, for each species, two populations that are native to habitats with contrasting patterns of temperature variability during growth. All plants were grown from cuttings or seeds in controlled environment chambers in an attempt to separate the effects of genotype from those of the growth environment.

2. THE RELATIONSHIP OF MINIMUM MICHAELIS CONSTANT TO HABITAT TEMPERATURE

Figure 16.2 compares two populations from each of three species. The populations compared are from the locations indicated in Table 16.1. The two populations of each species are native to a cool and a warm climate, respectively.

The relationship between apparent K_m and assay temperature (Figure 16.2) is similar in all three cases to that previously reported in many ectothermic animals (4, 8). The population native to the cooler habitat always exhibited the absolute minimum measured K_m for the species at any of the temperatures. The minimum K_m of all populations of the three species occurred at a temperature near the average habitat temperature during the growing season. Figure 16.3 plots the temperature at which the minimum K_m was measured for each population as a function of the average habitat temperature during the growing season. Although the data are few, there is a strong correspondence between the two values. The relationship, however, is based on climatological data collected in standard weather instrument shelters, and it is difficult to extrapolate accurately to plant tissue temperature.

We attempted to increase the resolution of such studies by quantifying daily and seasonal patterns of temperature fluctuation in the leaf canopy of plants of *P. glandulosa* (3). Leaf temperatures were directly measured in the field with fine wire thermocouples, and the entire seasonal pattern of tissue temperature was reconstructed from the field measurements and long-term weather records of 10 consecutive growing seasons. The K_m was measured at six assay temperatures over the 11–37°C range, which included the complete range of leaf temperatures commonly experienced in either location. The shape of the K_m response to assay temperature was closely related to the range of temperatures occurring during growth (Figure 16.2). The coastal population that experiences an average daily range in tempera-

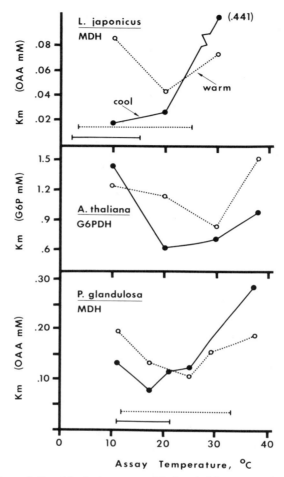

Figure 16.2. The relationship between population habitat temperature and the response of apparent Michaelis constant K_m to assay temperature for two populations of each of three plant species. For experimental growth conditions, see text and Ref. 3. The dashed and solid lines and open and closed circles are for the warm and cool populations, respectively. The horizontal lines are the ranges of minimum to maximum daily temperatures during growth in the native habitats of the warm (dashed line) and cool (solid line) habitats. *Abbreviations*: OAA, oxaloacetic acid; G6P, glucose-6-phosphate; MDH, malate dehydrogenase; G6PDH, glucose-6-phosphate dehydrogenase. Data for *Potentilla glandulosa* from Ref. 3, *Arabidopsis thaliana* from Teeri and Strang (unpublished), and *Lathyrus japonicus* from Ref. 17.

254

**Table 16.1. Native Habitats of the
Studied Populations of Three Species**

Lathyrus japonicus Willd.
  Cool: east coast Hudson Bay
  Warm: Michigan dunes
Arabidopsis thaliana (L.) Heynh.
  Cool: Dijon, France
  Warm: central Japan
Potentilla glandulosa Lindl.
  Cool: Pacific coast central California
  Warm: central Oregon

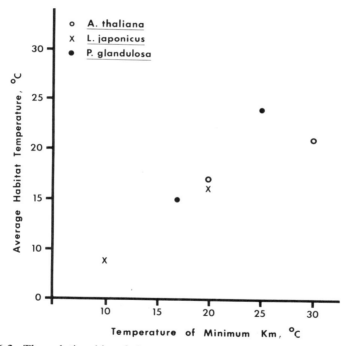

Figure 16.3. The relationship of the temperature at which the minimum apparent Michaelis constant K_m occurs and the average daily habitat temperature during growth of *Potentilla glandulosa, Arabidopsis thaliana,* and *Lathyrus japonica* grown from warm and cool populations. Same sources of data as for Figure 16.2.

ture of about 10°C had a less than two fold change in K_m over a temperature range of about 8°C. The inland population with an average daily temperature range of about 21°C had a less than two fold change over a range of 15°C.

3. THE TEMPERATURE INDEPENDENCE OF METABOLIC RATE

The fluctuation of K_m with changing assay temperature is less than two fold when measured over the range of probable tissue temperatures during growth. A less than two fold change in the K_m temperature-dependence curve has been interpreted (4, 9, 10) as providing the organism with a mechanism that makes the catalyzed reaction relatively independent of temperature fluctuation. Thus the rate of metabolism can exhibit temperature independence over this portion of the curve. Limited data (4, 8) from ectothermic animals suggest that the degree of temperature independence of metabolic rates of whole organisms is correlated with the type of K_m independence of temperature described above. There is still insufficient information to determine whether the correlation of the degree of temperature independence with the tissue temperature range is a general property of many plant enzymes. However the data for *P. glandulosa* (3) suggest that in at least some cases there may be a close correspondence. The greatest increase in the K_m of a single enzyme for a single individual appears when plants native to cool habitats are measured at high temperatures. It is highly unlikely that either the subarctic population of *L. japonicus* or the coastal population of *P. glandulosa* ever grows in nature at the respective upper assay temperatures of 40 and 37°C (Figure 16.2). Both populations exhibit a steep temperature dependence of K_m at these high assay temperatures. Similar large increases in K_m have been measured in fish native to cold water at assay temperatures much warmer than the environment, and are interpreted to indicate a major loss of catalytic and regulatory efficiency (10).

4. ACCLIMATION OF MICHAELIS CONSTANT TEMPERATURE DEPENDENCE

The degree to which the temperature dependence of K_m of an individual plant can be modified by growth temperature has not been extensively studied. In the case of *L. japonicus* the plants of the cool and warm populations were grown from seed at a single daytime temperature of 25°C. The K_m curves clearly are distinct, and at 25°C they did not acclimate to the extent that the two overlap. Plants of both populations of *P. glandulosa* were grown at both warm and cool day temperatures, and the temperature dependence of K_m did not acclimate (3). In the latter case there appeared to be a slight increase in K_m (i.e., inverse acclimation) when the plants were grown at a temperature considerably different from the normal habitat temperature. A few cases of thermal acclimation of the temperature dependence have been reported in fish (4). However some fish proteins appear not to respond to change in growth temperature (11).

It is evident from field data and controlled environment studies that in some plant species quantitative shifts in many proteins can occur in response to altered growth temperature (12, 13). Some proteins are detected only in certain restricted, usually very high or very low, temperatures and appear to be specific to a rather restricted set of growth conditions (12). The same pattern is seen for isozyme bands of a single enzyme, quantitative changes being elicited in most bands by simulated seasonal changes in temperature, and photoperiod, and some isozymes being expressed in restricted thermal ranges only (12).

In *Nicotiana tabacum,* DeJong (12) has demonstrated for malate (NAD) dehydrogenase from leaf tissue that either photoperiod or growth temperature can alter the relative staining intensity of the isozyme bands and the mobility of some of the bands. In addition, when grown at 8 h day length, one isozyme band was detected only at a high daytime growth temperature (35°C) and another band, which was detected at daytime temperatures of 15 and 25°C, was not evident at 35°C. When the plants were grown at 25°C and day length was varied, one band was detected only in long days of 18 h and not at 12 or 6 h day length. All the 7 MDH isozyme bands showed quantitative differences in staining intensity in response to changing temperature and day length. However several studies indicate that the degree of susceptibility of isozyme bands of a particular enzyme to environmental modulation is under genetic control. And in some species there is evidence of no quantitative or qualitative response of isozymes or K_m to a wide range of growth temperatures (3, 11).

Several kinds of evidence suggest that at least some thermal acclimation events are accompanied by concomitant changes in protein synthesis and degradation. Das and Prosser (14) measured the temperature response of ^{14}C-leucine incorporation into proteins of goldfish acclimated to high and low temperatures. When measured at low temperature (5°C) there was a greater maximal net synthesis of protein in tissues of low temperature (5°C) acclimated animals than in animals acclimated to 25°C. Depending on the particular tissue, these differences did not exist or were much smaller when both acclimation groups were measured at 25°C (14). There have been a number of observations of increased total protein content in vascular plants in response to low temperature (13, 15). In several angiosperm species proteins have been isolated from cold-hardened tissue, and these proteins have been demonstrated to protect thylakoids against frost damage (15). The protective proteins were most abundant in cold-hardened tissue. The time course of changes induced by low temperature in protein electrophoretic patterns has been studied for *Cynodon* species both in the field and in a controlled environment chamber (13). The appearance of new soluble protein bands, and an apparent overall increase in total soluble protein concentration, were strongly correlated with increased tolerance to subfreezing temperature. All these changes were only weakly expressed after exposure for 15 days to low temperature in a growth chamber, but by 30 days the chamber plants had become comparable to fully winter-hardened plants in the field. Sagisaka (16) similarly reported changes in amino acid metabolism induced by low temperature over a period of about one month.

The relationship between temperature-induced changes in tissue protein content and protein adaptation to altered temperature remains poorly understood. There may be several different mechanisms of protein adaptation at work in cases where changes in content occur. Some thermally induced proteins may function as protective agents (see, e.g., Ref. 15) at temperature extremes. In other cases the change in protein content may be directly related to altered protein functioning at the new temperature (e.g., temperature-induced isozymes of MDH) (4, 12). In some cases the enzyme apparently undergoes no structural or functional change in response to altered environmental temperature, as has been seen for MDH in some (3, 11), but not all, studies (12). Where no changes are seen, the population of molecules of a particular enzyme is presumed to have a fixed set of properties and, whenever temperature permits, the protein is capable of functioning.

5. SUMMARY

The data presently available concerning the shape of the response curve of K_m to assay temperature suggest that the degree of response is an adaptation to fluctuation of temperature. In the species studied, relatively low K_m values are observed over the range of normal tissue temperatures. There is at present no evidence that mature nonflowering plants have a different set of isozymes expressed during either the cool or the warm part of the day, or during brief periods of unusually warm or cool weather lasting only a few days during the growing season. There is evidence in *N. tabacum* and *Cynodon* species that there may be a change in protein structure and function at different seasons of the year. This alteration appears to be more highly correlated with resistance to injury induced by low temperature than to an acclimation *per se* of metabolic function. The available data from ectothermic animals and higher plants indicate that there is considerable variability among species in the thermal dependence of enzyme function. This variability may reflect different means of achieving metabolic adaptation to temperature fluctuation.

For the plant populations studied there is now evidence that within a species, genetic differentiation has occurred among different populations with regard to the sensitivity of K_m to change in temperature. The minimum apparent K_m of a particular population is usually measured near the mid-range of the native habitat temperature range that occurs during the growing season. In the three plant species studied, the absolute minimum apparent K_m for an enzyme is lower in populations native to cool habitats than in those native to warm habitats. Over the range of habitat temperatures likely to occur during growth, the apparent K_m of all studied populations varies less than two fold. This degree of relative thermal insensitivity may be related to the insensitivity of growth over the same habitat temperature range. The absolute maximum apparent K_m values are measured at high temperatures in plants native to cool habits, indicating a severe loss of catalytic efficiency at high temperatures. The study of the mechanism of environment-genotype interaction has only recently begun to focus on the functional properties of enzymes.

The data presented in this brief review suggest a number of possible functional adaptations of enzymes to environment. Many other possible interactions were not considered. It appears that investigations of the mechanisms of enzyme-environment interaction can offer considerable insight into the means by which plants are adapted to variability in environmental temperature. Such information may ultimately permit an increased ability to match genetically the temperature insensitivity of crop plants to the range of normal, as well as extreme, temperatures that occur at different locations on earth.

ACKNOWLEDGMENTS

I thank Drs. M. Feder and M. Peet for helpful discussion, Dr. J.-P. Simon for providing information on *Lathyrus japonicus,* and L. Strang and T. Teeri for technical assistance. This research was supported in part by National Science Foundation grant GB-41837, the Southeastern Plant Environmental Laboratories (NSF grant DEB 76-04150), and the DeKalb Foundation.

REFERENCES

1. W. M. Hiesey, *Am. J. Bot.,* **40,** 205 (1953).
2. J. A. Teeri, in H. H. Shugart, Ed., *Time Series and Ecological Processes,* SIAM, Philadelphia, 1977, p. 3.
3. J. A. Teeri and M. M. Peet, *Oecologia,* **34,** 133 (1978).
4. P. W. Hochachka and G. N. Somero, *Strategies of Biochemical Adaptation,* Saunders, Philadelphia, 1973.
5. R. Levins, *Evolution in Changing Environments,* Princeton University Press, Princeton, New Jersey, 1968.
6. G. N. Somero, *J. Exp. Zool.,* **194,** 175 (1975).
7. S. G. McNaughton, *Am. Nat.,* **106,** 165 (1972).
8. J. R. Hazel and C. L. Prosser, *Physiol. Rev.,* **54,** 620 (1974).
9. G. N. Somero, in C. L. Markert, Ed., *Isozymes,* Vol. II, *Physiological Function,* Academic Press, New York, 1975, p. 221.

10. G. N. Somero and P. S. Low, *Am. Nat.*, **111**, 527 (1977).
11. F. R. Wilson, G. S. Whitt, and C. L. Prosser, *Comp. Biochem. Physiol.*, **46B**, 105 (1973).
12. D. W. DeJong, *Am. J. Bot.*, **60**, 846 (1973).
13. D. L. Davis and W. B. Gilbert, *Crop Sci.*, **10**, 7 (1970).
14. A. B. Das and C. L. Prosser, *Comp. Biochem. Physiol.*, **21**, 449 (1967).
15. U. Heber and K. A. Santarius, in H. Precht, J. Christophersen, H. Heusel, and W. Larcher, Eds., *Temperature and Life*, Springer-Verlag, Berlin–Heidelberg–New York, 1973, p. 232.
16. S. Sagisaka, *Plant Physiol.*, **53**, 319 (1974).
17. J.-P. Simon, *Plant, Cell Environ.*, **2**, 23 (1979).

17 | Membrane Properties in Relation to the Adaptation of Plants to Temperature Stress

JOHN K. RAISON

Plant Physiology Unit, Division of Food Research, CSIRO, Sydney, New South Wales, Australia

JOSEPH A. BERRY and PAUL A. ARMOND

Department of Plant Biology, Carnegie Institution of Washington, Stanford, California

CARL S. PIKE

Department of Biology, Franklin and Marshall College, Lancaster, Pennsylvania

1. INTRODUCTION

Membranes appear to play a key role in both high and low temperature stress. At low, nonfreezing temperatures many plants, particularly those of tropical origin, experience metabolic dysfunction that leads to cell death after prolonged exposure (1). This syndrome is termed chilling injury, and it has many symptoms: leakage of ions from cells (2), accumulation of metabolic intermediates, such as ethanol and acetaldehyde (3), bleaching of chlorophyll (4), and loss of membrane integrity (4). All these symptoms

261

appear to be the result of a low-temperature-induced change in membrane structure. Many physiological processes of plants are also inhibited by high temperature (5). It is well known that proteins are denatured by high temperature (6), and there is evidence that high temperature might also disrupt membrane structure. For example, the irreversible inhibition of photosynthesis at high temperature correlates with the loss of photosystem 2 activity (7), the loss of photophosphorylation (8), and the change in the organization of chlorophyll in the membrane (9, 10): see also Chapter 15. The disruption of all these membrane-associated reactions occurs at about the same temperature in a given plant (5, 11). It seems reasonable, therefore, to postulate that all these high-temperature-induced effects are related to a perturbation of membrane structure and function by heat.

This chapter attempts to draw together some concepts of membrane structure, some information on the thermodynamics of the interactions that are thought to determine or maintain membrane structure, and some studies on the relationship between structure and function of biological membranes, in order to propose a hypothesis that deals with the disruptive effects of both high and low temperature on membranes. This model focuses on the molecular interactions between membrane components as they are affected by temperature. It is assumed that the manifold symptoms of chilling or heat damage to the intact plant are direct or indirect results of structural changes in the membrane; however these relationships are not considered in the model.

2. MEMBRANE STRUCTURE

The most commonly accepted structure of a biological membrane is that of a fluid, lipid bilayer with proteins either partially or fully embedded in the matrix (12, and Figure 17.1). This model was developed mainly from a

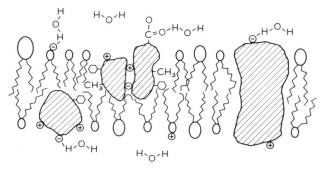

Figure 17.1. A schematic cross-sectional view of a membrane. The polar lipids are represented by the ovals (polar head groups) and wiggly lines (fatty acid chains). Integral proteins (shaded) appear either as a single unit or as an aggregate of two subunits. Hydrophilic regions of the protein and the polar head groups of the lipids are represented by the charged groups ⊕ and ⊖. The hydrophobic regions of the protein are represented by the —CH₃ and -○ groups.

consideration of the thermodynamic consequences of dispersing amphiphilic lipids in an aqueous environment. The structure is both stable and fluid.

Stability is a consequence of sequestering the hydrophobic alkyl chains of the membrane lipids to minimize the interactions of hydrophobic groups with water (Figure 17.1). In addition, the model postulates that hydrophilic interactions are maximized because the polar ionic and zwitterionic groups are in direct contact with the aqueous environment on either side of the bilayer.

Fluidity of the lipid bilayer is a consequence of the lack of strong attractive forces between the alkyl chains (13). The thermodynamic driving force for the formation of a bilayer and for the insertion of proteins into the bilayer is "a repulsive force acting on the separated elements [the aqueous and lipid phases] and not a preferential attraction" (13). When membrane lipids are in a fluid phase, there is no preference for any particular neighbors, and the bilayer can be formed equally well from mixed or pure amphiphilic lipids. Also, any large molecule, such as a protein, with a hydrophobic surface region can be incorporated into the bilayer without disrupting the stability. To this extent the lipid bilayer is a solvent for other lipophilic substances.

The three-dimensional structure of a protein is determined by the amino acid sequence and the environment. In an aqueous environment hydrophilic groups would be preferentially exposed at the surface of the protein, whereas in a lipid environment hydrophobic surface groups would be preferred. Some membrane proteins have portions of their amino acid sequences rich in hydrophobic side chains (14). Such molecules would probably fold so that a substantial part of the surface is hydrophobic, and such regions would associate with hydrophobic regions of the lipid bilayer. Hydrophilic surface regions could be on the portion of the protein exposed to the aqueous environment or to the polar regions of the lipids. Also, a hydrophilic group on the surface of one protein could bond to a hydrophilic group on the surface of another protein, thus linking two subunits by a hydrophilic bond in the hydrophobic interior of the bilayer. The folding of a protein may also allow some hydrophilic groups to bond to one another and form the core of the protein molecule (15). Proteins that are associated with membranes are amphiphilic, and it is reasonable to suppose that the structural arrangements of the proteins in the membrane are determined by bulk phase interactions of the protein with the bilayer. These interactions of proteins and lipids within a membrane are shown schematically in Figure 17.1.

The strength of the hydrophobic and hydrophilic interactions should determine the extent to which the protein is embedded in the membrane or exposed to the aqueous phase and the extent of the interaction of specific protein subunits. Also, the balance of these interactions should have more subtle effects on the association of protein subunits and on protein conformation. The catalytic function of the protein might therefore be affected if the balance between the forces acting on it were to change as a result of temperature change.

Studies of the temperature dependence of hydrophobic and hydrophilic interactions in defined chemical systems show that the strength of hydrogen bonds and electrostatic interactions within the water phase decrease as temperature increases (16). These are the forces that contribute to the hydrophilicity of a compound. In contrast, the strength of the repulsive interaction of hydrocarbon chains with water increases with increasing temperature. An amphiphilic substance should therefore associate more strongly with the lipid regions of the membrane at high temperature and with the aqueous phase at low temperature.

3. THE HYPOTHESIS

Extrapolating these arguments to proteins in the lipid bilayer of a membrane, it seems likely that in response to a temperature shift the proteins might move vertically in the plane of the bilayer: toward the aqueous interface at low temperature and toward the interior of the bilayer at high temperature. This type of movement would explain the instability of membrane proteins at both high and low temperatures, since under both conditions there would be considerable alteration in the forces maintaining the tertiary structure of the protein. Furthermore, a decrease in the strength of hydrophilic attractions at high temperature would weaken the forces linking the subunits of oligomeric membrane proteins.

4. STUDIES ON BIOLOGICAL MEMBRANES

Studies of temperature-induced changes in the thylakoid membranes of the blue-green alga *Anacystis nidulans* and of chloroplasts of the higher plant *Nerium oleander* have been conducted. We examined the effect of temperature on the physical properties of the lipid portions of the membranes, and we relate these studies to membrane structure in the case of *A. nidulans* or membrane function in the case of *N. oleander*. These studies provide some support for the hypothesis above.

Studies of the lipid properties were conducted using molecular probes containing a nitroxide spin label. Trace quantities of the probes were added to multilayer vesicles prepared by dispersing the extracted membrane polar lipids (galacto-, sulfo-, and phospholipids) in an aqueous buffer. Electron spin resonance spectra can be analyzed to determine either the partitioning of the probe between the aqueous and lipid phases or the motion of the probe, which is an indication of the fluidity of membrane lipids (17).

Figure 17.2 shows the effect of temperature on the partitioning of the amphiphilic spin label probe 2N8 between the lipid and aqueous phases of polar lipids from *A. nidulans* grown at 38°C. The fraction f of the spin label that is associated with the lipid phase increases with temperature. The change in partitioning is essentially due to a change in the balance between

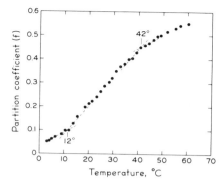

Figure 17.2. The partitioning of an amphiphilic spin label between the lipid and aqueous phases of a vesicle preparation as a function of temperature. Polar lipids were obtained by chromatography, using silicic acid, from a total lipid extract of the thylakoid membranes of *Anacystis nidulans* grown at 38°C. The lipids were dispersed (10 mg/ml) in 0.01 *M* Tris acetate buffer, pH 7.2, containing 5 m*M* EDTA, by brief sonication. A portion of this suspension (0.1 ml) was mixed with 0.1 nmol of the spin label 3-oxazolidenyloxy-2-methyl-2-octyl-4,4-dimethyl (2N8) and heated at 40°C for 5 min. The partitioning of the spin label between the lipid and aqueous phases was determined by electron spin resonance spectroscopy. Partitioning is expressed as *f*, the amount of label in the lipid phase divided by amount of label in both the lipid and aqueous phases.

the hydrophobic and hydrophilic interaction of that molecule with the lipid or the water phases and indicates that a net increase in the hydrophobic interactions between the probe and the membrane lipids occurs with increasing temperature, as predicted by the chemical studies. Furthermore, the distribution of the probe between lipid bilayer and aqueous phase is not a continuous function with temperature. Changes in the slope of *f* as a function of temperature at about 12 and 42°C indicate changes in molecular ordering; these temperatures are probably the limits of a phase transition for these lipids. In this instance the lipids are predominantly fluid above 42°C and predominantly solid below 12°C.

In addition to the spin label studies, fluorescence intensity measurements with the polyene fatty acid probe *trans*-parinaric acid provide information on lipid phase changes (18). As shown in Figure 17.3*a*, using membrane phospholipids there were marked changes in the slope of intensity as a function of temperature at about 12 and 42°C, in agreement with the spin label measurements on polar lipids. Fluorescence polarization measurements with *trans*-parinaric acid indicate the fluidity (19) of the *A. nidulans* phospholipid vesicles (Figure 17.3*b*). There was a marked change in the slope of the polarization ratio of the phospholipids as a function of temperature at about 42°C, indicative of a phase change, in agreement with the other measurements. The polarization ratio of *trans*-parinaric acid reaches a maximum

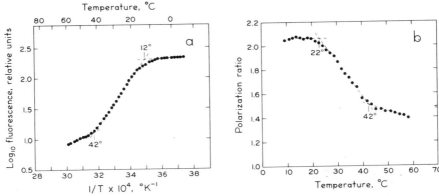

Figure 17.3. The effect of temperature on (*a*) the fluorescence intensity and (*b*) the polarization ratio of *trans*-parinaric acid in phospholipids from the thylakoid membranes of *Anacystis nidulans* grown at 38°C. A total lipid extract of the membranes was chromatographed on a silicic acid column. Phospholipids were obtained after sequential elution of pigments and galactolipids. The phospholipids were dispersed in 0.01 M Tris acetate buffer, pH 7.2, containing 5 mM EDTA and 25% v/v ethylene glycol, by brief sonication. For fluorescence analysis a suspension containing 400 μg of lipid per 3 ml of buffer, was labeled with 2.5 nmol of *trans*-parinaric acid. The sample was excited at 320 nm with polarized light, and fluorescence emission was measured at 420 nm, both perpendicular (I_1) and parallel (I_2) to the excitation light. The fluorescence intensity is I_2 and the polarization ratio is I_2/I_1.

when about 50% of the lipids are in the gel state (19). Here, this point occurs at about 22°C, several degrees higher than the 12°C at which the spin label studies suggest as the completion of the transition to the gel state.

Several thylakoid membrane functions were studied in *A. nidulans* grown at 38°C. Each function showed a characteristic increase in apparent activation energy below about 20 to 25°C (20). These changes thus occur within the temperature range of the phase transition.

When a phase change occurs there is an alteration in the molecular spacing and interactions of the fatty acid chains as seen by X-ray diffraction (21), by measurement of heat of fusion with differential scanning calorimetry (22), and by the sharp change in the temperature coefficient of motion of membrane lipids observed with both spin (17) and fluorescent (18, 19) probes. Because of the considerable heterogeneity of membrane lipids, the transition typically occurs over several degrees. Below the upper temperature limit of this phase transition there is a lateral separation of the fluid and gel phase lipids (23). This change in the molecular ordering of membrane lipids also apparently affects the strength of the hydrophobic and hydrophilic forces acting on the probe. If membrane proteins are also amphiphilic molecules, the hydrophilic and hydrophobic interactions of the protein with the lipid bilayer should change with temperature in much the same way as was observed with the spin label probe.

The changes in membrane structure that might be caused by these changes in the interactions that affect membrane proteins can be observed using the technique of freeze-fracture electron microscopy. Since the membrane is fractured along its hydrophobic zone, details of the internal membrane structure are revealed by this technique. Integral membrane proteins are seen as particles against the background matrix of lipids.

Examination of the protein particle distribution in the thylakoid membranes of *A. nidulans* confirms that shifts in temperature induce major changes in both the lipid-lipid and lipid-protein interactions in biological membranes. As shown in Figure 17.4*a*, membranes prepared and fixed at 38°C for freeze-fracture show an even distribution of protein particles with no extensive areas devoid of particles. Membranes fixed at 21°C show large particle-free regions (Figure 17.4*b*), and these regions are even more extensive in membranes fixed at 10°C (data not shown). The extent of these areas in thylakoid membranes, determined by analysis of many freeze-fracture electron micrographs, is shown in Table 17.1. At 21°C about 67% of the membrane is free of particles. This value is similar to the proportion of gel phase lipids, present at this temperature, determined by both *trans*-parinaric acid fluorescence polarization and spin label partitioning.

These observations are thus consistent with a low-temperature-induced lateral phase separation of lipids with the particle-free areas representing the gel phase lipids. The freedom from visible particles of the gel phase areas indicates that the phase change results in a major alteration in the protein-

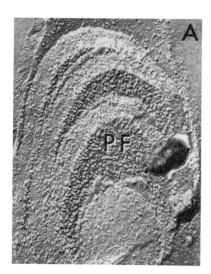
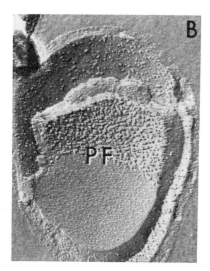

Figure 17.4. Fracture faces of thylakoid membranes of *Anacystis nidulans* grown at 38°C (× 42,000): (*a*) Membranes fixed with glutaraldehyde at 38°C before freeze-fracture. The PF fracture face (24) shows an even distribution of particles. (*b*) Membranes fixed at 21°C. The PF fracture face shows large particle-free regions clearly delineated from the particle-containing regions.

**Table 17.1. Particle Density on Fracture
Faces and Percentage of Particle-Free Areas
of Thylakoid Membranes of *Anacystis nidulans*
Grown at 38°C and Fixed at Three
Temperatures**

Values of particles/μm^2 are means ± one standard
deviation.

Fixation temperature (°C)	Particles/μm^2	Particle-free regions (%)
38	3693 ± 190	0
21	3804 ± 169	67
10	4278 ± 235	77

lipid interactions. The very small increase in particle density in fracture faces containing particles cannot account for the large increase in particle-free area at low temperatures and is inconsistent with a lateral movement of protein particles with the fluid phase lipids. The observations can, however, be explained by a vertical movement of some protein particles. These movements are indicated schematically in Figure 17.5 (24).

A vertical movement of protein particles is consistent with a decrease in the strength of the hydrophobic interactions and an increase in the strength of the hydrophilic interactions. Under these conditions the proteins capable of moving in the vertical plane would move toward the aqueous interface. If there is substantial lateral movement of protein particles with the fluid phase lipids, there must be some compensating change such that the number of particles in the fracture plane remains constant (Figure 17.5).

The change in lipid-protein interactions observed with *A. nidulans* membranes may also occur in other membranes; however the appropriate freeze-fracture studies have not been conducted. It is clear from studies with isolated lipids that phase transitions similar to that observed with *A. nidulans* occur in higher plant membranes. As shown in Table 17.2 the temperature of the transition varies among plant species and, in the case of *N. oleander*, with growth temperature. *Vigna radiata*, *Zea mays*, and *Phaseolus vulgaris* are classic examples of chilling sensitive plants. In these species the lipids begin to gel at 10 to 13°C. These phase transition temperatures have been determined by electron spin resonance spectroscopy of spin label probes and by fluorescence polarization using *trans*-parinaric acid. The techniques give comparable results. These plants exhibit a number of symptoms of chilling injury when exposed to temperatures below the phase transition temperature. Also shown are data obtained with the thermophilic C$_4$ plant *Tidestromia oblongifolia*. This plant has a photosynthetic apparatus that is quite stable at high temperatures (Chapter 15); however the lipids of this plant do not appear to gel until below 0°C. This result indicates that there

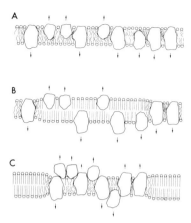

Figure 17.5. Model for the vertical displacement of membrane particles in *Anacystis nidulans*. The loss of particles from the fracture plane in the transition from a fluid phase (*a*) to a mixed-phase state (*b* and *c*) may involve different processes depending on whether the proteins remain in the less fluid portion of the membrane (*b*) or laterally migrate to the more fluid region of the membrane. If the proteins remain in their original position, integral membrane proteins that are not transmembranous (*b*, upper monolayer) could become undetectable at the fracture plane because of the change in bilayer thickness induced by the change of the physical state of the lipids. For such nontransmembranous proteins, approximately the same degree of surface exposure could be maintained. Transmembranous proteins would also require some increase in surface exposure to account for nondetectability at the fracture plane (*b*, lower monolayer). If the proteins are displaced to a more fluid region of the membrane (*c*), some proteins might be excluded from the fracture plane by the combined lateral pressure of the proteins and lipids. Integral membrane proteins may also partition between the two fracture faces in a different manner (arrows indicate possible fracture face affinities). The membrane models depicted in *b* and *c* are both consistent with areas of the membrane being devoid of particles, without substantial increases in the particle density of the remaining areas. From Ref. 24.

may be no necessary correlation between stability at high temperatures and sensitivity to low temperature chilling injury. *N. oleander,* an evergreen species that can be grown over a wide range of temperature, exhibits changes in the physical properties of its lipids in response to the contrasting thermal regimes of 20/15°C or 45/32°C day/night temperatures.

It is well known that certain bacteria can modify the lipids of their membranes in response to growth at different temperatures (25). The nearly complete adjustment of phase transition temperature and microviscosity with temperature has been termed "homeoviscous adaptation." The membranes of *N. oleander* do not change enough to result in complete homeoviscous adaptation, but they do change enough to partially compensate (perhaps one-third to one-half) for the difference in growth temperature.

For any given membrane the strength of the hydrophobic interactions

Table 17.2. Temperatures for the Thermal-Induced Structural Transition of Membrane Lipids from a Variety of Plants

The transition temperatures were determined from either spin labeling [lower temperature limit of the transition (17)] or fluorescence labeling [change in slope of a plot of the logarithm of intensity against reciprocal of absolute temperature (18)].

Species	Growth temperature (°C)	Transition temperature (°C)	
		Fluorescence	Spin labeling
Vigna radiata	27	—	15
Zea mays	30/20	12	13
Phaseolus vulgaris	30/20	11	10
Tidestromia oblongifolia	D.V.[a]	—	−1
Atriplex sabulosa	20/15	—	−1
Nerium oleander	20/15	−4	−5
Nerium oleander	45/32	7	4

[a]Grown in Death Valley, California. See Chapters 15 and 18 for temperature details.

would increase and that of the hydrophilic interactions would decrease at high temperature. The question may be asked: What effect would the changes in lipid properties of *N. oleander* induced by growth temperature (Table 17.2) have on the function of chloroplast membranes at high temperature?

An indication of the molecular organization of the chloroplast membrane can be gained from a measure of fluorescence yield of chlorophyll. Chlorophyll in solution is highly fluorescent. The fluorescence is quenched in the intact, undamaged chloroplast, because most of the excitation energy is trapped by the reaction centers. An increase in chlorophyll fluorescence suggests some disruption of the normally highly efficient transfer of energy in the chloroplast lamellae. As shown in Figure 17.6, for an intact leaf of oleander grown at 20/15°C the fluorescence yield increases at about 43°C, whereas for oleander grown at 45/32°C the increase occurs at 53°C. The temperature for the increase in chlorophyll fluorescence yield can be correlated with the temperature of irreversible damage to photosynthetic membranes. The result is a 10°C degree increase in the thermal stability of chloroplast membranes because of the increased growth temperature. We have examined some physical properties of membrane lipids of chloroplasts from these plants to determine whether there is a relationship between thermal stability and the properties of the lipids.

Figure 17.7 shows the motion of the spin label membrane probe 12NS in polar lipids from chloroplasts of *N. oleander* as a function of temperature. The motion of the probe, measured as the rotational time, decreases as temperature increases. Over the temperature range 20–50°C there is no

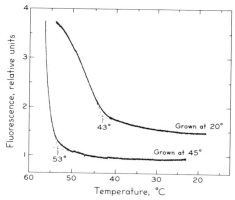

Figure 17.6. Change in the relative intensity of chlorophyll fluorescence in whole, attached *Nerium oleander* leaves as a function of temperature. Leaves from plants grown at the indicated temperatures were heated at about 1°C/min. The fluorescence from excitation by extremely low intensity light was measured continuously. The temperature of the fluorescence increase was determined as the point of intersection of the lines extending the two linear portions of the curves as shown.

change in the temperature coefficient of motion (the slope of the line), showing that there is no change in the molecular ordering of the lipids.

At a given temperature, the motion of the spin label in the polar lipids of *N. oleander* grown at 20°C is always faster than the motion in the lipids of the plant grown at 45°C. Since the motion of a spin label is related to membrane lipid viscosity (26), the membrane lipids of the 20°C grown plant are more fluid than the lipids of the plant grown at 45°C. If the thermal stability of the chloroplast membrane is related to membrane lipid fluidity, the fluorescence increase should occur at a temperature that corresponds to the same spin label motion and fluidity in the two lipid samples. As Figure 17.7 indicates, this is what is observed. For lipids from a plant grown at 20/15°C the motion at 43°C (the threshold for thermal damage) is 8.7×10^{-10} s, and for plants grown at 45/32°C the motion at 53°C is 8.8×10^{-10} s. A similar relationship is also observed with *Atriplex lentiformis* grown at different temperatures (results not shown). From these data it is concluded that the thermal stability of these photosynthetic membranes is related to the physical properties of the membrane lipids.

The fluorescence increase observed at high temperature is attributed to a dissociation of light-harvesting pigment-protein complexes from reaction center protein. In terms of the original model this dissociation would be expected if the forces linking these two proteins in the membrane were hydrophilic. With increasing temperature, the strength of the hydrophilic interactions decreases while that of hydrophobic interactions increases; thus proteins tend to associate more with the lipids (by hydrophobic interactions) than with one another (by hydrophilic interactions). Therefore the distance

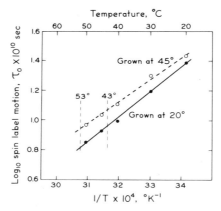

Figure 17.7. The motion of a spin label incorporated into the polar lipids of chloroplast membranes from *Nerium oleander* as a function of temperature. Plants were grown at 45/30 or 20/15°C day/night temperatures. The lipids were suspended as described for Figure 17.2 and labeled with 3-oxazolidinyloxy-2-(10-carbmethoxy decyl)-2-hexyl-4,4-dimethyl (12NS). Motion was calculated as described by Raison and Chapman (17) from five spectra at each temperature. The standard deviation for motion at 20°C was $\pm\ 0.2 \times 10^{-10}$ s, and at 50°C was $\pm\ 0.05 \times 10^{-10}$ s for both plants.

between the light-harvesting pigment protein and the reaction center protein will probably increase. Assuming no change in the amino acid composition of these two proteins, the loss of functional integrity at a common membrane lipid viscosity suggests that this property of the lipids is a useful indicator for assessing the critical point for the balance of hydrophobic and hydrophilic interactions in the membrane.

5. CONCLUSION

The loss of physiological function of membranes at both high and low temperatures can be explained in terms of a model based on a balance between the relative strengths of hydrophobic and hydrophilic interactions among proteins, lipids, and the aqueous environment. The predictions of the model are consistent with physical measurements of the properties of membrane lipids by three different techniques.

The model developed from physical studies of hydrophobic interactions as a function of temperature in *A. nidulans* can explain the breakdown in the functional properties of *N. oleander* membranes at high temperature in terms of molecular organization. Lipid viscosity can be used to predict the threshold temperature for thermal breakdown of chloroplasts of plants of the same species grown at different temperatures.

REFERENCES

1. J. M. Lyons, *Annu. Rev. Plant Physiol.*, **24**, 445 (1973).
2. A. Minchin and E. W. Simon, *J. Exp. Bot.*, **24**, 1231 (1973).
3. T. Murata, *Physiol. Plant.*, **22**, 401 (1969).
4. A. O. Taylor and A. S. Craig, *Plant Physiol.*, **47**, 719 (1971).
5. O. Björkman, J. Boynton, and J. Berry, *Carnegie Inst. Washington Yearb.*, **75**, 400 (1976).
6. V. Y. Alexandrov, *Cells, Molecules and Temperature*, Springer-Verlag, Berlin–Heidelberg–New York, 1977.
7. R. W. Pearcy, J. A. Berry, and D. C. Fork, *Plant Physiol.*, **59**, 873 (1977).
8. Y. Mukohata, in N. Nakao and L. Packer, Eds., *Organization of Energy Transducing Membranes*, University Park Press, Baltimore, 1973, p. 219.
9. U. Schreiber and P. A. Armond, *Biochim. Biophys. Acta*, **502**, 138 (1978).
10. U. Schreiber and J. A. Berry, *Planta*, **136**, 233 (1977).
11. P. A. Armond, U. Schreiber, and O. Björkman, *Plant Physiol.*, **61**, 411 (1978).
12. S. J. Singer and G. L. Nicolson, *Science*, **175**, 720 (1972).
13. C. Tanford, *Science*, **200**, 1012 (1978).
14. Y. Nakashima and W. Konigsberg, *J. Mol. Biol.*, **88**, 598 (1974).
15. S. J. Singer, in L. Bolis, K. Bloch, S. E. Luria, and F. Lynen, Eds., *Comparative Biochemistry and Physiology of Transport*, North-Holland, Amsterdam, 1974, p. 95.
16. D. Oakenfull and D. E. Fenwick, *Aust. J. Chem.*, **30**, 741 (1977).
17. J. K. Raison and E. A. Chapman, *Aust. J. Plant Physiol.*, **3**, 291 (1976).
18. L. A. Sklar, B. S. Hudson, and R. D. Simoni, *Proc. Nat. Acad. Sci. (US)*, **72**, 1649 (1975).
19. L. A. Sklar, G. P. Miljanich, and E. A. Dratz, *Biochemistry*, **18**, 1707 (1979).
20. N. Murata, J. H. Troughton, and D. C. Fork, *Plant Physiol.*, **56**, 508 (1975).
21. G. G. Shipley, in D. Chapman and D. F. H. Wallach, Eds., *Biological Membranes*, Academic Press, London–New York, 1973, p. 1.
22. D. Chapman, in D. Chapman and D. F. H. Wallach, Eds., *Biological Membranes*, Academic Press, London–New York, 1973, p. 91.
23. W. Kleemann, C. W. M. Grant, and H. M. McConnell, *J. Supramol. Struct.*, **2**, 609 (1974).
24. P. A. Armond and L. A. Staehelin, *Proc. Nat. Acad. Sci. (US)*, **76**, 1901 (1979).
25. M. Sinensky, *Proc. Nat. Acad. Sci. (US)*, **71**, 522 (1974).
26. A. D. Keith and W. Snipes, in H. A. Resing and C. G. Wade, Eds., *Magnetic Resonance in Colloid and Interface Science*, American Chemical Society, Columbus, Ohio, 1976, p. 426.

INTERACTION AND INTEGRATION OF ADAPTATIONS TO STRESS

V

In Parts I through IV the authors have been concerned with individual mechanisms of adaptation to stresses induced by water or by high temperature. In nature many mechanisms may play a part in the successful adaptation of a plant to its environment. As pointed out in Part I, water stress is usually accompanied by high temperatures; thus for plants to survive in arid regions, they must be able to withstand simultaneously both water deficits and high temperatures. The first two chapters in Part V discuss the mechanisms that play a part in the adaptability to water and high temperature stress of a desert environment.

In addition to the seasonally high temperatures and seasonally low availability of soil water, arid regions frequently have soils of poor nutrient status. Because water is the primary limiting resource, however, adding nutrients, particularly nitrogen, to crops in arid or semiarid regions can lead to reduced yields (1, 2). The nutrients promote vegetative growth, increase water use in the vegetative phase, and may deplete the water for the reproductive phase. Nevertheless, under a slightly more favorable soil water regime, nutrients may increase yield in otherwise similar environments. This important interaction between nutrients and water in arid and semiarid environments is the subject of Chapter 22.

The ultimate integration of all the adaptive mechanisms of a plant in a stressful environment is the yield and productivity in this environment. The influence of stress on the productivity and yield of plants is a feature of two of the chapters in this part. One discusses limitations to crop yield resulting from both water and high temperature stress in semiarid regions, and the second covers similar limitations to productivity arising from water stress in humid regions. Those familiar with only arid and semiarid regions tend to overlook the ability of water stress to limit productivity in temperate and tropical climates. Chapter 20 in this part and Chapter 26 in Part VI highlight the importance of water stress in crops and plants of the humid tropics.

Describing adaptation, Marshall wrote, "Adaptations are, in part, how the biological component of the ecosystem produces otherwise unpredictable results and makes the biological component of the ecosystem less easily understood and less frequently modeled than the physico-chemical component'' (3). Since adaptation is a characteristic of biological systems, incorporation of mechanisms of adaptation into models of natural and man-modified ecosystems is desirable and necessary. Chapter 23 describes such a model, albeit simple, that can be used to answer questions on the role of stomatal control of water loss in the adaptation to specified environments.

REFERENCES

1. K. P. Barley and N. A. Naidu, *Aust. J. Exp. Agric. Anim. Husb.*, **4**, 39 (1964).
2. R. A. Fischer and G. D. Kohn, *Aust. J. Agric. Res.*, **17**, 281 (1966).
3. J. K. Marshall, in J. K. Marshall, Ed., *The Belowground Ecosystem: A Synthesis of Plant-Associated Processes*, Dept. Range Sci. Ser. No. 26, Colorado State University, Fort Collins, 1977, p. 349.

18 | Seasonality and Gradients in the Study of Stress Adaptation

H. A. MOONEY

Department of Biological Sciences, Stanford University, Stanford, California

1. INTRODUCTION

To a large extent our knowledge of stress physiology has been derived from the study of plants that have direct economic importance. In fact, a great number of these plants belong to a single ecological group, the annuals. This has resulted in a somewhat narrow view of the multiplicity of adaptive characteristics that plants have evolved to cope with environmental stress.

In recent times in the study of environmental physiology there has been an increase in utilization of plants that originate from a variety of habitat types. This trend is likely to continue as we search for genetic stock that may have novel mechanisms for optimizing production under limiting water and nutrient regimes. Here I discuss approaches to the study of the mechanisms that native plants have evolved to cope with stress. In particular I treat the utilization of seasonality and environmental gradients for the study of the stress responses of plants in natural systems.

We can evaluate stress responses of plants in natural systems in terms of the reduction of potential production. Without stress, production should continually reach the theoretical limit. In nature, however, progression from favorable to unfavorable seasons in a single locality, or from benign to severe or stressful environments results in a reduction in plant productivity. This chapter indicates how we can use both seasonality and gradients to

279

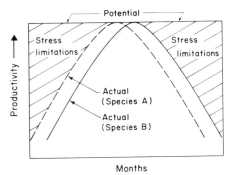

Figure 18.1. Potential and actual production of two co-occurring species. Potential production assumes no environmental limitations to production and a constant photoperiod. Various seasonal environmental limitations result in the actual production. As is characteristic of many communities, the component species are offset somewhat in their productive periods. Stress adaptations can be elucidated by comparing these offset species as well as by studying the seasonal response of either.

determine what particular mechanisms plants have evolved to cope with stress in both the short and long terms.

Virtually all organisms live in habitants in which environmental conditions change from less to more stressful during the course of the year (Figure 18.1). Studies of the seasonal metabolic response of an organism in its natural habitat, as well as comparisons of responses of co-occurring species, can be of considerable value in understanding the mechanisms by which plants adjust to stress.

In laboratory studies of stress physiology we generally consider a single species or genotype, then experimentally vary a single component of the environment. The effect of varying the environment, or inducing stress, can be quantified clearly in terms of any of a large number of metabolic responses of the test organism. In field studies, however, analysis is complicated by the possibility that two or more stress factors will vary simultaneously. Experimental manipulations can be used under field conditions to simplify analyses to a certain degree. For example, in environments where both water and temperature stress increase simultaneously, irrigation can remove one of these stresses. Generally, though, field studies are most useful in pointing to the most profitable direction and material for complementary studies under controlled environmental conditions.

The other research approach I draw from is the use of environmental gradients for the elucidation of the mechanisms that plants have utilized to enhance or maintain production under stress conditions.

Primary productivity is, of course, not uniformly high in all the world's ecosystems. Progressing along gradients, starting at the tropics, productivity decreases as temperatures become cooler or as moisture becomes less. It is obvious that the steepness of the decline in productivity would be much greater if there were no species replacements along these gradients (Figure 18.2). That is, along such gradients species evolve to use most efficiently the resources (or to operate under the stresses) that exist at any given point. Thus by examining replacements of ecotypes or species along such gradients, we can perceive the evolutionary options for stress adaptation.

2. ADAPTATIONS TO CHANGING SEASONS

Plants have evolved various physiological and morphological mechanisms to maintain productivity when the principal seasonally limiting factors are high temperature and low moisture. This section gives examples from our studies in Death Valley of some of these.

Death Valley is noted for its unusually hot summers. Daytime maxima can exceed 40°C for months on end. In the winter daytime maxima drop to below 20°C (Figure 18.3). Thus evergreen plants that inhabit this desolate region experience a wide range of temperatures. At different times of the year in Death Valley the mean daily maximum temperature is equivalent to June temperatures of localities ranging from coastal to interior California (Figure 18.3). It is only in the midsummer months that the climate is truly distinctive. Although annual rainfall is low on the valley floor, it can occur at any month of the year. Because of watershed effects, the amounts available in the soil after storms can be substantial in particular microsites. Thus Death Valley is a region of many climates, and in certain seasons the conditions for growth are similar to those found in localities not noted for climatic severity. How have the plants evolved to match the multiplicity of these environments?

In our studies of the photosynthetic capacity of Death Valley plants, we have noted what might be termed both "generalists" and "specialists." There are specialized species such as *Tidestromia oblongifolia*, which are

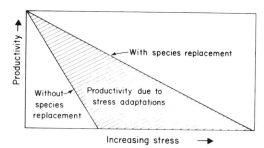

Figure 18.2. Theoretical effect of increasing environmental stress on community production with and without evolutionary adaptation or species (or ecotypic) replacements.

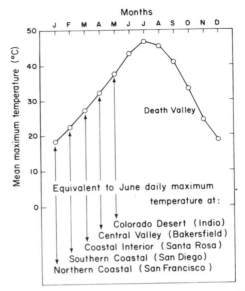

Figure 18.3. Mean maximum daily temperatures at Furnace Creek, Death Valley, California. The seasonal diversity of "growing season" climates in Death Valley is indicated by comparisons of June temperature values for other localities with equivalent months in Death Valley.

photosynthetically active only in the summer, and winter annual specialists such as *Camissonia claviformis*, which are present only in winter and spring (Figure 18.4). These species are thus nonoverlapping in the time they are photosynthetically active. As might be imagined, their thermal optima of photosynthesis are very different, with *C. claviformis* having an optimum near 20°C and that of *T. oblongifolia* exceeding 45°C (1).

Although these two species do not share activity times, they have a number of other properties in common. They have unusually high photosynthetic capacities and short leaf duration. These species are not subjected to great water stress under field conditions (Table 18.1) (this is particularly remarkable for *T. oblongifolia*, since it is active during a period of tremendous "evaporative demand"). Furthermore, the photosynthetic capacity of these species is greatly inhibited by water stress (Figure 18.5).

T. oblongifolia, the summer-growing herbaceous perennial, has evolved mechanisms to avoid moisture stress, such as establishment in favorable moisture microsites (2) and a system for rapid transport of water from the soil to the leaf so that water stress does not develop even under high evaporative demand (3). Although *T. oblongifolia* avoids water stress, it cannot avoid the high temperatures that occur during its growth period. Thus its principal stress-enduring mechanisms are related to efficient metabolic operation of high temperatures (4).

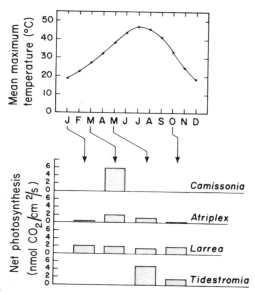

Figure 18.4. The seasonal maximum photosynthetic capacity for four species native to Death Valley. *Camissonia claviformis* is a winter annual, *Tidestromia oblongifolia* is a summer-active herbaceous perennial, and *Atriplex hymenelytra* and *Larrea divaricata* are evergreen perennials. Mean maximum daily temperatures are for Furnace Creek, Death Valley. From Ref. 1.

The two evergreen "generalist" species we have studied, *Larrea divaricata* and *Atriplex hymenelytra*, although photosynthetically active year-round have lower capacities at any given time than the specialists (Figure 18.4). *A. hymenelytra* and *L. divaricata* are both subjected to considerable water stress under natural conditions at some time during the year (Table 18.1). As would be expected, these species can tolerate considerable water stress (Figure 18.5) and, in fact, they apparently regulate their metabolism so that they operate at water potentials of −10 bars or so, even when fully watered.

It is tempting to speculate that the cost of being able to operate successfully at such low water potentials is the lack of capacity to do so with great efficiency at high potentials. An evolutionary argument has been proposed to explain this relationship (5). Plants that are subjected to environments where photosynthesis is normally diffusion-limited will maintain relatively low photosynthetic enzyme contents, hence will have comparatively low photosynthetic capacities.

Although both *A. hymenelytra* and *L. divaricata* are potentially photosynthetically active year-round if water is available, they face very different thermal regimes during the different seasons. There are a number of adaptive possibilities to adjust to changing temperatures or to avoid unfavorable

Table 18.1. Leaf Water Content, Dawn Leaf Water Potential, Mean Daily Leaf Osmotic Potential, and Mean Daily Leaf Turgor Potential of Four Death Valley Native Plants at Two Times of Year (From W. Bennert and H. A. Mooney, unpublished)

Species	Leaf water content (% dry weight)		Potentials (bars)						
			Total water		Osmotic		Turgor		
	March	August	March	August	March	August	March	August
Larrea divaricata	140	95	−11.0	−28.1	−29.6	−38.6	6.9	2.6
Atriplex hymenelytra	280	120	−14.9	−30.4	−32.9	−42.6	11.7	3.1
Camissonia claviformis	695	—	−4.1	—	−9.8	—	2.4	—
Tidestromia oblongifolia	—	290	—	−7.5	—	−19.5	—	5.6

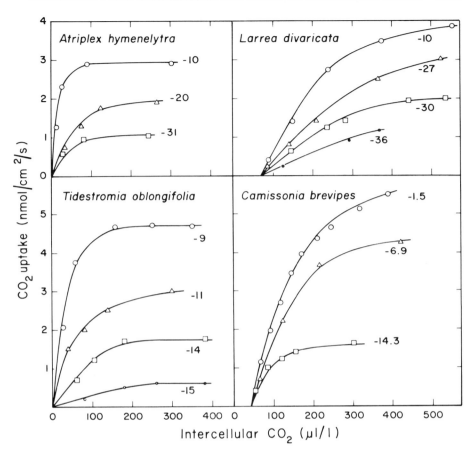

Figure 18.5. Relationship between net carbon dioxide (CO_2) uptake and intercellular CO_2 concentration at various leaf water potentials (in bars) of four species native to Death Valley. *Camissonia brevipes* is a winter annual, *Tidestromia oblongifolia* is a herbaceous perennial, and *Atriplex hymenelytra* and *Larrea divaricata* are evergreen perennials. Experiments were performed at 1600 $\mu E/m^2/s$ and 30°C on potted plants. Water stress was induced by withholding water. From Ref. 15 and unpublished data of Mooney.

temperatures (Figure 18.6). *C. claviformis*, for example, avoids high temperatures by growing in the winter and spring only. As pointed out in Chapter 15, *L. divaricata* adjusts to these different regimes by shifting its thermal optimum of photosynthesis (6). It can be shown that this shift in thermal optimum is adaptive by modeling the potential carbon gain of winter-acclimated plants to summer conditions, and vice versa (Figure 18.7). It should be noted that seasonal shifts in temperature-related photosynthetic optima are not necessarily always beneficial in terms of carbon gain (7).

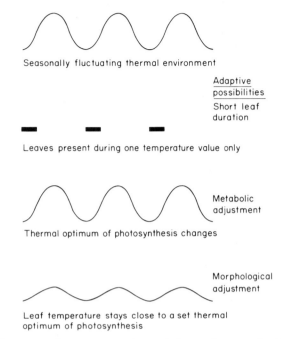

Figure 18.6. Diagrammatic representation of three photosynthetic adaptive possibilities in a thermally fluctuating environment.

A. hymenelytra does not show dramatic shifts in its temperature-related photosynthesis (6) but, like *Encelia farinosa* (8), utilizes changes in its leaf absorptive characteristics to maintain a leaf temperature that is near the thermal optimum of photosynthesis throughout the year. *E. farinosa* employs seasonally changing characteristics in surface hairs, and *A. hymenelytra* utilizes salt glands.

It is certainly evident that there are a great number of physiological and evolutionary options to cope with the stress conditions that occur in deserts like Death Valley, a few of which I have illustrated. Avoidance of stress is an evolutionary option that characterizes a number of Death Valley plants. The actual avoidance mechanisms are those that enable plants to gain sufficient carbon and to construct abundant seeds before the onset of stressful drought or high temperatures. The specific adaptive mechanisms are thus those relating to high carbon-gaining capacity, rapid developmental rates even at cool temperatures, and a trigger (probably a stress) that indicates at what time it is more beneficial in terms of ultimate reproductive output to switch from the carbon-gaining mode to the reproductive mode.

It is the evergreens such as *A. hymenelytra* and *L. divaricata* that have evolved mechanisms to operate under both temperature and water stress. The mechanisms to cope with temperature are primarily morphological in the case of *A. hymenelytra* and physiological in the case of *L. divaricata*.

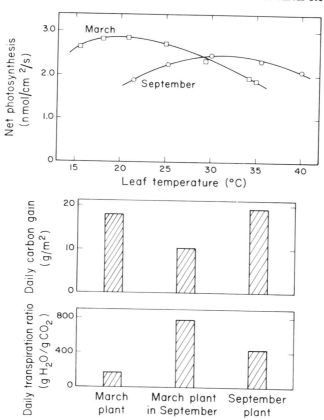

Figure 18.7. Temperature-related photosynthetic response of *Larrea divaricata* during March and September under water unlimited conditions in Death Valley, California (*top*) and the simulated daily carbon gain (*center*) and transpiration ratio (*bottom*) under the prevailing seasonal environmental conditions. The carbon gain and transpiration ratios are similar in both March and September, whereas an unacclimated plant, such as a plant with March photosynthetic characteristics under September climatic conditions, fixes less carbon than does an acclimated plant (*center*) and has a higher transpiration ratio (*bottom*). From Ref. 15 and simulations based on the photosynthetic model of Ehleringer and Miller (16).

It is likely, but not demonstrated, that the total productivity of the site is enhanced by mixtures of adaptive types, which together gather more of the available resources than would any individual adaptive type. Groups of species, growing in strongly seasonal climates, which have differing seasonal specializations, are no doubt more productive than groups without offset activities. This enhancement of site productivity is implied in Figure 18.1.

As an interesting footnote, it is most certainly true that desert plants such as those found in Death Valley have little competition for light. The striking offset in activity between species found there could be possible only under this circumstance. However it appears that there is competition for water among many species and, in fact, such competition has been proposed as a mechanism controlling spacing in such desert species as *L. divaricata* (9). It has been recently demonstrated that, indeed, competition for water exists between co-occurring *L. divaricata* and *Franseria dumosa* shrubs (10) as well as between *L. divaricata* shrubs (B. Mahall, personal communication).

In summary, our studies of the seasonal photosynthetic responses of a number of Death Valley plants have shown that species that co-occur in space do not necessarily do so in time. The mechanisms for coping with stress, or enhancing productivity, in a seasonally varying environment are numerous. Those species that endure the most extreme conditions apparently do so at the cost of faring relatively more poorly during the favorable seasons in comparison with species that avoid the stress periods.

3. GRADIENT

As noted earlier, the study of species replacement along environmental gradients can reveal evolutionary adaptive trends to particular stress conditions. This section describes the morphological and physiological trends found in species occupying habitats of varying aridity in Australia, California, and Chile.

3.1. Australia

As an example of the use of species replacements I utilize the genus *Eucalyptus*, which is remarkably diverse, containing more than 500 species. The distribution of the genus is centered in Australia, and it occurs in a wide range of climates, most of which have some seasonal arid periods. Although the genus is rich in species, the overall morphology of the plant is roughly constant. For example, the leaves of most of the species are clearly distinguishable as belonging to the genus *Eucalyptus*. Thus this genus would appear to be particularly well suited for studies of physiological mechanisms by which plants maintain productivity in progressively stressful environments.

We studied a series of *Eucalyptus* species in eastern Australia along a precipitation gradient from over 900 mm at the coast to less than 400 mm in the interior (11). At the moist end of the gradient the species are forest trees, and at the dry end, "mallee" types. Laboratory studies showed that photosynthetic capacity of these species is linearly related to leaf nitrogen content (Figure 18.8). Furthermore, it was shown that in nature leaves of these species had increasingly higher specific weights, hence more nitrogen per leaf area, with increasing habitat aridity. Thus it appears that the species of the arid sites reduce the amount of leaf or evaporative surface area per

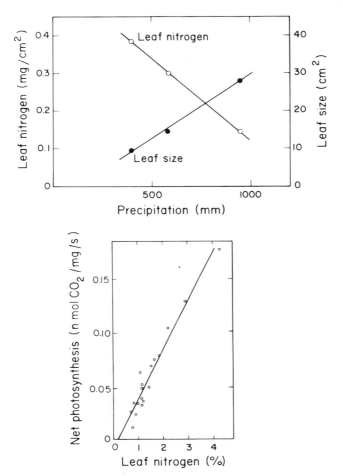

Figure 18.8. (*Top*): Mean leaf nitrogen content per unit leaf area and mean leaf size of a series of *Eucalyptus* species occurring along a precipitation gradient in eastern Australia. (*Bottom*): The relationship between photosynthetic capacity on a leaf weight basis and leaf nitrogen content. From Ref. 11.

volume of photosynthetic tissue. Hence potential production is not decreased, although there is a reduction in potential water use efficiency. This trend in leaf specific weight is possible because there is an increase in available light along the gradient. In the moistest sites light is limited in the dense forest, and for competitive reasons larger and thinner leaves are most appropriate.

Thus a simple change in the packaging of leaf tissue leads to plants that are adaptively appropriate for habitats differing in moisture availability. It should be noted, though, that each species is probably uniquely suited for its habitat. The species with the most efficient water use would do poorly in the light-limited habitat, because of light competition, and the most shade-adapted species would most assuredly do poorly in the arid habitat.

3.2. Chile and California

In studying adaptive variation along gradients it is important to select the gradients having as few environmental changes as possible. For example, along environmental gradients generated by elevation, both moisture and temperature generally change, making the assessment of the significance of any particular adaptive trait complex.

We have used aridity gradients in Chile and California to study adaptive patterns of plants to water stress. In both these regions there exist coastal latitudinal gradients where moisture regime changes rather dramatically, but because of the moderating influence of the cold coastal current, there is only a small change in thermal regime. Precipitation amounts extend to much lower values than those studied in the Australian gradient described above.

What sorts of adaptive pattern are seen along such gradients? In Chile (as well as in California) there is an obvious decrease in primary production, with decreasing precipitation going from the evergreen forests through evergreen scrub to a desert scrub. The most obvious shift in leaf morphological types, and presumably adaptive types along this gradient, is from evergreen to drought deciduous, to succulent photosynthetic tissue (Figure 18.9). At both extremes of the gradient the principal adaptive types endure the summer drought, which is relatively short at the wet end of the gradient and interminably long at the dry end. At the center of the gradient most of the dominant species are actually drought evaders. They avoid stress, but not at a cost of reduced production. These drought evaders have a higher photosynthetic capacity than the endurers when the water resource is unlimited (12). Miller (13) has modeled the environmental point at which it is more advantageous to shift from one adaptive mode to another to maintain optimum production along this gradient and has found a close fit with the natural pattern. Here, as noted earlier in the seasonal photosynthetic patterns in Death Valley, there is an inverse relationship between photosynthetic capacity and leaf duration. Species that have leaves that persist through a range of environmental conditions are less efficient photosynthetically during optimum conditions. This is an important relationship, which needs further study.

There is another important trend progressing along this aridity gradient among the woody plants that demonstrates the evolutionary interaction between plants and their physical, as well as biological, environment. Many of the forest species found in the moistest habitats have preformed terminal buds (Figure 18.10). Growth of new canopy tissue in this light-limited habitat can be rapid. In the community types that are drier, hence somewhat more

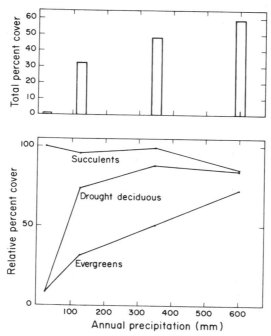

Figure 18.9. Total percentage cover of the woody plants and relative percentage cover of leaf types at four sites in Chile differing in annual precipitation. Data for the three moistest sites from Ref. 17 and the driest from Rundel (unpublished).

Figure 18.10. Branching patterns of plant species occurring along a gradient of moisture, hence of community density. *Left to right*: a branch of a tree of a moist closed forest, *Drimys winteri*; a shrub of mediterranean climate scrub, *Lithraea caustica*; and a shrub of the desert, *Oxalis gigantea*. Of these Chilean species, the *D. winteri* has preformed buds and terminal stem extension; *L. caustica* has terminal extension; and *O. gigantea* has leaves produced primarily on old wood.

291

open, growth of the shrubby species is still primarily terminal but not necessarily from preformed buds. Finally, in the driest and light-unlimited habitats many of the species produce leaves from old wood. At the driest site, then, leaf cost is the least of that at any of the sites, since the amount of support tissue produced is the least. Furthermore, the path length of water transport from root to leaf does not get progressively longer in the desert species as it does in the forest types.

3.3. Ecotypes

It is an unfortunate fact of history that the principal era of the study of the ecological and genetic variability of plants along environmental gradients ended in the 1960s before the development of readily accessible instrumentation for the measurement of environmental factors and plant metabolic responses under field conditions. Attention to the metabolic and structural variability between plant races occurring along environmental gradients gives us valuable information on the evolutionary response to stress.

I will not review here the abundant literature that demonstrates the genetically fixed specific adaptations of plants of local populations to the conditions prevailing in their habitat (14). Suffice it to state that it is clear that natural selection operates at a very fine level to produce genotypes that are most efficient at gathering the resources of the local habitat.

It is an evolutionary fact that no single genotype can be efficient in operating in a whole range of environmental conditions. For example, plants adapted to grow at high temperatures generally do poorly at low temperatures. Much of the work on ecotypes has demonstrated this clearly. There do exist, however, plants that are more generalist or plastic and can acclimate to a wide range of habitat conditions. As one example, I note *Heliotropium curassavicum*, which is a wide-ranging herbaceous perennial found in both North and South America in open, often saline habitats. Populations of this species occur in habitats as thermally dissimilar as Death Valley, California, where they are active even in the hot summer, and the beaches of Point Reyes, California, where the highest mean daily maximum summer temperatures rarely approach the mean daily maximum winter temperatures of Death Valley. Plants of both populations are very plastic in their temperature-related photosynthetic response. A coastal plant grown at the high temperatures characteristic of Death Valley in summer is metabolically indistinguishable from Death Valley natives grown at the same temperature (Figure 18.11). The reverse, however, does not seem to be true. But it appears that it is only in plants of naturally open or disturbed habitats, where there is generally little interspecific competition, that striking metabolic plasticity is obvious. We need more systematic studies on why certain species respond to new environments by a plastic response and others by a genetic ecotypic response.

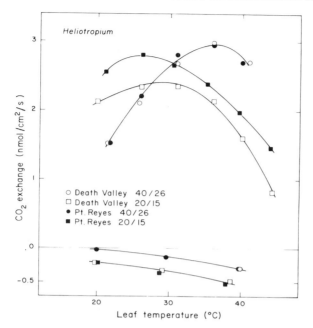

Figure 18.11. Temperature-related photosynthetic response of plants from populations of *Heliotropium curassavicum* originating from a hot desert climate (Death Valley) and a cool coastal climate (Point Reyes). Plants were grown at the day/night temperatures indicated (in °C) in a "phytocell" under full sunlight and with unlimited water and nutrients. From unpublished data of Mooney.

4. SUMMARY

Our present knowledge of the adaptive mechanisms that plants have evolved to cope with stress is based on examples from a rather small set of organisms. We can increase our understanding of the adaptive possibilities to stress by studying how plants optimize carbon gain under natural conditions when resources become limited (or stress ensues) either in time (seasonality) or space (gradients).

By studying plants in natural communities, we have the possibility of revealing not only what mechanisms plants use to cope with stress but also why they have evolved a particular adaptive pathway.

ACKNOWLEDGMENTS

Many of the examples described in this chapter grew out of studies supported by the National Science Foundation (grant DEB 75-19510).

REFERENCES

1. H. A. Mooney, O. Björkman, J. Ehleringer, and J. Berry, *Carnegie Inst. Washington Yearb.*, **75**, 410 (1976).
2. S. L. Gulmon and H. A. Mooney, *J. Ecol.*, **65**, 831 (1977).
3. J. H. Troughton, S. E. Camacho B., and A. E. Hall. *Carnegie Inst. Washington Yearb.*, **73**, 830 (1974).
4. O. Björkman, J. Boynton, and J. Berry, *Carnegie Inst. Washington Yearb.*, **75**, 400 (1976).
5. H. A. Mooney and S. L. Gulmon, in O. Solbrig, S. Jain, G. B. Johnson, and P. H. Raven, Eds., *Topics in Plant Population Biology*, Columbia University Press, New York, 1979, p. 316.
6. H. A. Mooney, J. Ehleringer, and O. Björkman, *Oecologia*, **29**, 301 (1977).
7. O. Lange, E.-D. Schulze, M. Evanari, L. Kappen, and U. Buschbom, *Oecologia*, **34**, 89 (1978).
8. J. R. Ehleringer, *Adaptive Significance of Leaf Hairs in a Desert Shrub*, Ph.D. thesis, Stanford University, Stanford, California, 1977.
9. S. R. J. Woodell, H. A. Mooney, and A. J. Hill, *J. Ecol.*, **57**, 37 (1969).
10. P. J. Fonteyn and B. Mahall, *Nature (London)*, **275**, 544 (1978).
11. H. A. Mooney, P. J. Ferrar, and R. O. Slatyer, *Oecologia*, **36**, 103 (1978).
12. H. A. Mooney and E. L. Dunn. *Am. Nat.*, **104**, 447 (1970).
13. P. C. Miller, in D. J. Horn, G. R. Stairs, and R. D. Mitchell, Eds., *Analysis of Ecological Systems*, Ohio State University Press, Columbus, Ohio, 1979, p. 179.
14. J. Heslop-Harrison, *Adv. Ecol. Res.*, **2**, 159 (1964).
15. H. A. Mooney, O. Björkman, and G. J. Collatz, *Carnegie Inst. Washington Yearb.*, **76**, 328 (1977).
16. J. R. Ehleringer and P. C. Miller, *Oecologia*, **19**, 177 (1975).
17. H. A. Mooney, J. H. Troughton, and J. A. Berry, *Carnegie Inst. Washington Yearb.*, **73**, 793 (1974).

19 | Leaf Morphology and Reflectance in Relation to Water and Temperature Stress

JAMES EHLERINGER

Department of Biology, University of Utah, Salt Lake City, Utah

1. INTRODUCTION

It is clear from previous ecological studies that plants growing in arid habitats tend to possess leaves that have more hairs (pubescence) than similar or related plants from more mesic habitats (1–3). This trend of increasing pubescence with aridity is seen repeatedly not only along geographical clines, but also within a single habitat as drought progresses in the growing season. Largely based on these observations, it is deduced that leaf hairs are an adaptive feature of plants occupying arid habitats (1, 4, 5). Although it may seem obvious that the leaf pubescence is a means of increasing plant fitness in arid habitats, and therefore is of adaptive value, the means by which leaf pubescence is adaptive are not so clear.

Several possibilities exist for the function of leaf pubescence in plants from arid habitats. These include (*a*) reduction of light absorption during conditions of high temperature and drought (6–8), (*b*) hindrance of the diffusion of gases across the leaf-air interface (9), and (*c*) reduction of predation by insects and larger herbivores (10). The function of leaf pubescence may differ among various plant species, but it would not be surprising to discover that in some cases pubescence served several functions.

With respect to high temperature and water stress, the presence of leaf

295

hairs is thought to modify leaf properties and physiological processes through several definable interactions (Figure 19.1). The pubescence may modify leaf absorption and boundary layer characteristics. A change in leaf absorptance will directly affect photosynthesis and leaf energy balance (leaf temperature). Indirectly, a change in leaf temperature will affect the rates of transpiration and photosynthesis. A change in the boundary layer may directly influence both photosynthesis and transpiration through its effect on the diffusion of carbon dioxide into and water vapor from the leaf surface. The boundary layer will directly affect leaf temperature by modifying the rate of heat transfer from the leaf. At the same time, the boundary layer will indirectly affect both photosynthesis and transpiration through leaf temperature as discussed previously. Thus it is clear that leaf pubescence has both direct and indirect effects on leaf physiological processes, and that the interactions among leaf hairs, environmental parameters, and physiological processes are tightly coupled. An understanding of the possible adaptive value of leaf hairs cannot be obtained without first quantifying the effects of the hairs on physiological processes.

2. INFLUENCE OF ENVIRONMENTAL FACTORS ON LEAF PUBESCENCE

Although the overall impact of leaf hairs on physiological activity is not yet understood for any system, at least one study has looked at the effects of hairs and other surface features on leaf spectral characteristics along en-

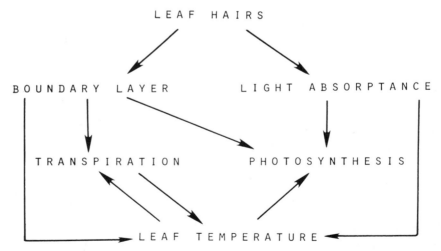

Figure 19.1. Interactions between leaf pubescence, energy exchange, and physiological processes. From Ref. 18, copyright by Springer-Verlag, Heidelberg, December 20, 1978.

vironmental clines of temperature and drought. Billings and Morris (11) compared reflectances of species from several communities and found that the reflectances of desert species were higher than those of plants from subalpine, pine forest, and shaded habitats. These investigators were comparing communities consisting of unrelated species and life forms, and several factors may have contributed to these reflectance differences.

Perhaps a less complicated situation for determining the value of leaf hairs to plants is to study the response(s) of a single species or of closely related species in a genus along an aridity gradient (see Chapter 18 for use of gradients in adaptation to stress). The arid land regions of the southwestern United States provide such an opportunity. From coastal California east to the deserts of the interior, there are many examples of genera with species distributed more or less allopatrically along a gradient of increasing aridity. As aridity increases along this gradient, species within a genus that have more pubescent leaves replace those that are less pubescent. These genera represent a number of families including Compositae, Labiatae, Polygonaceae, and Rhamnaceae, suggesting that the phenomenon is of broad ecological importance and is not restricted to a specific family.

One common genus along this aridity gradient, and perhaps representative of the leaf pubescence phenomena, is *Encelia* (Compositae). *Encelia* is a genus of suffrutescent shrubs with drought deciduous leaves. In terms of habitats, this genus extends from relatively mesic, warm coastal sites in southern California to extremely hot, arid desert sites in Arizona, California, and Mexico (Figure 19.2). At the coast we find the dark green, glabrous leaved *E. californica*. In the dry interior mountain slopes the light pubescent leaved *E. virginensis* occurs. At the drier lower elevations and throughout most of the desert *E. virginensis* is replaced by *E. farinosa*, a plant with white, extremely pubescent leaves. Aridity increases along a north-to-south coastal transect, and we find that *E. californica* is eventually replaced by the lightly pubescent leaved *E. asperifolia*. At the driest sites along this coastal transect *E. palmeri* with its densely pubescent leaves occurs.

Inspection of the spectral characteristics of leaves of *Encelia* species along an aridity transect makes it clear that the increased pubescence results in a decrease in leaf absorptance (Figure 19.3). Measurements of leaf absorptance between 400 and 700 nm are of interest because it is only these wavelengths that are useful for photosynthesis, and nearly 50% of the sun's irradiance and 80% of the solar radiation absorbed by leaves is in this wave band. The absorptance to solar radiation in the photosynthetically active wavelengths for leaves of *E. californica* is 85.2% (integral of solar irradiance and leaf absorptance spectra between 400 and 700 nm), a value similar to that observed for most green leaves (12). The light layer of pubescence on *E. virginensis* leaves reduces leaf absorptance to 75.9%. Leaves of *E. farinosa*, which have different amounts of leaf pubescence depending on the aridity of the site, have leaf absorptances substantially lower than the more glabrous species.

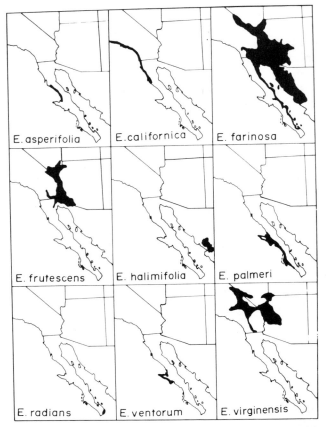

Figure 19.2. Distributions of species of *Encelia* in arid habitats of southwestern North America. *E. californica*, *E. frutescens*, *E. halimifolia*, *E. radians*, and *E. ventorum* are nonpubescent. *E. asperifolia* and *E. virginensis* are lightly pubescent, and *E. farinosa* and *E. palmeri* are heavily pubescent.

The absorptance spectra for *Encelia* in Figure 19.3 were obtained at a single point in time, April, the peak of the winter growing season. Measurements of absorptances on a seasonal basis reveal that leaf pubescence in *E. farinosa* varies during the growing season (Figure 19.4). Leaves produced early in the winter growth season have little pubescence and high leaf absorptances, whereas leaves produced in the spring and summer have more pubescence and lower leaf absorptances. The change in leaf absorptance of *E. farinosa* during the course of a season and the lack of change in *E. californica* are negatively correlated with air temperatures, suggesting that leaf pubescence might serve as a mechanism for modifying leaf energy balance and regulating leaf temperatures. The reasoning behind this is that as air temperatures increase, only two possibilities exist for reducing leaf temper-

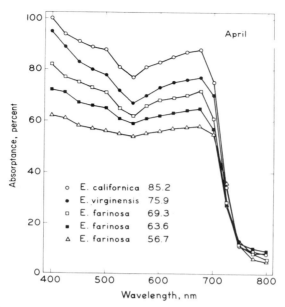

Figure 19.3. Absorptance spectra of intact leaves of *Encelia california, E. virginensis,* and *E. farinosa* along an aridity gradient during April. The values adjacent to each species represents the leaf absorptance to solar radiation between 400 and 700 nm.

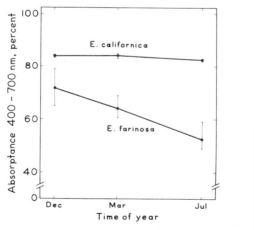

Figure 19.4. Seasonal courses of leaf absorptances (400 to 700 nm) for *Encelia farinosa* and *E. californica* from different sites. Based on data from Ref. 6.

atures: either an increase in transpiration rate is necessary or a decrease in energy absorbed is required. An increase in transpiration is very unlikely, given the aridity of these habitats. Thus one might expect that in going to more arid sites, leaves should exhibit lower leaf absorptances. This is indeed observed when the leaf absorptance and reflectance are plotted against the precipitation received during the growing season at different locations (Figure 19.5). These data show clearly that the leaf heat load is reduced as habitat aridity increases (decreasing precipitation means that less water is available for latent heat transfer from leaves).

As pointed out in Chapter 18, adaptation to drought and high temperature stresses by perennial desert plants should be viewed in two dimensions, since there is both a geographical or spatial component and a seasonal component. As Figures 19.3 to 19.5 indicate, environmental factors that result in high temperature or water stress responses in plants from one habitat early in a growing season may not appear until much later in the growing season in an adjoining habitat.

Development of leaf pubescence in *E. farinosa* appears to be a function of both drought and temperature stress. Two questions arise from the previous data: How does the plant sense water and temperature stress in its environment? Where are the sensors located? The evidence available suggests that these environmental sensors and control over pubescence occurs in the active apical meristems (13, 14). To investigate the question of response to water and temperature stresses, two sets of experiments were conducted, each aimed at eliminating one environmental stress to determine plant response to the other.

To eliminate temperature stress as a factor, *E. farinosa* plants were grown under greenhouse conditions. Plants were provided with a near constant aerial environment, and the response to drought was determined. As soil and

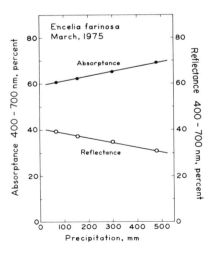

Figure 19.5. Leaf absorptance and reflectance (400 to 700 nm) of *Encelia farinosa* from various sites as functions of the precipitation received at each site during the growing season. Sites were initially chosen so that air temperatures at each site were similar. Based on data from Ref. 6.

leaf water potentials decreased, *E. farinosa* produced leaves that were progressively more pubescent, consequently having lower and lower leaf absorptances (Figure 19.6). At leaf water potentials near -10 bars leaves were lightly pubescent and leaf absorptance was about 78%. As leaf water potentials declined, leaf absorptance declined in a linear fashion. Below about -40 bars these plants did not produce new leaves; rather, they were drought deciduous and became inactive (14).

From both field and laboratory observations it appears that once a plant has been subjected to a mild water stress, removing the water stress does not result in the production of less pubescent leaves. Rather, the apical meristem continues to produce leaves with pubescence equal to the largest water stress encountered, until leaf water potentials have fallen below -40 bars (= dormancy?). Rewatering after a stress of -40 bars allows the apical meristem to "reset itself" and once again produce leaves with less pubescence and higher leaf absorptances.

The relationship between pubescence on *E. farinosa* leaves and temperature was determined with field-grown plants that received supplementary water throughout the year. The amount of leaf pubescence as measured by leaf absorptance was negatively related to mean maximum air temperature of the habitat (Figure 19.7). The variation in leaf absorptance due to temperature change, however, was much less than that due to water potential changes under the ranges of these factors normally found in nature. Although these plants were watered frequently, it is likely that to some extent leaf water potential covaried with air temperature. Nevertheless these results suggest

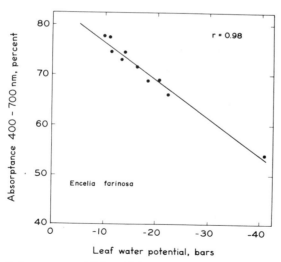

Figure 19.6. Leaf absorptance (400 to 700 nm) of *Encelia farinosa* as a function of midday leaf water potential. Plants were grown in tubs in a greenhouse and allowed to dry out slowly. From Ref. 14.

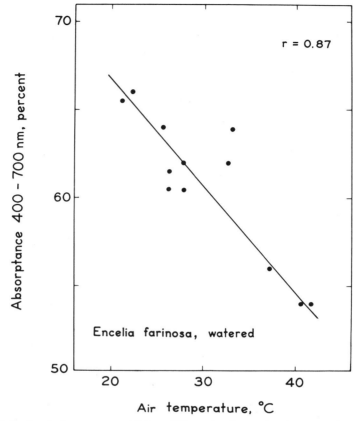

Figure 19.7. Leaf absorptance (400 to 700 nm) of *Encelia farinosa* that received supplemental water as a function of mean maximum air temperatures at field sites in Death Valley, California, and Deep Canyon, California. From Ref. 14.

that the environmental factors of drought and temperature both exert a strong influence over the development of leaf pubescence in *E. farinosa*.

3. PHYSIOLOGICAL EFFECTS OF LEAF PUBESCENCE

Our discussion thus far has centered around the general phenomenon of leaf pubescence in perennial plants from arid habitats, and in particular on leaf pubescence and environmental regulation of pubescence development in species of *Encelia*. I now concentrate on the effects that pubescence in *Encelia farinosa* has on leaf spectral characteristics, physiological processes, and carbon balance.

The presence of leaf hairs in *E. farinosa* greatly reduces absorptance by increasing leaf reflectance, not by increasing transmittance (Figure 19.5). Studies have shown that pubescence in *E. farinosa* is a blanket reflector between 400 and 700 nm. Consequently removing leaf pubescence from *E. farinosa* yields a leaf whose absorptance spectrum is virtually identical to that of *E. californica* (15). However over the entire solar spectrum the pubescence preferentially reflects near infrared radiation (700 to 3000 nm) (15). The heat balance of the leaf and the leaf temperature are influenced by leaf absorptance over the entire solar spectrum (400 to 3000 nm) rather than just the 400–700 nm band. The correlation between these two bands, however, is very tight (Figure 19.8). Whereas the 400 to 700 nm leaf absorptance for *E. farinosa* varies from 29 to 81%, the 400 to 3000 nm leaf absorptance varies from 15 to 46%. In comparison, *E. californica,* as is typical of green leaves, has a 400 to 700 nm leaf absorptance of 50%. These changes in 400 to 3000 nm leaf absorptance have a profound effect on leaf temperature. From energy balance calculations, a change in leaf absorptance from 50 to 15% will reduce midday leaf temperature by 5.1°C at a leaf conductance to water loss of 0.5 cm/s and typical summer conditions (air temperature 40°C). Solely as a result of this difference in leaf temperatures, the transpiration rate will decrease from 840 to 550 nmol/cm^2/s, a decrease of 34%.

In terms of photosynthesis, which is presumably the primary function of a leaf, pubescence may be necessary for the survival of *E. farinosa* under the hot, desert conditions of its native habitat. Figure 19.9 shows the temperature dependence of photosynthesis in *E. farinosa*. *Encelia farinosa* shows a temperature optimum for photosynthesis near 25°C and lacks the ability to acclimate the photosynthetic machinery to changes in growth conditions (16, 17). Consequently the temperature optimum remains near 25°C, and the upper limit of leaf temperatures is near 45°C. Pubescence (white leaf) allows *E. farinosa* to avoid potentially lethal temperatures and to photosynthesize close to the optimum temperature during summer conditions. Conversely, in

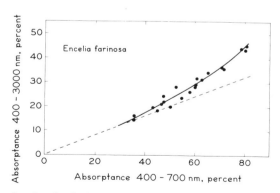

Figure 19.8. Total solar leaf absorptance (400 to 3000 nm) for *Encelia farinosa* plotted against the 400 to 700 nm leaf absorptance. From Ref. 15, copyright by Springer-Verlag, Heidelberg.

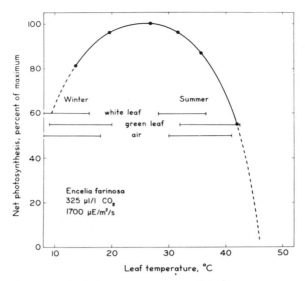

Figure 19.9. Daily ranges of air temperatures and leaf temperatures for green and white leaves of *Encelia farinosa* in the winter and summer and the temperature dependence of photosynthesis. Adopted from Ref. 18, copyright by Springer-Verlag, Heidelberg, December 20, 1978.

the winter the lightly pubescent form (green leaf) has a leaf temperature above air temperature and again closer to the temperature optimum of photosynthesis.

Our studies show, however, that pubescence in *E. farinosa* is not wholly advantageous. The presence of leaf pubescence decreases the photosynthetic rate by reflecting quanta that might otherwise be used in photosynthesis (Figure 19.10). Photosynthetic rates in these plants are quite high for C_3 plants and do not saturate at high light intensities. Because of this, pubescence affects photosynthesis at all light levels. When these photosynthetic data are replotted against absorbed rather than incident quantum flux, all differences between the curves are removed, showing that the pubescence serves primarily to reflect light and not to restrict diffusion of carbon dioxide across the boundary layer (18).

Several investigators have proposed that pubescence will increase the boundary layer resistance. If this effect on diffusion of heat and mass transfer is significant, the presence of hairs should modify photosynthesis, transpiration, and leaf temperature. Photosynthetic data presented previously suggested that this did not occur. It should be possible to estimate the additional boundary layer resistance arising from the presence of the hairs. I assume that the transfer of gases through the boundary layer is by

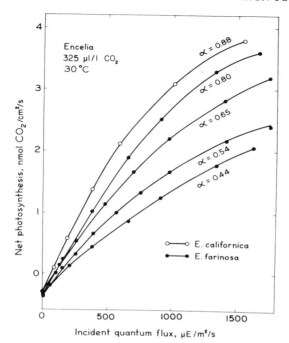

Figure 19.10. Relationship between net photosynthesis and incident quantum flux density for leaves of *Encelia farinosa* differing in pubescence and for *E. californica*. Alpha values represent leaf absorptance (400 to 700 nm) estimates. Thus the lower the alpha value, the greater will be the thickness of the pubescence layer. Adopted from Ref. 18, copyright by Springer-Verlag, Heidelberg, December 20, 1978.

diffusion, that is, there is no turbulent transfer, and I also assume no tortuosity effects on gas diffusion caused by the hairs. The increase in thickness of the boundary layer will then be equal to the thickness of the layer of leaf hairs. Knowing the increased thickness of the boundary layer, we can calculate the increase in boundary layer resistance as

$$r = \frac{L}{D} \tag{19.1}$$

where r is the additional boundary layer resistance for water vapor caused by the hairs, L is the thickness of the pubescent layer, and D is the diffusion coefficient for water vapor in air. The maximum expected increase in boundary layer resistance for a heavily pubescent *E. farinosa* leaf (0.36 mm thick pubescence) will be about 0.15 s/cm. A common boundary layer resistance will be approximately 0.3 s/cm. This means that under extreme conditions, pubescence may increase the boundary layer resistance by 50%. However even under heavy pubescence conditions, the hairs will have only a small effect on transpiration under natural conditions. This is because leaf resistances to water vapor are commonly greater than 5 s/cm when leaves are heavily pubescent (13). The effect of the hairs on the total resistance to transfer of carbon dioxide for photosynthesis will be even less than the minimal effect they have on transpiration.

The presence of leaf hairs has both advantages and disadvantages. The overall value of pubescence to *E. farinosa* was determined by Ehleringer and Mooney (18) by using computer simulations with a leaf energy budget–photosynthesis model. The model was designed to ask the following question: For a given combination of midday air temperatures and aridity, what value of all possible leaf absorptances will yield the highest photosynthetic rate? The simulations resulted in predictions that differing amounts of pubescence should be expected under different environmental regimes. Under mesic conditions (leaf conductance of 1 cm/s) and low to moderate air temperatures (20 to 40°C), the optimum pubescence level was the greenest possible leaf (leaf absorptance of 80%). However when air temperatures were above 40°C, even though conditions were mesic, a pubescent leaf had a higher rate of carbon gain than a nonpubescent leaf. Furthermore, as habitat aridity increased, the air temperature at which leaves are predicted to become pubescent decreased.

It is predicted that optimum leaf pubescence will vary with environmental conditions because there is a tradeoff between the decline in photosynthesis caused by an increased reflectance (see Figure 19.10) and the increase in photosynthesis caused by a decrease in leaf temperature (see Figure 19.9) (18). Because the photosynthetic machinery of *E. farinosa* does not acclimate to temperature changes, pubescence may become advantageous whenever leaf temperatures exceed the photosynthetic temperature optimum. Predictions from the model agree with field observations of pubescence levels by Ehleringer et al. (6), Ehleringer and Björkman (15), and Smith and Nobel (8).

Simulation results of the total daily carbon gain and transpiration loss by *E. farinosa* leaves under wet and dry conditions during the winter and summer seasons are in agreement with predictions from midday conditions by Ehleringer and Mooney (18). These simulations reveal that nonpubescent leaves are expected to have higher daily rates of carbon gain than are pubescent leaves, except when environmental conditions are hot and dry (Figure 19.11). Under the four environmental regimes considered in Figure 19.11, however, the pubescent leaves will have a lower daily water loss. As a consequence, the photosynthesis/transpiration ratio of pubescent leaves in the summer is predicted to be greater than that of glabrous leaves. With respect to natural selection, these simulation results and respective field observations would suggest that plant fitness in this perennial shrub may be maximized not by minimizing water loss, nor by maximizing the pho-

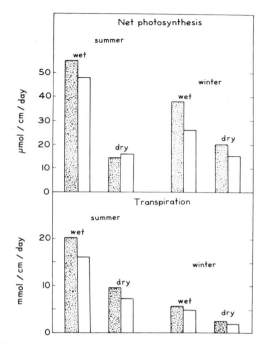

Figure 19.11. Daily rates of carbon gain and water loss for pubescent leaves (leaf absorptance 400 to 700 nm = 50%: open histograms) and for nearly glabrous leaves (leaf absorptance 400 to 700 nm = 80%: shaded histograms) of *Encelia farinosa* under wet (leaf conductance = 0.5 cm/s) and dry (leaf conductance = 0.2 cm/s) conditions in the winter (max/min air temperatures = 20/5°C) and summer (max/min air temperatures = 40/25°C) seasons.

tosynthesis/transpiration ratio, but rather through maximizing the carbon-gaining capacity of the leaf.

4. SUMMARY

I suggest that leaf pubescence has adaptive value to *E. farinosa* in its native desert habitats, because the hairs allow the leaf to gain a higher rate of carbon under arid conditions than the leaf could acquire without hairs, to avoid potentially lethal high leaf temperatures, and to lose less water daily, which allows the plant to extend its growth for a longer period into the drought. To a species that appears to be unable to acclimate its photosynthetic characteristics when faced with increasing air temperatures, the use of leaf hairs to produce a reflective surface is a viable alternative mechanism in adapting the plant to its environment.

REFERENCES

1. J. M. Coulter, C. R. Barnes, and H. C. Cowles, *A Text Book of Botany for Colleges and Universities*, Vol. 2, *Ecology*, American Book, New York, 1911.
2. A. F. W. Schimper, *Plant Geography Upon a Physiological Basis*, Clarendon Press, Oxford, 1903.
3. E. Warming, *Oecology of Plants: An Introduction to the Study of Plant Communities*, Oxford University Press, London, 1909.
4. H. Oppenheimer, in *Plant Water Relationships in Arid and Semi-arid Conditions: Reviews of Research*, UNESCO, Paris, 1960, p. 105.
5. F. Fritsch and F. Salisbury, *Plant Form and Function*, Bell, London, 1965.
6. J. Ehleringer, O. Björkman, and H. A. Mooney, *Science,* 192, 376 (1976).
7. W. K. Smith and P. S. Nobel, *Am. J. Bot.*, 65, 429 (1978).
8. W. K. Smith and P. S. Nobel, *Ecology*, 58, 1033 (1977).
9. J. T. Woolley, *Agron. J.*, 56, 569 (1964).
10. D. A. Levin, *Q. Rev. Biol.*, 48, 3 (1973).
11. W. D. Billings and R. J. Morris, *Am. J. Bot.*, 38, 327 (1951).
12. D. M. Gates, H. J. Keegan, J. C. Schleter, and V. R. Weidner, *Appl. Opt.*, 4, 11 (1965).
13. J. Ehleringer, *Adaptive Significance of Leaf Hairs in a Desert Shrub*, Ph.D. thesis, Stanford University, Stanford, California, 1977.
14. J. Ehleringer, in preparation.
15. J. Ehleringer and O. Björkman, *Oecologia*, 36, 151 (1978).
16. B. R. Strain and V. C. Chase, *Ecology*, 47, 1043 (1966).
17. J. Ehleringer and O. Björkman, *Plant Physiol.*, 62, 185 (1978).
18. J. Ehleringer and H. A. Mooney, *Oecologia*, 37, 183 (1978).

20 | Influence of Water Stress on the Photosynthesis and Productivity of Plants in Humid Areas

TADAYOSHI TAZAKI

Faculty of Science, Toho University, Chiba, Japan

KUNI ISHIHARA

Faculty of Agriculture, Tokyo University of Agriculture and Technology, Tokyo, Japan

TADAHIRO USHIJIMA

Faculty of Agriculture, Tokyo University of Agriculture and Technology, Tokyo, Japan

1. INTRODUCTION

Because of the high annual precipitation in Japan, water stress has until recently not been considered to be a limiting factor for plant growth. For example, the average annual precipitation in the Tokyo area is 1500 mm, the evaporation is only 1000 mm, and on average precipitation exceeds evaporation in every month of the year. However there is a period from the middle of July to the end of August, between the rainy *baiu* period of early summer and the typhoon season of September and October, when evapora-

309

tion often exceeds precipitation for a few weeks (1, 2). During this season rain-free periods of sufficient duration to damage crops often occur. The possibility of damage during this period is increased because the cloudiness and high humidity of the *baiu* period produces plants with shallow root systems and "shade" leaves that are very susceptible to injury when exposed to high radiation and low humidity. Furthermore, the summer droughts usually occur during the transition from vegetative to reproductive growth when plants are most susceptible to injury by water stress.

2. DAMAGE FROM SUMMER DROUGHT

Plants are seldom killed by drought in Japan, except on very sandy soils such as sand dunes. For example, a study of Japanese black pine (*Pinus thunbergii* Parl.) growing on a coastal dune area near Chigasaki showed that the seedlings emerged in April and grew well until the end of the *baiu* season, but nearly all died in midsummer (3). Their death was not caused by shading, poor mineral nutrition, salt deposition, or high soil salinity (4, 5), but by the lack of soil moisture. The available water in the top 30 cm of the dune soil was 18 mm, a small amount compared with that of a Kanto loamy soil of volcanic origin, which contains more than 90 mm in the same depth of soil (3). In the Kanto loam even small plants with shallow root systems were not killed in the summer of 1949, a severe drought year with no rain throughout August, whereas the pine seedlings growing in the Chigasaki sand had difficulty germinating and emerging. The seeds lying near the surface had difficulty germinating because of the dry surface soil, and seeds lying deep in the soil had difficulty emerging because of limited seed reserves. Thus the depth from which the seeds germinated and emerged was restricted to an area 1.5 to 3 cm from the soil surface (Figure 20.1). The seedlings that did

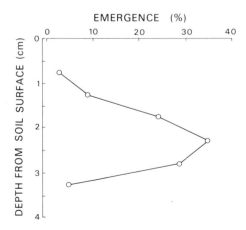

Figure 20.1. Relationship between seed depth and emergence of seedlings of *Pinus thunbergii* under a stand of pines in a dune area of Chigasaki. From Ref. 3.

emerge grew well until the beginning of July, but during July and August they suffered from drought. The vertical distribution of the soil water content in the sand during 3 rain-free weeks in July is shown in Figure 20.2. The last rain was on July 6 and the soil moisture was near field capacity (6%) throughout the soil profile. However 10 days after the last rain, on July 16, the water content was below the wilting percentage in the upper 5 cm of the soil, and after 3 weeks without rain the water content decreased near to or below the wilting percentage to a depth of 20 cm. Since the pine roots were less than 15 cm deep, all the seedlings were killed by the drying of the soil. Other herbaceous plants or shrubs growing under the pine stands survived the drought because their root systems were deeper than 20 cm.

3. EFFECTS OF WATER STRESS ON THE PHOTOSYNTHESIS AND PRODUCTIVITY OF MULBERRY

In Japan mulberry plants (*Morus alba*) are cultivated on slopes that are unsuitable for the cultivation of rice or upland crops. The mulberry plant is naturally a tree, but it is cultivated as a shrub for feeding to silkworms. The mulberry plants are pruned near to the ground in March to May to give five to ten new shoots: harvesting begins at the beginning of August when the shoots are about 2 m long. Mulberry is considered to be drought resistant, and it has a deep root system.

The plants used in the study reported here were growing in rows 160 cm

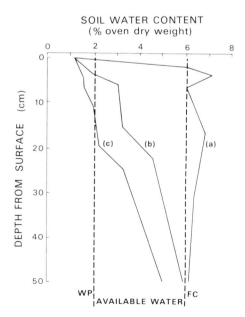

Figure 20.2. Vertical distribution of soil water content under a stand of pines in a dune area in Chigasaki during a rainless period, from July 7–27, 1949. Curve *a*, July 8; curve *b*, July 16; curve *c*, July 27; FC, field capacity; WP, wilting percentage. From Ref. 3.

apart, with plants 70 cm apart in the rows (6, 7). One field was near Tokyo, the other at Tōbu, Nagano, at an elevation of 960 m, where summer droughts often occur. The shrubs were 9 to 10 years old and had been pruned in the spring. The rate of photosynthesis of young leaves in the field at Nagano was measured by gas exchange techniques on clear sunny days, 2, 8, and 15 days after rain. The results are presented in Figure 20.3. It can be seen that as the soil water was depleted, the rate of photosynthesis decreased, especially in the afternoon.

A comparison was also made of the relative growth rates (RGR) and net assimilation rates (NAR) of young mulberry shoots at Tokyo and Nagano in both a wet year (1966) and a dry year (1969) (8). The results (Figure 20.4) indicate that even in the relatively wet year of 1966, beginning about 60 days after spring pruning, there was a greater decrease in RGR and NAR at Nagano than at Tokyo. In the dry year of 1969 the productivity was much higher in Tokyo than in Nagano, and this corresponds to a greater decrease in RGR and NAR at Nagano in 1969 compared to 1966. Although pruning occurs a month later in Nagano than in Tokyo because of the colder conditions at Nagano, the RGR and the NAR at Nagano were equal to or higher than those of Tokyo during the initial stages of growth, and it is likely that the low values of RGR and NAR during the summer are due to drought. The leaf area index at Nagano was also lower than at Tokyo, in the worst cases being only half that at Tokyo; thus the difference in dry matter production between Nagano and Tokyo was much larger than the difference in RGR and NAR (9). Thus we conclude that water stress restricts the productivity of mulberry even in Japan and even in a so-called drought resistant plant.

4. UNRESPONSIVE LEAVES IN MULBERRY

Another interesting feature found by Tazaki in 1951 was the anomalous transpiration of detached mulberry leaves. In normal leaves the transpiration rate decreased by 90% after detaching, whereas the transpiration of some leaves decreased by only 40 to 50%, and they were quickly killed by the loss of water (10). The high transpiration rates of these detached leaves was due to the imperfect closure of their stomata. The distinction between normal and unresponsive leaves is clear-cut. The change from a normal leaf to an unresponsive leaf is an aging phenomenon: loss of stomatal response appeared first in the oldest and lowest leaves on the stem at the end of June; by the beginning of August about half the leaves had become unresponsive, and by the middle of September all leaves were unresponsive except for those on the uppermost 25% of stem. This behavior was not peculiar to mulberry, but was common to several deciduous broad-leaved trees.

The change in photosynthetic rate in detached leaves was determined one autumn in both normal and the unresponsive leaves (Figure 20.5). In the normal leaf the photosynthetic rate decreased rapidly after detaching be-

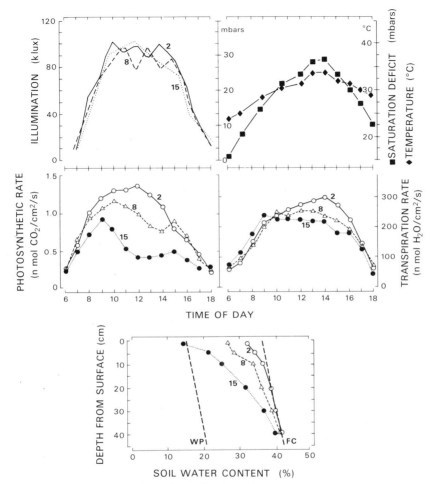

Figure 20.3. Diurnal variation of illumination, air temperature, vapor pressure (saturation) deficit, and rates of net photosynthesis and transpiration of young leaves in a montane mulberry field 2 (○, solid lines) (△, dashed lines), and 15 (●, dotted lines) days after rain. Vertical distribution of soil water content is also shown; FC, field capacity; WP, wilting percentage. Data for air temperature (●) and vapor pressure deficit (■) are for second day after rain only. Data for July 29 to August 10, 1969, at Tobu, Nagano, 200 km northwest of Tokyo, 960 m above sea level. From Refs. 6, 7.

cause the stomata closed rapidly. In the unresponsive leaves, although the photosynthetic rate immediately after detaching was lower than in normal leaves because the latter were younger, photosynthesis decreased slowly with loss of water. Since the stomata in the unresponsive leaves remained open after detaching, the study allowed us to look at the effect of water stress on the internal photosynthetic apparatus. The initial water content of

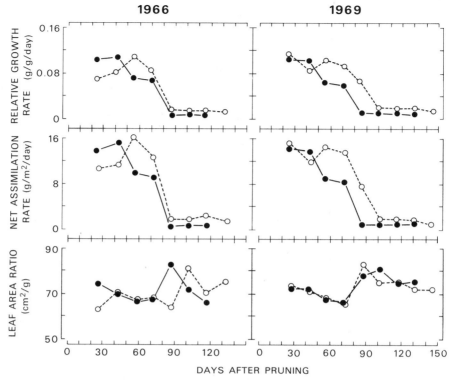

Figure 20.4. The relative growth rate, leaf net assimilation rate, and leaf area ratio of new mulberry shoots at Tokyo (○ – – – ○) and Nagano (●———●) in two different years. The plants were pruned on May 1 in Nagano and on April 1 in Tokyo; 1966 was a wet year and 1969 was a dry year. From Ref. 8.

the normal and unresponsive leaves was 200%. In the normal leaves in which the stomata closed, the photosynthetic rate was almost zero at a water content of 160%, whereas in the unresponsive leaves the photosynthetic rate was still positive even below the lethal range of 120% (Figure 20.5*b*), suggesting that the internal photosynthetic apparatus is much less sensitive to water stress than are the stomata. Since a considerable amount of water may be lost by transpiration from plants with leaves in which the stomata remain open, this property can have deleterious effects on mulberry under drought conditions.

5. WATER STRESS WITH ADEQUATE SOIL MOISTURE

Rice, which occupies nearly half the cultivated area of Japan, grows partly submerged for most of its life cycle. Nevertheless, slight wilting is often

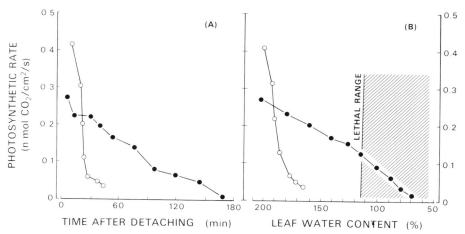

Figure 20.5. (*a*) Changes in rate of net photosynthesis after detaching and (*b*) the relationship between leaf water content and photosynthetic rate, in normal (○) and unresponsive (●) leaves of mulberry. The upper limit of lethal water range is also shown in *b*. From Ref. 11.

observed on fine days during midsummer. The diurnal variation in stomatal aperture, which reflects the water status of the leaf blade, was measured by the infiltration method, using a mixture solution of isobutanol and ethylene glycol, and the apertures were expressed as infiltration scores from 0 to 4 (12). The infiltration scores had a high positive correlation with the stomatal pore width measured by the microrelief method (12), the photosynthetic rate measured by gas exchange techniques (13), and stomatal conductance measured by diffusion porometry (14). On cloudy days the stomata opened slowly between sunrise and noon, were open maximally until 1500 h and thereafter closed rapidly (Figure 20.6). On fine days, on the other hand, the stomata opened in the morning to reach a maximum aperture at 0900 h and closed gradually throughout the rest of the day. This behavior was seen from seedlings to mature plants and suggests that even under partly submerged conditions the stomata were closed because of water stress on days when water absorption could not keep pace with transpiration (15). Stomatal closure after 0900 to 1100 h will cause photosynthesis to decrease, leading to poor utilization of solar energy. The same phenomenon was also found in maize (*Zea mays*) and sunflower (*Helianthus annuus*) grown in soils with adequate water (7).

Additional experiments on rice were based on the finding that when the nitrogen level of leaves is high, the stomata open widely in the morning (16), and when water absorption of roots is suppressed by the partial excision of the root system, the stomata close in the afternoon as a result of water stress (17). One week before the commencement of observations, some potted rice plants were given additional nitrogen fertilizer (high N treatment); soluble

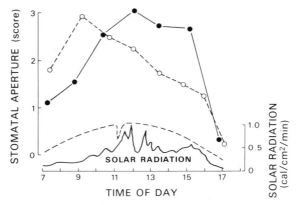

Figure 20.6. Diurnal variation of solar radiation and stomatal aperture of rice plants at the heading stage on a cloudy day (solid lines) and a sunny day (dashed lines). Cloudy day conditions: total solar radiation, 233 cal/cm²/day; maximum air temperature, 28.9°C; minimum relative humidity, 84%. Fine day conditions: total solar radiation, 508 cal/cm²/day; maximum air temperature, 31.2°C; minimum relative humidity, 67%. From Ref. 15.

starch was mixed with the soil to suppress water absorption of other plants (starch treatment) (18, 19), and a third (control) group was given no additional fertilizer or starch. The diurnal changes in the rates of transpiration, stomatal aperture, stomatal conductance, leaf water potential (ψ_{leaf}) measured with a thermocouple psychrometer, and xylem water potential (ψ_{xylem}) measured with a pressure chamber, were ascertained on clear, sunny days (20). Considerable differences in transpiration rate appeared between the control and high N treatment in the morning, with transpiration being higher in the high N treatment, whereas the starch treatment reduced transpiration below that in the controls during the afternoon. Both ψ_{xylem} and ψ_{leaf} were different: when ψ_{xylem} was -8 to -11 bars, ψ_{leaf} was only -4 to -6 bars. In another experiment using potted rice plants, ψ_{xylem} and ψ_{leaf} were measured simultaneously over a range of potentials on plants kept either in the dark to minimize transpiration or outdoors to maximize transpiration. In the dark ψ_{xylem} and ψ_{leaf} were similar at all water potentials, but outdoors ψ_{xylem}, measured with the pressure chamber, was 3 to 5 bars lower than ψ_{leaf}, measured with a psychrometer (Figure 20.7). Furthermore, the value of ψ_{xylem} became similar to that of ψ_{leaf} as the soil dried and transpiration stopped. From these results it is clear that the water status of the leaf mesophyll and the leaf xylem were not in equilibrium at rapid rates of transpiration (21). A similar phenomenon was observed in maize (22). We conclude that at rapid transpiration rates the determination of ψ_{leaf} with the pressure chamber can be misleading.

From the rates of transpiration mentioned above, it was inferred that the photosynthetic rates of plants given a high N treatment may be high during

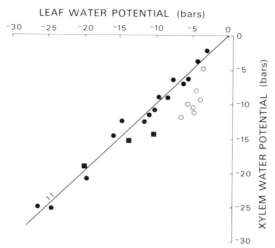

Figure 20.7. Relationship between leaf water potential, measured with a thermocouple psychrometer, and xylem water potential, measured with a pressure chamber, of leaf blades in rice: water potential of plants in dark room at various soil water contents (●); water potential of partly submerged plants in the sun (○); water potential of plants in the sun at various soil water contents (■). From Ref. 21.

the mornings when there is no water stress, whereas the plants given the starch treatment may have low photosynthetic rates during the afternoon because of water stress resulting from the restriction of water absorption by roots. The photosynthetic rates of intact leaves in the treatments was measured while changing the saturation deficit (vapor pressure deficit) inside the assimilation chamber (Figure 20.8). As we expected, the highest rates of photosynthesis were measured in high N plants at low saturation deficits, and photosynthesis decreased progressively as the saturation deficit increased: the photosynthetic rate of the plants treated with starch was similar to that in controls at low saturation deficits, but decreased more than in the controls as the saturation deficit of the air increased. Thus the effect of a high leaf N content was to increase the photosynthetic rate under no stress conditions, and the effect of restricting water uptake was to decrease net photosynthesis due to water stress at high irradiances and low humidities.

In rice, the effects of water stress during the day were found to differ with the growth conditions (24). Using potted rice plants, two treatments were established: (*a*) pots were buried in the interior of a paddy field at transplanting and transferred and buried at the southern border of the paddy on the day before measurement ("interior to border") and (*b*) pots were buried at the southern border from transplanting to the day of measurement ("border"). The diurnal change in stomatal behavior was measured simultaneously in both plots using the infiltration technique (Figure 20.9). Stomatal opening was similar in the early morning, but thereafter the stomatal aper-

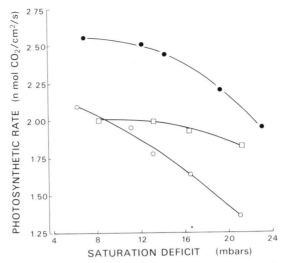

Figure 20.8. Relationship between vapor pressure (saturation) deficit and the rate of net photosynthesis of intact leaves of rice plants given high nitrogen (●), plants for which starch was added to the soil (○), and untreated controls (□). From Ref. 23.

tures of the "interior to border" plants were always less than those of the "border" plants. This suggests that partly submerged rice plants grown in the interior of a paddy field where self-shading occurs were more susceptible to water stress resulting from high solar radiation. Partly submerged rice plants growing in the interior of a paddy were then exposed to the same conditions as those growing along the southern border by harvesting several rows of plants on the south side of the experimental plants: again these were termed "interior to border" plants. The growth of the "interior to border" and the "border" plants was compared over intervals of 14 to 20 days at tillering, panicle formation, and heading. As was expected from the diurnal variation in stomatal aperture, both NAR and RGR were greater at each stage of growth in "border" than in "interior to border" plants (Table 20.1). Furthermore, the leaf area was greater and shoot and root weights were heavier in "border" plants than in "interior to border" plants. The root system of "border" plants was so well developed that the ratios of root weight to leaf area and of root to shoot were larger in "border" plants than "interior to border" plants (24).

From the foregoing observations it appears that rice plants grown at the border of a rice field are well adapted to their environment, having a well-developed root system, open stomata, little decrease in photosynthetic rate under water stress, and larger NAR and RGR. In contrast, rice plants growing in the interior of a rice field, with less solar radiation because of greater mutual shading and high humidity, were highly susceptible to water stress and were poorly adapted to a less humid environment.

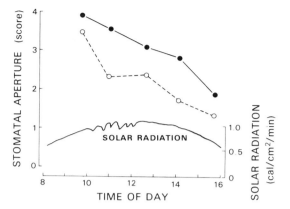

Figure 20.9. Diurnal variation of solar radiation and stomatal aperture in potted rice plants on the border (●——●) of a paddy field or moved from the interior to the border (○------○) at the panicle formation stage. Weather conditions: total solar radiation, 517 cal/cm²/day; maximum air temperature, 31.0°C; minimum relative humidity, 59%. From Ref. 24.

Table 20.1. Relative Growth Rates (RGR) and Net Assimilation Rates (NAR) of Rice at the Border of a Paddy (25).

Measured over intervals of 14 to 20 days at three stages of growth in plants grown at the border of a rice field ("border") or moved from the interior to the border of the field ("interior to border")

		Growth stage		
Rate	Position	Maximum tillering (20 days)	Panicle formation (14 days)[a]	Heading (17 days)
RGR	"Border"	82.5	23.6	35.1
(mg/g/day)	"Interior to border"	68.4	21.5	28.1
NAR	"Border"	7.85	2.87	5.69
(g leaf/m²/day)	"Interior to border"	7.01	2.58	4.42

[a]This period consisted of mostly rainy and cloudy days with an average of only 1 h of sunshine per day.

6. CONCLUSIONS

Japan is in a humid region in which precipitation considerably exceeds evaporation. Nevertheless, it has been shown that photosynthesis and dry matter production are reduced by water stress brought about both by the depletion of soil moisture during midsummer and, on clear sunny days, by high rates of transpiration outstripping water absorption even at high soil water contents. The effects in terms of the depression of photosynthesis and productivity may be small compared with the death and yield decreases of crops in arid or semiarid regions where severe drought frequently prevails. However in Japan, with its limited farming area and large population, agricultural production must be raised by raising the yield per unit area. In Japan, therefore, even a small decrease in photosynthesis and dry matter production is of concern. Furthermore, as the ratio of photosynthesis to transpiration decreases under water stress, the effects of water deficits on water use efficiency must be considered (26). To date, the effects of water stress in Japan have been disregarded: in fact, much work remains on the effect of water stress in humid regions.

REFERENCES

1. Annual Report of the Japan Meteorological Agency, Part 1, Tokyo, 1950–1976.
2. Daily meteorological surface observation of the Japan Meteorological Agency, Tokyo, 1967–1976.
3. T. Tazaki, *Jan. J. Bot.*, **17**, 239 (1960).
4. T. Tazaki, *Bot. Mag.*, **63**, 12 (1950).
5. M. Kadota and T. Tazaki, *Bull. Physiograph. Sci. Res. Inst., Univ. Tokyo*, **3**, 38 (1948).
6. T. Ushijima, M. Seto, and T. Tazaki, *Photosynthesis and Utilization of Solar Energy*, JIBP/PP—Photosynthesis Level III Group (1971), **1972**, p. 100.
7. T. Ushijima and T. Tazaki, *JIBP Synth., Tokyo*, **19**, 37 (1978).
8. T. Ushijima, unpublished data.
9. T. Ushijima, *J. Seric. Soc., Jan.*, **33**, 293 (1964).
10. T. Tazaki, *Bot. Mag.*, **72**, 68 (1959).
11. T. Tazaki and T. Ushijima, *Photosynthesis and Utilization of Solar Energy*, JIBP/PP—Photosynthesis Level III Group (1966–67), **1968**, p. 87.
12. K. Ishihara, T. Nishihara, and T. Ogura, *Proc. Crop Sci. Soc. Jan.*, **40**, 491 (1971).
13. K. Ishihara, R. Sago, T. Ogura, T. Ushijima, and T. Tazaki, *Proc. Crop Sci. Soc. Jan.*, **41**, 93 (1972).
14. K. Ishihara, T. Hirasawa, O. Iida, and T. Ogura, *Jan. J. Crop Sci.*, **48**, 319 (1979).
15. K. Ishihara, Y. Ishida, and T. Ogura, *Proc. Crop Sci. Soc. Jan.*, **40**, 497 (1971).
16. K. Ishihara, H. Ebara, T. Hirasawa, and T. Ogura, *Jan. J. Crop Sci.*, **47**, 664 (1978).
17. K. Ishihara, R. Sago, and T. Ogura, *Jan. J. Crop Sci.*, **47**, 499 (1978).
18. S. Mitsui, K. Kumazawa, and T. Ishihara, *J. Sci. Soil Anim. Fert. Jan.*, **24**, 45 (1953).
19. S. Mitsui and K. Kumazawa, *J. Sci. Soil Anim. Fert. Jan.*, **35**, 115 (1964).
20. K. Ishihara, T. Hirasawa, O. Iida, M. Kimura, and T. Ogura, *Jan. J. Crop Sci.*, **47** (Extra 1), 241 (1978).
21. K. Ishihara and T. Hirasawa, *Plant Cell Physiol.*, **19**, 1289 (1978).

22. T. Hirasawa and K. Ishihara, *Jan. J. Crop Sci.,* **48,** 557 (1979).
23. K. Ishihara, E. Kuroda, T. Hirasawa, and T. Ogura, unpublished data.
24. K. Ishihara, R. Sago, and T. Ogura, *Jan. J. Crop Sci.,* **47,** 515 (1978).
25. K. Ishihara, T. Hirasawa, H. Saito, and T. Ogura, unpublished data.
26. T. Tazaki, T. Ushijima, T. Seto, and Y. Masuda, *Photosynthesis and Utilization of Solar Energy*, JIBP/PP—Photosynthesis Level III Group (1970), **1970,** p. 39.

21 | Influence of Water Stress on Crop Yield in Semiarid Regions

R. A. FISCHER

Division of Plant Industry, CSIRO, Canberra, Australian Capital Territory, Australia

1. INTRODUCTION

In discussing the effects of water stress on crop yield, it is important in the context of the title of this book to try to emphasize adaptations that may increase yield in water-limited situations. Since the book encompasses the interests of both crop scientists and ecologists, we must remember that adaptations that are favorable for crop plants may not necessarily be favorable for plants in natural ecosystems. Crops are usually monogenotypic plant communities with intense intragenotypic competition (1); their overriding goal, or at least that of crop breeders and agronomists, is maximum yield per unit land area (other goals might be greater stability of yield, or better product quality). In a sense crop scientists are subjective teleologists (2); put bluntly, they are interested in manipulating plants to fill more bellies. On the other hand, in natural ecosystems intergenotypic competition is intense, with competitiveness or fitness probably the most important goal of natural selection. Competitiveness and maximum yield in pure stand may not be compatible goals (1). Harper (3) recently summarized the situation by pointing out that "plant breeding . . . is concerned to undo the results of

323

(natural) selection for selfish quantities of individual fitness and focus on the performance of populations.''

Despite this contrast, crop scientists should be able to learn from plants in natural ecosystems, hence from ecologists. On the one hand, the semiarid zone ecosystems may contain useful genes for adaptation to water-limited environments, either in the form of traits in wild relatives of present crops (4) or of new crop species (5). On the other hand, contemplation of natural ecosystems suggests there could be genes in present crop cultivars, which although desirable in their primitive progenitors, and retained unnoticed by mass selection, are now unnecessary or even limiting to yield [e.g., traits favoring competitiveness (1, 3)]. Finally, natural ecosystems may provide crop breeders with clues about useful breeding strategies and adaptive mechanisms. However such argument by analogy requires much caution.

This chapter emphasizes annual seed crops—the cereals, pulses, and oil seeds—which together provide most of man's food. The ensuing discussion of water stress in crops may seem simplistic, but bear in mind two major contrasts between crop science and ecology: namely, the preoccupation of crop scientists with the single selection goal of raising yield, and the relative simplicity of the crop community. My concern with adaptations is limited largely to genotypic differences that are expressed in response to water stress, so-called facultative adaptations. Interest in yield means that it is important for apparent adaptations at the level of the physiological process or plant part to be assessed in the context of final productivity per unit area of land. Often there are tradeoffs, as seemingly favorable adaptations are commonly accompanied by penalties in the broader context. Unfortunately there are few examples of the adaptive value of a trait having been convincingly established.

2. APPROACH TO WATER-LIMITED PRODUCTIVITY

Traditionally physiologists quantified the effects of water stress on plants and crops in terms of the soil water potential (ψ_{soil}); more recently plant water potential (ψ_{plant}) has been given a central role. This may provide a satisfactory basis for analyzing physiological processes on a short time scale, but I would argue that a better approach to overall productivity is to start with the limiting resource, in this case, water. As a rough guide to the extent to which water is limiting crop production in the semiarid zone, it has been suggested (6) that on average, actual crop evapotranspiration (ET_a) is from 0.2 to 0.5 of potential evapotranspiration (ET_p). Yield is correspondingly limited.

Water limits growth ultimately because water loss or transpiration is inextricably tied to carbon gain through photosynthesis. This approach views the semiarid zone crop as a system for harvesting scarce water: the crop trades acquired water for reduced carbon, which is converted into edible products. In the short term crop growth or dry matter accumulation is a question of transpiration (T) and photosynthesis per unit *transpiration*, which I define as transpiration efficiency (TE), as distinct from water use efficiency [i.e., dry matter accumulation per unit *evapotranspiration* (ET)] used elsewhere in this book. TE is the reciprocal of the familiar transpiration ratio. In the longer term, for example, the life cycle of a crop, this is still true; but additionally we must consider the proportion of assimilate allocated to each organ (leaves, roots, seeds, etc.). In this framework ψ_{plant} is clearly a secondary consideration. It is an indicator for the plant of the situation with respect to water and perhaps a trigger for changes in allocation, but not the ultimate limit to growth, which is water supply.

One example illustrates my point. Andrews and Newman (7) grew wheat plants in sealed pots with and without 60% root pruning, during a single 2 week drying cycle. Although the pruned plants showed lower transpiration rates initially, presumably because of lower ψ_{plant}, dry matter accumulation over the drying cycle was unaffected. The authors were surprised that pruning did not increase the sensitivity of the plants to drought; since pruning did not affect cumulative transpiration, however, nor were effects on TE likely, the result was to be expected.

The transpiration–transpiration efficiency–allocation framework, which has its origin in the studies of de Wit and colleagues (e.g., Ref. 8) has been discussed in detail elsewhere (6). My objective now is to use this framework to discuss water limitation and crop yield.

3. ENVIRONMENTAL CONTEXT

Water supply, that is, the water in the soil at sowing plus precipitation, sets an environmental limit over which the plant has no control. Water supply at any site varies greatly from year to year, but the temporal patterns of water stress seen by rainfed crops in the semiarid zone bear some relationship to the type of climate experienced. Adaptive strategies in turn are influenced by the probable timing of stress periods. The predominant occurrence of stress may be early in the crop cycle in some regions and growing seasons, in the middle stage of the crop in others, or in the later or seed-filling phase in yet others. This last-mentioned situation is probably the commonest and is given most attention here.

In contrast to natural ecosystems, the environment of crops can be modified by agronomic management. In particular, modern agronomy operates to reduce the possibility of early stress at the stage of germination and seedling establishment. The degree of water stress that may develop later can be influenced by fallowing, sowing date and density, fertilization practices, and so on. In addition, crop scientists often have the option of trading off the drought risk against other environmental hazards. For example, earlier flowering reduces the probability of drought during seed filling, but increases the probability of frost at flowering in many winter rainfall areas, and increases the probability of head molds in sorghum grown in summer wet seasons.

4. TRANSPIRATION

The major effect of environmental variation in water stress on the yield of a given crop is usually explained by variation in the total supply of water to the crop and is reflected in variation in crop evapotranspiration or more precisely crop transpiration (6). Because of water losses, largely through soil evaporation and sometimes in the form of available soil water in the profile at maturity, crop transpiration may be only half the total water supply or less, even in water-limited situations (6). It is worthwhile looking at some aspects of the crop that may increase crop transpiration as a fraction of water supply.

4.1. Ground Cover

Since soil evaporation depends largely on the radiation reaching the soil surface when it is wet, crops that reach full ground cover quickly in regions where rains are frequent, increase transpiration at the expense of soil evaporation. This effect is imitated by wheat sown at high density in a winter rainfall region (e.g., Refs. 9, 10; Table 21.1). At a given date in the spring (approximately anthesis) in the wetter years (1962, 1973) high plant density had produced more dry matter for little extra evapotranspiration or water use compared to low density. Except in the driest year, 1975, the marginal water use efficiency or extra dry matter divided by extra water use, was considerably greater than the prevailing TE of around 5 g/m²/mm (data not shown), because the extra growth and transpiration of the denser crops was largely at the expense of soil evaporation. Nevertheless a faster approach to full ground cover, with more early growth and some, albeit small, increase in water use inevitably reduces soil water available for transpiration during seed filling. The price of the extra early growth will be greatest and the compromise between early growth and early water use most critical where there is little soil evaporation to save—for example, when growth depends entirely on soil water stored at sowing, and when there is a low expectancy of precipitation during seed-filling.

4.2. Roots

Maximization of transpiration/water supply ratios for crops suggests at least two important roles for root systems that, as Passioura (11) points out, are not considered in short-term studies of the effects of the root system on ψ_{plant} and transpiration. First, how might roots compete against direct evaporation

Table 21.1. Effect of Low and High Plant Density on Evapotranspiration (ET) and Dry Matter Production (DM) by Wheat Crops from Sowing to Approximately Anthesis, as Influenced by Rainfall in a Mediterranean Climate

Plant density (per m²)	ET (mm)	DM (g/m²)	Marginal water use efficiency (g/m²/mm)[a]	Rainfall (mm)	Ref.
Tamworth (July 22–October 19, 1973)					
44	233	473	62	189	10
210	237	720			
Wagga Wagga (June 8–October 4, 1962)					
57	175	427	13	152	9
276	197	701			
Tamworth (August 1–October 23, 1975)					
44	206	500	6	115	10
297	238	684			

[a]Increase in dry matter per millimeter of increase in evapotranspiration.

for surface soil moisture? Whenever the topsoil is wet and ground cover incomplete, plants should draw all their transpiration from this layer. The observed concentration of roots in the topsoil would have this effect, but what happens when this layer dries to −15 bars or lower, and then is suddenly rewetted by rain?

The second important question involves the depth and extent to which the crop can dry the soil. Deeper rooting zones, and soil drying to lower levels of ψ_{soil} in heavier textured soils, increase the water available for transpiration, provided the deep soil water used is replenished between crops—but this does not always happen in semiarid cropping systems. Water used from deep in the profile is used efficiently because it is all transpired.

The maximum extent of soil drying in the root zone may depend more on the minimum ψ_{plant} that can be generated than on root density. In fact it has been suggested that plants have a higher root density than they need for maximal rates of water uptake (11, 12), and presumably for complete water extraction. Is high root density a carryover from wild progenitors that was desirable in competitive situations but is no longer useful? On the other hand, in some situations crops have been found to have moderate quantities of available water in the near subsoil at maturity, despite apparent water stress during seed-filling. Perhaps the very wet soil conditions common during the winter in mediterranean climates, where such observations have been made, inhibits deep rooting; also root pathogens or low subsoil nutrient content might be involved.

5. TRANSPIRATION EFFICIENCY

I adopted the transpiration–transpiration efficiency approach here because the trading of water for reduced carbon is basic to water-limited plant productivity. An added bonus of the approach is that the value of water (i.e., TE) is reasonably predictable, showing an inverse relationship to leaf-air vapor pressure difference or more crudely to potential evaporation, and being twice as high for species having C_4 pathways of photosynthesis compared to those with C_3 pathways. In the face of other environmental effects TE varies little (6, 8): the effects of water stress on TE appear to be no exception to this.

However satisfactory measurements of the effects of water stress on TE are limited. Most gas exchange studies have involved abnormally rapid drying and/or artificial control of leaf temperature. There is some evidence from gas exchange studies pointing to increases in TE with drying, such as that which causes leaf flagging or wilting in sunflower (13), and a small decrease in TE with slow drying in sorghum (14). Most of the results based on dry matter harvests of plants in controlled environments point to only small increases in TE in herbaceous C_3 plants with stress (6). Recent data of Gifford (15) are typical of these. Reductions to the water supply of wheat plants, such that overall transpiration was reduced by 70% of adequately watered controls, were associated with only a 20% increase in TE. Since the work was done at a relatively low level of total irradiance, even this may be an overestimation of what might happen in the field. The micrometeorological approach to this question in maize canopies under humid field conditions showed TE to be lower on a day when soil water stress was greater (16). Field data from serial harvests and soil sampling in wheat crops (10) suggest that water limitation may increase TE somewhat, but that the vapor pressure deficit, indicated by pan evaporation, is the dominant influence (Figure 21.1).

Obviously more field measurements of the effects of water stress on TE are needed, but these are difficult. Still it is puzzling that greater increases in TE do not occur under stress, especially when theory seems to suggest that there are mechanisms by which this could occur: the mechanisms suggested include leaf movement, increased leaf reflectance, prevention of the rise in residual resistance to carbon dioxide (CO_2) with stress, and stomatal closing at appropriate times, especially in C_3 species, perhaps in response to increased atmospheric vapor pressure deficit (17). However there may be penalties that have been overlooked: most of these facultative adaptations involve maintenance or improvement of TE at the expense of rates of transpiration and of photosynthesis. As Jones (18) has pointed out, when the

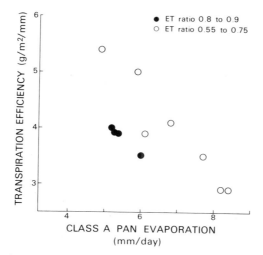

Figure 21.1. Relationship between transpiration efficiency and Class A pan evaporation rate for dense wheat crops at Tamworth, N.S.W., the lower range of evapotranspiration (ET) ratios indicating water stress. Adapted from Ref. 10.

maintenance respiration of plants is allowed for, these facultative responses appear less favorable.

6. ALLOCATION OF ASSIMILATE

The allocation of assimilate is the third major consideration in the approach I have adopted. My interest lies in changes in allocation to different tissue and organs induced by stress and their favorability for yield. Allocation changes induced by stress such as increased root/shoot ratio or altered leaf shape have implications for transpiration and TE. However this chapter discusses only several effects of stress on the allocation processes closely related to seed yield.

Water stress does not need to have any direct effect on short-term allocation patterns for apparent or indirect effects to arise in the long term. For example, stress during seed-filling will reduce harvest index (grain yield/total dry matter at maturity) simply because assimilate production during seed-filling is reduced. However it does seem that short-term allocation patterns can be affected by stress, and the effect may arise independently of stress effects on assimilate production. For example, allocation to leaf growth can be reduced by water stress before changes in leaf photosynthesis can be detected, and this is explained by the greater sensitivity of the former to falling leaf turgor potential.

6.1. Seed Number Bottleneck

In the case of grain crops the sensitivity of seed number to reduced ψ_{plant}, usually at some clearly definable stage of development around flowering, is a dominant allocation consideration. In the case of crops with a high degree of synchrony of development between tillers or branches (e.g., wheat, maize, sorghum), this effect can be disastrous.

I have studied seed number in wheat for which the period of greatest sensitivity to lowered ψ_{plant} is at about 7 days before anthesis at 19°C, when the sensitivity of seed number per spike is much greater than that of leaf senescence (19). Regarding the mechanism involved, part of the effect could be indirect and mediated through probable stress effects on assimilate production, because the inflorescence was growing rapidly in dry weight and was a major sink for assimilate when the stress had its greatest effect. Inflorescence dry weight, as estimated approximately by chaff dry weight at maturity (Table 21.2), was reduced by stress, but without data on assimilate allocation ratios during the actual stress period, the source of this reduction is unclear. Part of the stress effect on seed number, however, lies in seed number per unit inflorescence dry weight, as estimated approximately by chaff dry weight at maturity (Table 21.2), which appears to suggest a direct effect of stress on the growing inflorescence independent of assimilate supply. But even this is not certain because shading 7 days before anthesis leads to very similar effects on seed number, that is, reduced inflorescence weight and reduced seeds per unit inflorescence weight (Table 21.2). Effects of preanthesis water stress and shading on seed number tended to operate via reductions in competent florets per inflorescence at anthesis;

Table 21.2. Effect of Single Periods of Water Stress or Shading on Seed Number per Spike in Wheat in Pots (From Ref. 19 and R. A. Fischer, unpublished)

Stress period relative to anthesis (A) (days)	Terminal plant water potential	Value relative to control (%)[a]		
		Seed numbers	Chaff dry weight	Seed/chaff dry weight
No stress	>−8	(43)	(409)	(106)
Drought, A −22 to A −19	−20	69	59	118
Drought, A −16 to A −13	−19	68	84	81
Drought, A −9 to A −6	−22	38	69	54
Drought, A −2 to A +1	−27	79	75	106
Drought, A +6 to A +9	−30	84	83	101
Shading, A −22 to A −19	>−8	88	95	92
Shading, A −9 to A −6	>−8	58	79	75
Shading, A +6 to A +9	>−8	93	91	103

[a]Actual values of no stress control in parentheses: chaff dry weight in mg and seed/chaff dry weight in seeds/g.

grain set, or pollination and fertilization, which appears to depend on events at and after anthesis, was much less sensitive to lowered ψ_{plant} and to shading.

Although the exact mechanism by which stress reduces seed number may be unclear and may vary between species, its consequences in a teleological sense are probably similar. Harper (3) suggests evolution has favored homeostasis of seed size within most species because of the vital role of the seed in maintaining continuity between generations. In addition, man has selected for stability of seed size and in particular absence of shriveling. Sensitivity of seed number to water stress tends to ensure constancy of seed size by restricting seeds to be filled when assimilate for filling is likely to be limited. The opportunity for this continuous correlation of seed number with resources is greater with nonsynchronous flowering than with synchronous flowering typical of most determinate cereals.

Because shedding takes place before seed growth begins, the restriction of seed number appears to be a relatively efficient mechanism, with little dry matter involved in organs shed or terminated. Nevertheless the biggest effects of water stress on grain yield are usually associated with reductions in seed number. Does this effect represent a bottleneck, which can be eliminated or lessened, or is it a necessary risk-averting strategy? Insofar as seed crops appear to have sufficient photosynthetic capacity to fill more seeds than they form, especially when stress at flowering is followed by adequate water thereafter, but also when stress intensifies after flowering, it would seem that plants are unduly conservative. This might be considered another unwanted primitive trait. It would be interesting to relax selection pressure for plump, well-filled seeds and attempt to find genotypes whose seed number is less sensitive to water stress.

6.2. Seed-Filling and Atmospheric Stress

A second bottleneck in grain crops is the apparent sensitivity of seed-filling to brief periods of high atmospheric stress. This effect may have little to do with stress effects on assimilation. The consequence is shriveled grain and is well known with winter cereals in various regions of the world (referred to as hay curing in the United States, haying off in Australia, hot spell damage in Italy, and *zachvat* and *zapal* in Russia). Most reports claim that damage occurs with only brief periods of atmospheric stress and occurs even under apparent wet soil conditions, and that the greatest susceptibility arises about halfway between flowering and maturity, or about milk dough stage (e.g., Ref. 20). Despite its apparent importance, this phenomenon has received little attention from physiologists, and it is not even clear whether we are dealing with an effect of reduced ψ_{plant} or elevated plant temperature or both. It would seem that this reaction to stress has, teleologically speaking, no function; perhaps it is the response of the plant to a meteorological hazard insufficiently frequent in the evolutionary background of wheat to have encouraged any favorable adaptations.

I looked at this problem in wheat using a specially constructed torture chamber for simulating atmospheric stress. Wheat plants (cv. Gabo), grown in soil with abundant watering, were exposed to a severe atmospheric stress of 6 h at 10, 16, 24, or 31 days after anthesis. Table 21.3 shows the effects on final kernel weight averaged over all exposure dates, since the effect of date was simply a steady diminution in the depression of kernel weight the later the stress was applied, rather than showing more sensitivity at around the soft dough stage as suggested by others. Although the conditions were artificial, there is no doubt that the atmospheric stress was severe; the equivalent transpiration rate for treatment d in Table 21.3 was 2 mm/h, yet the rate was maintained unabated over the 6 h period, as attested by the 6 to 7°C depression in leaf and spike temperatures relative to air temperature throughout the exposure period. Since both plant water deficit and plant temperatures rose with atmospheric stress, it was not possible to attribute

Table 21.3. Effects of Artificial Atmospheric Stress on Relative Water Contents, Leaf and Spike Temperatures, Leaf Damage, and Kernel Weight in Wheat Grown in Pots (R. A. Fischer, unpublished)

Average of single 6 h stress period at either 10, 16, 24, or 31 days after anthesis.

	Control	Atmospheric stress			Atmospheric + soil stress	
	a	b	c	d	e	f
Air temperature (°C)	25	32	39	46	46	46
Relative humidity (%)	High	35	27	17	17	17
ψ_{soil} (bars)	>-1	>-1	>-1	>-1	>-2	>-3
At End of 6 h Stress						
Leaf relative water content (%)	96.0	91.5	92.4	81.4	67.0	<50
Grain relative water content (%)	94.6	93.0	91.2	89.5	87.4	84.8
Leaf temperature (°C)	—	30.0	32.2	39.0	43.2	46.2
Spike temperature (°C)	—	31.0	34.6	40.4	39.4	42.3
24 h Later						
Leaf damage (%)	0	0	0	0	0	50 to 75
At Maturity						
Final kernel weight (%)	100	97	95	87	83	72

the significant reductions in kernel weight with atmospheric stress alone (treatments *c* and *d*) to either one or other effect. When the atmospheric stress of treatment *d* was combined with soil stress (treatments *e* and *f*), potential transpiration rates were not maintained and kernel weight fell further, both plant water deficit and plant temperature rising.

A second experiment was attempted to resolve the water stress–temperature stress question. Stresses were applied at 7 days before anthesis or at 1, 12, or 20 days after, and nitrogen fertilizer treatments were also included (Table 21.4). Both soil and atmospheric stress at the latter two times reduced kernel weight; but since atmospheric stress (treatments *d, e, f*) reduced the kernel weight to a greater degree than soil water stress (treatment *c*), even though the leaf water deficits were greater in treatment *c* than in treatments *d, e,* and *f,* it is likely that the reduction in kernel weight with atmospheric stress arises from elevated temperatures, not from water deficits. Since plant water deficit always increased with such stress, however, we cannot entirely exclude it as a factor in atmospheric stress damage. Also in contradiction to the earlier comments, leaf area loss may be involved. The data in Table 21.4 also tend to confirm field observations that N fertilizer application, and presumably increased tissue N, increased the reduction in kernel weight with atmospheric stress, an effect associated with greater plant water deficit rather than higher plant temperature. Sections *B* and *C* in Table 21.4 show that although stress at 12 and 20 days did not affect seed number (data not shown), atmospheric stress alone (treatments *d* and *e*) at 7 days before anthesis reduced seed number less than soil stress alone (treatment *c*), the effects at this stage being positively related to leaf water deficits. However atmospheric stress at 1 day after anthesis had a greater effect on seed number than did soil stress, which at that stage of development had no effect on seed numbers even though the relative water content was low. This suggests that elevated plant temperature is the more deleterious factor for grain set. Obviously much more work needs to be done on yield reductions due to brief periods of severe atmospheric stress, since these constitute the sort of bottleneck to yield against which plant breeders may be able to make progress. Azzi (20) suggests interesting cultivar adaptations in this regard.

6.3. Seed-Filling and Stored Reserves

The influence of water stress on the mobilization of temporary reserves of assimilate to growing seeds has received some attention lately, in particular the mobilization of reserves present in the crop at anthesis. In wheat at anthesis water-soluble sugars can represent 5 to 7% of the total dry matter. It seems likely that this stem reserve of carbohydrate buffers seed-filling and grain yield against reductions in postanthesis photosynthesis caused, for example, by water shortage (21–25). There seems to be little doubt that the relative contribution to yield of reserves present at anthesis (i.e., reserves translocated to grain divided by total grain yield) increases with postan-

Table 21.4. Effect of Artificial Atmospheric Stress on Leaf Relative Water Content, Leaf and Spike Temperatures, Leaf Damage, and Grain Characteristics of Wheat Grown in Pots (R. A. Fischer, unpublished)

Single 6 h stress period at 7 days before anthesis or at 1, 12, or 20 days after anthesis.

	No atmospheric stress			Atmospheric stress		
	a	b	c	d	e	f
Air temperature (°C)	25	25	25	46	46	46
ψ_{soil} (bars)	>−1	>−1	>−14	>−1	>−1	>−2
N fertilizer	0	+	0	0	+	0

A. Average of Stress at 12 and 20 days After Anthesis

At End of Stress Cycle

Leaf relative water content (%)	98.3	96.7	60.3	90.0	81.0	89.6
Leaf temperature (°C)	~25	~25	~25	39.5	40.0	41.5
Spike temperature (°C)	~25	~25	~25	40.4	40.2	40.4

24 h Later

Leaf damage (%)	0	0	25	20	35	25

At Maturity

Final kernel weight (%)	100	95	91	82	67	83

B. Stress at 7 days Before Anthesis

Leaf relative water content (%)	98.7	97.0	63.8	91.8	86.6	66.2
Seed per spikelet (% of control)	100	91	69	79	91	44

C. Stress at 1 day After Anthesis

Leaf relative water content (%)	97.8	98.3	78.2	88.6	91.9	81.1
Seed per spikelet (% of control)	100	112	111	47	80	69

thesis water shortage. Whether the absolute contribution increases (i.e., g/m²) is not clear. Some data suggest that it does increase in wheat, barley, and maize (21, 22, 24, 25). However a study of drought during seed-filling in wheat and barley using repeated canopy labeling with $^{14}CO_2$ found no difference between stress treatments in the absolute amount of carbon reserves that was present at anthesis and subsequently translocated to the grain (23).

This can be seen from Figure 21.2, which gives the percentage of $^{14}CO_2$ in the grain at maturity and crop growth rate of stressed and unstressed plants: the absolute contribution of photosynthesis to the grain at any point in time is the product of the crop growth rate and proportion of $^{14}CO_2$ in the grain at maturity. The translocated reserves in this study amounted to 12% (irrigated crops) and 22% (droughted crops) of final grain yield or 8.5% (irrigated) and 11.5% (droughted) of total dry weight at anthesis (23).

Calculation of the reserves as a proportion of total dry weight at anthesis, at least for crops with fairly determinate flowering, is a useful guide in discussing this question. For example, in the maize experiment of Boyer and McPherson (21), the smaller relative depression with water stress of grain yield (reduced by 53% of control) compared to that of the total dry weight increment during grain-filling (reduced by 79% of control) is only partly explained by increased utilization for grain growth of reserves present at anthesis; the other factor is that in the control crop the total dry weight increment during grain-filling greatly exceeded grain yield, leaving considerable spare postanthesis assimilate. Reserves present at anthesis and subsequently translocated to the grain were only 18% of anthesis total dry weight under drought, a figure not so different from that of Bidinger et al. (23) for barley and wheat. In another example I calculated from published measurements of the loss of dry weight in nongrain tissue in a barley crop in a dry year (22) that the crop translocated 39% of total dry weight present at anthesis to the grain. This is so high as to seem very unlikely and suggests caution concerning the estimation of the preanthesis assimilate contribution to grain yield based, as is common, on dry weight changes of plant parts from anthesis to maturity, especially in field experiments entailing possible complications due to weight losses caused by weather and disease damage.

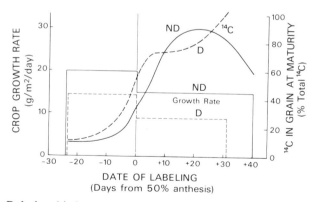

Figure 21.2. Relationship between crop growth rate (histograms) and ultimate fate of ^{14}C-labeled assimilate (curves), and date of labeling in the absence (ND) and presence (D) of water stress during grain filling. Mean of 2 wheat and 2 barley cultivars. Adapted from Ref. 23.

The question of reserves in crops with indeterminate or nonsynchronous flowering is complicated by the long time interval over which flowering extends; however it seems probable that reserves that accumulate before the onset of seed-filling can buffer yield against stress during seed-filling. This has been suggested for dryland soybeans (26) and oilseed rape (*Brassica napus* and *B. campestris*). It was concluded in the case of rape for a dry mediterranean environment that it is best to select for greater early growth before anthesis, when water is unlikely to be limiting, hence placing greater emphasis on seed-filling from the mobilizable reserves formed during early growth (27). A study in the same environment with lupin (*Lupinus angustifolius*), a legume with a determinate habit but nonsynchronous flowering, also indicated that the mobilization of stem reserves may be important in seed-filling (28).

It is interesting to note that the constitutive adaptation of accumulating large reserves by anthesis to guarantee seed-filling involves a clear penalty. Inflorescence growth and reserve accumulation probably compete for assimilate, so that seed number will suffer if large reserves are accumulated: under nonstress conditions, yield will be limited by seed number and the reserves will be unused, hence wasted. A more efficient strategy would seem to call for facultative adaptations leading to reduced seed number with maintenance or even expansion of reserves *only* when postanthesis stress is heralded.

6.4. Effect of Plant Height in Spring Wheat

Because tall wheat cultivars have been dominant in traditional dryland cropping, they might be expected to be less sensitive to water stress than the recent semidwarf cultivars that invariably contain the Norin-10 dwarfing genes and were largely selected under irrigation. The semidwarf habit represents a major constitutive change in allocation. A diverse set of tall (average height without stress 120 cm) and semidwarf (90 cm) wheats were compared on a heavy soil in the absence (ND) and presence (D) of a late drought that commenced just before anthesis, continued until maturity, and reduced the grain yield by 54% on average (Table 21.5). Despite the height differences (and other factors associated with height), the two groups did not differ under drought in yield or in most other traits shown in Table 21.5. However the change in relativity between the tall and semidwarf groups in going from nondrought (ND) to drought (D) showed some differences in yield, leaf water potential, leaf permeability, and the sensitivity of leaf permeability to leaf water potential (Table 21.5). There is much that could be discussed from the data in this table; my point here is that in a situation that might have led one to expect a big genotypic difference in response to stress, the net effect was small, at least in terms of the consequences for grain yield when exposed to late stress.

Table 21.5. Yield and Other Traits in Tall and Semidwarf Wheat Cultivars under Nondrought (ND) and Drought (D) Conditions in Northwest Mexico (29, 30)

Trait	Tall[a] ($n = 8$)	Semidwarf ($n = 23$)	Tall relative to semidwarf (%)	Change in relativity, D less ND (%)
Grain yield (g/m²)				
ND	447*	554	81	
D	225	232	97	+16
Total dry weight (g/m²)				
ND	1283	1322	97	
D	749	725	103	+6
Harvest index (%)				
ND	38**	46	83	
D	32*	35	91	+8
Kernel number ($\times\ 10^2$/m²)				
ND	106**	133	80	
D	66	74	89	+9
Kernel weight (mg)				
ND	42	42	100	
D	34	32	106	+6
Leaf water potential (bars)				
ND	−19.0*	−16.3	117	
D	−27.9	−28.3	99	−18
Leaf permeability (arbitrary units)				
ND	3.7	4.0	93	
D	1.5	1.3	115	+22
Sensitivity (per bar)[b]	0.08*	0.14		

[a]Significantly different from semidwarf at 5% (*) or 1% (**) level.
[b]Slope of linear relationship between leaf permeability and leaf water potential.

7. CONCLUSIONS

I have suggested that the effects of water limitations on crop yield in semiarid regions can be separated into the three major areas of water supply and transpiration, transpiration efficiency, and assimilate allocation. Yield reductions are mostly related to limitation of transpiration due to inadequate water supply, and to unfavorable effects on allocation, either from direct effects of lowered ψ_{plant} or indirect ones via reduced assimilate production at key stages. The greatest opportunity for the plant breeder to counter these effects probably lies with increasing transpiration as a fraction of water supply, and with yield "bottlenecks" that arise through the direct effects of stress on allocation, in particular on seed number. The former attempts to increase assimilation, and the latter attempts to alter favorably the distribution of assimilate within the bounds of a given total assimilate supply. Experience with supposedly drought-adapted wheat cultivars suggests that progress may be slow in such species.

REFERENCES

1. C. M. Donald, in K. W. Finlay and K. W. Shepherd, Eds., *Proceedings of the Third International Wheat Genetics Symposium*, Australian Academy of Sciences, Canberra, 1968, p. 377.
2. M. Grene, *The Understanding of Nature; Essays in the Philosophy of Biology*, Reidel, Dordrecht-Boston, 1974, pp. 207–227.
3. J. L. Harper, *Population Biology of Plants*, Academic Press, London–New York–San Francisco, 1977.
4. A. E. Hall, K. W. Foster, and J. G. Waines, in A. E. Hall, G. H. Cannell, H. W. Lawton, Eds., *Agriculture in Semi-Arid Environments*, Springer-Verlag, Berlin–Heidelberg–New York, 1979, p. 148.
5. J. A. Berry, *Science, 188*, 644 (1975).
6. R. A. Fischer and N. C. Turner, *Annu. Rev. Plant Physiol., 29*, 277 (1978).
7. R. E. Andrews and E. I. Newman, *New Phytol., 67*, 617 (1968).
8. H. van Keulen, C. T. de Wit, and H. Lof, in O. L. Lange, L. Kappen, E.-D. Schulze, Eds., *Water and Plant Life: Problems and Modern Approaches*, Springer-Verlag, Berlin–Heidelberg–New York, 1976, p. 408.
9. R. A. Fischer and G. D. Kohn, *Aust. J. Agric. Res., 17*, 255 (1966).
10. A. D. Doyle and R. A. Fischer, *Aust. J. Agric. Res., 30*, 815 (1979).
11. J. B. Passioura, in L. G. Paleg, and D. Aspinall, Eds., *The Physiology and Biochemistry of Drought Resistance*, Academic Press, New York–London–San Francisco, 1980, (in press).
12. E. I. Newman, in E. W. Carson, Ed., *The Plant Root and Its Environment*, University of Virginia Press, Charlottesville, 1974, p. 363.
13. H. M. Rawson, *Aust. J. Plant Physiol. 6*, 109 (1979).
14. M. M. Jones and H. M. Rawson, *Physiol. Plant., 45*, 103 (1979).
15. R. M. Gifford, *Aust. J. Plant Physiol., 6*, 367 (1979).
16. T. R. Sinclair, G. E. Bingham, E. R. Lemon, and L. H. Allen, *Plant Physiol., 56*, 245 (1975).
17. I. R. Cowan and G. D. Farquhar, in D. H. Jennings, Ed., *Integration of Activity in the Higher Plants*, Society for Experimental Biology, Symposium No. 31, Cambridge University Press, Cambridge, 1977, p. 471.
18. H. G. Jones, *J. Appl. Ecol., 13*, 605 (1976).
19. R. A. Fischer, in R. O. Slatyer, Ed., *Plant Response to Climatic Factors*, *Proceedings of the Uppsala Symposium, 1970*, UNESCO, Paris, 1973, p. 233.
20. G. Azzi, *Agricultural Ecology*, Constable, London, 1956.
21. J. S. Boyer and H. G. McPherson, *Adv. Agron., 27*, 1 (1975).
22. J. N. Gallagher, P. V. Biscoe, and R. K. Scott, *J. Appl. Ecol., 12*, 319 (1975).
23. F. Bidinger, R. B. Musgrave, and R. A. Fischer, *Nature (London), 270*, 431 (1977).
24. R. B. Austin, C. L. Morgan, M. A. Ford, and R. D. Blackwell, in *Plant Breeding Institute, Cambridge, 1977 Annual Report*, 1978, p. 136.

25. J. B. Passioura, *Aust. J. Plant Physiol.*, **3**, 559 (1976).
26. G. A. Constable and A. B. Hearn, *Aust. J. Plant Physiol.*, **5**, 159 (1978).
27. R. A. Richards and N. Thurling, *Aust. J. Agric. Res.*, **29**, 479 (1978).
28. E. A. N. Greenwood, P. Farrington, and J. D. Beresford, *Aust. J. Agric. Res.*, **26**, 497 (1975).
29. R. A. Fischer and R. Maurer, *Aust. J. Agric. Res.*, **29**, 897 (1978).
30. R. A. Fischer and M. Sanchez, *Aust. J. Agric. Res.*, **30**, 801 (1979).

22 | Interaction of Water Stress and Mineral Nutrition on Growth and Yield

A. N. LAHIRI

Division of Soil-Water-Plant Relationship, Central Arid Zone Research Institute, Jodhpur, Rajasthan, India

1. INTRODUCTION

Reviewers generally agree that it is debatable whether fertilizer application is beneficial under drought prone dry farming conditions (1, 2). Arnon (2) suggested that under conditions of limited soil moisture, nutrient deficiency may have adverse effects on plants and that addition of some fertilizer is warranted. On the other hand, it has also been pointed out that the fertilizer-induced increase in water use during the early vegetative period may have adverse effects by increasing water stress at critical stages. The reduced availability and utilization of nutrients under low soil moisture (3, 4) and reports of fertilizer-induced drought tolerance (5, 6) add complexity to the problem. The promotion of root growth by fertilizer application, which facilitates the extraction of water from deeper in the soil (1), and the increased absorption of phosphorus with the increased supply of soil phosphorus at a wide range of soil water potentials (7), also lend support to the argument that soil fertility should be maintained high under water-limited conditions. There are indications that turgor is not the only key to the reduction in growth arising from water deficits (8); some of the decrease in

341

growth and development of stressed plants may arise from nutrient deficiency (9). There is therefore need to evaluate the interaction of nutrient supply and water stress in the context of the type of droughts that occur in nature; we must also evaluate the conditions that modulate nutrient utilization of plants when there is a shortage of water.

2. INFLUENCE OF FERTILIZER ON GROWTH AND YIELD UNDER DIFFERENT CONDITIONS OF DROUGHT

Drought can have widely differing characteristics. Some idea of the different drought conditions and the consequent effects on the growth and yield of pearl millet (*Pennisetum typhoides*) was gained in a study undertaken over four successive years at Jodhpur (about 366 mm average annual rainfall; sandy loam soil with a pH of 7.5, 0.15% organic carbon, 8 kg/ha of available phosphorus as P_2O_5 and 180 kg/ha of available potassium as K_2O).

Figure 22.1 shows that during 1968 the soil water content gradually decreased from sowing to harvest. By 30 days after sowing 50% of the available moisture was removed, and continued removal reduced soil moisture to a very low level toward the end of the growing season. During 1969 very low precipitation kept the soil moisture of the major root zone below the lower level of availability; such a situation is often encountered in arid areas in years when the rains fail completely. In 1971 there was a rainfall of

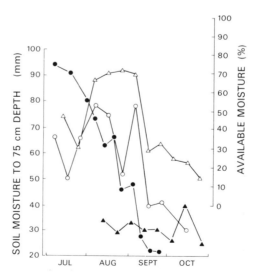

Figure 22.1. Seasonal change in water content and percentage available water in the upper 75 cm of soil under a pearl millet (*Pennisetum typhoides*) crop in 1968 (●), 1969 (▲), 1970 (△), and 1971 (○). The total annual rainfall and the rainfall in the cropping periods for the four years were 179 and 43 mm (1968), 93 and 76 mm (1969), 595 and 442 mm (1970), and 308 and 221 mm (1971), respectively.

308 mm, and soil moisture decreased below the 20% available level at an early stage of growth and again about 50 days after sowing. This situation may be experienced by plants at any stage of development, causing a reduction in vegetative growth and yield depending on the phase at which drought is experienced. Although 1970 was a relatively favorable year, and soil moisture was readily available for the first 50 days after sowing, beyond that the soil moisture sharply decreased to below the 30% available level. The data for dry matter production and yield during these years (Table 22.1) indicate that in 1968 when the soil water was principally utilized for vegetative growth, varietal and fertilizer effects on dry matter production were significant. Although the magnitude of the increases was small because of the general limitation of soil moisture during the vegetative period, there was a significant interaction between variety and fertilizer. However the grain yield was very low, and there was no significant effect of fertilizer on grain production. This situation is apparently akin to the condition in which all conserved soil moisture is utilized for vegetative growth: mitigation of this problem may be possible through identification of crop varieties having high root resistances and slow water use (10, 11). In 1969, when low pre-

Table 22.1. The Effects of Fertilizer Nitrogen and Varietal Differences on the Grain Yield and Dry Matter Production of Pearl Millet (*Pennisetum typhoides*) in Four Successive Years with Different Soil Water Availabilities (see Figure 22.1)

Treatments	Grain yield (kg/ha)				Dry matter yield (kg/ha)			
	1968	1969	1970	1971	1968	1969	1970	1971
Nitrogen (kg/ha)								
0	25	0.5	1680	760	820	66	5080	2370
20	22	0.5	1850	960	1170	70	5560	3170
40	26	0.6	2480	1220	1460	73	6860	3470
60	—	0.5	2310	1230	—	81	7020	3820
LSD								
$p = 0.05$	n.s.	n.s.	390	158	162	n.s.	730	434
$p = 0.01$	n.s.	n.s.	513	208	186	n.s.	960	571
Variety								
RSK	20	0.6	1940	860	1170	87	6720	3350
RSJ	39	0.6	1930	1060	1220	67	6860	3150
Ghana	35	0.5	1770	930	1670	73	5500	3080
Chadi	—	0.5	2000	950	—	83	4860	3030
HBI	4	0.5	2950	1420	540	51	6700	3370
LSD								
$p = 0.05$	11	n.s.	434	177	176	n.s.	812	n.s.
$p = 0.01$	15	n.s.	571	233	208	n.s.	1070	n.s.

cipitation kept the soil water contents low, dry matter production and grain yield were also reduced drastically, and variety and nitrogen had no significant effect on either variable. It is interesting to note, however, that in 1971 when the crop suffered from two cycles of stress, increasing the level of nitrogen up to 40 kg/ha significantly increased both the grain yield and dry matter production. Although the production was not high compared to that of 1970, the yields were reasonable for an arid environment. Even during 1970, stress occurred during seed maturation, but fertilizer significantly increased both dry matter production and grain yield.

Such large variations in soil water and growth are associated with large variations in water use and water use efficiency. Although the influence of variable climate on water use and water use efficiency, with particular reference to plants of the Indian arid zone, has been dealt with in greater detail elsewhere (12, 13), it can also be seen from the data presented in Table 22.2 for one variety of pearl millet grown in 1969 and 1970. In the year with low rainfall (1969) water use was approximately half of that in the wet year (1970), but the water use efficiency was drastically reduced.

Although the foregoing does not include all drought situations, it appears that fertilizer application under conditions of sporadic drought may have beneficial effects on plant growth, and adoption of techniques for soil moisture improvement and conservation may help under the other conditions of drought (14, 15).

3. INFLUENCE OF SOIL FERTILITY UNDER SPORADIC DROUGHT

Conditions of sporadic drought caused by erratic rainfall are frequently encountered not only in arid and semiarid regions, but also in areas of high rainfall (see Chapter 20). In this situation soil water stress does not prevail indefinitely, and plants pass through "wet" and "dry" phases. In view of the possible beneficial effects of fertilizers under these conditions, it is necessary here to discuss the mechanisms that may be involved.

3.1. Transpiration, Nutrient Uptake, and Metabolism During and After Sporadic Drought

It has been observed that plants grown under different levels and intervals of watering reduce water loss by reducing the area of individual leaves and/or the total leaf area of the plant by leaf death (16). However the rate of transpiration per unit leaf area seldom changes significantly. In fact, after a drought cycle, water stress induces a higher rate of transpiration: the increase in the transpiration rate is greater, the longer the cycle of stress. It therefore follows that the transpiration potential after a drought is dependent on the degree of stress to which the plants were subjected. Reports also indicate that water and nitrate follow uniform rates of uptake, and an increase in

Table 22.2. Grain Yields, Water Use, and Water Use Efficiency of Pearl Millet (*Pennisetum typhoides* var. HBI) at Four Levels of Nitrogen Fertilizer Use in a Year with High Rainfall (1970) and a Year with Low Rainfall (1969)

Nitrogen (kg/ha)	1969 (precipitation, 92.7 mm)			1970 (precipitation, 594.8 mm)		
	Grain yield (kg/ha)	Consumptive water use (mm)	Water use efficiency (kg/ha/mm)	Grain yield (kg/ha)	Consumptive water use (mm)	Water use efficiency (kg/ha/mm)
0	0.3	79	0.0038	1970	163	12.08
20	0.5	—	—	2500	171	14.62
40	0.6	81	0.0074	4070	174	23.39
60	0.5	75	0.0067	3280	199	16.48

345

transpiration will increase the rate of uptake of both water and nitrate (17). It may therefore be argued that the level of nitrate in the plant may return to normal when a favorable water regime follows a period of drought, provided the supply of nitrogen in the soil is not limiting. There is also evidence that photosynthesis returns to normal after a drought (18). Studies undertaken on three varieties of pearl millet indicated that the rate of nitrogen uptake remained virtually unaffected over a wide range of soil water potentials (ψ_{soil}). When ψ_{soil} was allowed to vary between field capacity and −1 bar, field capacity and −8 bars, and field capacity and −15 bars throughout the growing period (plants maintained close to field capacity served as controls), the rate of nitrogen uptake was reduced only when the soil water content was allowed to decrease to −15 bars (19). Therefore with a single cycle of drought, higher rates of water and nitrate uptake during the "wet" phase after drought may bring the nitrogen status of the plant back to normal levels.

Studies on the effects of nitrogen metabolism in pearl millet indicate that synthesis of proteins is enhanced on rewatering after a period of drought (20). Consequently the protein content at the postdrought stage was higher than that observed at the predrought stage. Thus we were not surprised to find that plants grown under a low moisture regime not only displayed a higher concentration of total nitrogen (presumably because of a decrease in dry matter), but also a higher concentration of protein nitrogen in the shoot as compared to plants grown under a favorable soil water regime (19). Also, the nitrogen concentration of the grain of plants grown under low soil water regime was higher than that in plants grown under the favorable soil moisture regime, even though the yield of the former was much smaller than the latter (19).

3.2. Nutritionally Induced Growth Under Sporadic Drought

The foregoing observations, coupled with the reports of the dependence of nitrate reductase activity on substrate concentration (21, 22), led us to speculate that a high availability of nitrogen, along with other nutrients, may induce greater plant vigor, which, in turn, should decrease the adverse effects of droughts of short duration. Sinha (23) has subsequently reported that the nitrate reductase activity decreases at wilting but subsequent irrigation immediately increases nitrate reductase such that often the level of nitrate reductase at the postdrought stage exceeds that of the predrought stage. Sinha (personal communication) further notes that improvement of the soil fertility (in this case an increase in the level of potassium with adequate supply of other nutrients) brings about a greater increase in the level of nitrate reductase after drought.

These observations support the hypothesis that increased mineral nutrition improves drought tolerance. Various measures of performance were examined in two varieties of pearl millet (var. RSK and HBI) in sand culture; only the level of nitrogen was varied, and a single cycle of stress ($\psi_{soil} = -15$

bars) was imposed to separate sets of plants at 10 day intervals from 20 to 60 days from sowing. The results showed that the reduction by drought in such characters as height, tiller number, leaf number, and yield may be substantially prevented by high soil nitrogen (6). This phenomenon has been found to exist in similar studies with wheat where the levels of nitrogen, phosphorus, and potassium were separately varied, and in the legumes *Cyamopsis tetragonoloba* (clusterbean), *Phaseolus aconitifolius* (moth bean), and *P. aureus* (mung bean), where phosphorus and potassium levels of the nutrient medium were altered separately. Figure 22.2 shows that the dry matter production and grain yield of clusterbean progressively and significantly improved under higher doses of phosphorus or potassium despite drought, and Figure 22.3 shows a similar effect in pearl millet given different nitrogen regimes. Drought decreased the absolute quantity of nitrogen in the pearl millet (Figure 22.4), but the level of nitrogen in the droughted plants given

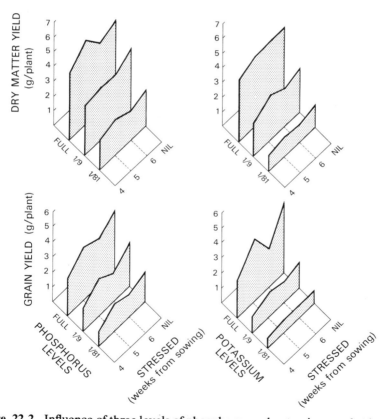

Figure 22.2. Influence of three levels of phosphorus and potassium on the dry matter yield and grain yield of clusterbean (*Cyamopsis tetragonoloba*) under stressed (at 4, 5, or 6 weeks from sowing) and unstressed (NIL) conditions. Effects due to drought, nutrient level and their interaction were significant at the 1% level in all cases.

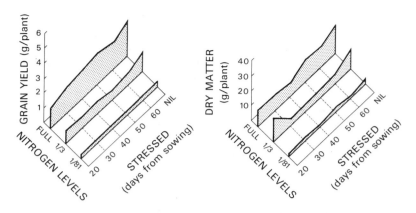

Figure 22.3. Influence of three levels of nitrogen on the dry matter production and grain yield of pearl millet (*Pennisetum typhoides*) under stressed (at 20, 30, 40, 50, or 60 days from sowing) and unstressed (NIL) conditions. Effects due to drought and nitrogen level were significant at the 1% level, but the interaction was significant for dry matter yield only.

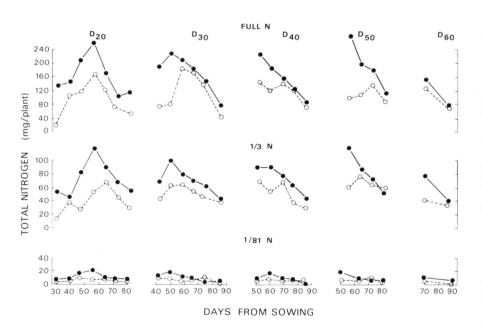

Figure 22.4. Post-drought total nitrogen per plant at various dates after sowing in pearl millet (*Pennisetum typhoides* var. RSK) grown at three levels of nitrogen and either unstressed (●—●) or stressed (○--○) at 20 (D_{20}), 30 (D_{30}), 40 (D_{40}), 50 (D_{50}), or 60 (D_{60}) days from sowing. From Ref. 6.

348

the full complement of nitrogen was comparable with that of undroughted control plants under only one-third the level of nitrogen, and the level of nitrogen in the droughted plants given only one-third the full dose of nitrogen was markedly higher than that in the control plants given only one-eighty-first of the full dose of nitrogen. It was also found that root growth of pearl millet was significantly increased at higher doses of nitrogen, even in the drought treatments. It is possible that enhanced uptake of nutrients during the "wet" phase and their use during the "dry" phase, when their availability and uptake are reduced, helped the plants to minimize the adverse effects of drought. The concept of storage and subsequent use of nutrients under drought is supported by the findings of Boatwright and Viets (24).

It has also been observed that at low values of ψ_{soil} progressive increases in soil fertility not only improve growth, but also increase the absolute quantities of nutrients in the plants. In wheat maintained at a ψ_{soil} of approximately -6 bars throughout the growing period (ψ_{soil} adjusted at 24 h intervals), increasing levels of soil nitrogen and phosphorus progressively promoted dry matter accumulation and also the absolute quantities of nitrogen, phosphorus, and potassium in the shoot (Table 22.3). The magnitude of the increase, however, became lower at increasingly lower values of ψ_{soil}.

Finally, investigations on wheat (var. Kalyansona) indicated that high levels of nutrition induced efficient enzyme activity and higher chlorophyll content even when desiccated (25). The results suggested that the water status of the tissue is not necessarily an index of the efficiency of metabolism. When desiccated, leaves of plants at high fertility lost more moisture than did leaves of plants raised under low fertility; but at the identical water status the enzyme activities and chlorophyll content remained higher in the leaves of plants raised under conditions of high fertility than in those grown under conditions of low soil fertility (Figure 22.5). We thus speculate that metabolic efficiency may depend more on the nutritional status than on the water content of the tissue, at least up to a critical level of water deficit.

Table 22.3. Shoot Dry Weight and Nitrogen, Phosphorus, and Potassium Contents in the Shoot at 115 days from Sowing in Wheat Grown in Soil with Different Levels of Added Nitrogen (N) and Phosphorus (P_2O_5) Fertilizer and Maintained Throughout at a Soil Water Potential of approximately -6 bars

Fertility status (kg/ha of N and P_2O_5)	Dry matter (g/plant)	Nitrogen (mg/plant)	Phosphorus (mg/plant)	Potassium (mg/plant)
0	0.45	6.5	0.13	1.8
40	0.70	9.0	0.50	4.4
80	1.15	15.0	1.06	9.4
120	2.65	34.0	2.25	19.0

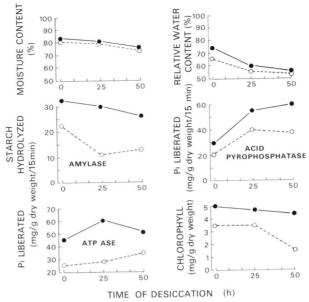

Figure 22.5. Changes during drying in water content, relative water content, chlorophyll content and activities of amylase (starch hydrolyzed), acid pyrophosphatase and ATPase (inorganic phosphate, i.e. Pi liberated) of detached leaves of wheat plants raised under high (●——●) or low (○--○) soil fertility. Redrawn from Ref. 25.

4. APPLICATION TO IRRIGATION MANAGEMENT

In the context of the foregoing, a question arises regarding the prospects of stabilizing the production of irrigated crops with a minimum of irrigation. This issue is particularly important in arid and semiarid areas, which generally have little irrigation water. It has been found that substantial production, associated with an improvement of water use efficiency, can be achieved in wheat (var. RS 31-1) by the optimization of nutrients under low ψ_{soil} (i.e., ψ_{soil} varying from -15 to -6 bars down the profile) (26). With the use of 80 kg/ha of nitrogen and phosphorus, a yield of about 2100 kg/ha was obtained with only 24.5 cm of irrigation water, whereas yield was reduced to 650 kg/ha with no added fertilizer and 40 cm of water (ψ_{soil} was -0.3 to -1 bar). Furthermore, Ram and Lahiri (27) used an economic analysis of yield data of wheat grown under both high and low levels of irrigation and various combinations of nitrogen and phosphorus levels to show that fertilizer application under low levels of irrigation can be economic and has the advantage that limited irrigation water can be spread over a larger area. Although the yield per unit area may be low, the total productivity of the region may increase severalfold. Singh (28) showed that 84 cm of water applied to 1 ha produced

5500 kg of wheat, whereas the same quantity spread over 3 ha produced 9100 kg of wheat. Similarly, 50 cm of water given to 1 ha yielded 1600 kg of sunflower, but the same amount of water spread over 2 ha produced a yield of 2700 kg of sunflower. Thus an improvement in the water use efficiency achieved by means of fertilizers may help to stabilize production in an area in which there is a shortage of irrigation water.

5. CONCLUSIONS

Where sporadic droughts prevail, it appears that periods of favorable soil moisture during plant development are vital for nutritionally induced stabilization of growth and yield. The body of data that suggests advantages of fertilizer application under dryland agriculture (29) possibly includes situations in which "wet" and "dry" phases exist during the growing period. Nevertheless, the benefits of fertilizers under such conditions may be smaller than in well-irrigated crops. When the soil water content is perpetually low—a condition commonly encountered in arid areas—the advantage of fertilizer application is admittedly uncertain. Where the soil water content gradually decreases during the growing season, several factors, such as initial water status of the soil profile, root growth, and moisture use by plants, may greatly influence the effect of fertilizers. Normally vegetative growth, rather than grain yield, is expected to be affected more by fertilizers in this situation. Thus strategies to stabilize production in arid and semiarid regions rest on the identification of the type of drought predominant in each area. Although the basic issue of the relative effect of tissue water status and nutrition on the growth reduction in stressed plants is difficult to resolve, the importance of nutrition in water-stressed plants cannot be completely ignored.

REFERENCES

1. F. G. Viets, Jr., *Adv. Agron.*, **14**, 223 (1962).
2. I. Arnon, in U. S. Gupta, Ed., *Physiological Aspects of Dryland Farming*, Oxford and IBH, New Delhi, 1975, p. 3.
3. F. G. Viets, Jr., in T. T. Kozlowski, Ed., *Water Deficits and Plant Growth*, Vol. 3, Academic Press, New York–San Francisco–London, 1972, p. 217.
4. J. E. Begg and N. C. Turner, *Adv. Agron.*, **28**, 161 (1976).
5. Anonymous, *Successful Farming's Soils Book*, 8th ed., Meredith, Des Moines, Iowa 1959.
6. A. N. Lahiri, S. Singh, and N. L. Kackar, *Proc. Indian Nat. Sci. Acad., Part B*, **39**, 77 (1973).
7. R. G. Fawcett and J. P. Quirk, *Aust. J. Agric. Res.*, **13**, 193 (1962).
8. Y. Vaadia and Y. Waisel, in R. M. Hagan, H. R. Haise, and T. W. Edmister, Eds., *Irrigation of Agricultural Lands*, Am. Soc. Agron., Madison, Wisconsin, 1967, p. 354.

9. R. O. Slatyer, in J. D. Eastin, F. A. Haskins, C. Y. Sullivan, and C. H. M. Van Bavel, Eds., *Physiological Aspects of Crop Yield,* Am. Soc. Agron. and Crop Sci. Soc. Am., Madison, Wisconsin, 1969, p. 53.

10. J. B. Passioura, *Aust. J. Agric. Res.*, **23**, 745 (1972).

11. J. B. Passioura, *Aust. J. Plant Physiol.*, **3**, 559 (1976).

12. A. N. Lahiri, *Ann. Arid Zone*, **14**, 135 (1975).

13. A. N. Lahiri, in P. L. Jaiswal, Ed., *Desertification and Its Control*, Indian Council of Agricultural Research, New Delhi, 1977, p. 225.

14. National Academy of Sciences, *More Water For Arid Lands: Promising Technologies and Research Opportunities*, National Academy of Sciences, Washington, DC, 1974.

15. H. S. Mann and A. N. Lahiri, *Proc. Indian Nat. Sci. Acad., Part B*, **45**, 1 (1979).

16. A. N. Lahiri and B. C. Kharabanda, *Proc. Nat. Inst. Sci. India, Part B*, **32**, 34 (1966).

17. G. Jensen, *Physiol. Plant.*, **15**, 791 (1962).

18. M. M. Ludlow and T. T. Ng, *Plant Sci. Lett.*, **3**, 235 (1974).

19. A. N. Lahiri and S. Singh, *Proc. Indian Nat. Sci. Acad., Part B*, **36**, 112 (1970).

20. A. N. Lahiri and S. Singh, *Proc. Nat. Inst. Sci. India, Part B*, **34**, 313 (1968).

21. T. E. Ferrari and J. E. Varner, *Plant Physiol.*, **44**, 85 (1969).

22. K. T. Glasziou, *Annu. Rev. Plant Physiol.*, **20**, 63 (1969).

23. S. K. Sinha, in L. G. Paleg and D. Aspinall, Eds., *The Physiology and Biochemistry of Drought Resistance*, Academic Press, New York–London–San Francisco, 1980 (in press).

24. G. O. Boatwright and F. G. Viets, Jr., *Agron. J.*, **58**, 185 (1966).

25. S. Kathju and A. N. Lahiri, *Plant Soil*, **44**, 709 (1976).

26. A. N. Lahiri, *Proceedings of the Symposium on Crop Response to Environmental Stresses*, Vivekananda Laboratory for Hill Agriculture, Almora, India, 1975, p. 13.

27. K. Ram and A. N. Lahiri, *Indian J. Agric. Res.*, **8**, 25 (1974).

28. S. D. Singh, *Trans. Indian Soc. Desert Technol. Univ. Cent. Desert Stud.*, **1**, 83 (1976).

29. G. Kemmler, *Plant Res. Dev.*, **5**, 70 (1975).

23 | Interaction and Integration of Adaptive Responses to Water Stress: The Implications of an Unpredictable Environment

H. G. JONES

East Malling Research Station, East Malling, Kent, U.K.

1. INTRODUCTION

Competition has led to the evolution of many mechanisms enabling plants to survive environmental stresses. For drought these include escape, tolerance of internal water deficits, and avoidance of internal water deficits by either maximizing water uptake or controlling water loss. Details of the various mechanisms have been mentioned by other contributors to this book (see Chapter 1), and reviewed elsewhere (1–5). Although it may be convenient for the ecologist or physiologist to consider the various mechanisms separately, most are not independent, since changing one character may require compensating changes in others. For example, changing the photosynthetic mechanism from C_3 to crassulacean acid metabolism involves alteration of stomatal behavior (6). Interestingly, each species tends to emphasize only one mechanism, so that most plants are readily classifiable into one of the groups mentioned (1), though combinations such as tolerant avoiders are known (3).

For agricultural crops, where normal evolutionary constraints do not apply, there seems no a priori reason for not combining several mechanisms

353

when breeding new cultivars. Similarly, the mechanisms favoring high agricultural productivity in drought are unlikely to be the same as those that have evolved in competitive situations (7). There is, however, little information available on how the different drought resistance mechanisms interact when in combination, and which ones are most likely to increase economic yields rather than genetic survival.

More important than the simple interactions among the different mechanisms is how they interact with the environment to influence yield. Most adaptations favoring survival when water is limited tend to reduce potential yield (e.g., Ref. 4). Even the observed responses of plants to stress—for example, reduction in growth and stomatal conductance—tend to reduce yield while helping to conserve water and minimize the stress. To what extent this phenomenon may be regarded as truly adaptive, or whether it is simply a biological example of Le Chatelier's principle (i.e., if a system in equilibrium is subjected to a stress, the system tends to react in such a way as to undo the effect of the stress), is a matter for debate. The former interpretation seems preferable, since although the general pattern of response to stress is similar in many plants, there are important differences in detail. Examples are the range in stomatal (8, 9) and developmental (10, 11) responses to drought.

This chapter discusses some problems of optimizing yield when water is likely to be limiting. Certain strategies for water use and drought avoidance and escape [the terminology of Levitt (3), not that in Chapter 1, is followed in this chapter] are considered in the context of the unpredictable nature of the environment and of the constraints set by the plant and its ability to sense the environment (especially water availability). Although similar techniques could be applied to the various tolerance mechanisms such as osmotic adaptation and desiccation tolerance, they would be difficult to quantify, since the "costs" involved in these adaptations are at present unknown (5). Presumably the costs are substantial, or else these mechanisms would be ubiquitous.

Before analyzing various strategies, we need to consider what is meant by drought resistance in terms of the "goals" to be achieved in different situations. In nature the primary goal of any organism is genetic survival, but for agricultural purposes the goals may be defined differently, generally on an economic basis. For example, in much advanced agriculture the goal is to maximize average economic yield over many years, whereas the goal of minimizing the chance of crop failure or of increasing yield stability may be more appropriate for subsistence agriculture.

The strategy required by the plant breeder to achieve the primary yield goal will depend on the plant material available, as well as on environment. Although in principle we are free to design plants *ab initio,* in practice we can modify existing material only to a limited degree, usually by incorporating characters that have arisen by natural selection in related genotypes. The constraints imposed by plant type may have important implications. For example, maximum average yield in an annual plant will be achieved by maximizing yield each year. In a perennial, however, there may be an optimum partitioning of dry matter into yield that is not equal to the maximum. Many fruit trees, for example, if allowed to crop too heavily in one year, go into a state of biennial bearing, producing a crop only in alternate years. Not only is total yield reduced compared with annual bearing, but economic yield is further reduced because fruit tend to be too small in the "on" years (12).

Since increased expression of most drought tolerance mechanisms reduces the potential maximum yield, there is usually an optimum level of expression for any situation. The major difficulty in determining the optimum is the unpredictability of the weather, though the capacity of the plant to sense and respond to changes in the environment and water supply is also important. Given a detailed knowledge of the environment and water availability during the course of a season, it is possible, at least in principle, to calculate retrospectively the precise tactics in terms of leaf and root development and stomatal behavior that would have been required to maximize yield. Unfortunately we do not have enough detailed physiological and developmental knowledge to do this exercise for any crop, though approximate answers could be obtained using a simple modeling approach.

The main problem, however, is that this approach can only be used a posteriori, since the optimum state at any instant depends on the probability of future rainfall and evaporative demand as well as on current environment. There have been few attempts (e.g., Ref. 13) to study quantitatively the implications for drought resistance of the uncertainty in timing and quantity of rainfall, though the problem is well known (7, 14–19). There has, however, been some success in determining optimum maturity dates for dryland crops in relation to climate and soil water-holding capacity (14, 19).

2. A STOCHASTIC MODEL TO INVESTIGATE PLANT BEHAVIOR

This section develops a stochastic modeling approach to compare a few types of behavior available to plants for achieving high yield in drought. In its simplest form the approach is based on the notion that the ideal behavior maximizes assimilation (therefore yield) for the amount of water *available*. This is more likely to be a general objective, where water is limiting, than maximizing water use efficiency (assimilation per water *used*), since maximizing water use efficiency does not necessarily make the best use of water and is a poor competitive strategy (15, 20, 21). The model considers the simple goal of maximizing mean annual assimilation (= vegetative yield). It is then used to compare stomatal behavior in different climates, assuming for simplicity that the climates differ only in the probability of rainfall.

Assume initially a crop that has developed its final leaf area and has a maximum growing period left of t days (e.g., delimited by extreme temperatures) with a certain amount of available water (W_0 = 5.5 mmol/cm²/day) stored in the soil profile. If the extra rainfall during the growing season is a random variable W (in mmol/cm²/day) whose probability density function is $f_W(w)$, the total amount of water available per day ($W^* = W_0 + W$) at a constant rate of use will have a probability density function $f_{W^*}(w^*)$. If the daily assimilation rate A (in mmol/cm²/day; initially assumed to be constant throughout a season) is given by any function $g(w^*)$, the mean A over many years is given by

$$\bar{A} = \int_{W_0}^{\infty} g(w^*)\, f_{W^*}(w^*)\, dw^* \tag{23.1}$$

The relationship between daily A and transpiration E used for the purposes of the present calculations is given by the solid line in Figure 23.1. This was calculated from data of Jones (15), which were assumed to be typical for a plant with a C_3 pathway of photosynthesis. The assumed probability density functions $f_{W^*}(w^*)$ for the two climates are shown in Figure 23.2. Each consists of a discrete probability of zero rainfall together with a portion of the normal distribution. Climate I represents a mediterranean or monsoonal climate at the end of the wet season; climate II is more maritime. We begin by assuming that all the rain falls early in the season. All calculations of $\bar{A}$ were done numerically using Equation 23.1, where probabilities of w^* were summed over intervals of 1 mmol/cm²/day.

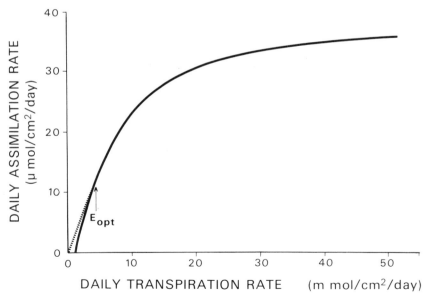

Figure 23.1. The assumed relationship between daily assimilation A and transpiration E, calculated from data of Jones (15; Figure 8, curve 2). The maximum ratio of A/E is obtained at E_{opt}.

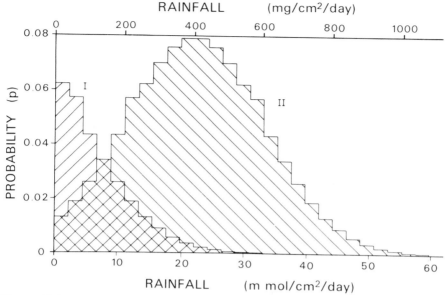

Figure 23.2. The probability density functions for rainfall in two climates. The probabilities of zero rainfall are 0.726 and 0.023 in climates I and II, respectively. The remaining parts of the distribution functions follow the normal distribution with mean $= -6.7$ and $\sigma = 22$ mmol/cm^2/day for climate I, and mean $= 22$ and $\sigma = 22$ mmol/cm^2/day for climate II. Rainfall is given in millimoles per square centimeter per day for consistency: the equivalent in milligrams per square centimeter per day is also given.

For any curve relating A and E, it can easily be shown that the maximum total seasonal assimilation will be achieved when daily transpiration is constant and equal to W^*, as long as $W^* \geqslant E_{opt}$ (where E_{opt} is the value of E giving the maximum water use efficiency—see Figure 23.1), and there are no points of inflexion in this region. If $W^* < E_{opt}$, however, total assimilation is maximal if E is again constant but equal to E_{opt} until all water is used (cf. Ref. 16). It follows from this that $g(w^*)$ for maximum $\bar{A}$ is given by the curve relating A to E for all $E \geqslant E_{opt}$, and by the straight line from the origin to the point on the curve corresponding to E_{opt} for all smaller values of E (see Figure 23.1).

3. RESULTS OF SIMULATION

The maximum mean total assimilation rate (yield) attainable with "ideal" stomatal behavior (determined retrospectively), such that each year the stomatal aperture is appropriate to give maximum A for the amount of rain in that year, is obtained by substituting the function $g(w^*)$ from Figure 23.1, into Equation 23.1. This value of $\overline{A}$ (Table 23.1, item 2) can be used as a basis for comparing other, nonideal, types of stomatal behavior, which have different functions $g(w^*)$. Water use and assimilation are expressed per day, so that results are independent of the absolute length of the growing season.

3.1. Constitutive or Nonresponsive Stomatal Behavior

One type of drought resistance involves permanent or constitutive adaptations. An example for stomata would be the limiting of the maximum leaf conductance by having few, small stomata (8, 9, 22), thus allowing conservation of limited water.

In the first case we consider, stomatal aperture is maintained at an appropriate value to give maximum assimilation if no rain falls (i.e., $W^* = W_0$). In this case $g(w^*)$ will be a constant, and total assimilation will also be constant, irrespective of rainfall (Table 23.1, item 1). Clearly the average difference between this and ideal behavior decreases as the probability of extra rain decreases. In fact, for the present example, the difference disappears as soon as the probability that $W^* > E_{opt}$ is zero (see Ref. 21).

The second alternative is the more optimistic approach of fixing the stomatal aperture appropriately for a certain amount of extra rain. In this case, although $g(w^*)$ will be a constant, total seasonal assimilation will vary, since the amount of water available may not sustain assimilation at this rate

Table 23.1. Effect of Stomatal Behavior on Mean Daily Assimilation in Different Climates and the Range Within which 95% of the Values Fall

Assimilation (μmol/cm^2/day) averaged over the whole season; W_0 was 5.5 mmol/cm^2/day (see Figure 23.2).

Behavior	Climate I	Climate II
All Rain Near Beginning of Season		
1. Stomata set for no extra rain	14.8 (14.8–14.8)	14.8 (14.8–14.8)
2. Ideal behavior	17.3 (14.8–23.0)	31.4 (1.5–35.7)
3a. Best fixed aperture	15.1 (14.1–23.0)	28.4 (7.7–35.7)
3b. Fixed rate of E to give 3a (mmol/cm^2/day)	7.2	22.8
4. Pessimistic responsive behavior	17.3 (14.8–23.0)	31.4 (1.5–35.7)
Uniform Probability of Rain Through Season		
5. Ideal behavior	—	26.5
6. Pessimistic responsive behavior	—	23.8
7a. Best fixed aperture	—	18.1
7b. Fixed rate of E to give 7a (mmol/cm^2/day)	—	8.3
8. Fixed rate with stomatal closure	—	21.8

for all the remainder of the season. The highest mean yields attainable, together with the constant E required are shown in items 3a and 3b of Table 23.1. Although it increases the chance of higher yields, this behavior also involves some risk (see Table 23.1), since if the required extra rain does not fall, some yield will be lost compared with the more conservative behavior above, if $E > E_{opt}$.

These two types of behavior are illustrated in Figure 23.3 (*a* and *b*). In the long run the best fixed aperture gives as much as 87 or 90% of the mean ideal

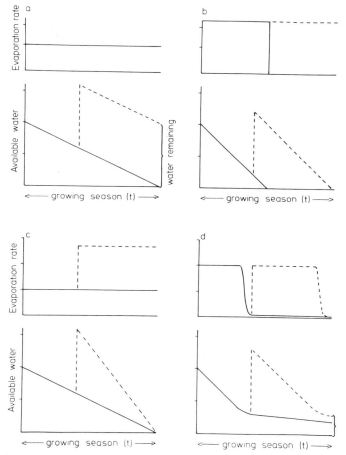

Figure 23.3. The main classes of stomatal behavior and their effects on water use. The solid lines represent the time course of evaporation rate (upper figure) or water remaining in the soil (lower figure) if no rain falls. The dotted lines illustrate the consequences of rain part way through the season. (*a*) Stomata fixed to use W_0. (*b*) Stomata set to use water faster than justified by W_0 (optimistic constitutive behavior). (*c*) Stomata respond to keep water use at current optimum (pessimistic responsive behavior). (*d*) Stomatal closure to prevent complete desiccation.

yield for the two climates, respectively, though there is a risk of very low yield, particularly with climate II (Table 23.1).

So far the assumption has been made that any extra rain falls early enough to be utilized efficiently. This is probably a reasonable simplification for our purposes, at least in climates where the rainfall probability falls off rapidly at the end of the wet season (23). For a more realistic analysis with climate II, however, it is necessary to take account of the timing of any rain, since equal amounts are generally more useful earlier in the season. To do this it was assumed for simplicity that any rain during the season falls at one time only, but that there is a uniform probability of this event occurring at any time during the season. In this case mean yield was estimated by graphical interpolation between values of $\overline{A}$, calculated as above for rain at particular times during the season.

Table 23.1 shows that introduction of uniform rainfall probability compared with all the rain near the start of the season reduces the maximum mean yield attainable with a fixed stomatal aperture from 28.4 to 18.1 μmol/cm²/day. There is a corresponding change in the optimal constant evaporation rate (Table 23.1). The low average yield in this situation arises because plants die when they have used all available water; therefore they cannot use any subsequent rain.

3.2. Pessimistic Responsive Stomatal Behavior

The other main type of behavior available consists of the plant sensing the amount of water available and responding appropriately to any inputs. Where the rain falls near the start of the season this response, with the stomata responding to maintain E and A at the current optimum, is identical to the ideal retrospective behavior (Table 23.1, items 2 and 4). With rain at other times, however, yield with this responsive behavior (Table 23.1, item 6) falls to 90% of that obtainable retrospectively (Table 23.1, item 5). The response of stomata maintaining E at the current optimum is illustrated in Figure 23.3c. This behavior may be regarded as "pessimistic," since the plants are always keeping enough reserves to last the season, and the only adjustments involve an increased aperture in response to rain.

3.3. Optimistic and Survival Stomatal Behavior

By analogy with the constitutive types of behavior already considered, it may be advantageous, even with responsive stomata, to use water more rapidly than is appropriate for the currently available water, in the expectation of more rain. The improvement in mean yield is, however, unlikely to be more than a few percent above that obtainable with the pessimistic responsive behavior discussed.

The other response not so far considered is the common one of stomatal closure to minimize desiccation damage (9), which is illustrated in Figure

23.3*d*. This ability reduces the risks associated with otherwise optimistic water use behavior and must have been of great selective advantage in evolution, since it is almost universal in higher plants. Its potential value even in our hypothetical crop is shown by noting that adding a stomatal closure response (preventing death) to an otherwise fixed aperture can improve the average yield in climate (Table 23.1, items 8 and 7a) from 18.1 to 21.8 μmol/cm^2/day. The latter value is 92% of the yield obtainable with the best pessimistic response, but has much less sophisticated requirements for environmental sensing.

3.4. Nonstomatal Responses

The simple models presented so far have indicated the advantages of responsive versus constitutive stomatal behavior in most climates. Variation in stomatal aperture, however, is only one of many responses affecting water use. These include changes in life cycle (3, 4, 7), root growth (24, 25), leaf angle (3, 26, 27), leaf anatomy (2, 3, 28), leaf area (3, 27, 29), and photosynthetic mechanism (6, 30).

Changes in leaf area provide a particularly important alternative to changing stomatal aperture, since at constant stomatal aperture, transpiration is approximately proportional to leaf area index (LAI) at values below about 3 (31). Changes in leaf area may permit full use of any extra rainfall while still allowing stomatal aperture to be maintained at the optimum for high A/E. The effects of leaf area changes in response to rain can be calculated in a similar way to that used in deriving Table 23.1. For example, if one assumes that LAI was originally 1 and that it changes ideally (up to a limit of 3) in response to any rainfall [with transpiration rate maintained optimal (= E_{opt}) throughout], the mean total assimilation increases from 17.3 μmol/cm^2/day with responsive stomata alone, to 18.7 μmol/cm^2/day.

It is worth noting that the leaf area response is inherently slower and less flexible than the stomatal response, and is therefore better suited for long-term adaptation, particularly since the large energy waste if leaves abscise before harvest makes it also intrinsically less reversible (15, 29). More rapid and reversible changes in effective leaf area can be achieved in some plants, especially in grasses, by leaf rolling (32 and Chapter 3).

4. DISCUSSION AND CONCLUSIONS

4.1. Sensitivity to Available Water

The preceding section noted that the general pessimistic response behavior requires that plants be able to measure accurately the total available water at any time. It is not clear, however, how this might be accomplished. The most obvious possibility is by way of the water potential of some part of the system (e.g., the shoot), but shoot water potential is a function not only of soil water but also of the rate of evaporation (see Ref. 33). However predawn water potentials, particularly in the roots, are less sensitive to the atmospheric environment. The second difficulty associated with shoot water potential as a measure of available water is that a small volume of wet soil near the surface will have the same effect on shoot water potential as would a large volume of wet soil. Individual measurement of water potential in each root and integration of the result, perhaps by production of abscisic acid (ABA) in proportion to stress at each point and subsequent translocation to the shoot, is a possibility. It is known that ABA is produced in roots in response to stress (34, 35) and that it can be translocated in the xylem (36). However the plant still has no means of determining the depth of soil with available water at an early stage in the season before roots have greatly extended. Cohen (20) has discussed the possible role of water potential gradients, soil carbon dioxide concentration, and diurnal temperature fluctuations as sensors of root depth as required for estimation of soil water.

In practice the more pessimistic water conservation behavior is rare (particularly in crop species), and most plants tend to use water faster (Figure 23.3d) than can be justified by the quantity available (29, 37). The lack of an appropriate sensing mechanism for available water probably explains the rarity of pessimistic responses in spite of their potential value. Constitutive water conservation responses, on the other hand, probably are rare because they have little selective advantage in natural competitive communities, but they are likely to be useful in agricultural situations where there is a high probability of suboptimal rainfall. In such a climate, selection for few, small stomata (22, 38, 39) or large root resistance (37) may well improve drought resistance.

Although most species close their stomata with increasing stress, there is a wide range in stomatal sensitivity to water stress (8, 9). Some workers have reported that a ''sensitive'' stomatal response to stress favors drought resistance (40, 41), but others say that the more drought resistant genotypes maintain their stomata open at lower leaf water potentials (42, 43). These differences may be interpreted in terms of the climate and degree of competition in the natural habitats, both water-saving and water-spending behavior being useful (3). For agricultural production the precise response required depends on whether continued assimilation is more important than survival in the particular agroeconomic situation, and is likely to differ in annuals and perennials. Further discussion can be found elsewhere in this volume (9).

4.2. Variability of Aerial Environment

An important simplification throughout this analysis has been the assumption of a constant aerial environment. In fact, it varies spatially, diurnally, and seasonally in an unpredictable as well as a systematic fashion. Such changes can be accommodated, at least retrospectively, when determining ideal stomatal behavior, but in practice random microclimatic variation raises difficulties similar to those already discussed for rainfall.

Some useful generalizations are, however, available. It has been shown (16, 21) that maximum assimilation for a given water use will be obtained if the stomata continually adjust to maintain $\partial E/\partial A = R'$, where R' is a constant (as long as the relationship between E and A is everywhere curved in the sense $\partial^2 E/\partial A^2 > 0$). Similarly, it can also be shown (16) that R' must be uniform among the different leaves of a plant. Although it might be thought that R' ought to be constant throughout a season, Cowan and Farquhar, Cowan (16, 21) argue that this is unlikely to be the case. One reason is that there may be a greater requirement for assimilation at certain critical stages of development. A more important reason, however, is the impossibility of determining an appropriate R' other than retrospectively, because of environmental unpredictability. The authors suggest that like the stomatal closure response discussed above, R' should decline as water availability decreases, to conserve water. In the present example the two responses are identical; but where microclimate changes during the day, it would probably be best to relate A and E using a constant R' rather than a constant conductance. The difference between the two results will be small except when stomata are nearly closed (16). Neither approach would easily maintain daily evaporation constant for long periods as assumed in the simulations given previously, but this should not affect the general conclusions reached.

4.3. Conclusions

The introduction pointed out that a plant can make a wide variety of active changes to adapt to stress. Interestingly, many of these (including production of leaf hairs, reduction in leaf size and number, altered flowering time, and stomatal closure) can be mimicked with exogenously applied ABA (44, 45). In fact there is increasing evidence that ABA may function as a growth regulator integrating plant responses to stress (46).

It is not possible to discuss all aspects of drought resistance behavior in this chapter, neither is there space to extend the model to consider non-vegetative crops. Many of these aspects are discussed by other contributors to this volume, and other important factors include the timing of the switch from vegetative to reproductive growth (13, 47, 48) and problems of assimilate partitioning in relation to water availability (7, 49), are covered elsewhere. The effects of rainfall variability on grain yield are even greater than they are on vegetative yield, because the length of the growing season (17) and distribution of assimilation during the season are so important.

Although we are a long way from a complete understanding of the interactions between the physiological, developmental, and morphological mechanisms that affect drought resistance, and how they are integrated to control yield in different climates, the stochastic modeling approach introduced here is a first step toward quantitative evaluation of the various types of behavior. This information should provide a basis for rationalizing breeding efforts and should help explain the types of drought tolerance observed in natural plant communities. The model is useful for comparing different types of behavior, even though the absolute values may not be realistic. It is likely that the most useful information will come from relatively simple models (such as those used here) operating over only one or two levels of organization, even though they neglect some second-order interactions in the short term and at the cellular and biochemical levels. Improved models may require better simulation of the distribution of rainfall sequences and amounts during the season (50).

REFERENCES

1. N. A. Maximov, *The Plant in Relation to Water*, Allen & Unwin, London, 1929.
2. J. Parker, in T. T. Kozlowski, Ed., *Water Deficits and Plant Growth*, Vol 1, Academic Press, New York–London, 1968, p. 195.
3. J. Levitt, *Responses of Plants to Environmental Stresses*, Academic Press, New York–London, 1972.
4. J. E. Begg and N. C. Turner, *Adv. Agron.*, **28**, 161 (1976).
5. N. C. Turner, in H. Mussell and R. C. Staples, Eds., *Stress Physiology in Crop Plants*, Interscience, New York–London, 1979, p. 343.
6. C. B. Osmond, *Annu. Rev. Plant Physiol.*, **29**, 379 (1978).
7. R. A. Fischer and N. C. Turner, *Annu. Rev. Plant Physiol.*, **29**, 277 (1978).
8. H. G. Jones, in H. Mussell and R. C. Staples, Eds., *Stress Physiology in Crop Plants*, Interscience, New York–London, 1979, p. 407.
9. M. M. Ludlow, Chapter 9, this volume.
10. J. F. Angus and M. W. Moncur, *Aust. J. Agric. Res.*, **28**, 177 (1977).
11. J. C. O'Toole and T. T. Chang, in H. Mussell and R. C. Staples, Eds., *Stress Physiology in Crop Plants*, Interscience, New York–London, 1979, p. 373.
12. L. B. Singh, *J. Hortic. Sci.*, **24**, 45 (1948).
13. D. Cohen, *J. Theor. Biol.*, **33**, 299 (1971).
14. R. O. Slatyer, in *Agroclimatological Methods: Proceedings of the Reading Symposium*, UNESCO, Paris, 1968, p. 73.
15. H. G. Jones, *J. Appl. Ecol.*, **13**, 605 (1976).
16. I. R. Cowan and G. D. Farquhar, in D. H. Jennings, Ed., *Integration of Activity in the Higher Plant*, Society for Experimental Biology, Symposium No. 31, Cambridge University Press, Cambridge, 1977, p. 471.
17. N. V. Kanitkar, *Dry Farming in India*, Manager of Publications, Delhi, 1944.
18. J. D. Bilbro, *Tex. Agric. Exp. Stn., Misc. Publ.*, **847** (1967).
19. I. Arnon, *Crop Production in Dry Regions*, Vols. 1, 2, Hill, London, 1972.
20. D. Cohen, *Isr. J. Bot.*, **19**, 50 (1970).
21. I. R. Cowan, *Adv. Bot. Res.*, **4**, 117 (1977).
22. H. G. Jones, *J. Exp. Bot.*, **28**, 162 (1977).
23. P. Meigs, in *Reviews of Research on Arid Zone Hydrology*, UNESCO, Paris, 1953, p. 203.

24. R. S. Russell, *Plant Root Systems: Their Function and Activity in the Higher Plant*, Cambridge University Press, Cambridge, 1977.
25. H. M. Taylor and B. Klepper, *Adv. Agron.*, **30**, 99 (1978).
26. J. E. Begg and B. W. R. Torssell, in R. L. Bieleski, A. R. Ferguson, and M. M. Cresswell, Eds., *Mechanisms of Regulation of Plant Growth*, Bulletin 12, Royal Society of New Zealand, Wellington, 1974, p. 277.
27. J. E. Begg, Chapter 3, this volume.
28. P. S. Nobel, Chapter 4, this volume.
29. J. B. Passioura, in I. F. Wardlaw and J. B. Passioura, Eds., *Transport and Transfer Processes in Plants*, Academic Press, New York–San Francisco–London, 1976, p. 373.
30. C. C. Black, *Annu. Rev. Plant Physiol.*, **24**, 253 (1973).
31. J. L. Monteith, *Principles of Environmental Physics*, Edward Arnold, London, 1973.
32. R. B. Austin and H. G. Jones, *Plant Breeding Institute, Cambridge, 1974 Annual Report*, 1975, p. 20.
33. P. G. Jarvis, in D. A. de Vries and N. H. Afgan, Eds., *Heat and Mass Transfer in the Biosphere*, Vol. I, *Transfer Processes in the Plant Environment*, Scripta, Washington, DC, 1975, p. 369.
34. M. L. Barr, *Plant Physiol. (Suppl.)*, **51**, 47 (1973).
35. D. C. Walton, M. A. Harrison, and P. Cote, *Planta*, **131**, 141 (1976).
36. G. V. Hoad, *Planta*, **124**, 25 (1975).
37. J. B. Passioura, *Aust. J. Agric. Res.*, **23**, 745 (1972).
38. A. K. Dobrenz, L. N. Wright, A. B. Humphrey, M. A. Massengale, and W. R. Kneebone, *Crop Sci.*, **9**, 354 (1969).
39. K. E. Miskin and D. C. Rasmusson, *Crop Sci.*, **10**, 575 (1970).
40. O. Stocker, S. Rehm, and H. Schmidt, *Jahrb. Wiss. Bot.*, **91**, 278 (1973).
41. P. S. Nobel, *Oecologia*, **27**, 117 (1977).
42. C. Y. Sullivan, in K. L. Larson and J. D. Eastin, Eds., *Drought Injury and Resistance in Crops*, Spec. Publ. No. 2, Crop Sci. Soc. Am., Madison, Wisconsin, 1971, p. 1.
43. R. G. Henzell, K. J. McCree, C. H. M. Van Bavel, and K. F. Schertz, *Crop Sci.*, **16**, 660 (1976).
44. B. V. Milborrow, *Annu. Rev. Plant Physiol.*, **25**, 259 (1974).
45. S. A. Quarrie and H. G. Jones, *J. Exp. Bot.*, **28**, 194 (1977).
46. H. G. Jones, *Nature (London)*, **271**, 610 (1978).
47. G. W. Paltridge and J. V. Denholm, *J. Theor. Biol.*, **44**, 23 (1974).
48. J. V. Denholm, *J. Theor. Biol.*, **52**, 251 (1975).
49. A. E. Hall, K. W. Foster, and J. G. Waines, in A. E. Hall, G. H. Cannell, and H. W. Lawton, Eds., *Agriculture in Semi-Arid Environments*, Springer-Verlag, Berlin–Heidelberg–New York, 1979, p. 148.
50. T. A. Buishand, *J. Hydrol.*, **36**, 295 (1978).

BREEDING AND SELECTION FOR ADAPTATION TO STRESS | VI

Successful breeding for improved drought tolerance usually involves breeding for short-duration crops, that is, breeding for drought escape. Although this may reduce yields in favorable years, it increases yield stability over all years. Agronomists have long realized that genetic gains might be made more rapidly and predictably if desirable morphological and physiological attributes could be identified and selected for in the parents and their progenies. Breeders of wheat and rice have achieved large gains in yield by deliberately selecting for particular photoperiodic responses and canopy architectures (1, 2). To date few physiological or morphological characters have

been used in breeding programs for dought tolerance. Hurd (3) describes a program in which
deep-rooting characters were selected and used in
successfully breeding a wheat for semiarid conditions.

The preceding parts, particularly Parts II, III, and
IV, have identified the morphological and physiological adaptations to water and heat stress. Part
V discussed the importance of these characters
on yield and productivity. This part stresses the
possibility of utilizing these characters in breeding programs. Since diversity is the tool with
which the plant breeder must work, our knowledge of the diversity of adaptive mechanism is
first assessed in Chapters 24, 25, and 26, and attempts at utilizing the characters in breeding programs are reported in Chapters 25, 26, and 27.

REFERENCES

1. J. Bingham, *Agric. Prog.*, **44**, 30 (1969).
2. P. R. Jennings, *Science,* **186**, 1085 (1974).
3. E. A. Hurd, *Agric. Meteorol.,* **14**, 39 (1974).

24 | Differences in Adaptation to Water Stress Within Crop Species

J. M. MORGAN

New South Wales Department of Agriculture, Agricultural Research Centre, Tamworth, New South Wales, Australia

1. INTRODUCTION

For the purpose of discussion of differences in adaptation to water stress within crop species, the role of water within the plant and the responses of plant tissue to water deficits may be viewed in terms of intracellular or protoplasmic responses to changes in extracellular water potential (i.e., in xylem and cell walls) and factors influencing the level of extracellular water potential. Included in the first category are responses that tend to maintain cellular function and integrity by maintaining turgor, and therefore water content. The second category includes the responses that tend to regulate water potential *per se* by regulation of water loss from the soil. This distinction arises from the view of water flow in the plant being driven by potential gradients (1, 2). The bulk of water flow is from the soil to leaf by way of the conducting elements, the rate of change of water content being small compared with the magnitude of flux through the plant. Thus water does not flow through the protoplasm of most cells; rather, the water contents and potentials of the protoplasm adjust to changes in the water potential of the cell walls and xylem vessels. Although in many instances these categories may be treated as distinct, feedbacks do occur in the leaf stomata, where the turgor of guard cells may affect the rate of water loss, hence leaf water potential, and in the roots, where the rate of elongation of growing root tips, which is driven by turgor pressure, may affect the rate of absorption of water, hence plant water potential. Both these factors are important in the regulation of water flow. In some species reduction in leaf area by leaf rolling may also be important in controlling water loss and also reflects changes in leaf turgor. These feedbacks highlight the interdependence of the various factors affecting the performance of a plant during periods of stress, which makes the formulation of simple criteria for selecting drought tolerant genotypes difficult. The relative importance of these factors undoubtedly depends on the type of plant and its environment. Thus characteristics that tend to ration water supply, such as root resistance (3), may be more significant for crops that grow largely on stored water, whereas osmotic adjustment could improve adaptation to both sporadic and protracted stresses.

Although touched upon in previous reviews (4, 5), this chapter emphasizes cellular mechanisms that lead to the maintenance of water content and turgor; much of the discussion reflects a personal view rather than an exhaustive review of the subject.

2. SYMPLASTIC RESPONSES TO WATER STRESS

2.1. Turgor Maintenance

Cell Wall Elasticity. During periods of water stress the water potential ψ of the leaf cell walls and xylem sap declines, maintaining the transpiration stream. Initially a potential gradient is established between the cell walls and protoplasm which may be reduced in one of several ways. If positive turgor exists, there may be either (a) an outflow of water and reduction in turgor and osmotic potential or (b) an increase in the osmotic potential π, due to increase in the amount of solute. It is possible also that both may occur to varying degrees. Where turgor potential has been reduced to zero, water is lost, causing a reduction in π, though the rate of water loss reflects the water adsorption properties of the cell constituents. In situation a the rate of change of turgor depends largely on the elastic properties of the cell walls (6):

$$dP = \epsilon \frac{dV}{V} \tag{24.1}$$

where P is the turgor potential, ϵ the modulus of elasticity, and V the cell volume. Cells with a high modulus of elasticity may be expected to show only a small decrease in water content per unit change in water potential; hence more water is retained in the protoplasm during periods of stress. As well as elasticity, the amount of solute in the protoplasm at a given water potential, usually indicated by the value of π at full or zero turgor, may affect the change in water content with reduction in ψ. The relationship between ψ, ϵ, π, and cell volume (which reflects water content) may be described by the following relation (7):

$$\frac{1}{V_0} \frac{dV}{d\psi} = \frac{1}{\epsilon + \pi_0}$$

(24.2)

where the subscript 0 indicates the value at zero turgor.

Differences in both ϵ and π_0 (or π_t, the osmotic potential at full turgor) among different species have been found (8), but there seems to have been little exploration of these parameters within crop species. Although ϵ was not measured, Johnson and Brown (9) provided evidence of differences in π_t between *Agropyron* hybrids. The values of π_t varied from -12.6 to -15.1 bars. Although these differences were not large and consequently may not be expected to be significant in a field situation, rankings based on these criteria generally agreed with field performances (presumably based on growth or yield).

Osmotic Adjustment. Variations in ϵ and π_t affect the magnitude and rate of change of turgor potential with reduction of water potential, but the turgor potential may still decline linearly with fall in water potential (i.e., water is lost from the protoplasm). As water is withdrawn from the cell, π changes by concentration effects. In an ideal osmotic system this effect is described approximately by:

$$\pi = \frac{RWC_0 \pi_0}{RWC}$$

(24.3)

where RWC is the relative water content. This formulation is useful in discriminating between changes in π that are due to concentration effects and those that may be attributed to changes in the amount of solute in the protoplasm (10).

Such changes in π commonly occur in fungi and algae (11) and also in crop plants (10, 12, also Chapter 7). In most instances in which osmotic adjustment has been observed, the change in π was sufficient to equal or exceed the change in ψ (10, 13) resulting in maintenance of turgor potential and water content. Because of the importance of these factors in expansion growth and metabolism, these processes would be maintained as well. Not only are growth and function maintained over a greater range of values of ψ than when osmotic adjustment is absent, but also the amount of water retained at low levels of ψ is increased (14), and this should lead to improvement in the critical water potential for survival. Although this mechanism of adaptation could be of considerable significance as a means of improving the adaptation of crop plants to water stress, little is known of the extent of variation within crop species, or of the degree to which it is influenced by fluctuations in the environment (i.e., of the basic physiology). Some information on the pattern of the response of π to ψ during osmotic adjustment, and of differences between wheat genotypes has been obtained by studying plants in pots during drying cycles (10). The type of response characterizing osmotic adjustment in leaves is shown in Figure 24.1, where it is contrasted with that due to concentration effects (Figure 24.2). In this study osmotic adjustment was virtually absent from the leaves of some genotypes, but extensive in others. Where it occurred turgor was fully maintained, and in fact increased slightly, over the range of leaf water potential (ψ_{leaf}) from 0 to -15 bars, and did not reduce to 0 until a ψ_{leaf} of about -35 bars was reached. In genotypes without osmotic adjustment, however, there was a gradual decline in turgor from the onset of stress with zero turgor being reached at a ψ_{leaf} of about -20 to -25 bars. In this situation the presence of osmotic adjustment would have allowed the leaves to function unimpaired down to a ψ_{leaf} of -15 bars, and at reduced turgor over a greater range of ψ_{leaf} than where it was absent.

Genotypes with osmotic adjustment usually had higher levels of water retention. For example, at ψ_{leaf} of -30 bars, the relative water content of *Triticum spelta*, a genotype showing osmotic adjustment, was 0.78 compared with a value of 0.50 for a *T. durum* that did not show adjustment.

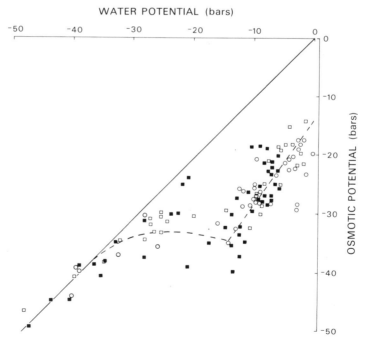

Figure 24.1. Relationship between leaf water potential and leaf osmotic potential for genotypes showing osmotic adjustment: □, *Triticum aestivum*; ○, *T. dicoccum*; ■, *T. spelta*. From Ref. 10.

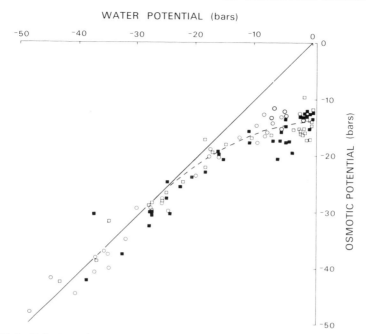

Figure 24.2. Relationship between leaf water potential and leaf osmotic potential for genotypes without osmotic adjustment: □, *Triticum durum*; ○, *T. aestivum*; ■, *T. boeticum*. From Ref. 10.

Although the data were scattered, there was some suggestion that the buildup in osmotic solutes was greater after anthesis than before (14). This is discussed at greater length below with respect of field-grown plants.

With species other than wheat, there appears to be little information on differences among genotypes. Jones and Turner (15) found no difference in osmotic adjustment between two sorghum cultivars, though differences in water retention observed by Blum (16) may well have been due to osmotic adjustment.

Although the observation of differences among genotypes grown under glasshouse conditions is encouraging, the expression of these differences in a field situation has yet to be confirmed. Too little is known about the mechanism of osmotic adjustment in crop plants to permit appreciation of the effects on it of a fluctuating environment, of the kind often experienced in the field. Also, the basis of the observed genotypic differences is unknown, and without this knowledge it is difficult to appreciate how these might interact with a field environment. Thus if the solute used in adjustment consists largely of sugar molecules (17), the extent of adjustment may be influenced by availability, which may in turn depend on requirements in other parts of the plant. Genotypes could then differ because of differences in the pattern of distribution of sugars rather than in inherent ability to degrade larger polysaccharide molecules in response to stress.

2.2. Difficulties Encountered in Studying Turgor Maintenance

Although more exhaustive studies have yet to be made, some preliminary studies of field behavior suggest that some genotypes that responded to water stress by osmotic adjustment in the glasshouse experiment (10) showed evidence of this in the field. Comparison of field and glasshouse measurements is difficult because of the difference in the pattern and extent of drying cycles. Figure 24.3 shows the changes in leaf turgor potential P_{leaf}, leaf osmotic potential π_{leaf}, and ψ_{leaf} for the genotype *T. spelta* over a period of several weeks in the field. The samples for measurement of ψ_{leaf} and π_{leaf} were made between 1000 and 1200 h in conditions of bright sunshine and cloud; measurements were made using thermocouple psychrometers. This particular set of data is interesting in that there was little change in ψ_{leaf} after an initial decline from -18 to about -25 bars. P_{leaf} (calculated as the difference between

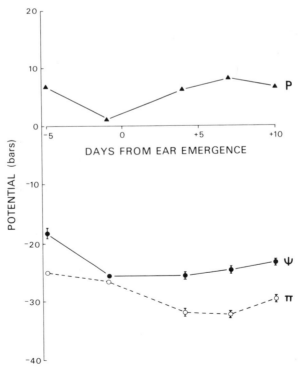

Figure 24.3. Changes in water potential ψ, osmotic potential π, and turgor potential P, with time during the period around ear emergence for leaves of *Triticum spelta*.

ψ_{leaf} and π_{leaf}) declined with the initial fall in ψ_{leaf}, but it subsequently increased to a value of about 7 bars, which exceeds the initial value of 6 bars measured when ψ_{leaf} was only -18 bars. Following this response there was little further change in either π_{leaf} or ψ_{leaf}. It seems from these results that an increase in solute (or turgor) occurred in response to the initial decrease in ψ_{leaf}. Analysis of the response of π_{leaf} to RWC suggested that the initial change in P_{leaf} and π_{leaf} was due to concentration effects, so that osmotic adjustment did not begin until P_{leaf} was almost zero. This contrasts with the observations of osmotic adjustment made on plants grown in the glasshouse (10), where osmotic adjustment occurred over the range of ψ_{leaf} from 0 to -15 bars. It would seem, then, that in the field situation a considerable degree of pre-stressing was a prerequisite for osmotic adjustment. Jones and Turner (15) also provide evidence of increase in π_{leaf} following stressing cycles in the leaves of glasshouse-grown sorghum plants.

Although these field data may well indicate a need for preconditioning, it is also possible that the capacity for osmotic adjustment varies with ontogeny. The results of Sionit and Kramer (18) indicate a buildup of solutes during the period between flowering and pod filling in nonstressed soybeans, though this may have been in response to a small decline in ψ after flower induction.

The results for the genotype *T. dicoccum* also indicate a general decline in π_{leaf} with change in ontogeny from jointing to the grain-filling stage (Figure 24.4), though as in previous instances these changes could also be interpreted as responses to reductions in ψ_{leaf}. For example, during the period of

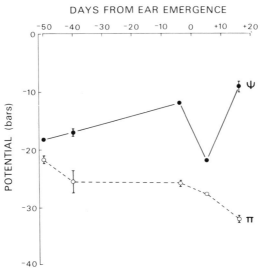

Figure 24.4. Changes in leaf water potential ψ and leaf osmotic potential π with time for *Triticum dicoccum* grown in the field.

11 days after anthesis π_{leaf} fell from -28 to -32 bars while ψ_{leaf} increased from -22 to -9 bars. This response, however, followed a sharp fall in ψ_{leaf} just prior to anthesis and could have been a reaction similar to that of *T. spelta*.

These results clearly indicate the need for caution in the timing of measurements made to differentiate genotypes with respect to osmotic adjustment in leaves of plants in the field. Until more is understood about possible changes in osmotic adjustment with ontogeny of the plant, or leaf age, and with fluctuations in the environment, several sample times should be used when examining differences between genotypes. Also differences in the behavior of plants in the field compared with the glasshouse suggest the need for caution in extrapolating to the field situation results obtained in the glasshouse. However several genotypes showed osmotic adjustment in both the glasshouse and field, and this does suggest that glasshouse studies may be of some value. There may also be merit in sampling as wide a range of genetic variability as possible (10, 19). Fischer and Sanchez (19) measured π_{leaf} and ψ_{leaf} in a range of modern bread and durum wheats over a number of sampling dates. Differences between genotypes in turgor potential failed to reach statistical significance, but the variation in values at low ψ_{leaf} was encouragingly large. Indeed, the values of turgor potential of 6 and 5 bars at values of ψ_{leaf} of -27.8 and -29.6 bars for 'Penjamo 62' and 'Tall Z-Bouteille' were of the order of magnitude found in genotypes showing osmotic adjustment (10). On the other hand, most other genotypes had almost zero turgor at these levels of ψ_{leaf}, similar to genotypes without osmotic adjustment (10). Differences in turgor potential of a similar order of magnitude were also found in a field study covering a wide range of genotypes (Morgan, unpublished).

Apart from leaves, there is evidence that osmotic adjustment occurs in roots (20), in expanding hypocotyls (13), and in the wheat inflorescence during spike expansion (14, 21) and possibly during grain growth (14). Of these, only the study of the wheat inflorescence (14) included a range of genotypes. In this work and in the experiment of Barlow et al. (21), osmotic adjustment always occurred in the developing spike. Thus genotypes differed only with respect to the extent of adjustment of the osmotic potential. Osmotic adjustment in the developing spike differs from that occurring in leaves in several respects. First the level of turgor maintained is usually considerably less, that in the spike being about 5 bars compared with about 15 bars in leaves, though this does not seem to affect water retention. Second, water retention may be influenced by the phase following the main adjustment of osmotic potential. In both leaves and spikelets, the response of the osmotic potential to changes in water potential seems to consist of three phases: (*a*) an increase in the number of solute molecules and maintenance of turgor (i.e., osmotic adjustment), (*b*) maintenance or reduction in the number of solute molecules and reduction in turgor, and (*c*) maintenance of the number of solute molecules and zero turgor. However in the spike for the range of values of water potential of 0 to −40 bars, the analysis of the relationship of $\log RWC$ with $\log \pi$ (Figure 24.5) suggests that in some genotypes (*T. durum, T. dicoccoides*) the changes in osmotic potential in the second phase may be less than would be expected if changes were only due to concentration effects. This may indicate continued osmotic adjustment, but at a reduced rate. The factors that seemed to be important in maintaining high RWC were the level of osmotic potential reached during the first phase, during which there was little change in RWC, the slope of the response of $\log \pi$ to $\log$ RWC in the second phase, and the RWC at which osmotic adjustment began.

Genotypes differed with respect to each of these factors (Figure 24.5), and consequently with respect to levels of RWC at low levels of water potential. Thus in *T. durum,* although osmotic adjustment did not commence until RWC was approximately 0.9, there was no change in RWC until the osmotic potential was −20 bars, and only a slow change thereafter. In *T. aestivum,* on the other hand, although osmotic adjustment commenced at RWC of about 0.95, the extent of osmotic adjustment was only −16 bars, and the change in RWC with osmotic potential was greater than in *T. durum.* Thus the values of RWC at a water potential of −20 bars ($\pi = -25$ bars) were 0.83 and 0.64 for *T. durum* and *T. aestivum,* respectively. No measurements of growth or survival were made in this study, but it seems from comparisons of leaves and apices that the ability to survive and recover from severe water stress is associated with water retention (21). On this basis we might expect these genotypes to differ in ability to survive water

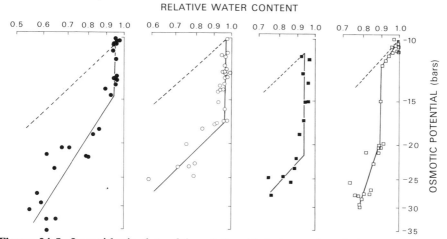

Figure 24.5. Logarithmic plots of the relationship between the relative water content and osmotic potential of spikelets during spikelet growth for the genotypes *Triticum dicoccoides* (●), *T. aestivum* (○), *T. dicoccum* (■), and *T. durum* (□). The broken lines represent Equation 24.3.

stress. On the other hand, growth is likely to be slowed by reduction in turgor potential (12).

Since there were differences in the levels of water potential at which turgor began to decline, it is likely that there were also differences in the water potential at which growth rate was reduced. Both apical growth and spikelet formation were adversely affected by levels of water potential less than -10 bars in 'Heron' wheat (21). Even though the apex was younger, this is close to the value at which turgor began to change in *T. dicoccoides* (about -12 bars), but below that of *T. dicoccum* (about -17 bars). These genotypes form a small sample of possible variation, and differences larger than these may occur.

There is also evidence that osmotic adjustment may occur in the postanthesis spikelet (14), though the data were very variable. Such adjustment usually occurred in genotypes that showed osmotic adjustment in leaves (e.g., *T. spelta, T. dicoccum*). In other genotypes there was no evidence of osmotic adjustment, but rather negative turgors developed at low water potential. Moreover, the levels of RWC were higher at low water potential in genotypes showing osmotic adjustment.

In view of the significance of continued spike and grain growth during periods of water stress to grain number and size, the extent of osmotic adjustment at both these stages could be of importance in discriminating between genotypes. Indeed this may be singularly more important than leaf responses to stress, particularly since the emerged ear may contribute substantially to its own requirements for assimilates (22) or may utilize assimilates stored in other parts of the plant.

3. FACTORS AFFECTING WATER POTENTIAL

The level of the water potential in the plant ψ_{plant} is generally considered to be affected by the water potential of the soil ψ_{soil}, the concentrations of water in the atmosphere c_a and at the evaporating surface in the leaf c_l, and the resistances to water flow in the pathway between soil and air (23). The way in which these relate may be expressed in very simplistic and qualitative terms and assuming isothermal, steady state conditions by (2)

$$\psi_{plant} = \psi_{soil} - RE \qquad (24.4)$$

where R is the resistance to water flow in soil and plant, the evaporation rate E is proportional to $(c_l - c_a)/(r_l + r_a)$, and r_l and r_a are the leaf and boundary layer resistances, respectively: however, it should be recognized that the strict form of this relationship depends on canopy architecture (24).

Most of these variables may be altered either directly, as in the case of r_l, or indirectly as in the case of leaf geometry, which affects r_a, by changes of cultivar or species. Soil water potential may be affected by variations in patterns of crop development and the extent and geometry of the root system (3). The level of water potential in turn affects cell water content and turgor, though the extent of this influence depends on the degree of turgor regulation, discussed in the preceding section.

3.1. Stomatal Responses

Reference was made earlier to feedback between turgor responses in the leaf and the control of ψ_{leaf} through stomatal resistance. The turgor of the guard cells should be affected by the turgor of the leaf tissue (25). However leaf turgor may be buffered against changes in ψ_{leaf} by regulatory mechanisms such as osmotic adjustment. Unfortunately most attempts to assess genotypic differences in stomatal response to water stress have used ψ_{leaf} as a measure of stress (16, 19, 26). The possible difficulty with this approach is that the relationship between ψ_{leaf} and turgor may vary both with genotype (10) and with environment (15). It may be more meaningful to use either turgor or RWC, which is associated wih changes in turgor, to evaluate differences in stomatal response between genotypes, though these measurements on leaf tissue may still be remote from the water relations of the guard cells that directly affect stomatal resistance.

It is possible then that genotypes with high levels of osmotic adjustment may also maintain low leaf resistances to lower water potentials than genotypes without osmotic adjustment (Chapter 9). This may explain instances in which the more drought resistant genotypes were those that had lower leaf resistances at low ψ_{leaf} (16, 27). Thus although very responsive stomata may be desirable for plants to survive long periods of drought, this mechanism may not be highly desirable in relatively fast-growing, short life cycle crop plants such as sorghum or wheat (see Chapter 23). Particularly for crops that rely on rainfall during the growing season, maintenance of growth during intermittent stress may be desirable.

3.2. Soil Water

One extensively utilized method of maintaining water potential involves the alteration of phenology by exploiting differences in photoperiod response between genotypes (5). This approach probably has led to considerable improvements in adaptation to water stress (28). It is possible that in a particular environment there are strategies of growth and development that improve the yield of the crop when water is limiting. For example, higher yields of wheat in New South Wales have resulted from cultivars that may be sown late enough to avoid late frosts but develop rapidly enough to substantially avoid water stress in late spring. Because of variations in weather, however, sowing time cannot always be controlled, and under these circumstances it may be possible to increase yield by altering developmental patterns by varying the genotype and seeding density, thereby regulating water use before anthesis, hence supply during grain growth (29).

The possibility that yield may be improved by rationing water supply before anthesis, not by control of plant development *per se* but by changing resistance to water flow in the root, was presented by Passioura (3). By judicious control of root morphology he was able to show that an increase in resistance to flow produced in this way could conserve water before anthesis, hence improving supply during the subsequent grain-filling stage when stress is more usual. It is possible that root resistance may be altered genetically by selecting genotypes with main xylem vessels of smaller diameter, reduced number of seminal axes, and increased frequency of split xylem vessels (30). Differences in vessel diameter and the occurrence of split vessels have been found within wheat species, and these factors appear to be sufficiently heritable to attempt selection of isolines for comparative studies (30).

3.3. Variations in Water Potential Within the Plant

The foregoing strategies influence the water potential of the spikelets during flowering and grain growth by regulating the water supply to the plant. However during periods of water stress the water potential of the spikelet may be substantially greater than that of the leaf (26). Indeed, at low leaf water potentials, the difference may be as much as 20 bars (26). It is therefore conceivable that the grain may continue growing and the spikelets photosynthesizing well beyond the wilting point of the leaf, and after cessation of leaf photosynthesis. The extent to which the spikelet continues functioning depends on the extent of turgor maintenance as well as the extent to which its water potential differs from that of the leaf.

Under these circumstances we might expect the photosynthetic area of the ear to be important, particularly since there may be an insufficient supply of assimilates if photosynthesis of the flag leaf has been reduced (22). This may be why genotypes with awns have often yielded better than awnless types during periods of water stress (31, 32), but the same when water was not limiting. In a field study with isolines with and without awns,

Table 24.1. Spikelet and Leaf Water Potentials for Four Wheat Genotypes (14)

Values are means ± one standard error

Genotype	Water potentials (bars)	
	Leaf	Spikelet
T. spelta	−31.3 ± 2.1	−17.0 ± 1.5
T. aestivum cv. Condor	−31.2 ± 0.8	−21.5 ± 1.9
T. durum	−31.3 ± 3.5	−31.6 ± 2.0
T. aestivum cv. Capelle Desprez	−31.3 ± 2.0	−25.3 ± 1.8

Evans et al. (33) found that grain yield was reduced by stress after anthesis by 20% in the awnless lines but by only 11% in the awned lines. Since the stress was sufficiently severe to cause pronounced leaf rolling, it is probable that photosynthesis had stopped in the leaf, but continued in the ear, because of higher water potential.

Genotypes may differ in ability to maintain the water potential of the ear. For example, the results in Table 24.1, from an experiment by Morgan (14), compare four genotypes at the same ψ_{leaf} under controlled conditions. Differences between ear and leaf water potential varied from 0 bars in *T. durum* to 14.3 bars in *T. spelta*.

4. CONCLUSIONS

From the foregoing discussion it is possible to make several observations relevant to the selection for drought tolerance in crop plants. First, the place of individual mechanisms in the overall system of the plant and its environment should be considered. Within the plant it is important to appreciate the way in which different mechanisms interact with each other and the significance of their role in the functioning of the plant. Interactions between mechanisms may be very important, as in the instance of stomatal responses and osmotic adjustment. In assessing interactions it may be useful to view the system in terms of symplastic responses and factors controlling the plant water potential.

Second, the type of environment may be important in determining which mechanisms are of greatest importance. Thus where crop growth depends almost entirely on stored water, mechanisms such as increased root resistance, which ration water use to improve supply during and after flowering, may be significant. Mechanisms such as osmotic adjustment, which extend the range of water potential over which the plant can function, are probably advantageous during periods of intermittent or protracted stress. Similarly, strategies that tend to improve partitioning, such as high harvest index, may improve yield in both stressed and nonstressed environments, particularly when this is achieved at the expense of stem rather than roots.

There may be considerable scope for improvement in wheat by focusing more attention on the water relations of the developing spike and the ear during grain growth. Factors of importance are the maintenance of high water potential during stress, the maintenance of turgor and water content, and in the emerged ear, increase in supply of assimilates, either from increased photosynthetic area (awns) or from reserves.

REFERENCES

1. I. R. Cowan and F. L. Milthorpe, in T. T. Kozlowski, Ed., *Water Deficits and Plant Growth,* Vol. 1, Academic Press, New York–London, 1968, p. 137.

2. W. R. Gardner, in T. T. Kozlowski, Ed., *Water Deficits and Plant Growth,* Vol. 1, Academic Press, New York–London, 1968, p. 107.
3. J. B. Passioura, *Aust. J. Agric. Res.,* **23,** 745 (1972).
4. L. H. May and F. L. Milthorpe, *Field Crop Abstr.,* **15,** 171 (1962).
5. J. E. Begg and N. C. Turner, *Adv. Agron.,* **28,** 161 (1976).
6. W. Wenkert, E. R. Lemon, and T. R. Sinclair, *Ann. Bot. (London),* **42,** 295 (1978).
7. J. R. Philip, *Plant Physiol.,* **33,** 264 (1958).
8. J. Warren Wilson, *Aust. J. Biol. Sci.,* **20,** 349 (1967).
9. D. A. Johnson and R. W. Brown, *Crop Sci.,* **17,** 507 (1977).
10. J. M. Morgan, *Nature (London),* **270,** 234 (1977).
11. J. A. Hellebust, *Annu. Rev. Plant Physiol.,* **27,** 485 (1976).
12. T. C. Hsiao, *Annu. Rev. Plant Physiol.,* **24,** 519 (1973).
13. R. F. Meyer and J. S. Boyer, *Planta,* **108,** 77 (1972).
14. J. M. Morgan, *J. Exp. Bot.,* **31,** (in press) (1980).
15. M. M. Jones and N. C. Turner, *Plant Physiol.,* **61,** 122 (1978).
16. A. Blum, *Crop Sci.,* **14,** 691 (1974).
17. H. Kauss, *Plant Biochem. (Tiflis),* **13,** 119 (1977).
18. N. Sionit and P. J. Kramer, *Plant Physiol.,* **58,** 537 (1976).
19. R. A. Fischer and M. Sanchez, *Aust. J. Agric. Res.,* **30,** 801 (1979).
20. E. L. Greacen and J. S. Oh, *Nature (London) New Biol.,* **235,** 24 (1972).
21. E. W. R. Barlow, R. Munns, N. S. Scott, and A. H. Reisner, *J. Exp. Bot.,* **28,** 909 (1977).
22. L. T. Evans and H. M. Rawson, *Aust. J. Biol. Sci.,* **23,** 245 (1970).
23. R. O. Slatyer, *Plant-Water Relationships,* Academic Press, London–New York, 1967.
24. I. R. Cowan, *Q. J. R. Meteorol. Soc.,* **94,** 523 (1968).
25. R. O. Slatyer, in L. T. Evans, Ed., *Environmental Control of Plant Growth,* Academic Press, New York–London, 1963, p. 33.
26. J. M. Morgan, *Aust. J. Plant Physiol.,* **4,** 75 (1977).
27. H. G. Jones, *J. Agric. Sci.,* **88,** 267 (1977).
28. N. C. Turner, in H. Mussell and R. C. Staples, Eds., *Stress Physiology in Crop Plants,* Interscience, New York–London, 1979, p. 343.
29. A. D. Doyle and R. A. Fischer, *Aust. J. Agric. Res.,* **30,** 815 (1979).
30. R. A. Richards and J. B. Passioura, *Aust. CSIRO Div. Plant Ind. Ann. Rep.,* **1977,** p. 123.
31. F. J. Grundbacher, *Bot. Rev.,* **29,** 366 (1963).
32. G. J. Vervelde, *Neth. J. Agric. Sci.,* **1,** 2 (1953).
33. L. T. Evans, J. Bingham, P. Jackson, and J. Sutherland, *Ann. Appl. Biol.,* **70,** 67 (1972).

25 | Genetic Variability in Sorghum Root Systems: Implications for Drought Tolerance

WAYNE R. JORDAN

Texas Agricultural Experiment Station, Temple, Texas

FRED R. MILLER

Texas Agricultural Experiment Station, Texas A&M University, College Station, Texas

1. INTRODUCTION

Successful crop production in temperate climates depends heavily on the capability of the environment to supply water. Droughts, as meteorological events, are commonplace: they may be either seasonal or temporal, and the agricultural consequences may be either beneficial or harmful. Mechanisms by which crops resist the deleterious consequences of drought are as varied as the patterns of drought themselves (1–4). Defining drought as a historical meteorological event or sequence of events (see Chapter 1) implies that once a crop is planted, the option to avoid drought is no longer viable. We must recognize that drought, as a meteorological circumstance, may have little, if any, consequence in crop production. Rather, our main concern with drought is that it acts to set in motion a sequence of events

383

resulting in reduced soil water availability, increased tissue water deficits, and disturbance of processes that contribute to yield of marketable products.

This chapter examines the utilization of stored soil water by annual crop plants and describes one approach that may lead to enhanced drought resistance resulting from postponement of tissue dehydration. This approach is illustrated with recent results from studies with grain sorghum (*Sorghum bicolor* L. Moench).

2. UTILIZATION OF STORED SOIL WATER

2.1. Drought Patterns and Soil Water Availability

Examples of soil water availability patterns associated with the growth of a summer annual crop such as cotton or grain sorghum in three typical temperate dryland production areas are illustrated in Figure 25.1. The patterns

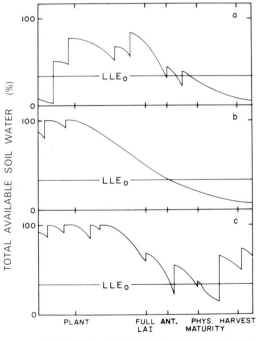

Figure 25.1. General aspects of soil water availability during growth and development of a grain sorghum crop in temperate dryland regions. (*a*) Limited rainfall concentrated during summer growing season. (*b*) Distinct wet-dry seasonal pattern. (*c*) Basic wet-dry seasonal pattern with dry season receiving irregular rainfall. LLE_0 represents the lower limit of soil water availability that will support water loss at potential rates (i.e., $ET/ET_0 < 1.0$ below LLE_0).

suggest the nature of the resistance mechanisms required to tolerate the consequences of either seasonal or temporal drought. The pattern illustrated by Figure 25.1a would be typical for the Rolling Plains region of Texas and portions of the Great Plains where rainfall is limited in quantity and seasonal distribution. Since most of the rainfall occurs during the summer cropping season, planting is delayed until sufficient moisture is present to assure successful germination and establishment. Seldom does sufficient rainfall occur to fill the soil profile completely before planting. Satisfactory growth during the season requires timely rainfall because of the absence of deep stored soil moisture. The crop frequently experiences tissue water deficits because of a paucity of rainfall coupled with a high evaporative demand, and reduced rates of vegetative development and wilting or leaf rolling are commonly observed. Cultivars capable of successful production in this region must possess drought tolerance at the seedling stage as well as at later stages (see Chapter 27). The primary root must be capable of rapid proliferation in both depth and surface area because frequent and severe drying of the seedbed may restrict development of lateral or crown roots. Lateral or crown roots that do develop may be required to grow through relatively dry layers of soil to reach moist regions deeper in the profile. Because mechanical resistance to root growth increases dramatically as soil water content decreases (5), these roots must be capable of exerting considerable "growth pressure," presumably through alterations in root diameter or increased osmotic pressure in root cells (6). Since the wetted portion of the profile seldom extends below 1 m, the capability to form an exceptionally deep root system is not required. Of greater importance is the ability to achieve high root length density in the wetted regions of the profile to allow efficient and thorough extraction.

Regions typified by distinct wet-dry seasons are illustrated in Figure 25.1b. A crop planted near the end of the wet period must be capable of achieving final production on stored soil moisture only. The simplest way to accomplish this goal involves matching total crop water use to total soil water availability. The proper choice of maturity genotype is an important consideration in this approach (7). Alternate mechanisms may allow conservation of water during vegetative growth so as to have more water available during later, more critical stages (8). Conservation, used in this context, is achieved through a reduction in the rate of canopy development. Crop evapotranspiration (ET) is closely related to leaf area until an effective, full canopy is reached (9); therefore more water remains in the profile for use at later, more critical periods (8, 10). In this situation the rate of leaf area development is reduced by increased plant water deficits resulting from the inability of roots to transport water in quantities required to match evaporative demand. These resistances may lie in either the radial or axial transport paths within roots, or result from low root densities.

The third pattern (Figure 25.1c) combines aspects of both Figures 25.1a and 25.1b and is typical for most of the Great Plains area, of which our location at Temple, Texas, is a part. Winter rainfall is usually sufficient to fill the soil profile completely (11). Spring rainfall normally maintains high available soil water for 30 to 40 days after planting, but optimum production requires additional summer rainfall. These summer rainfall events are unpredictable in both occurrence and magnitude; therefore the profile gradually dries and may result in development of plant water deficits near anthesis for grain sorghum and from early to peak flowering for cotton. These deficits occur even though appreciable soil water is available at soil depths below 50 cm (9). Cultivars successful in this situation must be capable of achieving high root densities below 50 cm, and they must be able to absorb water from rewetted surface layers rapidly following rainfall. The latter aspect may depend on the rapid regeneration of roots near the surface, which were lost as the soil dried (12).

2.2. Soil Water Depletion by a Grain Sorghum Crop: An Example

General aspects of the soil water distribution with depth are given in Figure 25.2 for Houston Black Clay soil and a grain sorghum crop. Total water availability and water transmission characteristics of this soil are well known

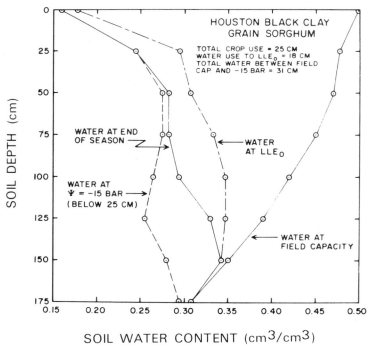

Figure 25.2. Distributions of soil water content with depth during the growing season for a grain sorghum crop growing on Houston Black Clay at Temple, Texas. For definition of LLE_0 see legend for Figure 25.1; ψ is the soil water potential.

(9, 13). At Temple, Texas, the profile is underlain with a dense calcareous stratum (caleche) that apparently restricts root penetration below about 1.7 m. At crop maturity essentially all available water above 75 cm is removed, and approximately 6 cm of available water (above -15 bars) remains in the lower 100 cm of the profile. A more significant feature of these curves is that only about 18 cm of water is removed by the growing crop before actual ET falls below potential evapotranspiration (ET_0) (see curve LLE_0 and Ref. 9). Since ET/ET_0 drops below 1.0 at soil water contents represented by the LLE_0 curve, this may be taken as the point at which soil resistance in the entire root zone becomes an important component of the total flow resistance. Leaf water potentials fall as the total resistance increases, and leaf rolling and stomatal closure may result (14). The distribution of soil water content (θ) with depth at this time depends not only on water transmission characteristics of the soil, but on rooting characteristics of the crop as well. Similar distributions of θ at LLE_0 are found for commercial grain sorghum hybrids and cotton cultivars. We believe the root density L_v for current crop cultivars is insufficient at depths below 50 cm to allow extraction of sufficient quantities of water from this region of the soil profile to meet the evaporative demand.

2.3. Root Proliferation and Soil Water Extraction: Predictions from Theory

The problem as stated by Hsiao et al. (6) is central to a definition of desired root properties, that is, ". . . how extensively must a soil volume be permeated by roots for the effective use of the stored water? Or, how and by what are the limits set for the distance water can be transported through the soil to the roots at a rate sufficient to prevent plant water stress?'' It is obvious that an answer to this and similar queries must be based on a recognition that soil and plant properties act in concert to influence the flow of water through the system. The theoretical analysis outlined below was used to examine soil and plant parameters in the flow path, to define desired root length densities.

Appropriate equations describing the flow of water through both soil and plant pathways have been derived (15–17). The treatment by Hsiao et al. (6) and Reicosky and Ritchie (18) is followed, modified as deemed suitable for the specific case of grain sorghum and Houston Black Clay soil. The main assumptions on which this model is based are that the roots act as uniform sinks for water uptake and are parallel and evenly spaced in the soil at a given L_v, that water moves only radially from the soil to the root surface, that the soil hydraulic conductivity is constant in the radial path, and that transport is steady state (6, 15). The basic flow equation is given by:

$$q = \frac{\psi_s - \psi_l}{R_s + R_r + R_c} \tag{25.1}$$

where ψ_s and ψ_l are water potentials in bulk soil and leaf, respectively, and R_s, R_r, and R_c are resistances associated with flow from the bulk soil to the root surface, from the root surface to the root xylem, and from the root xylem to the point of evaporation in the leaf, respectively. The term R_c has been called the "connection resistance" (W. S. Meyer and J. T. Ritchie, personal communication). Flow, q, is given in cubic centimeters of water transported per second per square centimeter of land area; that is, it has units of cm/s.

Resistance to flow in the soil pathway R_s is given by:

$$R_s = \frac{ln\ (1/r^2\pi L_v)}{4\pi L_v \Delta Z}\ \frac{1}{K} \qquad (25.2)$$

where r is the average root radius (cm), L_v is the average root length density (cm of root/cm³ of soil) in a specified soil layer of thickness ΔZ (cm), and π is 3.1416. The hydraulic conductivity of the soil K is expressed as cm²/bar/s and R_s has units of bar s/cm.

Root resistance R_r, in bar s/cm, is assumed to represent the total resistance of the root system with all roots acting as parallel resistors as given by:

$$R_r = \frac{R'_r}{L_v \Delta Z} \qquad (25.3)$$

where R'_r is the resistance per unit length of root with units of bar s/cm. Recent studies with grain sorghum have shown R'_r to be flow dependent (W. S. Meyer and J. T. Ritchie, personal communication), with this dependence of the form:

$$R'_r = \frac{R'_r\ (min)}{1\ -\ exp\ (-Bq/L_v \Delta Z)} \qquad (25.4)$$

where B is a scaling factor; R'_r (min) is the minimum resistance per unit length of root, found to be approximately 2.0×10^5 bar s/cm². Based on a comparison of water potentials of exposed and covered leaves, where flow was zero, a substantial resistance was shown to occur in the liquid flow path between the root xylem and leaves. This additional "connection" resistance R_c, in bar s/cm, may be described by:

$$R_c = 0.9R_r \qquad (25.5)$$

The equations above were solved numerically for L_v using $r = 0.01$ cm (19), and $K = f(\psi_s)$ determined from Ritchie et al. (9). In solution, q, ΔZ, ψ_l, and ψ_s are specified, and L_v was determined to meet the specified flow.

Results of this analysis are shown in Figure 25.3 for a flow rate of 0.8 mm/h, typical for the peak rate on a high ET day in midsummer, and ΔZ of 1.25 m representing the portion of the profile between 50 and 175 cm. Referring to Figure 25.2, the average ψ_s between 50 and 175 cm when LLE_0 is reached is approximately -1.5 bars. Since ET begins to deviate from

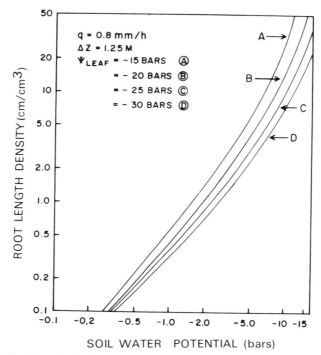

Figure 25.3. Predicted relationship between root length density and soil water potential in a 1.25 m layer of soil (ΔZ) and given a flow rate q of 0.8 mm/h at four leaf water potentials ψ_{leaf}.

ET_0 at this time, the analysis suggests that the average value of L_v within the zone must be near 1.0. LLE_0 is usually reached after the crop reaches full leaf area (i.e., during boot stage, anthesis, or early grain-filling period). If L_v were increased during this period, water could be supplied at rates sufficient to meet ET_0 and tissue dehydration could be delayed, perhaps until the crop had passed through this period so critical to yield (20). Of course it must be recognized that rapid depletion of soil water during the critical period increases the risk of severe plant water deficit later if rainfall does not occur. The yield reduction in this situation may not be severe, provided the cultivar can translocate a large portion of the preanthesis dry matter to the developing grain (6).

The result that a flow of 0.8 mm/h cannot be met with $L_v < 1$ at ψ_l of -15 bars does not imply that water is not being extracted from the zone. If lower flows are specified, as in the case of cloud cover or changes in intercepted radiation, required quantities can be supplied from drier soil without altering L_v (Figure 25.4). For example, with $L_v = 1$, a flow of 0.8 mm/h will be met

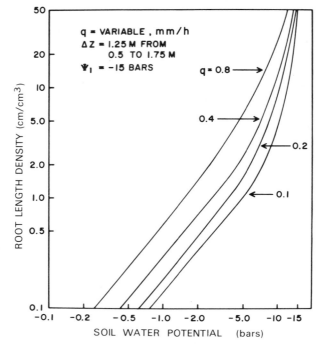

Figure 25.4. Predicted relationship between root length density and soil water potential at various levels of evaporative demand (i.e., flow rate q) for plants with a leaf water potential ψ_l of -15 bars. ΔZ is the 1.25 m soil layer between 0.5 m and 1.75 m.

only at ψ_s above about -1.5 bars, but a flow of 0.1 mm/h may be met from a soil with an average ψ_s of -15 bars. These analyses demonstrate the sensitivity of the entire system to change in only one component part of the chain. Relatively large benefits are predicted from morphological adaptations such as leaf rolling, which reduce intercepted radiation (see Chapter 3), and consequently reduce required flow, but little benefit may result from a reduction in leaf water potential alone.

Predictions based on the model presented above raise interesting questions regarding the benefit derived from stomatal adjustment to soil water deficit, one manifestation of osmotic adjustment (21–23, see also Chapters 7, and 9). Relatively large reductions in ψ_l (from -15 to -30 bars) are needed to maintain a flow of 0.8 mm/h if ψ_s is decreased by only 1 bar, from -1.5 to -2.5 bars, at an L_v of 1.0 (Figure 25.3). Since it is well known that the root/shoot ratio increases in plants subjected to prolonged soil water deficits, it appears likely that the major benefit derived from osmotic adjustment is continued root growth necessary for exploration of large volumes of soil where ψ_s is high.

3. AN ASSESSMENT OF GENETIC VARIABILITY IN SORGHUM ROOT SYSTEMS

To improve a given trait through crop breeding, variability in the trait must exist and must be identifiable within the germ plasm pool, and heritability of the trait must be high enough to afford a reasonable chance of success in developing the desired type. We screened a wide range of breeding material for root system characteristics that differ appreciably from a widely used breeding line, 'RTx2536.' Initially we used a solution culture technique (24, 25) to assess growth characteristics of both shoots and root, but more recently we have monitored the rooting patterns of a limited number of genotypes in the field using root observation chambers (26) and the minirhizotron technique of Waddington (27) as described by Bohm (28) and modified by Arkin (G. F. Arkin, personal communication).

3.1. Results from Solution Culture Studies

Evaluation of available genetic material for differences either in root system morphology or in the balance between shoot and root growth is in progress. We are using two approaches in these investigations. First, detailed studies have made use of well-defined genetic model systems involving isogenic cultivars differing only in height (29) and maturity loci (24, 25, 30). The second approach involves screening a large number of cultivars from diverse genetic backgrounds (31). Similar screening approaches have been carried out on maize (32), wheat (33), and soybean (34).

Comparisons with Isogenic Cultivars. Previous work with a range of species demonstrated that close coordination exists between shoot and root growth (35, 36). We considered that cultivars with well-defined specific shoot differences in a uniform genetic background would serve as model genetic systems to investigate root growth and development. Two nearly isogenic maturity lines were tested (24, 30), an early maturing cultivar (60 SM-$ma_1ma_2ma_3ma_4$) and a late one (100M-$Ma_1Ma_2Ma_3Ma_4$). Crown roots appeared earlier and root length increased at a more rapid rate in the early (60 SM) cultivar. Increased root growth rates of '60 SM' were associated with more rapid leaf area development. The relation between shoot and root growth, expressed as the ratio of leaf area to root length, was similar for both cultivars, demonstrating that early maturity was associated with greater organ growth rates in general. Later field studies with root observation chambers showed that the additional time spent in the vegetative stage by the late isoline allowed production of greater leaf area per plant and greater root densities throughout the profile (Blum, unpublished results). Therefore, with regard to maturity isolines in the milo genetic background, the balance between shoot and root growth appears to be maintained regardless of maturity, and differences in root growth rates or rooting patterns reflects a difference in leaf area resulting from the difference in time spent in vegetative growth.

Isolines differing in height loci were also studied using cultivars from a kafir and milo background (29). Evidence from studies on the hormonal modification of stem growth in other crops, and from earlier studies with sorghum (37), suggested that variations in height may be associated with content or action of gibberellins. The effect of height genes on root growth of these sorghums was unknown, although Bower (38) was unable to detect differences in rooting patterns in height isolines of 'RS610' hybrid sorghum. Our studies also failed to detect appreciable differences in root systems or in the balance between shoot and root growth, even though height differences of nearly 1 m developed. A similar comparison between semidwarf and tall winter red wheats was recently presented by Cholick et al. (39).

Thus from our detailed studies using well-defined genetic model systems we failed to detect root variants that corresponded to shoot variants. However most of the isogenic comparisons made to date with sorghum have been restricted to a very narrow sample of the total germ plasm.

Results with Standard and Recently Converted Lines. A range of diversity in root characteristics does exist within sorghum cultivars. A screening program using hydroponic culture has led to the conclusion that the balance between shoot and root growth, and in the distribution of growth within the root system of many of the older parental lines developed in Texas breeding programs before the late 1960s, are similar, whereas considerable variability exists in recently converted lines (31, 40). Leaf area to root length (LA/RL) and shoot to root (S/R) dry weight ratios are taken as standard expressions of the balance of growth between shoot and root. In an earlier study of 30 lines, seven of the eight entries with LA/RL ratios more than 30% higher than 'RTx2536,' the standard of comparison, also had higher S/R ratios. Of 53 lines tested thus far, 19 (36%) and 16 (30%) had LA/RL and root length/root volume (RL/RV) ratios outside the range expected for 'RTx2536.' Examples of these variations appear in Figure 25.5.

How valuable are data on root length, root volume, or root weight from studies such as these in relation to drought tolerance under field conditions? One recent report (41) addresses these questions specifically in the case of sorghum, although the selection of genotypes included in the study was based on drought tolerance of seedlings. Of 10 entries representing a range of drought tolerance, the group regarded as possessing highest levels of tolerance based on field reaction had consistently higher root weights, greater root volumes, and lower S/R ratios. Those considered to be intermediate or susceptible to drought were not separable on these bases. These findings are similar in principle to an earlier report on root development of tolerant and susceptible sorghum strains growing in a natural field situation (42). Because of an unexpected drought during the normal rainy season (kharif), eight strains were separable into two groups, one that wilted and

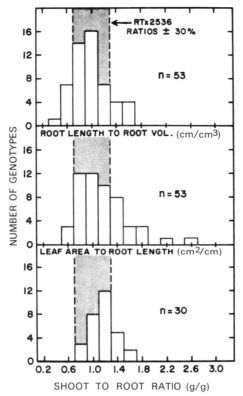

Figure 25.5. Frequency distributions for the ratios of root dry weight to shoot dry weight, leaf area to root length, and root length to root volume obtained from grain sorghum cultivars grown in hydroponics. Ratios for all cultivars are expressed relative to those for the standard genotype 'RTx2536.' Ratios differing from 'RTx 2536' (set equal to 1.0) by more than 30% are significantly different.

one that did not. When observed near the end of the drought (40 days after planting), roots of resistant strains had penetrated an average of nearly 15 cm deeper than susceptible strains. As a group, resistant strains also had more primary and secondary roots, greater root weights, and lower S/R ratios. Thus deep soil penetration associated with a greater partition of dry matter to roots appears to enhance drought tolerance in sorghum in a manner similar to that for wheat (33).

3.2. Results from Field Studies

Based on agronomic yield performance and on earlier root studies, we tested 10 cultivars in the field in 1977. Root growth patterns were followed throughout the season using the minirhizotron technique (cf. Section 3). The soil was the Houston Black Clay described earlier in this chapter.

Considerable variation in root growth rates and rooting patterns was found among the 10 cultivars. Examples of this variability are shown in Figure 25.6 for three dates. Patterns exhibited by 'RTx2536' are taken as representative for older Texas lines in general. Water remaining in the profile above −15 bars is shown in Figure 25.6a (from Figure 25.2) as a general example for a common hybrid. These rooting profiles represent growth during a long steady drying cycle during which water was extracted from progressively greater depths. Rooting patterns differed with respect to maximum L_v, depth at which maximum L_v was observed, and distribution of L_v with depth. The dynamics of root growth and disappearance are well illustrated by all three examples. Of the three genotypes, 'SC0 056-14' produced the deepest, most extensive rooting pattern; 'BTx622' produced a relatively intense, shallow rooting pattern, and L_v below 1 m for 'RTx2536' was intermediate between 'BTx622' and 'SC0 056-14.' In general, L_v was lower for 'RTx2536' compared with the other two. Soil water measurements were not made in 1977; therefore water depletion patterns associated with these rooting patterns may be only surmised. It is anticipated that 'SC0 056-14' may be better able to use deep stored soil moisture than either 'RTx2536' or 'BTx622.'

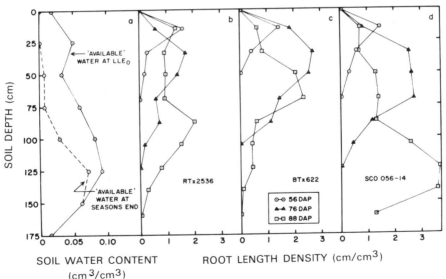

Figure 25.6. Relationship between the distribution of available soil water with soil depth (from Figure 25.2) (a), and between root length density and soil depth, for 'RT x2536' (b), 'BTx622' (c), and 'SC0 056-14' (d), determined at 56, 76, and 88 days after planting (DAP). Root length densities were calculated from minirhizotron observations.

3.3. Analytical Analyses of Rooting or Root Activity Patterns

Comparisons of rooting patterns among genotypes is difficult and often leads to subjective comparisons or tremendous investments in labor and money to secure sufficient samples to allow statistical analysis. Use of the minirhizotron technique greatly reduces the expense of data collection, but the problem of analysis remains. We are currently using an analysis of rooting patterns that can be represented by three descriptors as illustrated in Figure 25.7. For any time, a rooting pattern is described by the intensity I, the depth to maximum intensity $\bar{Z}$, the distribution of roots in the profile D. In this instance D is taken as the soil depth above which 80% of the total L_v exists. Although suitable for our data, this technique would be unsatisfactory for rooting patterns in which a linear or exponential decrease in L_v from the surface is normal (6, 43–45). However even in these cases it may be possible to use the minirhizotron technique to describe and compare root activity patterns in terms of water or nutrient uptake (12, 46–53).

An alternate way to analyze the root distribution pattern would be to consider the soil and root resistances to water uptake as given by Equations 25.2 and 25.3 in Section 2.3. This treatment would integrate the two variables of soil water distribution in relation to the associated L_v within the profile. Hence one could calculate the total theoretical resistance from L_v and soil water data, and confirm it from ψ_l and ET measurements. This approach also provides a method for comparing results from different soils.

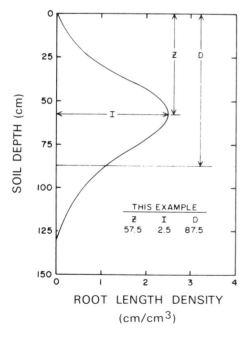

Figure 25.7. Illustration of three descriptors that define rooting or root activity patterns in soils to allow statistical comparisons among genotypes or treatments at any single time: I represents maximum root length density or root activity, $\bar{Z}$ is the soil depth at which I exists, and D is the soil depth above which 80% of the root length density or root activity occurs.

4. CONCLUDING REMARKS

A case for enhanced drought tolerance of summer crops through more efficient extraction of soil water stored deep in the profile is presented. Both theoretical analysis and actual rooting patterns of present-day grain sorghum cultivars substantiate the hypothesis that quantities of water required to match high evaporative demand cannot be supplied from deep in the profile because of low L_v. A modest increase in L_v from 1 to 2 cm/cm^3 would allow flow to match high evaporative demands until the soil dried to about -3 bars below 50 cm. An increase in L_v of this magnitude would allow the crop to maintain a high plant water status until approximately 24 cm of stored water has been extracted, rather than 18 cm, which now appears normal. Since the additional stress-free period would occur near anthesis, the overall effect of this relatively small amount of additional "freely available" water may be magnified because of the sensitivity of the crop to water deficits at this growth stage. Once the early grain-filling period is past, plant water deficits induced by high evaporative demands and depleted soil water reserves may have little deleterious effect on yields, provided preanthesis photosynthate can be mobilized to fill developing grain (6, 54).

An increase in L_r from 1 to 2 cm/cm^3 requires that a substantial additional investment be made by the crop in root biomass. Assuming a soil layer of 1.25 m and a dry matter content for roots equivalent to 20% of root volume, an increase in L_r of 1 cm/cm^3 requires at least 800 kg/ha more dry matter be partitioned to the root system prior to anthesis. Whether this may be accomplished in sorghum without concomitant reductions in grain yield remains to be demonstrated, but it certainly appears to be possible based on results with wheat (33).

Our efforts to identify sources of genetic variability in sorghum root systems have met with moderate success, although much remains to be done. Only a very small portion of the more than 17,500 lines in the World Collection of Sorghums has been studied. Our efforts are currently directed toward only the lines released from the Sorghum Conversion Program (40, 55) because these are the only exotic materials now suitable for rapid incorporation into commercial breeding programs. Our findings that older lines, selected and developed in Texas under similar conditions, all have similar root and shoot growth patterns, suggest the need for plant breeders to select parental lines under rather specific conditions for optimum performance.

Identification of sources of potentially superior rooting characteristics is only the first step in a crop improvement program. Little is known about heritability of deep root growth, extensive lateral branching, or efficient water or nutrient uptake. Zobel (56) estimates from mutation studies with tomato that 10% of radiation-induced mutations affect root growth separate from shoot growth, and another 20% affect both shoot and root. No such results are known for sorghum, but we recently initiated a similar mutation program in an attempt to produce isolines or near isolines that differ only in rooting characteristics. Only in this manner will it be possible to define accurately the contributions of specific root characters to drought tolerance. Under present breeding systems, claims for drought tolerance stemming from any single character may be clouded by the complexity of other interacting mechanisms, which may also impart some degree of tolerance as suggested by Blum (57).

ACKNOWLEDGMENTS

We acknowledge the assistance of Dr. R. G. C. Smith, Department of Agronomy and Soil Science, University of New England, Armidale, N.S.W., Australia, in predicting the root length densities in Figures 25.3 and 25.4.

REFERENCES

1. J. Levitt, *Responses of Plants to Environmental Stresses*. Academic Press, New York–London, 1972.
2. M. M. Caldwell, in O. L. Lange, L. Kappen, and E.-D. Schulze, Eds., *Water and Plant Life: Problems and Modern Approaches*, Springer-Verlag, Berlin–Heidelberg–New York, 1976, p. 63.
3. R. A. Fischer and N. C. Turner, *Annu. Rev. Plant Physiol.*, **29**, 277 (1978).
4. T. C. Hsiao and E. Acevedo, *Agric. Meteorol.*, **14**, 59 (1974).
5. H. M. Taylor, in E. W. Carson, Ed., *The Plant Root and Its Environment*, University of Virginia Press, Charlottesville, 1974, p. 271.
6. T. C. Hsiao, E. Fereres, E. Acevedo, and D. W. Henderson, in O. L. Lange, L. Kappen, and E.-D. Schulze, Eds., *Water and Plant Life: Problems and Modern Approaches*, Springer-Verlag, Berlin–Heidelberg–New York, 1976, p. 281.
7. A. Blum, *Agron. J.*, **62**, 333 (1970).
8. J. B. Passioura, *Aust. J. Agric. Res.*, **23**, 745 (1972).
9. J. T. Ritchie, E. Burnett, and R. C. Henderson, *Agron. J.*, **64**, 168 (1972).
10. P. J. Salter and J. E. Goode, *Crop Responses to Water at Different Stages of Growth*, Comm. Agric. Bur., Farnham Royal, 1967.
11. D. E. Kissel, J. T. Ritchie, and C. W. Richardson, *Tex. Agric. Exp. Stn., Misc. Publ.*, **1201** (1975).
12. H. M. Taylor and B. Klepper, *Agron. J.*, **65**, 965 (1973).
13. C. W. Richardson and J. T. Ritchie, *Trans. ASAE*, **16**, 72 (1973).
14. J. T. Ritchie and W. R. Jordan, *Agron. J.*, **64**, 173 (1972).
15. W. R. Gardner, *Soil Sci.*, **89**, 63 (1960).
16. R. A. Feddes and P. E. Rijtema, *J. Hydrol.*, **17**, 33 (1972).
17. M. N. Nimah and R. J. Hanks, *Soil Sci. Soc. Am. Proc.*, **37**, 522 (1973).
18. D. C. Reicosky and J. T. Ritchie, *Soil Sci. Soc. Am. J.*, **40**, 293 (1976).
19. C. Hackett, *Aust. J. Biol. Sci.*, **26**, 1211 (1973).
20. R. G. Lewis, E. A. Hiler, and W. R. Jordan, *Agron. J.*, **66**, 589 (1974).

21. K. J. McCree, *Crop Sci.*, **14**, 273 (1974).
22. K. W. Brown, W. R. Jordan, and J. C. Thomas, *Physiol. Plant.*, **37**, 1 (1976).
23. J. E. Begg and N. C. Turner, *Adv. Agron.*, **28**, 161 (1976).
24. A. Blum, G. F. Arkin, and W. R. Jordan, *Crop Sci.*, **17**, 149 (1977).
25. A. Blum, W. R. Jordan, and G. F. Arkin, *Crop Sci.*, **17**, 153 (1977).
26. G. F. Arkin, A. Blum, and E. Burnett, *Tex. Agric. Exp. Stn., Misc. Publ.*, **1386** (1978).
27. J. Waddington, *Can. J. Bot.*, **49**, 1850 (1971).
28. W. Böhm, *Z. Acker-Pflanzenbau*, **140**, 282 (1974).
29. S. A. Wright, *The Effect of Height Genes and Gibberellic Acid on Root and Shoot Development of* Sorghum bicolor *(L.) Moench*, M.S. thesis, Texas A&M University, College Station, 1978.
30. J. R. Quinby, *Sorghum Improvement and the Genetics of Growth*, Texas A&M University Press, College Station, 1974.
31. W. R. Jordan, F. R. Miller, and D. E. Morris, *Crop Sci.*, **19**, 468 (1979).
32. T. A. Kiesselbach and R. M. Weihing, *Agron. J.*, **27**, 538 (1935).
33. E. A. Hurd, *Agric. Meteorol.*, **14**, 39 (1974).
34. C. D. Raper, Jr., and S. A. Barber, *Agron. J.*, **62**, 581 (1970).
35. R. Brouwer, in F. L. Milthorpe and J. D. Ivins, Eds., *The Growth of Cereals and Grasses*, Butterworths, London, 1965, p. 153.
36. R. S. Russell, *Plant Root Systems: Their Function and Interaction with the Soil*, McGraw-Hill, New York, 1977.
37. P. W. Morgan, F. R. Miller, and J. R. Quinby, *Agron. J.*, **69**, 789 (1977).
38. J. T. Bower, *A Comparison of Root Systems of Two Isogenic Height Lines of Hybrid Grain Sorghum*, M.S. thesis, University of Nebraska, Lincoln, 1972.
39. F. A. Cholick, J. R. Welsh, and C. V. Cole, *Crop Sci.*, **17**, 637 (1977).
40. J. C. Stephens, F. R. Miller, and D. T. Rosenow, *Crop Sci.*, **7**, 396 (1967).
41. A.-E. M. Nour and D. E. Weibel, *Agron. J.*, **70**, 217 (1978).
42. S. Bhan, H. G. Singh, and A. Singh, *Indian J. Agric. Sci.*, **43**, 828 (1973).
43. H. M. Taylor and B. Klepper, *Adv. Agron.*, **30**, 99 (1978).
44. W. C. Mayaki, L. R. Stone, and I. D. Teare, *Agron. J.*, **68**, 532 (1976).
45. I. D. Teare, E. T. Kanemasu, W. L. Powers, and H. S. Jacobs, *Agron. J.*, **65**, 207 (1973).
46. E. L. Greacen, P. Ponsana, and K. P. Barley, in O. L. Lange, L. Kappen, and E.-D. Schulze, Eds., *Water and Plant Life: Problems and Modern Approaches*, Springer-Verlag, Berlin–Heidelberg–New York, 1976, p. 86.
47. B. K. Kaigama, I. D. Teare, L. R. Stone, and W. L. Powers, *Crop Sci.*, **17**, 555 (1977).
48. T. L. Lavy and J. D. Eastin, *Agron. J.*, **61**, 677 (1969).
49. J. W. McClure and C. Harvey, *Agron. J.*, **54**, 457 (1962).
50. F. S. Nakayama and C. H. M. Van Bavel, *Agron. J.*, **55**, 271 (1963).
51. R. W. Pearson, in E. W. Carson, Ed., *The Plant Root and Its Environment*, University of Virginia Press, Charlottesville, 1974, p. 247.
52. R. W. Rickman, R. R. Allmaras, and R. E. Ramig, *Agron. J.*, **70**, 723 (1978).
53. J. D. C. Vega, *Comparative Dynamics of Root Growth and Subsoil Water Availability in Unirrigated Corn and Sorghum*, Ph.D. thesis, University of California, Davis, 1972.

54. K. S. Fischer and G. L. Wilson, *Aust. J. Agric. Res.*, **22**, 33 (1971).
55. J. W. Johnson, D. T. Rosenow, F. R. Miller, and K. F. Schertz, *Tex. Agric. Exp. Stn. Prog. Rep.*, **2942**, 46–57 (1971).
56. R. W. Zobel, in J. G. Torrey and D. T. Clarkson, Eds., *The Development and Function of Roots*, Academic Press, New York–San Francisco–London, 1975, p. 261.
57. A. Blum, in H. Mussell and R. C. Staples, Eds., *Stress Physiology in Crop Plants*, Interscience, New York–London, 1979, p. 429.

26 | Adaptation to Water Stress in Rice

P. L. STEPONKUS and J. M. CUTLER

Department of Agronomy, Cornell University, Ithaca, New York

J. C. O'TOOLE

The International Rice Research Institute, Los Baños, The Philippines

1. INTRODUCTION

Rice (*Oryza sativa* L.) is a major food grain, supplying one-third of the world's population with more than half their calories and nearly half their protein. It is estimated that 48% of the world's 141 million hectares of rice is cultured in rainfed fields where inadequate water at one growth stage or another limits yield. The development of new cultivars with greater drought resistance and the potential to increase and stabilize yields for this rainfed sector is of significant concern. The great diversity of the world's rainfed rice-growing environments and the traditional plant types adapted to them (1) suggest that rice germ plasm contains numerous adaptive mechanisms that collectively enable the crop to cope with water deficits. Compared to the other major cereals, however, little information is available on the response of rice to water deficits. Such information is essential to the advancement of plant breeding efforts to improve rice yield potential and stability under rainfed cultivation.

Drought resistance is the result of numerous morphological, anatomical, and physiological characteristics, both constitutive and inducible, which

401

interact to allow for the maintenance of growth and developmental processes under edaphic and climatic conditions generally termed "drought." Following Levitt (2), such characteristics have been categorized as drought escape (evasion), drought avoidance, or drought tolerance mechanisms (see also Chapter 1). Such categories may serve as useful guides, but they are somewhat provocative when a strict and parochial usage of the terms is attempted. The perspective from which a given characteristic is viewed greatly influences whether it is deemed an avoidance or a tolerance mechanism. In addition, usage of the terms has fostered the concept that drought resistance of a given species may be ascribed to a single characteristic. Although dramatic examples of a single mechanism being responsible for increased drought resistance can be cited in extremely xerophytic species, drought resistance of most mesophytic crop species cannot be ascribed to a single mechanism. On the contrary, drought resistance of crop species should be considered to be a complex of interacting mechanisms, all contributing to the observed plant performance.

Drought resistance is a manifestation of the extent of coincidence over the growing season between the plant's water requirements for growth and development and its actual water status. The compatibility of the plant with its environment is ultimately determined by the tolerance of the plant, manifested in growth and developmental processes, to internal plant water deficits. The internal deficits are a function of the balance between water uptake and transpirational water loss, which is modified by internal water storage and resistance to water movement through the plant (Figure 26.1). Soil water availability is influenced by various mechanisms associated with water uptake (root depth, density, and permeability), whereas climatic factors determining evaporative demand are ameliorated by water conservation mechanisms (stomatal and cuticular resistance, leaf rolling, color and pubescence, and overall canopy architecture). These factors combine to determine *what* the plant water status will be in relation to the prevailing environment. *Whether* that level of plant water status is detrimental is a function of the tolerance of growth and developmental processes to water deficits. Furthermore, different stages of growth and development may vary in sensitivity to water deficits with respect to both the level of plant water status required for unimpaired plant performance and the extent of water deficits spanning the spectrum between unimpaired and complete impairment of plant performance (Figure 26.1). Because plants are more often than not under conditions of an internal water deficit, it is implied that there is some basal level of tolerance to this water deficit. Such tolerance may be measured either biologically or economically. Knowledge of *how* plant water status is transduced into plant performance (growth and developmental processes) is necessary for the complete understanding of how a plant responds to drought. Hsiao and co-workers (3, 4) have discussed the various possibilities, but the mechanism or mechanisms of transduction have not been clearly identified.

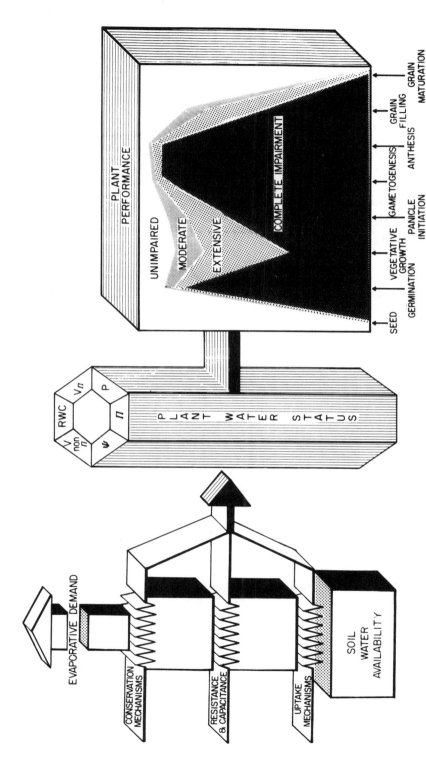

Figure 26.1. A schematic representation of the integration of various plant characteristics that influence the plant water status and the subsequent transduction into plant performance as a function of the edaphic and climatic environment: ψ = total water potential; π = osmotic potential, P = turgor potential, RWC = relative water content, V_π = osmotic volume, $V_{non\pi}$ = nonosmotic volume.

403

Finally, adaptation to drought is the result of both long-term phylogenetic evolution of different ecotypes within a species (constitutive) and the short-term adaptations in response to environmental stimuli that are elicited within a single life cycle (facultative). The latter inducible mechanisms, which depend on phylogenetic adaptation a priori, may allow for both amelioration of external factors that influence the internal plant water status and amelioration of the plant's tolerance to internal water deficits and provide a dynamic aspect of the functions in Figure 26.1.

Our discussion of adaptive mechanisms uses examples obtained from rice, first to consider plant characteristics that influence the relationship between the external environment and the internal water status of the plant, and second to consider how the internal water status relates to plant performance.

2. SYNCHRONIZATION OF PLANT ONTOGENY WITH THE HYDROLOGICAL ENVIRONMENT

The probability that a favorable water supply will coincide with the rice crop's water requirements for growth and development may be maximized by selection for characteristics such as early maturity or photoperiodic cuing of sensitive reproductive stages. Natural selection under the wide range of climatic conditions where rainfed rice has been cultured has provided a wide spectrum of genetic diversity with respect to these characteristics (5). The characteristics, however, have not been extensively utilized in improving drought resistance of rainfed rice. Krishnamurthy et al. (6) selected cultivars suitable for the drought prone region of north central India by considering grain yield in relation to days to flowering. Grain yield was maximum in cultivars requiring 70 days or less to flowering and rapidly declined in longer duration cultivars.

Another means by which the coincidence of sensitive reproductive stages with periods of high rainfall probability can be maximized is through photoperiodic control of flowering. However the use of photoperiod sensitivity in breeding programs was set back by the advent of the nonphotoperiodic sensitive modern rice cultivars developed in the 1960s. These first generation modern cultivars were suited to a limited range of hydrological conditions, principally those with good water control (i.e., low probability of flood or drought). Currently there is renewed interest in photoperiod sensitivity, and requirements for successful rainfed varieties to cope with problems of rainfall deficiency or distribution are being reexamined (7, 8).

3. MINIMIZATION OF INTERNAL WATER DEFICITS

Initial efforts in drought resistance breeding relied heavily on visual scoring of plant performance under drought conditions (9, 10), and this method is currently being utilized in international rice testing programs (11). Subsequent efforts to establish relationships between visual scoring criteria and plant water status indicate that characteristics associated with the maintenance of high leaf water potential (ψ_{leaf}) have been selected for in traditional cultivars adapted to drought prone regions. O'Toole and Moya (12) illustrated the high correlation between visual drought scoring and maintenance of ψ_{leaf}. A more extensive survey of more than 1500 rices has shown that upland cultivars, and hybrids with an upland parent from Asia, Africa, or South America that were rated superior by visual scoring criteria, also maintained relatively higher predawn ψ_{leaf} (Figure 26.2). Several plant characteristics can serve to minimize the development of internal water deficits resulting from reduced soil water availability, or increased atmospheric evaporative demand or both. An assessment of the significance and extent of genetic diversity of individual characteristics that contribute to

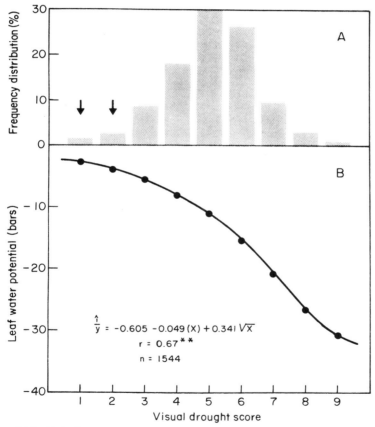

Figure 26.2. Relationship between visual drought score and predawn leaf water potential for 1544 cultivars and breeding lines evaluated in the drought screening greenhouse at the International Rice Research Institute during 1977. (*a*) Frequency distribution of visual drought scores (arrows indicate upland adapted cultivars and hybrids with at least one upland parent). (*b*) Relationship of predawn leaf water potential (Y) to visual drought score (X).

maintenance of higher ψ_{leaf} is required to advance plant breeding efforts and to allow for more complete exploitation of the available genetic resources.

3.1. Root System Characteristics

Chang and Vergara (13) and O'Toole and Chang (1) reviewed earlier work that compared the root systems of upland- and lowland-adapted cultivars. When compared to lowland cultivars, traditional upland cultivars have a deeper rooting habit and thicker roots. Soil injection of radioactive phosphorus has confirmed the superiority of traditional upland cultivars in West Africa for rooting depth and activity at 1 m (14).

Extensive characterization of rice root systems at the International Rice Research Institute has shown great diversity among rice cultivars on the basis of rooting depth and density when grown in greenhouse root boxes and in upland field conditions (15–18). In these reports the root/shoot ratio and the deep root/shoot ratio (root fraction below 30 cm) were adopted as a means of comparing cultivars. These ratios were highly correlated with field drought scoring results (16) and suggest that the high correlation between visual scoring and maintenance of high ψ_{leaf} (Figure 26.2, and Ref. 12) is in part attributable to the deep and prolific root systems of traditional upland cultivars.

On the basis of this study and previous visual scoring results, two cultivars with contrasting root systems, but similar phenology, were chosen to further consider varietal differences in soil water utilization. The water status (ψ_{leaf}) of the two cultivars grown in a deep (1 m) aerobic soil under greenhouse conditions was determined during a 30 day period following cessation of irrigation. When both cultivars—'Kinandang Patong,' an upland cultivar with a high deep root/shoot ratio, and 'IR28,' a hybrid selected for irrigated culture—were exposed to the same soil and atmospheric conditions, they differed greatly in maintenance of high ψ_{leaf} (Figure 26.3). Both cultivars exhibited a declining trend in midday ψ_{leaf} after cessation of irrigation, but the dawn measurements of ψ_{leaf} indicate that 'Kinandang Patong' rehydrated to higher ψ_{leaf} during the night throughout the drying period. The dawn ψ_{leaf} is considered to be a good estimate of the maximum ψ_{soil} encountered by the roots (19). Concurrent measurements of stomatal resistance and leaf rolling, which are discussed later, indicated that the observed differences in ψ_{leaf} were not attributable to differences in water conservation mechanisms. Thus it may be assumed that the upland cultivar 'Kinandang Patong' possesses a superior root system for absorption of soil moisture from lower depths, which allows rehydration during the night hours and relatively higher midday ψ_{leaf}.

3.2. Plant Capacitance and Conductance

Increased capacity to conduct water from the soil-root interface to the leaves may also be an adaptive mechanism that results in maintenance of relatively

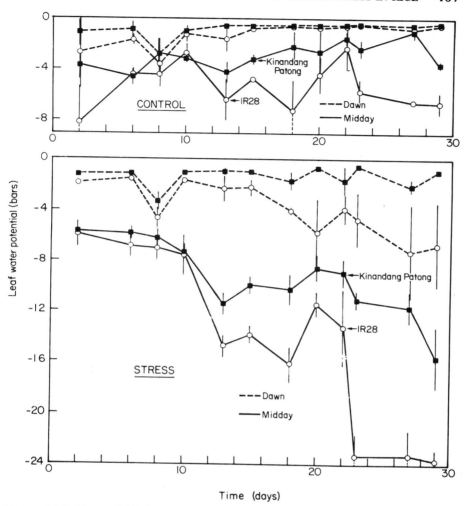

Figure 26.3. Dawn (0500 h) and midday (1300 h) values of leaf water potential of two rice cultivars ('Kinandang Patong' and 'IR28') subjected to well-watered control and water stress treatments during a 30 day period (51). Bars denote ± one standard error of the mean.

high ψ_{leaf}. Indirect evidence of differences in conductance by the vascular systems among cultivars has been observed among rices grown in irrigated fields in the dry season under conditions of high irradiance (424 cal/cm²/day) and high vapor pressure deficits (18 to 21 mbars). Table 26.1 illustrates the differences in ψ_{leaf} measured between 1000 and 1400 h among 15 cultivars or lines of rice. Adaxial stomatal resistances were similar for all cultivars (1.5 to 2.0 s/cm), and all plants were grown under submerged conditions (10

Table 26.1. Comparison of Mean Daily Minimum Leaf Water Potentials of 15 Rice Cultivars Adapted to Different Hydrological Conditions When Grown in Irrigated Plots During the 1977 Dry Season at the International Rice Research Institute, The Philippines

Means followed by a common line are not significantly different at the 5% level as indicated by Duncan's multiple range test.

Cultivar	Origin	Mean daily minimum leaf water potential (bars)
M1-48	Hybrid upland	−16.3
Salumpikit	Traditional upland	−17.4
Kinandang Patong	Traditional upland	−18.1
Dular	Dual upland-lowland	−18.9
C-22	Hybrid upland	−19.8
IR442-2-58	Lowland hybrid	−20.1
IR30	Lowland hybrid	−20.2
IR1561-228-3-3	Lowland hybrid	−20.3
IR36	Lowland hybrid	−20.6
IR2071-105-9-1	Lowland hybrid	−21.0
IR20	Lowland hybrid	−21.1
IR26	Lowland hybrid	−21.1
IR8	Lowland hybrid	−21.4
IR2035-117-3	Lowland hybrid	−21.6
IR5	Lowland hybrid	−22.3

cm of water). Assuming, for liquid phase transport, that:

$$\psi_{\text{leaf}} = \psi_{\text{soil}} - (\text{flux } r_p) \qquad (26.1)$$

where ψ_{soil} is the soil water potential and r_p is the resistance to water flow from soil to leaf (20), and assuming nonlimiting soil moisture conditions and equivalent transpiration rates among the cultivars, the cultivars with higher ψ_{leaf} have lower resistances or higher conductances to water flow through the plant. It is interesting to note that the five cultivars with the highest ψ_{leaf} are from rainfed upland habitats; the other 10 are high yielding hybrids selected in irrigated culture.

3.3. Water Conservation Mechanisms

Thus far we have dealt with mechanisms that result in maintenance of relatively high ψ_{leaf} by absorbing and conducting water in the liquid phase of the soil-plant-atmosphere continuum. Other mechanisms function to inhibit water loss from aerial plant organs in the vapor phase of the continuum. The influence of these mechanisms on maintaining relatively high ψ_{leaf} or in retarding its decrease is dependent on their responsiveness to internal stress.

Stomatal closure has received considerable attention (3, 21, 22) as a plant response to internal water stress, and it is generally agreed that stomatal control of water loss is the principal plant response that retards the decrease of ψ_{plant}. Rarely, however, has the role of altered leaf form or shape been considered along with stomatal closure, even though leaf rolling, a common observation in grass species, is intuitively associated with decreased transpiration (23, 24). Assuming that decreased exposure of transpirational surface area affects water loss (25) and that rolling alters the leaf's boundary layer resistance (26), the rolling of leaves may be a significant additional means of inhibiting water loss from grass species (see Chapter 3).

In the previously discussed experiment, which addressed varietal differences in maintenance of ψ_{leaf} (Figure 26.3), concurrent measurements of stomatal resistance and leaf rolling were made. In both cultivars leaf diffusive resistance and leaf rolling changed simultaneously relative to ψ_{leaf}, and the adaxial stomata of both cultivars were less responsive than was the abaxial stomata to decreasing ψ_{leaf}. However as the leaf's bulliform tissues lose turgor, the adaxial leaf surface is rolled inside; thus the adaxial stomata may be responding to the modified microclimate rather than to the ψ_{leaf}. Although the effects of leaf rolling to transpiration and net photosynthesis have not been experimentally demonstrated, it should be noted that leaf rolling is a dynamic, scalar characteristic, the varying degrees of which may well serve as an effective moderator of leaf water status with minimum reduction in net photosynthesis.

Cuticular resistance represents an additional significant factor inhibiting water loss from aerial plant organs. Yoshida and de los Reyes (27) found a wide range of cuticular resistance values along rice cultivars that were lower than those measured for corn and sorghum. This trait has also been found to exhibit some facultative plasticity. Both upland and lowland cultivars significantly increased cuticular resistance in response to a slow drying treatment (1). Differences in cuticular waxes of rice cultivars have been shown using scanning electron microscopy (1). Quantitative analysis of chloroform-extracted cuticular waxes by gas liquid chromatography indicated that the amount of epicuticular wax per unit leaf surface area in rice (0.04 mg/dm^2) is considerably less than that reported for other cereals: 1.5 mg/dm^2 for barley (28), 2.0 mg/dm^2 for sorghum (29), 2.5 mg/dm^2 for wheat (30).

It can be concluded from the discussion above that several adaptive mechanisms result in the maintenance of relatively higher ψ_{leaf}. Those that influence water uptake and conductance may be separated from those that influence water loss even though they share a common objective, because (a) the former involve absorption and conductance of soil moisture, whereas the latter involve inhibition of water loss from the shoot; (b) the former may maintain ψ_{leaf} without significant internal stress development, whereas the latter are generally observed as a response to some level of internal stress; and (c) the two possibilities may present different consequences for productivity because of their contrasting effects on leaf gas exchange.

In dealing with adaptive mechanisms that function to minimize the degree of internal plant water stress, our discussion has relied heavily on ψ_{leaf} as an estimate of plant water status, as have several recent reviews and reports on drought resistant mechanisms in crop species (1, 12, 31, 32). Although ψ_{leaf} is readily measured and frequently is used to indicate the level of plant water status, plant water status may be described by several other parameters (Figure 26.1). Hsiao et al. (4) cautioned that ψ_{leaf} "is not a good indicator of water status from the viewpoint of plant responses." The question of which parameter best describes plant water status as it influences growth and developmental phenomena requires an understanding of how plant water status is transduced into plant performance. The complete exploitation of genetic diversity will require evaluating mechanisms that ameliorate internal stress as well as those minimizing it.

4. MITIGATION OF INTERNAL WATER DEFICITS

The preceding discussion has considered the obvious constitutive adaptations that interact to determine the internal plant water status relative to the external environment. However many of these constitutive adaptations exhibit facultative plasticity and are influenced by the water stress history of the plant. Such plasticity, which is important when one considers the transduction of plant water status into plant performance (Figure 26.1), is manifested in the long-appreciated phenomenon of drought hardening (33, 34). Hardening, or drought conditioning, implies that preexposure to a sublethal level of stress results in decreased sensitivity to a subsequent stress. A variety of morphological and developmental alterations contribute to this reduced sensitivity, including leaf senescence and abscission, reduction in leaf size, increase in the root/shoot ratio, and delay in developmental events.

A variety of anatomical alterations are also observed in drought-hardened plants, but their significance is poorly understood. Concomitant with these morphological and anatomical alterations are physiological changes that allow growth and developmental processes to continue despite the development of soil water deficits. Such inducible mechanisms would then allow for amelioration of reduced soil water availability and increased consumptive requirements. Additionally, it is possible that drought conditioning could alter the tolerance of various growth and developmental processes to internal water deficits.

Collectively, such alterations are manifestations of the facultative plasticity and provide for a dynamic interaction between the external environment and plant performance rather than the static situation that might be inferred from Figure 26.1. The following discussion treats such facultative plasticity as the capacity for adaptive drought tolerance.

One component of adaptive drought tolerance that has received considerable attention is the capacity for osmotic adjustment, that is, the reduction of osmotic potential (see Chapter 7). A result of osmotic adjustment is an improved capacity to maintain turgor at lower plant water potentials and reduced water content (35, 36). Although osmotic adjustment has been considered to be a tolerance mechanism—that is, uptake of water can be maintained to lower soil water potentials—it may also be viewed as an internal water deficit avoidance or postponement mechanism—that is, loss of turgor is deferred to lower plant water potentials. A fundamental manifestation of osmotic adjustment is the maintenance of turgor under conditions of reduced water content (3, 4).

Several upland-adapted rice cultivars were investigated with respect to manifestations and mechanisms of adaptive drought tolerance. Plants were conditioned by exposure to several cycles of soil drying where the dawn ψ_{plant} was allowed to reach -10 to -14 bars before rewatering. The responses of these plants were compared to those of continuously well-watered plants when both were subsequently exposed to a challenge stress period. During this time ψ_{leaf} and its components, leaf turgor potential (P_{leaf}), and leaf osmotic potential (π_{leaf}) were monitored as indicators of water status and physiological adjustment. The relationship between P_{leaf} and π_{leaf} and relative water content (RWC) are illustrated for conditioned and control plants in Figure 26.4. All plants showed P_{leaf} declining rapidly with declining RWC, illustrating that much of the decline in ψ_{leaf} is absorbed by the P component. Conditioned plants had higher values of P_{leaf} (3 to 5 bars in all varieties examined) than did control plants at a given RWC. The π_{leaf} of conditioned plants was 5 bars more negative than that of control plants at any given RWC (Figure 26.4b). During dehydration the π_{leaf} of all plants declined slowly with decreasing leaf RWC, as expected from the concentrating influence of dehydration.

Additional evidence for osmotic adjustment in leaves of drought-conditioned rice plants is provided by the relationship between P_{leaf} and ψ_{leaf} as presented in Figure 26.5. For all varieties examined, conditioned plants had higher levels of P_{leaf} than did control plants throughout the range of measured water potentials. The intercept of these slopes of fitted regression lines varied significantly between conditioned and control plants of any variety. The differences ranged from 3 to 5 bars among the varieties examined without any significant difference among the varieties. Note, however, that the zero P_{leaf} was achieved at -15 bars in control plants, but was deferred to -20 bars in conditioned plants.

The data in Figures 26.4 and 26.5 demonstrate the capacity for osmotic adjustment in rice leaves, which was observed in several cultivars in response to a conditioning stress period. This physiological adjustment allows for achievement of lower plant water potentials and permits the plant to maintain gradients for water flow and to extract water from the soil at lower ψ_{soil} while simultaneously maintaining positive P_{leaf}. Thus it can be viewed as a tolerance mechanism from a decreased soil water perspective but as an avoidance or postponement mechanism from a cellular turgor perspective. However in neither case does osmotic adjustment allow for increased tolerance of growth and developmental processes to internal water deficits. Any

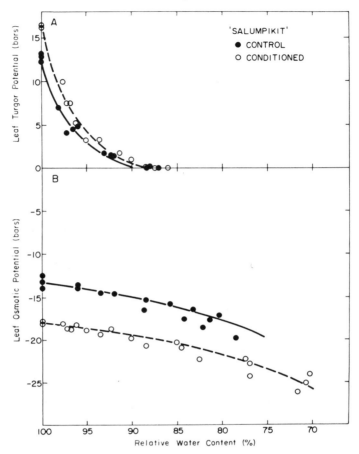

Figure 26.4. Relationship between (*a*) leaf turgor potential and relative water content (RWC) and (*b*) leaf osmotic potential and RWC for leaves of plants of the cultivar 'Salumpikit' that had been conditioned by several cycles of stress or in unconditioned controls.

such alteration would have to be manifested as a decreased sensitivity of plant performance to plant water deficits.

A major void in the study of plant water relationships is an understanding of the factors responsible for the primary transduction between plant water status and growth and developmental processes (Figure 26.1). Turgor potential has been viewed as an obvious candidate for such a role (3, 4) because growth and development are a concatenation of cell division, enlargement, and differentiation in the meristematic regions of the plant and are ultimately related to cell turgor (3). Cell expansion requires the maintenance of turgor above a threshold level (37–39), and rates of organ enlargement are

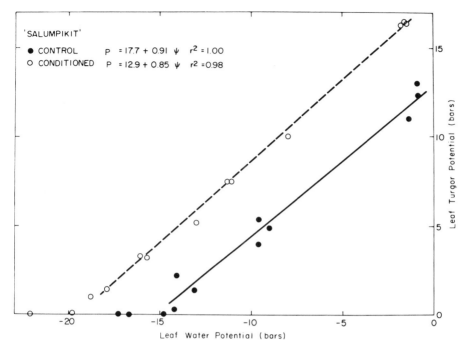

Figure 26.5. The relationship between leaf turgor potential P and leaf water potential ψ for leaves of conditioned and control plants of the cultivar 'Salumpikit.'

extremely sensitive to changes in turgor (40, 41). Cell division may be tightly coupled to cell expansion (42, 43), and organ differentiation and development are ultimately dependent on the harmonious synchronization of cell expansion and division. In fact, Slatyer has generalized that "deleterious effects of water deficits are most pronounced in tissues and organs which are in stages of most rapid growth and development" (44).

Although turgor may be a primary transducer for the effects of water deficits on plant growth and development, direct measurements of P_{leaf} are difficult. Only recently have techniques for the direct measurement of P_{leaf} been reported (45) and applied to higher plants (46). In view of the high sensitivity of leaf enlargement to changes in turgor, Boyer and McPherson (21) and Hsiao and Acevedo (47) suggested that leaf elongation rate be used as a criterion for the evaluation of drought sensitivity.

Recent work has focused on the adaptive nature of osmotic adjustment from the perspective of maintaining turgor at lower ψ_{plant}. Such studies have usually culminated in the analysis of water potential and its various components at various tissue water contents without considering the influence on plant performance (Figure 26.1). A further step would be to question whether drought hardening merely involves induction of cellular mechanisms to maintain turgor (osmotic adjustment, changes in cell wall elasticity, changes in tissue water distribution) or whether it alters the sensitivity of growth and developmental processes to turgor levels.

To address this question, leaf elongation rates (LER) of control and conditioned plants were monitored as an index of plant performance during a challenge stress period. The conditioning regime affected the maximum LERs differently, depending on the cultivar. In some cultivars, such as '63-83,' the conditioning regime did not have any influence on the maximum LERs; in other cultivars such as 'Kinandang Patong' there was a slight decrease in maximum LERs of conditioned plants; in still other cultivars such as 'Salumpikit' conditioned plants had slightly higher maximum LERs. This increased maximum LER was transient and presumably analogous to the compensatory growth phenomenon observed in plants relieved from water deficits (41, 48). However in all cultivars the daily LER decreased very steeply with declining predawn ψ_{leaf}. An example for 'Salumpikit' is given in Figure 26.6. It is readily apparent that LER provides a sensitive and easily measured indication of plant performance in relation to water status.

Water stress conditioning resulted in leaf elongation being maintained at

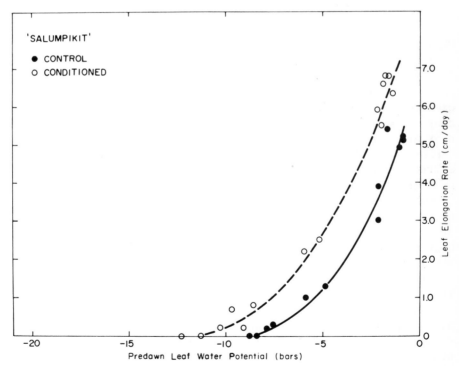

Figure 26.6. The relationship between leaf elongation rate and predawn leaf water potential for leaves of conditioned and control plants of the cultivar 'Salumpikit.'

lower predawn ψ_{leaf}. For 'Salumpikit' (Figure 26.6), LER declined to zero at predawn water potentials of -8.0 and -11.5 bars for control and conditioned plants, respectively. For another variety, '63-83,' this difference was particularly striking, with LER declining to zero at water potentials of about -5 and -10 bars, respectively. It can be concluded that this adaptive mechanism renders the plant more tolerant to reduced ψ_{leaf}. However a comparison of the relationship between LER and ψ_{leaf} does not provide any insight into whether this increase comes about by way of avoidance of a loss of turgor due to osmotic adjustment or whether the turgor threshold required for the maintenance of leaf elongation is altered. By comparing the relationship between P_{leaf} and ψ_{leaf} (Figure 26.5) at levels where LER is zero (from Figure 26.6), it can be seen that P_{leaf} of zero LER was 6.3 bars in control plants and 7.3 bars in conditioned plants. The conditioning does not appear to influence significantly the turgor threshold required for growth. A comparison of the ψ_{leaf} and P_{leaf} at which there is a 50% reduction in LER provides a comparison of the sensitivity of LER to water deficits. For control plants a 50% reduction in LER occurs at ψ_{leaf} of -3.0 bars and a P_{leaf} of 10.3 bars, whereas in conditioned plants the ψ_{leaf} was -4.2 bars and P_{leaf} was 14.0 bars (Figure 26.5). Thus from the relationship of LER and P_{leaf} the conditioned plants have a higher turgor requirement for growth. However at similar levels of

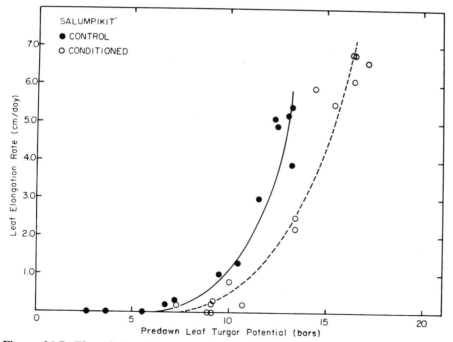

Figure 26.7. The relationship between leaf elongation rate and predawn leaf turgor potential for leaves of conditioned and control plants of the cultivar 'Salumpikit.'

ψ_{leaf} the conditioned plants have a smaller reduction in LER; that is, they are less sensitive to a reduction in ψ_{leaf}.

These extrapolations are with LER data and ψ_{leaf} measurements made during the development of a water stress when the plant could be adapting (49). To take into account the capacity for adaptation of the control plants during the challenge stress, the relationship of LER to P_{leaf} during a challenge stress was determined directly (Figure 26.7). Under such conditions the P_{leaf} threshold values for leaf elongation are approximately 6.0 and 7.0 bars for the control and conditioned plants, respectively. The turgor required for maintenance of 50% maximal LER was 12 bars for control plants (compared with 10.3 bars, above) and was at 14 bars for the conditioned plants (same as previously): the lack of any increase in the adaptive capacity is considered elsewhere (50). Although the conditioned plants appear to have a larger turgor requirement for maximal growth, these turgor levels were achieved when control and conditioned plants were at equal ψ_{leaf}. Hence it appears that conditioning renders LER less sensitive to changes in turgor above the turgor threshold. Thus adaptation appears to both increase P_{leaf} at a given ψ_{leaf} because of decreases in π_{leaf} by osmotic adjustment, but it also alters the sensitivity of one growth process, leaf elongation, to changes in P_{leaf}.

5. CONCLUSIONS

Though not commonly associated with the problems of drought, a significant proportion of the world's rice production is subject to the vagaries of rainfall, which results in extensive limitations of crop production. The development of rice cultivars with increased drought resistance is currently a major research goal. Despite an enormous reservoir of genetic material—no less than 100,000 cultivars in the two cultivated species, which embodies a broad spectrum of genetic diversity—little is known about the physiological bases for drought resistance in rice. As a result, current screening techniques rely almost exclusively on visual scoring criteria. Future progress in rice breeding for drought resistance is dependent on the identification of the constitutive and facultative plant characteristics that interact to determine both the plant water status in response to alterations in the edaphic and climatic environment and the impact of the resultant plant water status on the growth and development of rice.

This preliminary analysis of drought resistance in rice indicates that singular approaches, whether they be from an edaphic or climatic perspective, plant water status perspective, or plant performance perspective, should be integrated to ensure achievement of the maximum exploitation of germ plasm, which is necessary for the development of rice cultivars with improved drought resistance.

ACKNOWLEDGMENTS

This is Cornell University Department of Agronomy Series Paper No. 1280.

REFERENCES

1. J. C. O'Toole and T. T. Chang, in H. Mussell and R. C. Staples, Eds., *Stress Physiology in Crop Plants*, Interscience, New York–London, 1979, p. 373.
2. J. Levitt, *Responses of Plants to Environmental Stresses*, Academic Press, New York–London, 1972.
3. T. C. Hsiao, *Annu. Rev. Plant Physiol.*, **24**, 519 (1973).
4. T. C. Hsiao, E. Acevedo, E. Fereres, and D. W. Henderson, *Philos. Trans. R. Soc. London, Ser. B*, **273**, 479 (1976).
5. T. T. Chang and H. I. Oka, in S. Yoshida, Ed., *Climate and Rice*, International Rice Research Institute, Los Baños, 1976, p. 87.
6. C. Krishnamurthy, S. V. S. Shastry, and W. H. Freeman, *Oryza*, **8**, No. 2 (Suppl.), 47 (1971).
7. H. I. Oka, *Phyton (Buenos Aires)*, **11**, 153 (1958).
8. B. S. Vergara and T. T. Chang, *Int. Rice Res. Inst. Tech. Bull.*, **8** (1976).
9. T. T. Chang, G. C. Loresto, and O. Tagumpay, *Sabrao J.*, **6**, 9 (1974).
10. G. C. Loresto, T. T. Chang, and O. Tagumpay, *Philipp. J. Crop Sci.*, **1**, 36 (1976).
11. International Rice Research Institute, *Standard Evaluation System for Rice*, International Rice Research Institute, Los Baños, 1975.
12. J. C. O'Toole and T. B. Moya, *Crop Sci.*, **18**, 873 (1978).
13. T. T. Chang and B. S. Vergara, in International Rice Research Institute, *Major Research in Upland Rice*, International Rice Research Institute, Los Baños, 1975, p. 71.
14. F. N. Reyniers and T. Binh, in I. W. Buddenhagen and G. J. Persley, Eds., *Rice in Africa*, Academic Press, London–New York–San Francisco, 1978, p. 278.
15. International Rice Research Institute, *Annual Report for 1974*, International Rice Research Institute, Los Baños, 1975.
16. International Rice Research Institute, *Annual Report for 1975*, International Rice Research Institute, Los Baños, 1976.
17. International Rice Research Institute, *Annual Report for 1976*, International Rice Research Institute, Los Baños, 1977.
18. International Rice Research Institute, *Annual Report for 1977*, International Rice Research Institute, Los Baños, 1978.
19. R. O. Slatyer, *Plant-Water Relationships*, Academic Press, London–New York, 1967.
20. M. R. Kaufmann and A. E. Hall, *Agric. Meteorol.*, **14**, 85 (1974).
21. J. S. Boyer and H. G. McPherson, *Adv. Agron.*, **27**, 1 (1975).
22. J. E. Begg and N. C. Turner, *Adv. Agron.*, **28**, 161 (1976).
23. H. R. Oppenheimer, in *Plant-Water Relationships in Arid and Semi-Arid Conditions: Reviews of Research*, UNESCO, Paris, 1960, p. 105.

24. J. Parker, in T. T. Kozlowski, Ed., *Water Deficits and Plant Growth*, Vol. 1, Academic Press, New York–London, 1968, p. 195.

25. D. M. Gates, *Annu. Rev. Plant Physiol.*, **19**, 211 (1968).

26. I. R. Cowan and F. L. Milthorpe, in T. T. Kozlowski, Ed., *Water Deficits and Plant Growth*, Vol. 1, Academic Press, New York–London, 1968, p. 137.

27. S. Yoshida and E. de los Reyes, *Soil Sci. Plant Nutr. (Tokyo)*, **22**, 95 (1976).

28. P. von Wettstein-Knowles, in R. A. Nolan, Ed., *Barley Genetics II*, Washington State University Press, Pullman, 1971, p. 146.

29. A. Ebercon, A. Blum, and W. R. Jordan, *Crop Sci.*, **17**, 179 (1977).

30. S. Kumar, *Indian Phytopathol.*, **27**, 508 (1974).

31. A. Blum, *Crop Sci.*, **14**, 361 (1974).

32. A. Blum, in H. Mussell and R. C. Staples, Eds., *Stress Physiology in Crop Plants*, Interscience, New York–London, 1979, p. 429.

33. R. Gauss, *Am. Breed. Mag.*, **1**, 209 (1910).

34. I. I. Tumanov, *Planta*, **3**, 391 (1927).

35. M. M. Jones and N. C. Turner, *Plant Physiol.*, **61**, 122 (1978).

36. J. M. Cutler and D. W. Rains, *Physiol. Plant.*, **42**, 261 (1978).

37. J. A. Lockhart, *J. Theor. Biol.*, **8**, 264 (1965).

38. E. L. Greacen and J. S. Oh, *Nature (London) New Biol.*, **235**, 24 (1972).

39. P. B. Green, R. O. Erickson, and J. Buggy, *Plant Physiol.*, **47**, 423 (1971).

40. J. S. Boyer, *Plant Physiol.*, **43**, 1056 (1968).

41. E. Acevedo, T. C. Hsiao, and D. W. Henderson, *Plant Physiol.*, **48**, 631 (1971).

42. R. Cleland, *Annu. Rev. Plant Physiol.*, **22**, 197 (1971).

43. A. Poljakoff-Mayber, in A. Poljakoff-Mayber and J. Gale, Eds., *Plants in Saline Environments*, Springer-Verlag, Berlin–Heidelberg–New York, 1975, p. 97.

44. R. O. Slatyer, in J. D. Eastin, F. A. Haskins, C. Y. Sullivan, and C. H. M. Van Bavel, Eds., *Physiological Aspects of Crop Yield*, Am. Soc. Agron., Crop Sci. Soc. Am., Madison, Wisconsin, 1969, p. 53.

45. E. Steudle, V. Zimmermann, and U. Luttge, *Plant Physiol.*, **59**, 285 (1977).

46. D. Hüsken, E. Steudle, and U. Zimmermann, *Plant Physiol.*, **61**, 158 (1978).

47. T. C. Hsiao and E. Acevedo, *Agric. Meteorol.*, **14**, 59 (1974).

48. W. Wenkert, E. R. Lemon, and T. R. Sinclair, *Agron. J.*, **70**, 761 (1978).

49. K. W. Shahan, J. M. Cutler, and P. L. Steponkus, *Agron. Abstr.*, p. 85, 1978.

50. J. M. Cutler, K. W. Shahan, and P. L. Steponkus, *Crop Sci.*, **20**, (in press) (1980).

51. J. C. O'Toole and R. T. Cruz, *Plant Physiol.*, **65**, 428 (1980).

27 | Improvement of Perennial Herbaceous Plants for Drought-Stressed Western Rangelands

DOUGLAS A. JOHNSON

USDA–SEA–AR, Crop Research Laboratory, Utah State University, Logan, Utah

1. INTRODUCTION

As demands on United States rangeland increase, improved forage plants and range management techniques should be used to achieve optimum returns. Inasmuch as most rangeland in the western part of the country lies within the arid or semiarid climatic zones (1), plants growing in these areas are affected seriously by drought stress (2). Consequently improved range plants must have drought resistance mechanisms, adaptations that allow plants to grow and survive in areas subjected to periodic water deficits. In an effort to provide improved forages for the western rangeland, the U.S. Department of Agriculture (USDA) Agricultural Research in cooperation with Utah State University, has assembled a forage and range research team at Logan, Utah. The team consists of a cytogeneticist, a plant physiologist, a range scientist, and two plant breeders.

419

This chapter deals with approaches that have been useful or hold promise in isolating superior plant materials for water-limited areas: these approaches are discussed with particular reference to the USDA Range and Forage Program at Logan. Plant adaptations to drought are classically categorized into avoidance and tolerance mechanisms (3–6). Avoidance adaptations allow the plant to escape drought stress, and mechanisms of tolerance enable the plant to either postpone or withstand dehydration (see Chapter 1). Adaptations to drought stress include both morphological and physiological mechanisms. These numerous adaptations are discussed in Parts II and III.

In production of annual crops attention is usually directed toward growth and yield rather than toward survival (7). However perennial species are usually seeded on rangelands; consequently survival as well as production is important. The perennial habit increases the complexity of breeding and selection. With perennials a portion of current production is diverted for storage reserves for subsequent regrowth the following season. Also, in range plant production, where the economic end product is not primarily grain, a considerably smaller investment of current assimilate is required for seed production. Seed production is necessary only to ensure new genetic combinations and to provide seed for seedling establishment in disturbed or adjacent habitats. Thus care should be used in making blanket comparisons between adaptations in herbaceous annuals and adaptations that might be beneficial to semiarid perennials.

Drought tolerance adaptations involving dehydration tolerance theoretically would be most advantageous in forage plants where leaf production is important. Dehydration tolerance adaptations should allow range plants to produce maximum leaf growth at a given water potential. Dehydration postponement mechanisms, such as increased root growth or increased stomatal sensitivity, theoretically would be less desirable than tolerance mechanisms because they usually develop at the expense of top growth. In practice, however, postponement mechanisms, such as extensive root growth and transpirational control, are also important for successful adaptation to water-limited environments. Thus improved range plants probably have a combination of dehydration postponement and tolerance mechanisms that enable them to survive periods of severe drought, but to continue active growth during periods of less severe drought.

2. CHARACTERIZING THE ENVIRONMENT

Hanson (8), Reitz (9), Boyer and McPherson (10), and Fischer and Turner (11) emphasized the importance of thoroughly characterizing the particular drought situation. Not only is total precipitation important, but also its distribution. Begg and Turner (12) stressed that intensity and duration of the water deficit are important in determining the particular drought adaptations that are most appropriate for a particular species. Boyer and McPherson (10) emphasized the importance of timing in considering lack of water. In environments where drought is unpredictable or sporadic, plants with superior performance in one season may exhibit inferior performance in another. Consequently the potential for success in providing improved plant materials is probably greatest in environments where drought occurs predictably during the same part of the growing season each year.

The areas to which our plant improvement program is directed generally receive less than 380 mm annual precipitation, most of it during late fall, winter, and spring (Table 27.1). Because high evaporative losses during the summer months make the already low summer precipitation even less effective, production on most of the western rangeland relies heavily on stored winter–spring moisture.

3. DEFINING THE SELECTION CRITERIA

The effect of drought stress on range plants is particularly pronounced during germination, emergence, and early seedling growth (14, 15). McGinnies (16), Knipe and Herbel (17), and Wilson et al. (18) reported that low soil moisture delays germination, reduces total germination, and decreases seedling growth of several range species. Planting failures should be minimized by the development of plant materials with superior establishment characteristics.

Inasmuch as stored winter–spring moisture is so important in determining productivity of Intermountain rangelands, early spring growth is an essential feature for improved plant materials. This adaptation, one of drought escape, allows the plant to grow and mature during the portion of the season when water is most readily available. This early growth benefits the live-

Table 27.1. Long-Term Precipitation Normals for Representative Range Areas in the Western United States Categorized According to 3 Month Periods (13)

Climatic region	Precipitation normals, 1941–1970 (mm)				
	January through March	April through June	July through September	October through December	Total annual
Southeast Oregon	72	75	25	77	248
Central Plains, Idaho	73	73	26	75	246
Wind River, Wyoming	33	137	54	43	268
Northeastern Nevada	73	93	41	75	282
Western Utah	49	64	47	55	216

stock producer because winter feed is often scarce and the summer ranges are not yet available for grazing. An integral part of successful seedling growth, particularly in semiarid areas, involves early root initiation and elongation. Rapid root extension is critical in stand establishment because it allows the seedling to compete successfully both with other species and with evaporative drying for the rapidly diminishing moisture of the soil surface (19).

If winter moisture is in short supply, range plant growth may depend almost entirely on precipitation during April, May, and June. In such years seed must germinate on a relatively small amount of soil moisture. Germination under low soil moisture is important so that the seedlings can immediately take advantage of the spring rains. After emergence, the seedlings may not receive additional precipitation for another 10 to 20 days. Thus another important range plant characteristic involves the ability of seedlings to survive desiccating conditions and resume growth at the end of the drought period.

Because of the extremely low precipitation and high evaporative demands during July, August, and September, range forage plants generally become dormant. Thus capacity to withstand summer drought is also important. Plant materials should also be able to respond to precipitation in late fall, when evaporative demands are lower.

4. ASSEMBLING A BROAD GENETIC BASE

Genetic advance in a plant improvement program critically hinges on assembling a diverse collection of germ plasm, which can be comprised of released varieties, experimental strains from other breeders, collections from old plantings, and plant introductions. This diverse germ plasm serves as a breeding pool and provides a broad genetic base for obtaining variation for particular characteristics. Hurd (20) emphasized that large populations are required because improvement in drought resistance probably involves many genes having small effects that are difficult to measure. Hanson (8) also stressed the importance of diverse germ plasm pools in increasing the genetic diversity for selection and in improving the frequency of genes for potentially valuable plant characteristics.

5. DEVELOPING SCREENING TECHNIQUES

After a broad-based germ plasm pool has been assembled, plant improvement generally involves screening this large source population to find the plants that have the desired combination of characters. Levitt (21) and Cooper (22) stressed the importance of reliable screening tests as an integral component of a plant improvement program. Hanson (8) emphasized that progress in plant breeding has been impeded by the lack of appropriate screening procedures.

Plant screening techniques should (a) assess plant performance at the critical developmental stage, (b) be completed in a relatively short period of time, (c) use relatively small quantities of plant materials, and (d) be capable of screening large populations. Although many techniques are available for examining plant water relationships, most of these are too laborious and time-consuming for use in a plant improvement program. In addition many techniques for evaluating plant water relations measure dynamic plant characteristics that change daily or throughout the season.

Numerous possibilities also exist for screening anatomical or morphological characteristics related to drought resistance. Many researchers have examined the association between individual morphological characters and drought resistance. However their results often conflict and document the many interactions that these characters have with the environment. Wright (14), Moss et al. (23), and Boyer and McPherson (10) state that essentially none of these associations has proved useful as a reliable indicator of drought resistance. Consequently Wright (24) recommended that until the fundamentals of drought resistance characteristics are known more precisely, plant improvement programs will have to utilize screening techniques that are based on plant response rather than on specific plant characteristics. Moss et al. (23) also suggested that selection criteria for improving plant performance under drought may be a necessary compromise between impossibly complex measurements and the more convenient, rapid screening techniques.

Roy and Murty (25) and Mederski and Jeffers (26) have suggested that selection under optimum growth conditions might yield breeding lines that also perform well under water stress conditions. In this case improved response to drought is apparently an unidentified component of stability in performance over different environments. However high yielding varieties under conditions of adequate moisture are not always high yielding under moisture stress (27–29). Boyer and McPherson (10) emphasized that although assessments under optimum conditions may sometimes be successful, to screen for characteristics that become apparent only during drought, selection must be done under desiccating conditions. In the latter case potentials for growth and for drought resistance are controlled by separate genetic entities. Thus after drought resistance components have been identified, they must be incorporated into high yielding cultivars.

Breeding lines that use water efficiently in a dry environment may not do as well as other lines under more favorable water conditions. This is because tradeoffs exist regarding plant responses in different environments (30). Therefore selecting plants for wide adaptability may be selecting for mediocrity (9). As a result, the most promising route for plant improvement under drought stress probably involves selection under water-limiting conditions.

5.1. Field Methods

Field performance of plant lines is generally regarded as the standard by which plant response under drought stress is assessed. However field trials require large expenditures of time and money. In addition, semiarid areas are notorious for their fluctuations in environmental conditions from year to year and from site to site. Consequently screening under range conditions can be a time-consuming, uncertain endeavor (24). However several recently developed techniques might help reduce the vagaries of climatic fluctuations in semiarid field situations.

Rainout Shelters. One new method involves the use of rainout shelters (31–34), which mechanically cover field plots, thus excluding rainfall from them. Modern designs of rainout shelters are electronically controlled and automatically shelter the plot area from incoming precipitation. Besides excluding rainfall, the shelters tend to minimize disturbance to other environmental parameters such as irradiation, temperature, and wind. As such, rainout shelters provide a field setting with strict control over water stress conditions. A disadvantage of shelters is that the area covered is relatively small (12 × 12 m) for large-scale plant breeding applications. If selections can be made at the seedling stage, however, this may not be a serious drawback.

Line Source Sprinklers. A second field technique that seems promising for plant screening involves a line source sprinkler installation (35). Sprinklers are spaced so closely along an irrigation line that water distribution is essentially constant along any line parallel to the sprinkler line. A continuous water variable is obtained by applying water so that the amount of water application decreases linearly from a high rate at the point of origin to zero at some point a given distance from the sprinkler line. Thus the system produces a water application pattern that is uniform along the length of the irrigation line and continuously, but uniformly (under windless conditions), variable at right angles to the sprinkler line. This field installation is ideal for evaluating the response of range seedlings to different water levels, an especially critical characteristic for successful seedling establishment on western rangeland. One potential disadvantage is that since this technique cannot exclude precipitation, the results are affected by fluctuating precipitation. However in much of the Intermountain West experiments can be run during July, August, and September, when precipitation is typically low.

Infrared Photography. Another field selection procedure that holds promise for screening potential breeding lines involves infrared photography. Blum (36) and Blum et al. (37) used infrared photography to screen sorghum genotypes for dehydration postponement in the field. Their work was based on the premise that infrared reflectance of plant leaves is associated with leaf hydration. Large field nurseries of sorghum were screened for leaf dehydration levels by photometric inspection of aerial infrared photographs. Although this procedure depends on precise timing of canopy development and is affected by physiological conditions other than hydration of the plant, it could be used for evaluating large numbers of breeding lines.

5.2. Laboratory Methods

Inasmuch as field screening often requires large expenditures of time and money, and because fluctuating weather conditions make field comparisons among years extremely difficult, reliable laboratory or greenhouse screening techniques are desirable. Laboratory procedures allow independent control of environmental conditions and isolation of the direct and indirect effects of drought.

Extensive Root Growth. A notable exception to the limited applicability of plant characters to drought resistance was reported by Hurd (20, 38–40). In that wheat breeding program selection for extensive root systems was successful and resulted in the release of high-yielding, drought tolerant cultivars (41–43). Hurd and co-workers concluded that a rapidly penetrating root system with an extensive network of primary and secondary roots is essential for high wheat yields in a semiarid environment. In some semiarid environments where plant growth must rely on water stored in the soil, conservative water behavior has been suggested as a desirable trait. Passioura (44) demonstrated that water could be conserved during the vegetative stage by increasing the resistance of the root system of wheat to water flow by means of reducing the numbers of seminal roots. This allowed greater water availability during heading and resulted in seed yields that were higher than those from plants with normal root systems. However in forage production, where leaf rather than grain production is critical, water conservation for later growth stages is probably of less concern. Therefore a rapidly penetrating and extensive root system probably should be of primary concern in programs to develop improved perennial herbaceous cultivars for semiarid areas.

Controlled Environment Chamber. Controlled environment chambers have proved particularly successful in the identification of drought resistant seedlings. Wright and Streetman (45) extensively reviewed the historical aspects of atmospheric and soil drought testing. With a growth chamber designed by Wright (46), important range grass species were evaluated for seedling drought resistance (24, 47, 48). Seedlings were exposed in a chamber programmed to represent field drought conditions prevalent during establishment. After regular watering, seedlings were rated for recovery. This screening and selection work, which was summarized by Wright (15), has resulted in the release of an improved cultivar of *Eragrostis curvula* (Schrad.) Nees that is well adapted for seeding arid range sites in the Southwest (49).

This technique of withholding water and observing subsequent recovery of seedlings is appealing because of its simplicity and capacity to evaluate large populations. However as Begg and Turner (12) have cautioned, the rate of development of stress during a drying cycle should be carefully considered. In particular, restriction of plant roots in small containers can accelerate the rate of stress development. Thus Begg and Turner (12) recommended using plant containers that allow for more realistic root development and consequently a more gradual transition from mild to severe stress. Careful consideration should also be given to the length of recovery for the particular species and to the severity of stress (50, 51). Additionally, caution should be used in extrapolating the results obtained with seedlings to mature plants.

Emergence Under Drought Stress. Germination and emergence under drought stress are key factors in the establishment of many crops, particularly range forage plants (14). Several techniques have been used to assess germination and early seedling growth under water stress conditions. A commonly used procedure involves the germination of seeds on a blotter containing an osmotic solution of known solute potential. Solutions of mannitol, sodium chloride, and polyethylene glycol have been used to provide a range of water potentials; however different osmotica may have specific effects on germination, independent of water potential. Consequently techniques in which seed is in direct contact with the osmoticum may confound the effects of drought stress with the direct effect of the osmotic solution on the seed.

A procedure used by Kaufmann (52), as modified by Johnson and Asay (53), has been successfully used in a plant improvement program. This technique is similar to those proposed by Painter (54), Zur (55), Cox and Boersma (56), and Tingey and Stockwell (57) in that a semipermeable membrane separates the osmotic solution from the germination medium, but allows water diffusion across the membrane. Direct seed and osmoticum contact is avoided, and by use of soil as a germination medium this procedure closely simulates contact between seed surfaces and soil water. With this procedure, Kaufmann and Ross (58) obtained germination data that closely correlated with results from field tests. Johnson and Asay (53) demonstrated that this method provides a realistic procedure for examining seedling emergence of range grass species under drought stress, and at the same time allows assessments of large breeding populations.

Plant Turgor. Osmotic adjustment and subsequent maintenance of plant turgor has received increasing attention as evidenced by recent reviews by Cram (59), Hellebust (60), Zimmermann (61), and Turner and Jones (Chapter 7). Work by Noy-Meir and Ginzburg (62), Sánchez-Díaz and Kramer (63),

Cheung et al. (64), Tyree (65), Johnson and Brown (66), and Roberts and Knoerr (67) suggests that analysis of plant turgor might be useful for predicting plant performance under drought stress. Examination of plant turgor response is especially appealing because maintenance of turgor integrates both water supply and water loss mechanisms of the plant. The hypothesis is that plants that maintain relatively high turgor over a wide range of water stress exhibit relatively greater growth over that same stress range.

Methods for determining turgor within plants generally involve psychrometric analysis (68, 69), pressure chamber determination (64, 70), and pressure probe assessment (71, 72). The advantages and disadvantages of these methods are discussed by Zimmermann (61). Most of the presently available techniques to measure plant turgor are too laborious for use in a plant improvement program, at least during the initial selection phases. Furthermore, environments in which plants are grown can influence turgor response (73) and threshold turgors for growth (74). Thus further studies are required that more precisely define the interactions of turgor with plant growth and water regulation. If these inadequacies are overcome, however, turgor response might hold promise for screening plant materials for improved performance under drought stress.

Other Laboratory Methods. Desiccation tolerance and heat tolerance are often positively correlated (75, 76). Inasmuch as desiccation tolerance screening methods are more time-consuming than are heat tolerance techniques, heat tolerance tests have been used to indicate desiccation tolerant breeding lines. Sullivan (77) has used an electroconductivity technique modified from Dexter et al. (78) to assess heat tolerance among sorghum breeding lines. Sullivan et al. (79) showed that sorghum lines with greatest field drought resistance also had the highest heat tolerance. This technique has been used to sample 1600 plants in one day (80).

Another heat tolerance technique that has been used to indicate drought resistance involves chlorophyll stability. In this technique, termed "chlorophyll stability index," the colorimetric readings of the chlorophyll extract from heated and unheated leaf samples are compared. Kaloyereas (81), Murty and Majumder (82), and Kilen and Andrew (83), working with loblolly pine, rice, and maize, respectively, found high correlations between chlorophyll stability index and field drought resistance. In pearl millet, however, Fanous (84) found a poor correlation between this index and drought resistance.

Proline accumulation during periods of water deficit has been reported in many plant species (85–89). Singh et al. (90) indicated that barley varieties that accumulated more proline tended to survive extreme water stress better and grew more rapidly during recovery than other varieties. Rao and Asokan (91) also found that drought resistant varieties of sugarcane accumulate more proline than susceptible ones. Blum and Ebercon (92) reported a significant positive relation between free proline accumulation and ability to recover after the relief of stress in sorghum. However Hanson et al. (93) working with two barley cultivars demonstrated that a simple, positive correlation between proline accumulation and drought resistance may not always hold true. Therefore they suggested that proline accumulation should not be used as a positive index of drought resistance and cautioned that selection for high proline accumulation might even result in selection for drought susceptibility (see Chapter 12).

6. APPLICATION OF SCREENING TECHNIQUES

Hurd (40) emphasized that after a diverse germ plasm pool has been assembled and appropriate selection techniques developed, a plant breeding program involves successive generation cycling to accumulate the desired additive genes. The genetic advance R expected in selection for a trait such as drought resistance can be expressed as:

$$R = i\sigma_p h^2 \tag{27.1}$$

where i is the intensity of selection, σ_p is the phenotypic standard deviation, and h^2 is the heritability in the narrow sense of the particular characteristic (94).

This equation can be presented as in Figure 27.1. The mean drought resistance of the base population is represented by line A. By use of selection procedures in the base population, individuals with superior performance can be identified with their mean drought resistance represented by line C. Consequently a selection differential between lines A and C has been identified and is governed both by the phenotypic variation present in the base population and the intensity of selection applied by the screening procedure. However because the heritability of the desired characteristic is

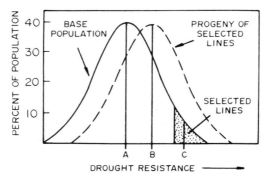

Figure 27.1. Diagrammatic representation of expected genetic advance from selection for a trait such as drought resistance (40): A = mean of base population, C = mean of selected lines, and B = mean of progenies of selected lines.

BREEDING SCHEME

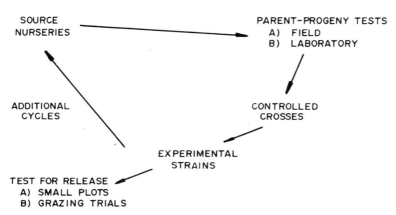

Figure 27.2. A representative plant breeding sequence used in the United States Department of Agriculture Range and Forage Program at Logan, Utah. Field and laboratory screening procedures are used to identify superior performance in lines of the diverse source population. These lines form the parentage of experimental strains that are tested for possible varietal release.

less than unity, only a portion of the selection differential will actually be achieved in the subsequent generation. Therefore the mean drought resistance of the subsequent generation will be somewhere between the mean of the base population and the mean of the selected portion of the base population. More detailed discussions of selection and genetic advancement are contained in Falconer (94) and Allard (95).

The overall sequence representing our range plant improvement program is diagrammed in Figure 27.2. The laboratory and field screening is an integral part of the parent and progeny testing phase of our program and data from these evaluations are incorporated into a general selection index. Superior lines form the parentage of experimental strains that are tested in small plots and grazing trials. Based on these tests, the plant material is either released or included in additional breeding cycles.

7. SUMMARY AND CONCLUSIONS

A plant improvement program for supplying plant material with superior performance in a water-limited environment should first characterize the drought in the particular area. Not only must total precipitation be considered, but also its distribution, timing, and predictability. Second, important adaptive mechanisms for resisting drought should be considered, and the most important selection criteria for evaluating superiority or inferiority of the breeding lines should be identified. After a broad genetic base of plant material has been assembled, screening techniques must be developed that can (a) assess plant performance at the critical developmental stages, (b) be completed quickly, (c) use small quantities of plant materials, and (d) screen large populations. Finally, the selection techniques must be incorporated into a plant breeding program for meaningful genetic advance. Each of these steps is discussed with particular emphasis toward perennial herbaceous plant improvement for the semiarid rangelands of the western United States.

ACKNOWLEDGMENTS

I gratefully thank Drs. K. H. Asay, R. W. Brown, M. M. Caldwell, D. R. Dewey, H. H. Wiebe, and R. E. Wyse for helpful comments on the manuscript. Appreciation is also extended to C. J. Andreasen for bibliography assistance and to C. C. Lemon for secretarial assistance.

REFERENCES

1. P. Meigs, in *Reviews of Research on Arid Zone Hydrology*, UNESCO, Paris, 1953, p. 203.
2. R. W. Brown, in R. E. Sosebee, Ed., *Rangeland Plant Physiology*, Range Sci. Ser. No. 4, Soc. Range Manage., Denver, 1977, p. 97.
3. H. L. Shantz, *Ecology*, **8**, 145 (1927).
4. J. Levitt, *Annu. Rev. Plant Physiol.*, **2**, 245 (1951).
5. J. Levitt, *Responses of Plants to Environmental Stresses*, Academic Press, New York–London, 1972.
6. L. H. May and F. L. Milthorpe, *Field Crop Abstr.*, **15**, 171 (1962).
7. C. Y. Sullivan and J. D. Eastin, *Agric. Meteorol.*, **14**, 113 (1974).
8. A. A. Hanson, in V. B. Youngner and C. M. McKell, Eds., *The Biology and Utilization of Grasses*, Academic Press, New York–San Francisco–London, 1972, p. 36.
9. L. P. Reitz, *Agric. Meteorol.*, **14**, 3 (1974).
10. J. S. Boyer and H. G. McPherson, *Adv. Agron.*, **27**, 1 (1975).
11. R. A. Fischer and N. C. Turner, *Annu. Rev. Plant Physiol.*, **29**, 277 (1978).
12. J. E. Begg and N. C. Turner, *Adv. Agron.*, **28**, 161 (1976).
13. *Monthly Averages of Temperature and Precipitation for State Climatic Divisions 1941–1970*, U.S. Dept. Comm. NOAA Environmental Data Service, Climatography of the United States No. 85, Asheville, North Carolina, 1973.
14. L. N. Wright, in K. L. Larson and J. D. Eastin, Eds., *Drought Injury and Resistance in Crops*, Spec. Publ. No. 2, Crop Sci. Soc. Am., Madison, Wisconsin, 1971, p. 19.
15. L. N. Wright, in R. S. Campbell and C. H. Herbel, Eds., *Improved Range Plants*, Range Symp. Series No. 1, Soc. Range Manage., Denver, 1975, p. 3.
16. W. J. McGinnies, *Agron. J.*, **52**, 159 (1960).

17. D. Knipe and C. H. Herbel, *J. Range Manage.*, **13**, 297 (1960).
18. A. M. Wilson, J. R. Nelson, and C. J. Goebel, *J. Range Manage.*, **23**, 283 (1970).
19. G. A. Harris, in J. K. Marshall, Ed., *The Belowground Ecosystem: A Synthesis of Plant-Associated Processes*, Range Sci. Series No. 26, Colorado State University, Fort Collins, 1977, p. 93.
20. E. A. Hurd, in K. L. Larson and J. D. Eastin, Eds., *Drought Injury and Resistance in Crops*, Spec. Publ. No. 2, Crop Sci. Soc. Am., Madison, Wisconsin, 1971, p. 77.
21. J. Levitt, in *Forage Plant Physiology and Soil-Range Relationships*, Spec. Publ. No. 5, Am. Soc. Agron., Madison, Wisconsin, 1964, p. 57.
22. J. P. Cooper, *Report Welsh Plant Breeding Station, Aberystwyth, for 1973*, 1974, pp. 95–102.
23. D. N. Moss, J. T. Woolley, and J. F. Stone, *Agric. Meteorol.*, **14**, 311 (1974).
24. L. N. Wright, *Crop Sci.*, **4**, 472 (1964).
25. N. N. Roy and B. R. Murty, *Euphytica*, **19**, 509 (1970).
26. H. J. Mederski and D. L. Jeffers, *Agron. J.*, **65**, 410 (1973).
27. G. W. Burton, in *Research on Water*, Spec. Publ. Series No. 4, Am. Soc. Agron., Madison, Wisconsin, 1964, p. 95.
28. E. A. Hurd, *Agron. J.*, **60**, 201 (1968).
29. V. A. Johnson, S. L. Shafer, and J. W. Schmidt, *Crop Sci.*, **8**, 187 (1968).
30. G. H. Orians and O. T. Solbrig, *Am. Nat.*, **111**, 677 (1977).
31. M. L. Horton, *Iowa Farm Sci.*, **17**, 16 (1962).
32. E. A. Hiler, *Trans. ASAE*, **12**, 499 (1969).
33. I. D. Teare, H. Schimmelpfennig, and R. P. Waldren, *Agron. J.*, **65**, 544 (1973).
34. G. F. Arkin, J. T. Ritchie, M. Thompson, and R. Chaison, *Tex. Agric. Exp. Stn. Misc. Publ.*, **1199** (1975).
35. R. J. Hanks, J. Keller, V. P. Rasmussen, and G. D. Wilson, *Soil Sci. Soc. Am. J.*, **40**, 426 (1976).
36. A. Blum, *Z. Pflanzenzucht.*, **75**, 339 (1975).
37. A. Blum, K. F. Schertz, R. W. Toler, R. I. Welch, D. T. Rosenow, J. W. Johnson, and L. E. Clark, *Agron. J.*, **70**, 472 (1978).
38. E. A. Hurd, *Euphytica*, **18**, 217 (1969).
39. E. A. Hurd, *Agric. Meteorol.*, **14**, 39 (1974).
40. E. A. Hurd, in T. T. Kozlowski, Ed., *Water Deficits and Plant Growth*, Vol. 4, Academic Press, New York–San Francisco–London, 1976, p. 317.
41. E. A. Hurd, L. A. Patterson, D. Mallough, T. F. Townley-Smith, and C. H. Owen, *Can. J. Plant Sci.*, **52**, 687 (1972).
42. E. A. Hurd, T. F. Townley-Smith, D. Mallough, and L. A. Patterson, *Can. J. Plant Sci.*, **53**, 261 (1973).
43. E. A. Hurd, T. F. Townley-Smith, L. A. Patterson, and C. H. Owen, *Can. J. Plant Sci.*, **52**, 689 (1972).
44. J. B. Passioura, *Aust. J. Agric. Res.*, **23**, 745 (1972).
45. L. N. Wright and L. J. Streetman, *Ariz. Agric. Exp. Stn. Tech. Bull.*, **143** (1960).
46. L. N. Wright, *Ariz. Agric. Exp. Stn. Tech. Bull.*, **148** (1961).
47. L. N. Wright and G. L. Jordan, *Crop Sci.*, **10**, 99 (1970).
48. L. N. Wright and S. E. Brauen, *Crop Sci.*, **11**, 324 (1971).
49. L. N. Wright, *Crop Sci.*, **11**, 939 (1971).

50. H. M. Laude, in K. L. Larson and J. D. Eastin, Eds., *Drought Injury and Resistance in Crops*, Spec. Publ. No. 2, Crop Sci. Soc. Am., Madison, Wisconsin, 1971, p. 45.

51. A. Corleto and H. M. Laude, *Crop Sci.*, **14**, 224 (1974).

52. M. R. Kaufmann, *Can. J. Bot.*, **47**, 1761 (1969).

53. D. A. Johnson and K. H. Asay, *Crop Sci.*, **18**, 520 (1978).

54. L. I. Painter, *Agron. J.*, **58**, 459 (1966).

55. B. Zur, *Soil Sci.*, **102**, 394 (1966).

56. L. M. Cox and L. Boersma, *Plant Physiol.*, **42**, 550 (1967).

57. D. T. Tingey and C. Stockwell, *Plant Physiol.*, **60**, 58 (1977).

58. M. R. Kaufmann and K. J. Ross, *Am. J. Bot.*, **57**, 413 (1970).

59. W. J. Cram, in U. Lüttge and M. G. Pitman, Eds., *Encyclopedia of Plant Physiology*, New Series, Vol. 2, Part A, Springer-Verlag, Berlin–Heidelberg–New York, 1976, p. 284.

60. J. A. Hellebust, *Annu. Rev. Plant Physiol.*, **27**, 485 (1976).

61. U. Zimmermann, *Annu. Rev. Plant Physiol.*, **29**, 121 (1978).

62. I. Noy-Meir and B. Z. Ginzburg, *Aust. J. Biol. Sci.*, **22**, 35 (1969).

63. M. F. Sánchez-Díaz and P. J. Kramer, *J. Exp. Bot.*, **24**, 511 (1973).

64. Y. N. S. Cheung, M. T. Tyree, and J. Dainty, *Can. J. Bot.*, **53**, 1342 (1975).

65. M. T. Tyree, in M. G. R. Cannell and F. T. Last, Eds., *Tree Physiology and Yield Improvement*, Academic Press, London–New York–San Francisco, 1976, p. 329.

66. D. A. Johnson and R. W. Brown, *Crop Sci.*, **17**, 507 (1977).

67. S. W. Roberts and K. R. Knoerr, *Oecologia*, **28**, 191 (1977).

68. C. F. Ehlig, *Plant Physiol.*, **37**, 288 (1962).

69. R. W. Brown, *Agron. J.*, **68**, 432 (1976).

70. M. T. Tyree and H. T. Hammel, *J. Exp. Bot.*, **23**, 267 (1972).

71. E. Steudle and U. Zimmermann, *Z. Naturforsch. B*, **26**, 1302 (1971).

72. D. Hüsken, E. Steudle, and U. Zimmermann, *Plant Physiol.*, **61**, 158 (1978).

73. D. A. Johnson, *Crop Sci.*, **18**, 945 (1978).

74. J. A. Bunce, *J. Exp. Bot.*, **28**, 156 (1977).

75. J. Levitt, *The Hardiness of Plants*, Academic Press, New York, 1956.

76. C. Y. Sullivan, N. V. Norcio, and J. D. Eastin, in A. Muhammed, R. Aksel, and R. C. von Borstel, Eds., *Genetic Diversity in Plants*, Plenum, New York, 1977, p. 301.

77. C. Y. Sullivan, in N. G. P. Rao and L. R. House, Eds., *Sorghum in Seventies*, Oxford and IBH, New Delhi–Bombay–Calcutta, 1972, p. 247.

78. S. T. Dexter, W. E. Tottingham, and L. F. Graber, *Plant Physiol.*, **7**, 63 (1932).

79. C. Y. Sullivan, J. D. Eastin, and E. J. Kinbacher, *Univ. Nebraska, Farm Ranch Home Q.*, **15**, 7 (1968).

80. C. Y. Sullivan, in K. L. Larson and J. D. Eastin, Eds., *Drought Injury and Resistance in Crops*, Spec. Publ. No. 2, Crop Sci. Soc. Am., Madison, Wisconsin, 1971, p. 1.

81. S. A. Kaloyereas, *Plant Physiol.*, **33**, 232 (1958).

82. K. S. Murty and S. K. Majumder, *Curr. Sci.*, **11**, 470 (1962).

83. T. C. Kilen and R. H. Andrew, *Agron. J.*, **61**, 669 (1969).

84. M. A. Fanous, *Agron. J.*, **59**, 337 (1967).

85. A. R. Kemble and H. T. MacPherson, *Biochem. J.*, **58,** 46 (1954).
86. N. M. Barnett and A. W. Naylor, *Plant Physiol.*, **41,** 1222 (1966).
87. D. G. Routley, *Crop Sci.*, **6,** 358 (1966).
88. G. Pálfi and J. Jahász, *Plant Soil,* **34,** 503 (1971).
89. R. P. Waldren, I. D. Teare, and S. W. Ehler, *Crop Sci.*, **14,** 447 (1974).
90. T. N. Singh, L. G. Paleg, and D. Aspinall, *Aust. J. Biol. Sci.*, **26,** 65 (1973).
91. K. C. Rao and S. Asokan, *Sugar J.*, **40,** 23 (1978).
92. A. Blum and A. Ebercon, *Crop Sci.*, **16,** 428 (1976).
93. A. D. Hanson, C. E. Nelsen, and E. H. Everson, *Crop Sci.*, **17,** 720 (1977).
94. D. S. Falconer, *Introduction to Quantitative Genetics*, Ronald Press, New York, and Oliver and Boyd, Edinburgh, 1960.
95. R. W. Allard, *Principles of Plant Breeding*, Wiley, New York, 1964.

SUMMARY AND SYNTHESIS | VII

28 | Adaptation of Plants to Water and High Temperature Stress: Summary and Synthesis

1. STRESS TERMINOLOGY

J. Levitt

Department of Plant Biology, Carnegie Institution of Washington, Stanford, California

The terminology proposed in Chapter 1 was not accepted by all authors, some preferring to remain with that of Levitt (1). Indeed, all stress terminology should be based on two principles:

1. A terminology for water stress, or for any other stress, must not be developed by itself. It is necessary first to develop a terminology applicable to all environmental stresses, even those induced by the organismal environment.

2. Stress is a physiological concept, and all physiology is based on physics and chemistry. Therefore, just as water potential terminology proved successful only when it was based on principles of thermodynamics, so also environmental stress terminology can succeed only if based on principles of mechanics.

437

Some biologists have not been able to accept a physical concept of stress as a basis for the terminology of environmental stress. Whatever their arguments, they are too late; for we are in the space age, and the mechanical concept of stress is already a part of environmental terminology. Thus gravitational stress is measured by the number of G's to which an organism is exposed when sent into space. Are we, then, to propose a second kind of stress, and teach our students and colleagues that they must accept two mutually exclusive definitions for stress? The answer of course is a resounding NO!

Consequently we must base *all* stress terminology on the following equation from mechanics:

$$M = \frac{\text{stress}}{\text{strain}} \qquad (28.1)$$

where M is the modulus of elasticity. But M has been used for so many other quantities, and the true meaning of "modulus of elasticity" is really the resistance to the stress, since the larger M is, the smaller is the strain per unit stress. For environmental stress, therefore, the defining equation is:

$$R = \frac{\text{stress}}{\text{strain}} \qquad (28.2)$$

where R is the resistance to stress.

How can this resistance be increased? Equation 28.2 gives two possibilities: *stress avoidance,* that is, a decrease (or partial avoidance) of stress by erecting some kind of barrier between it and the living cells of the organism, and *stress tolerance*, or full exposure of living cells to stress, but the organism must tolerate the stress without suffering an injurious strain. The three quantities are related as follows:

Stress resistance	
Stress avoidance	Stress tolerance

There may be several ways available to a plant for avoiding a specific stress. Stress tolerance, however, can be of two main kinds, regardless of the nature of the stress. The resistant organism may tolerate the stress due to possession of *strain avoidance,* a decrease in the strain produced per unit stress, or *strain tolerance,* an ability to undergo without injury a strain that is fatal to a less resistant organism. These relations can be illustrated by the simple case of linear strain in a physical body (Figure 28.1). Strain tolerance in its turn may be of two kinds. The strain may be fully reversible, or it may be repaired.

The early agriculturists were intuitively aware of this physical nature of environmental stress. This is why they coined the term "hardening" for the process of adapting to stress. The strain-avoiding plant, for instance, "hardens," therefore suffers less strain per unit stress, similar to the thickening of the wire in Figure 28.1, which makes it more rigid, thus less yielding to the

I. Stress injury

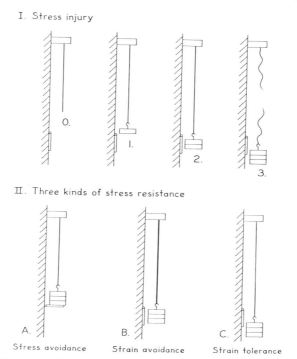

II. Three kinds of stress resistance

A.   B.   C.

Stress avoidance   Strain avoidance   Strain tolerance

Figure 28.1. Diagrammatic representation of stress injury and three types of stress resistance in plants.

stress. The strain-tolerating plant is able to undergo without injury a strain that would injure an intolerant plant by, for example, an annealing process that increases its elastic extensibility. However the term "acclimation" has come into more common use and may be accepted in place of "hardening." Strictly speaking, however, *acclimation* can only be used for climatic stresses, not for radiation stress, salt stress, or toxic ion stress.

The four kinds of stress resistance just discussed are related as shown in Figure 28.2, and all have been identified in plants adapted to stress. Therefore correct use of these terms is absolutely essential for a clear understanding of stress injury and resistance.

2. MORPHOLOGICAL ADAPTATIONS TO WATER STRESS

H. H. Wiebe

Department of Biology, Utah State University, Logan, Utah

The morphological adaptations of two plant organs, leaves and roots, are considered in detail in this book. Developments at the apex in response to

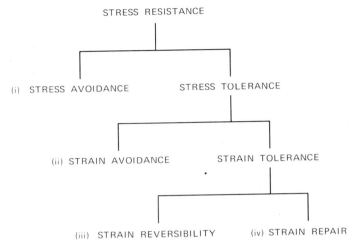

Figure 28.2. Relationships between the four basic types of stress resistance in plants.

water stress, discussed in Chapter 13, can be considered to be morphological adaptations but are not discussed here.

The successful colonization of terrestrial environments by higher plants is believed to have occurred largely as a result of the evolution of a wide range of morphological adaptations associated with the conservation of water (Chapter 3). The morphological adaptations to water stress may be classified as (*a*) permanent, including the production of smaller leaves, leaves with changed anatomy, and the senescence or shedding of older leaves, or (*b*) temporary, involving changes in leaf angle or orientation. A temporary change may take the form of a passive wilting response as in sunflowers and cotton, or the active parahelionastic movements evident in a number of leguminous species. Leaf rolling or folding is a common response in grasses. These mechanisms all reduce the radiation load on water-stressed leaves and may thereby prevent overheating or firing when there is little water available to dissipate energy as latent heat. An important feature of these temporary adaptations of leaves to water stress is their reversibility and rapid recovery on relief of stress, perhaps without serious yield reduction.

A more permanent adaptation to water stress is that of changes in leaf anatomy discussed in Chapter 4, which quantifies the effect of leaf anatomy as the ratio of internal mesophyll surface area divided by the leaf surface area, the A^{mes}/A ratio. High light during leaf development markedly increases A^{mes}/A, and water stress, salinity stress, and high temperature also increase this anatomical feature. An increase in A^{mes}/A leads to an increase in water use efficiency, thus the chapter provides an explanation for the influence of leaf anatomy on water use efficiency.

Turning now to roots, it is clear from Chapters 6 and 25 that selection criteria for root adaptations of annual crops may differ with climate and soil type. A deep root system would seem best in a climatic region in which the soil is recharged to considerable depth during the winter and rains during the growing season are infrequent. In contrast, a shallower system with a higher root length density would seem to be optimum for a climate in which rains, even though limited, occur during the growing season and the deep soil is usually not recharged during the winter. After a sorghum crop in relatively deep Blackland soils in Central Texas, Jordan and Miller (Chapter 25) reported that the water content of soils below 1 m was still near field capacity. Models incorporating root length density and depth predicted that increased root densities in deeper soils could increase the total water available to a sorghum crop by up to 33%. Even more important, dehydration of the crop could be delayed, perhaps beyond the critical anthesis stage. Thus a slight postponement of stress may pay large dividends in yield. Studies indicate that substantial differences in root growth rate, in depth of penetration, and in root length density profiles exist among cultivars of sorghum and soybean. It thus appears that ample root variability exists to permit considerable choice in crop breeding and selection.

Authors differed in their evaluation of the importance of root density. Taylor (Chapter 6) suggested that selection for root densities greater than 0.2 to 0.3 cm/cm^3 was not profitable in cotton and soybean, whereas Jordan and Miller (Chapter 25) concluded that increasing the root density at soil depths below 50 cm in sorghum from 1 to 2 cm/cm^3 would allow significantly more extraction of water from a drying soil. These differences in interpretation and conclusion may arise from the different soil types for which the plants were selected. Studies have suggested that sorghum plants with high root densities at depth do better than those with low root densities at depth in a heavy clay soil, but not in a sandy loam soil.

The concept that xylem or root longitudinal resistance is negligible continues to be challenged. Potential energy decreases of 2.6 bar/cm have been reported, and reports of resistances of 0.1 to 0.01 bar/cm (10 and 1 bar/m, respectively) are common. Yet others fail to find any measurable xylem resistance. It may be that this matter should be reinvestigated and perhaps considered in crop selection. Indeed our models, which too often imply that a crop consists of many plants, each with one root, stem, and leaf, may be inadequate. Models based on the resistance of parallel, merging pathways or channels would be more realistic.

The mechanism by which drought influences the shoot/root ratio by limiting shoot growth more than root growth apparently has not been studied. Assuming that elongation is a function of turgor, then changes in the relative turgors of elongating regions of the shoot and root could account for the altered shoot/root ratios, but no data seem to exist. This would also imply a high xylem resistance, which is entirely plausible in developing xylem near the elongating zones. Otherwise, the water potentials of root and shoot elongating regions would be nearly the same. Allocation of assimilates shifts toward the roots during drought. This does not necessarily imply that assimilates promote root growth, but rather that under drought conditions root growth is less impaired than stem growth; therefore roots are relatively greater assimilate sinks.

Although Taylor (Chapter 6) found some soil conditions in which root length density was practically uniform throughout the soil, it usually falls off more or less exponentially with depth. One might ask the evolutionary reason for root length density of most plants falling off with depth. Even if we allow for manipulation by man, we are generally referring to varieties that have been selected for yield, often in arid regions and for thousands of years. If these have a particular root length density profile, is this not in itself a hint that these characteristics have been among those unconsciously selected because they contribute to yield? Root length density profiles of native plants or weeds are influenced by long-term natural selection; it might be instructive to study the root systems of some of the more successful and obnoxious annual weeds associated with the crop of interest.

Root systems of perennial native desert and chaparral species are discussed in Chapter 5. Although it might appear reasonable for a root system, once developed, to be maintained for the life of a shrub or tree, there are many data indicating that the finer feeder roots are quite transient and are continuously replaced. The turnover may involve a third or half of the total root system annually. The figure for the feeder roots may be much higher because the large roots are the perennial ones, which contribute the most to the biomass, whereas the fine roots contribute to the root length density. Root regeneration may involve the bulk of the annual energy budget of some desert shrubs, but significant sloughing and regeneration has been reported for species of humid regions as well. Apparently there is some biological advantage in sloughing roots after a period of function and continually replacing them. Perhaps the advantage is in random reexploration of the available soil volume; perhaps less energy is used in building new roots than in maintaining old ones. Maintenance involves such energy-intensive mechanisms as suberization, which though it may increase root longevity by protecting against decay microorganisms, also decreases the absorptive capacity of the roots.

The energy costs of root growth are in fact referred to several times throughout the book. Diversion of assimilates to feeder roots reduces the quantity available for agricultural yield. The roots may repay the investment by giving a limited yield at lower soil moistures or by enabling survival. Although drought reduces yields generally, there are reports that varieties selected for yield under drought conditions do not yield as well as other varieties during wetter years. If a variety could be bred that yielded best under all conditions, it would be expected to capture the market. It may be that the crossover point comes during extremely dry years, when it is not profitable to harvest the crop. A challenging problem for modelers would be to determine the optimum plan for energy investment in roots versus yield, given the average and range of rainfalls for a region. Modelers might also suggest the optimum variety for farmers to plant, given climate, its variability, and economic and human cultural factors.

3. PHYSIOLOGICAL ADAPTATIONS TO WATER STRESS

J. S. Boyer

USDA–SEA–AR, Department of Botany and Agronomy, University of Illinois, Urbana, Illinois

Part III, which discusses the physiological adaptations to water stress, indicates that plants are capable of internal osmotic adjustment using various metabolites (Chapters 7, 9, and 13). This is a considerable advance from the concept that osmotic adjustment involved primarily the uptake of ions by plants in saline soils. Of particular significance is the metabolic control of osmotic adjustment, which suggests that plants respond according to a genetically programmed system that has as the net result the continuation of germination, the preservation of growth, and the alteration of stomatal behavior in the face of dry conditions. With these findings, we should be able to select for this behavior, as the chapter by Morgan (Chapter 24) elegantly attests. At this time, however, it is clear that we know little about how plants accomplish osmotic adjustment. Do solutes accumulate in cells simply because they are being less rapidly metabolized, or are solutes directed to organs that are important to survival under dry conditions?

Although osmotic adjustment is a plant characteristic that preserves growth in an adverse environment, adaptation to dry conditions also operates in other directions. The desiccation tolerance of certain higher plants, described by Gaff in Chapter 14, is a good example. He shows that survival in extreme environments involves cessation of growth rather than preservation of growth. Furthermore, he demonstrates that severe selection pressure can markedly improve the ability of higher plants to survive. Thus man ought to be able to select for plants having superior survival abilities in dry environments. It is disturbing to find, however, that this character is associated with markedly slower growth in mesic environments. We need to know whether this tradeoff represents a fundamental biological principle, or whether it is observed only in the extreme cases cited in Chapter 14.

The biochemical adaptations and possible hormonal control of development under dry conditions are addressed by Stewart and Hanson, and Aspinall in Chapters 11 and 12, which demonstrate some of the large biochemical changes that take place as plants dehydrate, particularly in the proline content of the tissue. As each chapter points out, however, we need to know whether these changes contribute to the ability of plants to grow in dry areas.

Chapter 8 deals with the diverse behavior of stomata in different species and the structural modifications that mediate some of these responses. This work is an example of the large amount of information that can be obtained from plants adapted to widely different habitats. In a related chapter, Osmond (Chapter 10) speculates that stomatal closure may decrease the recycling of carbon dioxide at low water potentials, resulting in damage to the photosynthetic apparatus from reductant that is generated in the light under dry conditions. I think that this is a real possibility, but I doubt that excess reductant is the primary cause of losses in chloroplast activity: the losses are too rapid, they can occur in the dark, and they frequently do not involve any changes in chlorophyll content of the chloroplasts. However the effects of excess reductant could be tested easily by varying the oxygen concentration around leaves as they dehydrate, and I hope that these experiments will soon be done.

Where then should we direct our research in the future? It is clear from this book that the field of plant water relations has made a great deal of progress in the 15 years since reliable methods of measuring plant water status became available. We can now identify and begin to understand the diverse ways in which plants adapt to dry conditions. However we need to determine which adaptations are advantageous to the plant, a point made by several contributors. In my opinion the *only* way to accomplish this goal is to evaluate the impact of adaptations on the reproductive success of the plant, whether individually or in plant communities. Reproductive success is the sole criterion for evolutionary success in natural communities and is the primary concern of agriculture. I therefore urge the use of experiments that evaluate reproductive success of adaptations to dry conditions in research programs in the future.

Another important type of research is the identification of regulatory mechanisms controlling adaptation. Once adaptations having value to the plant have been identified, it is essential to understand the mechanisms by which the degree of adaptation is controlled. Given an understanding of the regulatory mechanisms, we have a vastly improved chance of being able to manipulate plants to our own ends, either genetically or through environmental means.

4. ADAPTATION TO HIGH TEMPERATURE STRESS

J. R. McWilliam

Department of Agronomy & Soil Science, University of New England, Armidale, New South Wales, Australia

It is difficult to provide an adequate quantitative definition of heat stress in plants, since this response depends on a number of factors including the thermal adaptation of the particular species, the duration of the exposure to high temperature, and the activity or stage of growth of the exposed tissue. For psychrophiles, plants adapted to cold environments, temperatures above

15°C may represent a heat stress, whereas at the other extreme, the thermophiles, which grow at temperatures above 30°C, the lower limit for heat stress could be as high as 45°C. Furthermore, the upper thermal limit for plant survival is usually lower for growing than for resting organisms, and it is influenced by many factors, including the level of irradiance, the degree of tissue hydration, which is critical in seeds, the extent of prior hardening, and the duration of exposure (since heat damage varies exponentially with time).

Heat stress is a major factor influencing the productivity and adaptation of wild and cultivated plants, especially when temperature extremes coincide with critical stages of plant development. The temperature of the aerial portions of the plant under these conditions may rise above or remain below the environmental temperature depending on the plant's ability to transfer excess heat to the environment. This relationship between plant temperature and ambient temperature can be very important for plant adaptation in hot semiarid environments, where leaf temperatures may exceed the 45–55°C range usually accepted as the upper temperature limit for most plants. Even higher temperatures can be experienced at the soil surface, which can represent a major constraint for the establishment of young seedlings. As Chapter 21 shows, it is often difficult to separate the discrete effects of heat stress from those imposed by water stress, since under conditions of elevated temperature these two components are inevitably confounded.

Plants adapt to high temperature by means of heat avoidance or heat tolerance, and good examples of both kinds of adaptation are given in this book. Ehleringer (Chapter 19) describes an elegant example of adaptation for heat avoidance in *Encelia* involving leaf pubescence, which by conferring greater reflectance, reduces leaf temperature and permits this species to survive the high temperatures in its native habitat. An interesting feature of this adaptation is that the ability to develop pubescence is triggered by exposure to severe water stress (−40 bars) and the actual elongation of the leaf hairs takes place at quite low leaf water potentials.

Work with the winter dominant species *Tidestromia oblongifolia* reported by Björkman et al. (Chapter 15) provides an equally good example of a heat tolerant desert species that is well adapted to temperatures in excess of 50°C during the growing season. Such plants are often poorly adapted to cool environments by contrast to other evergreen desert perennials such as *Larrea divaricata*, which grow over a much wider range of thermal environments and show good evidence of temperature acclimation in such processes as photosynthesis.

The underlying mechanisms that are responsible for the ability of plants to adapt to high temperatures can be expressed at the morphological level as described by Ehleringer in *Encelia* (Chapter 19), at the process level as with the irreversible inhibition of photosynthesis and loss of photophosphorylation discussed by Björkman et al. (Chapter 15), or at the biochemical and molecular level as reviewed by Teeri (Chapter 16) and Raison et al. (Chapter 17). In most cases the actual mechanisms are quite complex and probably operate at all levels, such that the combined effect is responsible for the general adaptive response to heat stress.

For thermotolerant species that survive by enduring the temperature stress, adaptation appears to be bound up with the thermostability and continued activity of key molecules such as proteins and nucleic acids and the associated membranes involved in metabolism at elevated temperatures. Further evidence for this view is provided in the chapter by Björkman et al. (Chapter 15) with respect to some of the enzymes associated with photosystem 2 and noncyclic photophosphorylation in heat-adapted desert species such as *Tidestromia oblongifolia*. These authors also report differences in the *in vitro* heat stability of a number of stroma and extrachloroplastic photosynthetic enzymes from heat- and cold-adapted plants and suggest that this may contribute to the greater heat tolerance of the former. Thermostability of enzymes *in vitro* may not represent the true situation in the intact leaf, since there is evidence that in addition to greater heat stability, thermotolerant species may also possess more thermostable protein-synthesizing machinery, which can replace enzymes denatured at high temperature.

These differences suggest that adaptive changes may occur in the structure or function of these proteins. Some evidence to support this view is provided by Teeri (Chapter 16), who reports that for two enzymes (malate dehydrogenase and glucose-6-phosphate dehydrogenase) at least one kinetic parameter, the substrate-binding Michaelis constant K_m, has a temperature optimum that reflects the thermal environment of the plant habitat. Also the temperature range over which the K_m is close to maximum reflects the temperature range to which the species or varieties are adapted. He found no evidence of short-term acclimation to temperature in terms of K_m or any form of isozyme mechanism: however this does not rule out the possibility of acclimation in other kinetic properties or in other significant rate-limiting enzymes that are more relevant for plant adaptation.

Despite ample evidence for the denaturation of active proteins at high temperature, there is also growing evidence that high temperatures causes the disruption of membrane (lipid) structure. Raison et al. (Chapter 17) suggest that many of the high-temperature-induced effects that have been described in plants may be directly related to a perturbation of membrane structure and function by heat. They also present evidence to suggest that variation at this level could help explain acclimation and adaptation to extremes of temperature.

The concept of the fluidity of a membrane bilayer, with a change in state from largely gel to gel-liquid to liquid with progression up the temperature scale is now generally accepted. Raison et al. (Chapter 17) propose that the control of this condition resides in the balance between the hydrophilic and hydrophobic forces acting on the vertical movement of the protein in the plane of the bilayer. This suggests that at high temperatures, protein movement toward the interior of the bilayer due to an increase in hydrophobic interactions causes considerable alteration in the forces maintaining the tertiary structure, resulting in severe membrane instability and loss of integrity. Data from electron spin resonance studies and evidence from the sudden increase in the fluorescence of chloroplast membranes observed at high temperature, thought to be associated with the separation of the light-harvesting pigment-protein complex and the protein at the reaction center, support this hypothesis. This work may well provide a valuable tool to assess the inherent adaptability of plants to temperature extremes.

In many respects our understanding of how plants respond and adapt to high and low temperature extremes is more advanced than our current knowledge of the responses to water stress. The reason for this, apart from the greater precision and relative ease of application of temperature stress, is probably the greater emphasis given in recent years to the study of the basic molecular and biochemical processes involved in temperature stress.

The chapters dealing with aspects of adaptation to heat stress provide valuable background material and document further examples of specific high temperature adaptation. We are rapidly approaching the stage at which this information will permit us to develop more effective screening techniques, such as the fluorescence probe described earlier, that will facilitate selection for greater heat tolerance in susceptible crop plants.

5. INTERACTION AND INTEGRATION OF ADAPTATIONS TO STRESS

J. T. Ritchie

USDA–SEA–AR, Grassland, Soil, & Water Research Laboratory, Temple, Texas

The interaction and integration of the adaptive responses of plants to stress is complex and difficult. As is pointed out in several chapters including Chapter 21, it is often difficult to separate the effects of high temperature stress from those of water stress at high temperatures. Moreover, adaptations that reduce water stress and improve water use efficiency may also reduce high temperature stress, and vice versa. Ehleringer in Chapter 19 provides an example of how pubescence in *Encelia* reduces heat stress and also reduces water stress, and Mooney in Chapter 18 refers to *Atriplex hymenelytra*, in which highly angled leaves and salt excretion reduce heat stress and water stress. Mooney et al. (2) have calculated that the reduced absorptance of leaves resulting from salt crystallization on the leaf surface and dehydration reduces the leaf temperature by 4 to 5°C and reduces transpiration by about 20% under conditions similar to those encountered in Death Valley in the summer. Leaves angled at 70° to the horizontal reduce leaf temperatures by a further 2°C and also reduce transpiration by an additional 7% compared with horizontal leaves. Since photosynthesis is saturated at relatively low quantum flux densities, it is concluded that the reduced absorption of light and the angling of leaves will have little effect on the rate of photosynthesis and will improve the water use efficiency (2).

Hsiao et al. (3) give some idea of the complexity of responses and feedback of water stress alone on the processes leading to crop yield. It is therefore not surprising that no author attempted to integrate the interactions of water stress and high temperature stress and the influence of adaptations to these stresses on plant productivity and crop yield. Jones (Chapter 23) describes a simple model that can be used to determine the influence of various stomatal behavior patterns on yield in climatic situations characterized by unpredictable rainfall. This must be seen as a first attempt at a difficult subject, but one that will undoubtedly grow more useful and fruitful as more data become available on the regulatory mechanisms and physiological implications of adaptations to water and high temperature stress.

The many small effects of water stress and high temperature stress on the morphological and physiological processes and the adaptations to stress are ultimately evident as an effect on the productivity and yield of the plant or crop. Fischer and Turner (4) suggested that plant productivity in arid and semiarid regions can be considered to be a function of the total transpiration, the water use efficiency, and the partitioning of the assimilate. Using a similar approach, Fischer (Chapter 21) suggests that the factors limiting crop yield under water-limited situations can be divided into water supply and transpiration, transpiration efficiency, and assimilate allocation: this separation provides a specific mechanism for evaluating stress as related to production. Furthermore each factor can be evaluated in terms of genetic characterization of plants and in management of plants. Genetics and management can be combined to provide a production system that best fits the environmental context of the climate as pointed out by Fischer, namely, early, middle, late, or irregular stresses. The supply of water to a crop, through the stored soil water component or through the transpiration rate, can modify the level of stress that plants encounter within any of the environments influenced by various climate cycles in the world.

In attempting to maximize transpiration relative to soil evaporation by the various strategies suggested in Chapter 21, such as manipulation of the ground cover and more root extraction near the surface when the soil is wet, it is important to recognize the tradeoff between soil evaporation and transpiration that exists when the vegetative cover of the soil is only partial. When the soil surface is wet following rains or irrigation, most of the radiation reaching the soil is repartitioned into latent heat. As such the ground surface temperature stays relatively cool, and transpiration is approximately proportional to the radiation intercepted by plants. When the soil surface is dry, however, much of the radiation reaching the soil is repartitioned into sensible heat, causing elevated temperatures in the lower parts of the plant canopy and a rather strong increase in transpiration over that expected from radiation interception. This is sometimes called the "clothesline effect." Because rainfall frequency differs greatly among regions and at different times of year, and because soil hydraulic properties vary considerably in ability to transport water to the surface in response to drying by soil evaporation, caution should be used in attempting to couple transpiration with production as implied in the calculation of yield or dry matter as a function of water use. This sort of analysis is qualitatively helpful in the evaluation of marginal production systems, but water use efficiency does not fill bellies.

The generality of the transpiration efficiency concept, used effectively by Fischer (Chapter 21) and others (5) in certain situations, can be questioned. The use of a variable such as pan evaporation to normalize transpiration efficiency obtained in several environments differing in saturation deficit, as suggested in Chapter 21, certainly aids in making transpiration efficiency more widely applicable. Nevertheless several criticisms can be leveled against the widespread adoption of this variable. One major disadvantage is that it does not account for the known differences in sensitivity of plant processes to water stress, namely, that cell expansion is more sensitive to stress than are processes controlled by stomata, such as photosynthesis and transpiration. This difference in sensitivity should decrease transpiration efficiency; instead there is the slight increase reported in Chapter 21. Probably the duration of the differential effect of stress on plant processes will determine its impact on the transpiration efficiency. Since low levels of nutrients adversely influence biomass accumulation more than transpiration, questions can also be raised concerning the influence of available nutrients on the transpiration efficiency. Water stress also causes a greater proportion of the carbon to be allocated to roots; if the aboveground biomass only is considered, this will decrease the transpiration efficiency, but if the total biomass is considered it might have no effect.

The emphasis by Fischer (Chapter 21) on the effects of the allocation of assimilates on seed number and size is commendable. A majority of agronomic experiments have overlooked this point by reporting only yield, giving little insight into the major factors affecting yield, especially with regard to water deficits.

The chapter by Tazaki et al. (Chapter 20) is timely in that it points out the importance of short and irregular water shortages on production in humid regions such as Japan, where soil water storage is often low. Japan is only one example of an area in which research on plant water stress has been neglected because the traditional climatic evaluation of monthly precipitation and evaporation indicates that there should be no water deficits. There are, however, large untapped areas of temperate and tropical regions of the world in which these short droughts, not evident from traditional climatic indices, drastically reduce production. The southeastern United States and the Cerrados in Central Brazil are examples. Natural soils in these areas are low in some essential plant nutrients and have either shallow topsoil or toxicity to root growth at depths below the plough layer. It is common to find plants able to extract no more than 20 to 80 mm of soil water before being severely affected by water stress, as in the case of the pine seedling study reported by Tazaki et al. (Chapter 20). With a decreasing water supply available for irrigation in the arid and semiarid regions of the world, production in the humid and subhumid regions will become more important. Understanding how to cope with water deficits of short duration will be a major challenge.

6. BREEDING AND SELECTION FOR ADAPTATION TO STRESS

6.1. GENETIC IMPROVEMENT OF DROUGHT ADAPTATION

A. Blum

Agricultural Research Organization, The Volcani Center, Bet Dagan, Israel

As mentioned in Section 1 of this chapter, authors have differed in their attempt to define stress and resistance to stress in physiological and ecological terms. I suggest that an agronomic definition will be useful to bring together the individual physiological processes and the integrated plant responses leading to economic yield. The genetic improvement of crops involves the engineering of new cultivars tailored to meet given specifications. The highest possible yield for a given set of conditions (i.e., the potential yield) is a major specification. An agronomic definition of drought stress must consider yield. Thus drought stress is in evidence if yield is reduced below the potential level as a result of a water deficit at any period of growth. The magnitude of drought injury can be measured by the ratio of actual to potential yield. Drought stress or injury therefore need not involve complete or massive yield loss. Thus at a given level of water deficit, drought resistance is manifested and measured as a relatively small reduction in actual yield below the potential yield.

Since, however, potential yield changes with genotype, the use of the actual ratio of yield to potential yield is not helpful in the genetic evaluation of a range of genotypes to stress. Thus the plant breeder recognizes "stability" of yield performance as a measure of the amplitude of variation between potential and actual yield of a given genotype across changing environments. Although stability is a crude statistic that can be affected by the potential yield level, it is a valuable yardstick for comparison among cultivars of common genetic germ plasm. Stability is calculated from yield data obtained from repeated testing over many locations and years. If the water regime is an important variable among the testing sites, a stable cultivar will possess some unidentified drought resistance mechanism that will buffer it against an excessive yield reduction at the drier end of the testing spectrum.

The identification of these unidentified drought resistance mechanisms, as well as any other attributes of drought resistance induced by selection pressure, is invaluable for the development of nonempirical breeding methods. A simple example for sorghum follows. An open-pollinating population of adapted grain sorghum (cross-pollinating by means of a male sterility gene) was submitted to selection under both severe dryland conditions and irrigated conditions in Israel. Selection was performed repeatedly for a larger weight of grain per panicle. After five selection cycles it was shown that under conditions of drought stress, selection for larger panicle weight caused a correlated genetic shift toward earliness in flowering (and maturity) by about 10 days. This range of earliness was thereafter used successfully as an additional selection criterion in developing new sorghum cultivars for dryland conditions.

Additional studies have shown that earliness in sorghum was associated with reduced water use during most stages of growth as a result of a smaller leaf area and root density per plant in early genotypes. Earliness is also negatively correlated with potential yield. Therefore there is a tradeoff between potential yield and drought adaptation that is probably very delicately balanced for any given stress environment.

The relationships between potential yield and drought resistance must stem from some basic physiological interactions. For example, the smaller size of various plant organs, tissues, and cells appear to constitute a drought adaptive attribute. Zimmermann (6) points out that smaller cells are better adjusted osmotically. However yield is in itself a manifestation of a larger organ size. Thus potential yield may be undermined by a genetic shift toward smaller cells and meristem size to improve drought adaptation. The complex of size, potential yield, and drought adaptation has many interesting facets that are beyond the scope of this summary. It points, however, to a probable balance in the integrated plant system between the morphological buildup of potential yield and its impact on drought adaptation. Thus it is possible that a further improvement of potential yield will involve a corresponding decrease in drought resistance.

From the pioneering studies by Briggs and Shantz (7) to the recently developed plant production models, total production of plant biomass has been shown to be correlated linearly with water use. For a given crop plant and a normalized environment (including crop management), this relationship appears to hold little variation (see Chapter 21). The question this raises is whether the biomass/water use ratio (i.e., the water use efficiency) can be genetically increased, especially at the lower range of water use. In cereals, improved grain yield has been genetically obtained over a range of regimes of water use by partitioning more of the total biomass into the grain: an increase in total biomass per unit of water use has not been recorded for these crops. However the total biomass/water use ratio was genetically improved in various forage crops. For example, in Israel total dry matter (aboveground) production per unit of water use was almost doubled by moving from open-pollinated varieties to F_1 hybrids in forage sorghum growing on 160 to 290 mm of water: the possibility of roots balancing this differential is very remote. With grain cereals, when selection pressure was applied to grain yield (not total biomass), the correlated physiological response led to increased grain production at the expense of straw (and root?) production. However when selection pressure was applied to total plant production, as in forage crops, total biomass (roots neglected) was genetically improved for varying levels of water use.

Improvement of potential yield appears to result in relatively similar improvements in yield under conditions of limited water supply; this applies to both grain and total plant production. As pointed out earlier, however, serious physiological deficiencies in drought adaptation may be expected as a result of further increases in the genetic yield potential. An alternative pathway for improving yield under conditions of drought stress is being tried in some breeding programs. This is the selection for maximum yield under conditions of stress, even though this is a subpotential yield. However selection for subpotential yield is inefficient. Since yield is a complex trait of low heritability, the greater the reduction in yield below the potential level, the less efficient is the selection for yield. Therefore the genetic improvement of yield under conditions of drought stress requires dissociation from yield as a major genetic selection criterion. Additional physiological selection criteria are required. These criteria, together with yield, may improve the probability of securing the most productive genetic recombinants for conditions of drought stress. On the physiological plane, these selection criteria constitute some very relevant and definite drought resistant mechanisms commonly grouped into "drought escape" and "drought tolerance" categories. Together with various collaborators we have explored the question of what might constitute physiological selection criteria for drought resistance in sorghum breeding. These are reported in detail elsewhere (8).

6.2. BREEDING FOR DROUGHT RESISTANCE

F. Bidinger

International Crops Research Institute for the Semi-Arid Tropics, Hyderabad, Andhra Pradesh, India

Drought resistance has been and remains a problem area in crop breeding. Conventional "resistance breeding" strategy is not applicable in most environments because it is very difficult to establish uniform and repeatable drought resistance screening nurseries for field selection. Biochemical or physiological research on mechanisms of adaptation to stress, as reported in Part III and IV, has demonstrated the value of certain of these mechanisms, but in most cases we lack proof that breeding directly for these mechanisms will result in improved crop yield or yield stability under stress. Many plant breeders are unwilling to undertake breeding for physiological mechanisms without this type of evidence. Few physiologists have undertaken to provide such evidence, partly, at least, because the only conclusive way of so doing is to breed for the mechanisms in question and to test the results.

As pointed out in Section 6.1 of this chapter, the options remaining to plant breeders are either to concentrate on high overall yield potential, on the assumption that this will provide some benefit under stress as well, or to rely on multilocation testing of advanced materials to provide some exposure to stress. Neither option, however, gives resistance to drought stress any significant priority in the breeding program, and neither is likely to provide solid evidence that it is, in fact, possible to make a significant improvement in drought resistance through breeding.

Despite individual successes (9), the possibility of breeding for drought resistance has not been conclusively demonstrated. The question of whether the common correlation between dry matter production and (evapo)transpiration represents a fixed relationship, or whether greater productivity per unit of available water can be achieved in similar plant materials through selection as suggested in Section 6.1 of this chapter, remains unanswered. There are theoretical arguments that an increase in dry matter per unit transpiration may be possible under stress, based on the relative effects of stress on resistance to carbon dioxide and water vapor exchange in crops (4), but the field data to critically test this hypothesis appear not to exist.

We can also ask whether breeding for high yield potential and breeding for drought resistance are compatible objectives. A tradeoff between yield potential and resistance could occur if resistance involved adaptations such as greater dry matter distribution to roots compared with aerial parts, or maintenance of greater stem carbohydrate reserves, which could limit grain yields in nonstress conditions (see Section 6.1 of this chapter). Recent work suggests that such differences may occur between dwarf bread wheat cultivars bred under maximum yield conditions, and taller "drought resistant" bread wheats selected under dryland conditions (10). These conclusions were based on the mean performance of the two classes. However within-group variations in the stress resistance index were nearly of the same magnitude as between-group variations, which suggests that the general pattern is not inevitable. It may be that the particular strategy of a breeding program, especially the selection emphasis on drought resistance in comparison to that on yield potential, will determine whether such tradeoffs will occur.

Setting aside the possibility of theoretical limitations to breeding for improved drought resistance, what are the realistic alternatives to approach the problem? Although they can be formulated in a variety of ways, in general, they range from attempting to build the case for breeding for physiological adaptations (i.e., a physiological rather than a breeding option) to developing ways of controlling stress environments to make direct selection for genotype performance in stress a more effective approach. The four chapters in Part VI illustrate these options.

The chapters by Morgan (Chapter 24), Jordan and Miller (Chapter 25), and Steponkus et al. (Chapter 26) represent the physiological approach, that is, looking within the target crop for differences among individuals that may be important in conditioning productivity under stress. It is a long-term approach, involving the determination of whether significant genetic variation exists for a particular character or response, the demonstration that differences in the character or response are related to differences in genotypes by stress interactions, and finally, the demonstration that breeding for a candidate response to stress is both possible and effective. It is not the type of approach with significant appeal to many plant breeders and probably will have to be carried out largely by physiologists. Yet there may be no other way of determining whether physiological responses or adaptations to stress can be used as the basis of a plant breeding program.

The opposite approach is exemplified by Johnson (Chapter 27) in his discussion of the breeding of range grasses. This is the traditional approach to plant breeding—direct selection for performance under the problem conditions—but here it is extended by the use of controlled environments and special field techniques to expose breeding materials to a number of the important aspects of drought stress. This is an important contribution: if it is difficult to identify resistance to stress as a general phenomenon in field testing, it may be possible to build up resistance by working on parts of the problems that can be dealt with more readily. The team approach utilized in this project is particularly well suited to this alternative.

A third approach is the use of physiological responses to stress as final selection criteria among otherwise acceptable lines (8). This approach, as noted earlier, does not place major emphasis on drought resistance, but it has two attractive features. First, it offers a way to use physiological knowledge of adaptations or responses to stress without the burden of proof required in direct breeding for resistance mechanisms. Second, it is compatible with the possibilities and limitations of most ongoing breeding programs; that is, it could be added to a breeding program with a minimum of additional resources and without necessitating a compromise with objectives such as disease resistance and yield. Selection/evaluation for potential drought resistance adaptations at the final stage of a breeding program almost certainly is not the most effective approach in the long term, since there may be little variability for resistance traits remaining by the final stage of testing. But this approach is a useful way to begin to utilize existing knowledge of plant adaptation to stress, while such mechanisms are too poorly understood to form the basis of a program breeding directly for stress resistance.

7. CONCLUDING REMARKS

The ability to adjust to the environment is one of the most important but frustrating attributes of plant behavior. It is important because this adjustment permits plants to colonize diverse environments, which has immense practical significance. It is frustrating because of the plasticity and variability of plant response, which makes it hard to identify responses having adaptive value and difficult to repeat experiments. To be a scientist studying adaptation is an exciting, but humbling, experience. This book has attempted to identify the various morphological and physiological mechanisms of adaptation to water and high temperature stress and to evaluate their importance in the light of survival and productivity. Additionally, it has assessed their importance in breeding and selection for stress resistance or tolerance. The amalgamation of the expertise of ecologists, agronomists, plant physiologists, and plant breeders has led to a view and synthesis that would not have been possible if only the individual disciplines had considered the adaptation of plants to water and high temperature stress.

It is clear from this book that our understanding of the physiological basis and physiological consequences of morphological and physiological adaptive mechanisms to water and high temperature stress has increased very significantly in the past two decades. Furthermore, our understanding of the physiology of high temperature stress is more advanced than our physiological understanding of water stress. Even so, our understanding of the mechanisms of adaptation is insufficient to be able to predict the interaction and integration of the various mechanisms that may be operating in a plant or that may be bred into a plant. Presumably this is because we still are not able to assess the importance of one particular physiological or morphological mechanism relative to another. The various morphological or physiological mechanisms have been identified and assessed in widely differing germ plasms. Until several mechanisms of adaptation can be assessed in a similar genetic background, the relative importance of the various mechanisms will be open to debate. Thus the suggestion in Chapter 3 that it may be more fruitful to search for morphological than for biochemical adaptation to water stress, and the conclusion in Chapter 5 that leaf adaptations rather than root adaptations appear to be chiefly responsible for the success of a species in a water-stressed environment, will remain as unsubstantiated hypotheses until the adaptive mechanisms can be tested in a similar genetic background.

The book has also highlighted the need to assess the various mechanisms of adaptation to stress through their impact on the reproductive success of the plant or crop. Plant breeders have long recognized this, and as pointed out in Section 6.2 of this chapter, the majority will likely remain unconvinced about the value of a particular mechanism until its value can be demonstrated in terms of improved yield under stress conditions. Survival of stress is a secondary assessment of the advantage of a mechanism of adaptation: survival must ultimately confer reproductive success on a plant or community if it is to be advantageous. It has been pointed out elsewhere that mechanisms of survival, although important in natural ecosystems, may be of lesser value in a managed ecosystem such as commercial crop production (4, 11).

In Chapter 2 Ritchie emphasized the importance of research on plant stress. We conclude this book on a similar note. Gauss wrote in 1910, "Let no one look with indifference upon the possibility . . . of acclimatizing valuable crop species to arid regions, or underestimate the magnitude of the achievement. So vast an achievement would rank with the discovery of a new continent in its enlargement of the sources of human subsistence'' (12). Although areas in which water and nutrition can be readily controlled have seen significant yield increases in recent years as a result of the Green Revolution, this "revolution" has had little impact in areas of dryland farming in which water is the major constraint to yield. Smaller yield increases must be expected in rainfed agricultural systems; nevertheless increases from improvement in yield as a result of breeding greater drought tolerance and adaptation to stress will be significant. Progress to date notwithstanding, there is a need for greater emphasis on the understanding of and breeding for adaptation to water and high temperature stress in the future.

REFERENCES

1. J. Levitt, *Responses of Plants to Environmental Stresses*, Academic Press, New York–London, 1972.
2. H. A. Mooney, J. Ehleringer, and O. Björkman, *Oecologia*, **29**, 301 (1977).
3. T. C. Hsiao, E. Fereres, E. Acevedo, and D. W. Henderson, in O. L. Lange, L. Kappen, and E.-D. Schulze, Eds., *Water and Plant Life: Problems and Modern Approaches*, Springer-Verlag, Berlin–Heidelberg–New York, 1976, p. 281.
4. R. A. Fischer and N. C. Turner, *Annu. Rev. Plant Physiol.*, **29**, 277 (1978).
5. H. van Keulen, C. T. de Wit, and H. Lof, in O. L. Lange, L. Kappen, and E.-D. Schulze, Eds., *Water and Plant Life: Problems and Modern Approaches*, Springer-Verlag, Berlin–Heidelberg–New York, 1976, p. 408.
6. U. Zimmermann, *Annu. Rev. Plant Physiol.*, **29**, 121 (1978).
7. L. J. Briggs and H. L. Shantz, *J. Agric. Res.*, **3**, 1 (1914).
8. A. Blum, in H. Mussell and R. C. Staples, Eds., *Stress Physiology in Crop Plants*, Interscience, New York–London, 1979, p. 429.
9. E. A. Hurd, *Agric. Meteorol.*, **14**, 39 (1974).
10. R. A. Fischer and R. Maurer, *Aust. J. Agric. Res.*, **29**, 897 (1978).
11. N. C. Turner, in H. Mussell and R. C. Staples, Eds., *Stress Physiology in Crop Plants*, Interscience, New York–London, 1979, p. 343.
12. R. Gauss, *Am. Breeders Mag.*, **1**, 209 (1910).

Author Index

Page numbers in **boldface** refer to literature cited in end-of-chapter lists.

Species Index

Subject Index

475

Disclosure

ALSO BY MICHAEL CRICHTON

Fiction

The Andromeda Strain
The Terminal Man
The Great Train Robbery
Eaters of the Dead
Congo
Sphere
Jurassic Park
Rising Sun

Non Fiction

Five Patients
Jasper Johns
Electronic Life
Travels

DISCLOSURE

A Novel by
Michael Crichton

LONDON NEW YORK SYDNEY TORONTO

This edition published in 1994 by BCA
by arrangement with Century, Random House UK Limited

Copyright © Michael Crichton 1993

CN 3866

Printed in England by Clays Ltd, St Ives plc

For Douglas Crichton

It shall be an unlawful employment practice for an employer: (1) to fail or refuse to hire or to discharge any individual, or otherwise to discriminate against any individual with respect to his compensation, terms, conditions or privileges of employment because of such individual's race, color, religion, sex, or national origin or (2) to limit, segregate, classify his employees or applicants for employment in any way which would deprive or tend to deprive any individual of employment opportunities or otherwise adversely affect his status as an employee, because of such individual's race, color, religion, sex, or national origin.

Title VII, Civil Rights Act of 1964

Power is neither male nor female.

Katharine Graham

DISCLOSURE

MONDAY

FROM: DC/M
ARTHUR KAHN
TWINKLE/KUALA LUMPUR/MALAYSIA

TO: DC/S
TOM SANDERS
SEATTLE (AT HOME)

TOM:

CONSIDERING THE MERGER, I THOUGHT YOU SHOULD GET THIS AT HOME AND NOT THE OFFICE:

TWINKLE PRODUCTION LINES RUNNING AT 29% CAPACITY DESPITE ALL EFFORTS TO INCREASE. SPOT CHECKS ON DRIVES SHOW AVG SEEK TIMES IN 120–140 MILLISECOND RANGE WITH NO CLEAR INDICATION WHY WE ARE NOT STABLE AT SPECS. ALSO, WE STILL HAVE POWER FLICKER IN SCREENS WHICH APPEARS TO COME FROM HINGE DESIGN DESPITE IMPLEMENTATION OF DC/S FIX LAST WEEK. I DON'T THINK IT'S SOLVED YET.

HOW'S THE MERGER COMING? ARE WE GOING TO BE RICH AND FAMOUS?

CONGRATULATIONS IN ADVANCE ON YOUR PROMOTION.

ARTHUR

Tom Sanders never intended to be late for work on Monday, June 15. At 7:30 in the morning, he stepped into the shower at his home on Bainbridge Island. He knew he had to shave, dress, and leave the house in ten minutes if he was to make the 7:50 ferry and arrive at work by 8:30, in time to go over the remaining points with Stephanie Kaplan before they went into the meeting with the lawyers from Conley-White. He already had a full day at work, and the fax he had just received from Malaysia made it worse.

Sanders was a division manager at Digital Communications Technology in Seattle. Events at work had been hectic for a week, because DigiCom was being acquired by Conley-White, a publishing conglomerate in New York. The merger would allow Conley to acquire technology important to publishing in the next century.

But this latest news from Malaysia was not good, and Arthur had been right to send it to him at home. He was going to have a problem explaining it to the Conley-White people because they just didn't—

"Tom? Where are you? Tom?"

His wife, Susan, was calling from the bedroom. He ducked his head out of the spray.

"I'm in the shower!"

She said something in reply, but he didn't hear it. He stepped out, reaching for a towel. "What?"

"I said, Can you feed the kids?"

His wife was an attorney who worked four days a week at a downtown firm. She took Mondays off, to spend more time with the kids, but she was not good at managing the routine at home. As a result, there was often a crisis on Monday mornings.

"Tom? Can you feed them for me?"

"I can't, Sue," he called to her. The clock on the sink said 7:34. "I'm already late." He ran water in the basin to shave, and lathered his face. He was a handsome man, with the easy manner of an athlete. He touched the dark bruise on his side from the company touch football game on Saturday. Mark Lewyn had taken him down; Lewyn was fast but clumsy. And Sanders was getting too old for touch football. He was still in good shape—still within five pounds of his varsity weight—but as he ran his hand through his wet hair, he saw streaks of gray. It was time to admit his age, he thought, and switch to tennis.

Susan came into the room, still in her bathrobe. His wife always looked beautiful in the morning, right out of bed. She had the kind of fresh beauty that required no makeup. "Are you sure you can't feed them?" she said. "Oh, nice bruise. Very butch." She kissed him lightly, and pushed a fresh mug of coffee onto the counter for him. "I've got to get Matthew to the pediatrician by eight-fifteen, and neither one of them has eaten a thing, and I'm not dressed. Can't you please feed them? Pretty please?" Teasing, she ruffled his hair, and her bathrobe fell open. She left it open and smiled. "I'll owe you one . . ."

"Sue, I can't." He kissed her forehead distractedly. "I've got a meeting, I can't be late."

She sighed. "Oh, all right." Pouting, she left.

Sanders began shaving.

A moment later he heard his wife say, "Okay, kids, let's go! Eliza, put your shoes on." This was followed by whining from Eliza, who was four, and didn't like to wear shoes. Sanders had almost finished shaving when he heard, "Eliza, you put on those shoes and take your brother downstairs right now!" Eliza's reply was indistinct, and then Susan said, "Eliza Ann, I'm talking to you!" Then Susan began slamming drawers in the hall linen closet. Both kids started to cry.

Eliza, who was upset by any display of tension, came into the bathroom, her face scrunched up, tears in her eyes. "Daddy . . . ," she sobbed. He put his hand down to hug her, still shaving with his other hand.

"She's old enough to help out," Susan called, from the hallway.

"Mommy," she wailed, clutching Sanders's leg.

"Eliza, will you *cut it out.*"

At this, Eliza cried more loudly. Susan stamped her foot in the hallway. Sanders hated to see his daughter cry. "Okay, Sue, I'll feed them." He turned off the water in the sink and scooped up his daughter. "Come on, Lize," he said, wiping away her tears. "Let's get you some breakfast."

He went out into the hallway. Susan looked relieved. "I just need ten minutes, that's all," she said. "Consuela is late again. I don't know what's the matter with her."

Sanders didn't answer her. His son, Matt, who was nine months old, sat in the middle of the hallway banging his rattle and crying. Sanders scooped him up in his other arm.

"Come on, kids," he said. "Let's go eat."

When he picked up Matt, his towel slipped off, and he clutched at it. Eliza giggled. "I see your penis, Dad." She swung her foot, kicking it.

"We don't kick Daddy there," Sanders said. Awkwardly, he wrapped the towel around himself again, and headed downstairs.

Susan called after him: "Don't forget Matt needs vitamins in his cereal. One dropperful. And don't give him any more of the rice cereal, he spits it out. He likes wheat now." She went into the bathroom, slamming the door behind her.

His daughter looked at him with serious eyes. "Is this going to be one of those days, Daddy?"

"Yeah, it looks like it." He walked down the stairs, thinking he would miss the ferry and that he would be late for the first meeting of the day. Not very late, just a few minutes, but it meant he wouldn't be able to go over things with Stephanie before they started, but perhaps he could call her from the ferry, and then—

"Do I have a penis, Dad?"

"No, Lize."

"Why, Dad?"

"That's just the way it is, honey."

"Boys have penises, and girls have vaginas," she said solemnly.

"That's right."

"Why, Dad?"

"Because." He dropped his daughter on a chair at the kitchen table, dragged the high chair from the corner, and placed Matt in it. "What do you want for breakfast, Lize? Rice Krispies or Chex?"

"Chex."

Matt began to bang on his high chair with his spoon. Sanders got the Chex and a bowl out of the cupboard, then the box of wheat cereal and a smaller bowl for Matt. Eliza watched him as he opened the refrigerator to get the milk.

"Dad?"

"What."

"I want Mommy to be happy."

"Me too, honey."

He mixed the wheat cereal for Matt, and put it in front of his son. Then he set Eliza's bowl on the table, poured in the Chex, glanced at her. "Enough?"

"Yes."

He poured the milk for her.

"No, Dad!" his daughter howled, bursting into tears. "*I* wanted to pour the milk!"

"Sorry, Lize—"

"Take it out—take the milk out—" She was shrieking, completely hysterical.

"I'm sorry, Lize, but this is—"

"I wanted to pour the milk!" She slid off her seat to the ground, where she lay kicking her heels on the floor. "Take it out, take the milk out!"

His daughter did this kind of thing several times a day. It was, he was assured, just a phase. Parents were advised to treat it with firmness.

"I'm sorry," Sanders said. "You'll just have to eat it, Lize." He sat down at the table beside Matt to feed him. Matt stuck his hand in his cereal and smeared it across his eyes. He, too, began to cry.

Sanders got a dish towel to wipe Matt's face. He noticed that the kitchen clock now said five to eight. He thought that he'd better call the office, to warn them he would be late. But he'd have to quiet Eliza first: she was still on the floor, kicking and screaming about the milk. "All right, Eliza, take it easy. Take it easy." He got a fresh bowl, poured more cereal, and gave her the carton of milk to pour herself. "Here."

She crossed her arms and pouted. "I don't want it."

"Eliza, you pour that milk *this minute.*"

His daughter scrambled up to her chair. "Okay, Dad."

Sanders sat down, wiped Matt's face, and began to feed his son. The boy immediately stopped crying, and swallowed the cereal in big gulps. The poor kid was hungry. Eliza stood on her chair, lifted the milk carton, and splashed it all over the table. "Uh-oh."

"Never mind." With one hand, he wiped the table with the dish towel, while with the other he continued to feed Matt.

Eliza pulled the cereal box right up to her bowl, stared fixedly at the picture of Goofy on the back, and began to eat. Alongside her, Matt ate steadily. For a moment, it was calm in the kitchen.

Sanders glanced over his shoulder: almost eight o'clock. He should call the office.

Susan came in, wearing jeans and a beige sweater. Her face was relaxed. "I'm sorry I lost it," she said. "Thanks for taking over." She kissed him on the cheek.

"Are you happy, Mom?" Eliza said.

"Yes, sweetie." Susan smiled at her daughter, and turned back to Tom. "I'll take over now. You don't want to be late. Isn't today the big day? When they announce your promotion?"

"I hope so."

"Call me as soon as you hear."

"I will." Sanders got up, cinched the towel around his waist, and headed upstairs to get dressed. There was always traffic in town before the 8:20 ferry. He would have to hurry to make it.

He parked in his spot behind Ricky's Shell station, and strode quickly down the covered walkway to the ferry. He stepped aboard moments before they pulled up the ramp. Feeling the throb of the engines beneath his feet, he went through the doors onto the main deck.

"Hey, Tom."

He looked over his shoulder. Dave Benedict was coming up behind him. Benedict was a lawyer with a firm that handled a lot of high-tech companies. "Missed the seven-fifty, too, huh?" Benedict said.

"Yeah. Crazy morning."

"Tell me. I wanted to be in the office an hour ago. But now that school's out, Jenny doesn't know what to do with the kids until camp starts."

"Uh-huh."

"Madness at my house," Benedict said, shaking his head.

There was a pause. Sanders sensed that he and Benedict had had a similar morning. But the two men did not discuss it further. Sanders often wondered why it was that women discussed the most intimate details of their marriages with their friends, while men maintained a discreet silence with one another.

"Anyway," Benedict said. "How's Susan?"

"She's fine. She's great."

Benedict grinned. "So why are you limping?"

"Company touch football game on Saturday. Got a little out of hand."

"That's what you get for playing with children," Benedict said. DigiCom was famous for its young employees.

"Hey," Sanders said. "I scored."

"Is that right?"

"Damn right. Winning touchdown. Crossed the end zone in glory. And then I got creamed."

At the main-deck cafeteria, they stood in line for coffee. "Actually, I would've thought you'd be in bright and early today," Benedict said. "Isn't this the big day at DigiCom?"

Sanders got his coffee, and stirred in sweetener. "How's that?"

"Isn't the merger being announced today?"

"What merger?" Sanders said blandly. The merger was secret; only a handful of DigiCom executives knew anything about it. He gave Benedict a blank stare.

"Come on," Benedict said. "I heard it was pretty much wrapped up. And that Bob Garvin was announcing the restructuring today, including a bunch of new promotions." Benedict sipped his coffee. "Garvin is stepping down, isn't he?"

Sanders shrugged. "We'll see." Of course Benedict was imposing on him, but Susan did a lot of work with attorneys in Benedict's firm; Sanders couldn't afford to be rude. It was one of the new complexities of business relations at a time when everybody had a working spouse.

The two men went out on the deck and stood by the port rail, watching the houses of Bainbridge Island slip away. Sanders nodded toward the house on Wing Point, which for years had been Warren Magnuson's summer house when he was senator.

"I hear it just sold again," Sanders said.

"Oh yes? Who bought it?"

"Some California asshole."

Bainbridge slid to the stern. They looked out at the gray water of the Sound. The coffee steamed in the morning sunlight. "So," Benedict said. "You think maybe Garvin won't step down?"

"Nobody knows," Sanders said. "Bob built the company from nothing, fifteen years ago. When he started, he was selling knockoff modems from Korea. Back when nobody knew what a modem was. Now the company's got three buildings downtown, and big facilities in California, Texas, Ireland, and Malaysia. He builds fax modems the size of a dime, he markets fax and e-mail software, he's gone into CD-ROMs, and he's developed proprietary algorithms that should make him a leading provider in education markets for the next century. Bob's come a long way from some guy hustling three hundred baud modems. I don't know if he can give it up."

"Don't the terms of the merger require it?"

Sanders smiled. "If you know about a merger, Dave, you should tell me," he said. "Because I haven't heard anything." The truth was that Sanders didn't really know the terms of the impending merger. His work involved the development of CD-ROMs and electronic databases. Although these were areas vital to the future of the company—they were the main reason Conley-White was acquiring DigiCom—they were essentially technical areas. And Sanders was essentially a technical manager. He was not informed about decisions at the highest levels.

For Sanders, there was some irony in this. In earlier years, when he was based in California, he had been closely involved in management decisions. But since coming to Seattle eight years ago, he had been more removed from the centers of power.

Benedict sipped his coffee. "Well, I hear Bob's definitely stepping down, and he's going to promote a woman as chairman."

Sanders said, "Who told you that?"

"He's already got a woman as CFO, doesn't he?"

"Yes, sure. For a long time, now." Stephanie Kaplan was DigiCom's chief financial officer. But it seemed unlikely she would ever run the company. Silent and intense, Kaplan was competent, but disliked by many in the company. Garvin wasn't especially fond of her.

"Well," Benedict said, "the rumor I've heard is he's going to name a woman to take over within five years."

"Does the rumor mention a name?"

Benedict shook his head. "I thought you'd know. I mean, it's your company."

On the deck in the sunshine, he took out his cellular phone and
called in. His assistant, Cindy Wolfe, answered. "Mr. Sanders's
office."

"Hi. It's me."

"Hi, Tom. You on the ferry?"

"Yes. I'll be in a little before nine."

"Okay, I'll tell them." She paused, and he had the sense that she was
choosing her words carefully. "It's pretty busy this morning. Mr. Garvin
was just here, looking for you."

Sanders frowned. "Looking for me?"

"Yes." Another pause. "Uh, he seemed kind of surprised that you
weren't in."

"Did he say what he wanted?"

"No, but he's going into a lot of offices on the floor, one after another,
talking to people. Something's up, Tom."

"What?"

"Nobody's telling me anything," she said.

"What about Stephanie?"

"Stephanie called, and I told her you weren't in yet."

"Anything else?"

"Arthur Kahn called from KL to ask if you got his fax."

"I did. I'll call him. Anything else?"

"No, that's about it, Tom."

"Thanks, Cindy." He pushed the END button to terminate the call.

Standing beside him, Benedict pointed to Sanders's phone. "Those
things are amazing. They just get smaller and smaller, don't they? You
guys make that one?"

Sanders nodded. "I'd be lost without it. Especially these days. Who
can remember all the numbers? This is more than a telephone: it's my
telephone book. See, look." He began to demonstrate the features for
Benedict. "It's got a memory for two hundred numbers. You store them

by the first three letters of the name." Sanders punched in K-A-H to bring up the international number for Arthur Kahn in Malaysia. He pushed SEND, and heard a long string of electronic beeps. With the country code and area code, it was thirteen beeps.

"Jesus," Benedict said. "Where are you calling, Mars?"

"Just about. Malaysia. We've got a factory there."

DigiCom's Malaysia operation was only a year old, and it was manufacturing the company's new CD-ROM players—units rather like an audio CD player, but intended for computers. It was widely agreed in the business that all information was soon going to be digital, and much of it was going to be stored on these compact disks. Computer programs, databases, even books and magazines—everything was going to be on disk.

The reason it hadn't already happened was that CD-ROMs were notoriously slow. Users were obliged to wait in front of blank screens while the drives whirred and clicked—and computer users didn't like waiting. In an industry where speeds reliably doubled every eighteen months, CD-ROMs had improved much less in the last five years. DigiCom's SpeedStar technology addressed that problem, with a new generation of drives code-named Twinkle (for "Twinkle, twinkle, little SpeedStar"). Twinkle drives were twice as fast as any in the world. Twinkle was packaged as a small, stand-alone multimedia player with its own screen. You could carry it in your hand, and use it on a bus or a train. It was going to be revolutionary. But now the Malaysia plant was having trouble manufacturing the new fast drives.

Benedict sipped his coffee. "Is it true you're the only division manager who isn't an engineer?"

Sanders smiled. "That's right. I'm originally from marketing."

"Isn't that pretty unusual?" Benedict said.

"Not really. In marketing, we used to spend a lot of time figuring out what the features of the new products were, and most of us couldn't talk to the engineers. I could. I don't know why. I don't have a technical background, but I could talk to the guys. I knew just enough so they couldn't bullshit me. So pretty soon, I was the one who talked to the engineers. Then eight years ago, Garvin asked me if I'd run a division for him. And here I am."

The call rang through. Sanders glanced at his watch. It was almost midnight in Kuala Lumpur. He hoped Arthur Kahn would still be awake. A moment later there was a click, and a groggy voice said, "Uh. Hello."

"Arthur, it's Tom."

Arthur Kahn gave a gravelly cough. "Oh, Tom. Good." Another cough. "You got my fax?"

"Yes, I got it."

"Then you know. I don't understand what's going on," Kahn said. "And I spent all day on the line. I had to, with Jafar gone."

Mohammed Jafar was the line foreman of the Malaysia plant, a very capable young man. "Jafar is gone? Why?"

There was a crackle of static. "He was cursed."

"I didn't get that."

"Jafar was cursed by his cousin, so he left."

"What?"

"Yeah, if you can believe that. He says his cousin's sister in Johore hired a sorcerer to cast a spell on him, and he ran off to the Orang Asli witch doctors for a counter-spell. The aborigines run a hospital at Kuala Tingit, in the jungle about three hours outside of KL. It's very famous. A lot of politicians go out there when they get sick. Jafar went out there for a cure."

"How long will that take?"

"Beats me. The other workers tell me it'll probably be a week."

"And what's wrong with the line, Arthur?"

"I don't know," Kahn said. "I'm not sure anything's wrong with the line. But the units coming off are very slow. When we pull units for IP checks, we consistently get seek times above the hundred-millisecond specs. We don't know why they're slow, and we don't know why there's a variation. But the engineers here are guessing that there's a compatibility problem with the controller chip that positions the split optics, and the CD-driver software."

"You think the controller chips are bad?" The controller chips were made in Singapore and trucked across the border to the factory in Malaysia.

"Don't know. Either they're bad, or there's a bug in the driver code."

"What about the screen flicker?"

Kahn coughed. "I think it's a design problem, Tom. We just can't build it. The hinge connectors that carry current to the screen are mounted inside the plastic housing. They're supposed to maintain electrical contact no matter how you move the screen. But the current cuts in and out. You move the hinge, and the screen flashes on and off."

Sanders frowned as he listened. "This is a pretty standard design, Arthur. Every damn laptop in the world has the same hinge design. It's been that way for the last ten years."

"I know it," Kahn said. "But ours isn't working. It's making me crazy."

"You better send me some units."

"I already have, DHL. You'll get them late today, tomorrow at the latest."

"Okay," Sanders said. He paused. "What's your best guess, Arthur?"

"About the run? Well, at the moment we can't make our production quotas, and we're turning out a product thirty to fifty percent slower than specs. Not good news. This isn't a hot CD player, Tom. It's only incrementally better than what Toshiba and Sony already have on the market. They're making theirs a lot cheaper. So we have major problems."

"We talking a week, a month, what?"

"A month, if it's not a redesign. If it's a redesign, say four months. If it's a chip, it could be a year."

Sanders sighed. "Great."

"That's the situation. It isn't working, and we don't know why."

Sanders said, "Who else have you told?"

"Nobody. This one's all yours, my friend."

"Thanks a lot."

Kahn coughed. "You going to bury this until after the merger, or what?"

"I don't know. I'm not sure I can."

"Well, I'll be quiet at this end. I can tell you that. Anybody asks me, I don't have a clue. Because I don't."

"Okay. Thanks, Arthur. I'll talk to you later."

Sanders hung up. Twinkle definitely presented a political problem for the impending merger with Conley-White. Sanders wasn't sure how to handle it. But he would have to deal with it soon enough; the ferry whistle blew, and up ahead, he saw the black pilings of Colman Dock and the skyscrapers of downtown Seattle.

DigiCom was located in three different buildings around historic Pioneer Square, in downtown Seattle. Pioneer Square was actually shaped like a triangle, and had at its center a small park, dominated by a wrought-iron pergola, with antique clocks mounted above. Around Pioneer Square were low-rise red-brick buildings built in the early years of the century, with sculpted façades and chiseled dates; these buildings now housed trendy architects, graphic design firms, and a cluster of high-tech companies that included Aldus, Advance Holo-Graphics, and DigiCom. Originally, DigiCom had occupied the Hazzard Building, on the south side of the square. As the company grew, it expanded into three floors of the adjacent Western Building, and later, to the Gorham Tower on James Street. But the executive offices were still on the top three floors of the Hazzard Building, overlooking the square. Sanders's office was on the fourth floor, though he expected later in the week to move up to the fifth.

He got to the fourth floor at nine in the morning, and immediately sensed that something was wrong. There was a buzz in the hallways, an electric tension in the air. Staff people clustered at the laser printers and whispered at the coffee machines; they turned away or stopped talking when he walked by.

He thought, Uh-oh.

But as a division head, he could hardly stop to ask an assistant what was happening. Sanders walked on, swearing under his breath, angry with himself that he had arrived late on this important day.

Through the glass walls of the fourth-floor conference room, he saw Mark Lewyn, the thirty-three-year-old head of Product Design, briefing some of the Conley-White people. It made a striking scene: Lewyn, young, handsome, and imperious, wearing black jeans and a black Armani T-shirt, pacing back and forth and talking animatedly to the blue-suited Conley-White staffers, who sat rigidly before the product mock-ups on the table, and took notes.

When Lewyn saw Sanders he waved, and came over to the door of the conference room and stuck his head out.

"Hey, guy," Lewyn said.

"Hi, Mark. Listen—"

"I have just one thing to say to you," Lewyn said, interrupting. "Fuck 'em. Fuck Garvin. Fuck Phil. Fuck the merger. Fuck 'em all. This reorg sucks. I'm with you on this one, guy."

"Listen, Mark, can you—"

"I'm in the middle of something here." Lewyn jerked his head toward the Conley people in the room. "But I wanted you to know how I feel. It's not right, what they're doing. We'll talk later, okay? Chin up, guy," Lewyn said. "Keep your powder dry." And he went back into the conference room.

The Conley-White people were all staring at Sanders through the glass. He turned away and walked quickly toward his office, with a sense of deepening unease. Lewyn was notorious for his tendency to exaggerate, but even so, the—

It's not right, what they're doing.

There didn't seem to be much doubt what that meant. Sanders wasn't going to get a promotion. He broke into a light sweat and felt suddenly dizzy as he walked along the corridor. He leaned against the wall for a moment. He wiped his forehead with his hand and blinked his eyes rapidly. He took a deep breath and shook his head to clear it.

No promotion. Christ. He took another deep breath, and walked on.

Instead of the promotion he expected, there was apparently going to be some kind of reorganization. And apparently it was related to the merger.

The technical divisions had just gone through a major reorganization nine months earlier, which had revised all the lines of authority, upsetting the hell out of everybody in Seattle. Staff people didn't know who to requisition for laser-printer paper, or to degauss a monitor. There had been months of uproar; only in the last few weeks had the tech groups settled down into some semblance of good working routines. Now . . . to reorganize again? It didn't make any sense at all.

Yet it was last year's reorganization that placed Sanders in line to assume leadership of the tech divisions now. That reorganization had structured the Advanced Products Group into four subdivisions— Product Design, Programming, Data Telecommunications, and Manufacturing—all under the direction of a division general manager, not yet appointed. In recent months, Tom Sanders had informally taken over as DGM, largely because as head of manufacturing, he was the person most concerned with coordinating the work of all the other divisions.

But now, with still another reorganization . . . who knew what might happen? Sanders might be broken back to simply managing DigiCom's production lines around the world. Or worse—for weeks, there had been persistent rumors that company headquarters in Cupertino was going to take back all control of manufacturing from Seattle, turning it over to the individual product managers in California. Sanders hadn't paid any attention to those rumors, because they didn't make a lot of sense; the product managers had enough to do just pushing the products, without also worrying about their manufacture.

But now he was obliged to consider the possibility that the rumors were true. Because if they were true, Sanders might be facing more than a demotion. He might be out of a job.

Christ: out of a job?

He found himself thinking of some of the things Dave Benedict had said to him on the ferry earlier that morning. Benedict chased rumors, and he had seemed to know a lot. Maybe even more than he had been saying.

Is it true you're the only division manager who isn't an engineer?

And then, pointedly:

Isn't that pretty unusual?

Christ, he thought. He began to sweat again. He forced himself to take another deep breath. He reached the end of the fourth-floor corridor and came to his office, expecting to find Stephanie Kaplan, the CFO, waiting there for him. Kaplan could tell him what was going on. But his office was empty. He turned to his assistant, Cindy Wolfe, who was busy at the filing cabinets. "Where's Stephanie?"

"She's not coming."

"Why not?"

"They canceled your nine-thirty meeting because of all the personnel changes," Cindy said.

"What changes?" Sanders said. "What's going on?"

"There's been some kind of reorganization," Cindy said. She avoided meeting his eyes, and looked down at the call book on her desk. "They just scheduled a private lunch with all the division heads in the main conference room for twelve-thirty today, and Phil Blackburn is on his way down to talk to you. He should be here any minute. Let's see, what else? DHL is delivering drives from Kuala Lumpur this afternoon. Gary Bosak wants to meet with you at ten-thirty." She ran her finger down the call book. "Don Cherry called twice about the Corridor, and you just got a rush call from Eddie in Austin."

"Call him back." Eddie Larson was the production supervisor in the Austin plant, which made cellular telephones. Cindy placed the call; a moment later he heard the familiar voice with the Texas twang.

"Hey there, Tommy boy."

"Hi, Eddie. What's up?"

"Little problem on the line. You got a minute?"

"Yes, sure."

"Are congratulations on a new job in order?"

"I haven't heard anything yet," Sanders said.

"Uh-huh. But it's going to happen?"

"I haven't heard anything, Eddie."

"Is it true they're going to shut down the Austin plant?"

Sanders was so startled, he burst out laughing. *"What?"*

"Hey, that's what they're saying down here, Tommy boy. Conley-White is going to buy the company and then shut us down."

"Hell," Sanders said. "Nobody's buying anything, and nobody's selling anything, Eddie. The Austin line is an industry standard. And it's very profitable."

He paused. "You'd tell me if you knew, wouldn't you, Tommy boy?"

"Yes, I would," Sanders said. "But it's just a rumor, Eddie. So forget it. Now, what's the line problem?"

"Diddly stuff. The women on the production line are demanding that we clean out the pinups in the men's locker room. They say it's offensive to them. You ask me, I think it's bull," Larson said. "Because women never go into the men's locker room."

"Then how do they know about the pinups?"

"The night cleanup crews have women on 'em. So now the women working the line want the pinups removed."

Sanders sighed. "We don't need any complaints about being unresponsive on sex issues. Get the pinups out."

"Even if the women have pinups in *their* locker room?"

"Just do it, Eddie."

"You ask me, it's caving in to a lot of feminist bullshit."

There was a knock on the door. Sanders looked up and saw Phil Blackburn, the company lawyer, standing there.

"Eddie, I have to go."

"Okay," Eddie said, "but I'm telling you—"

"Eddie, I'm sorry. I have to go. Call me if anything changes."

Sanders hung up the phone, and Blackburn came into the room. Sanders's first impression was that the lawyer was smiling too broadly, behaving too cheerfully.

It was a bad sign.

P hilip Blackburn, the chief legal counsel for DigiCom, was a slender man of forty-six wearing a dark green Hugo Boss suit. Like Sanders, Blackburn had been with DigiCom for over a decade, which meant that he was one of the "old guys," one of those who had "gotten in at the beginning." When Sanders first met him, Blackburn was a brash, bearded young civil rights lawyer from Berkeley. But Blackburn had long since abandoned protest for profits, which he pursued with singleminded intensity—while carefully emphasizing the new corporate issues of diversity and equal opportunity. Blackburn's embrace of the latest fashions in clothing and correctness made "PC Phil" a figure of fun in some quarters of the company. As one executive put it, "Phil's finger is chapped from wetting it and holding it to the wind." He was the first with Birkenstocks, the first with bell-bottoms, the first with sideburns off, and the first with diversity.

Many of the jokes focused on his mannerisms. Fussy, preoccupied with appearances, Blackburn was always running his hands over himself, touching his hair, his face, his suit, seeming to caress himself, to smooth out the wrinkles in his suit. This, combined with his unfortunate tendency to rub, touch, and pick his nose, was the source of much humor. But it was humor with an edge: Blackburn was mistrusted as a moralistic hatchet man.

Blackburn could be charismatic in his speeches, and in private could convey a convincing impression of intellectual honesty for short periods. But within the company he was seen for what he was: a gun for hire, a man with no convictions of his own, and hence the perfect person to be Garvin's executioner.

In earlier years, Sanders and Blackburn had been close friends; not only had they grown up with the company, but their lives were intertwined personally as well: when Blackburn went through his bitter divorce in 1982, he lived for a while in Sanders's bachelor apartment in

Sunnyvale. A few years later, Blackburn had been best man at Sanders's own wedding to a young Seattle attorney, Susan Handler.

But when Blackburn remarried in 1989, Sanders was not invited to the wedding, for by then, their relationship had become strained. Some in the company saw it as inevitable: Blackburn was a part of the inner power circle in Cupertino, to which Sanders, based in Seattle, no longer belonged. In addition, the two men had had sharp disputes about setting up the production lines in Ireland and Malaysia. Sanders felt that Blackburn ignored the inevitable realities of production in foreign countries.

Typical was Blackburn's demand that half the workers on the new line in Kuala Lumpur should be women, and that they should be intermingled with the men; the Malay managers wanted the women segregated, allowed to work only on certain parts of the line, away from the men. Phil strenuously objected. Sanders kept telling him, "It's a Muslim country, Phil."

"I don't give a damn," Phil said. "DigiCom stands for equality."

"Phil, it's their country. They're Muslim."

"So what? It's our factory."

Their disagreements went on and on. The Malaysian government didn't want local Chinese hired as supervisors, although they were the best-qualified; it was the policy of the Malaysian government to train Malays for supervisory jobs. Sanders disagreed with this blatantly discriminatory policy, because he wanted the best supervisors he could get for the plant. But Phil, an outspoken opponent of discrimination in America, immediately acquiesced to the Malay government's discriminatory policy, saying that DigiCom should embrace a true multicultural perspective. At the last minute, Sanders had had to fly to Kuala Lumpur and meet with the Sultans of Selangor and Pahang, to agree to their demands. Phil then announced that Sanders had "toadied up to the extremists."

It was just one of the many controversies that surrounded Sanders's handling of the new Malaysia factory.

Now, Sanders and Blackburn greeted each other with the wariness of former friends who had long since ceased to be anything but superficially cordial. Sanders shook Blackburn's hand as the company lawyer stepped into the office. "What's going on, Phil?"

"Big day," Blackburn said, slipping into the chair facing Sanders's desk. "Lot of surprises. I don't know what you've heard."

"I've heard Garvin has made a decision about the restructuring."

"Yes, he has. Several decisions."

There was a pause. Blackburn shifted in his chair and looked at his hands. "I know that Bob wanted to fill you in himself about all this. He came by earlier this morning to talk to everyone in the division."

"I wasn't here."

"Uh-huh. We were all kind of surprised that you were late today."

Sanders let that pass without comment. He stared at Blackburn, waiting.

"Anyway, Tom," Blackburn said, "the bottom line is this. As part of the overall merger, Bob has decided to go outside the Advanced Products Group for leadership of the division."

So there it was. Finally, out in the open. Sanders took a deep breath, felt the bands of tightness in his chest. His whole body was tense. But he tried not to show it.

"I know this is something of a shock," Blackburn said.

"Well," Sanders shrugged. "I've heard rumors." Even as he spoke, his mind was racing ahead. It was clear now that there would not be a promotion, there would not be a raise, he would not have a new opportunity to—

"Yes. Well," Blackburn said, clearing his throat. "Bob has decided that Meredith Johnson is going to head up the division."

Sanders frowned. "Meredith *Johnson?*"

"Right. She's in the Cupertino office. I think you know her."

"Yes, I do, but . . ." Sanders shook his head. It didn't make any sense. "Meredith's from sales. Her background is in sales."

"Originally, yes. But as you know, Meredith's been in Operations the last couple of years."

"Even so, Phil. The APG is a technical division."

"You're not technical. You've done just fine."

"But I've been involved in this for years, when I was in Marketing. Look, the APG is basically programming teams and hardware fabrication lines. How can she run it?"

"Bob doesn't expect her to run it directly. She'll oversee the APG division managers, who will report to her. Meredith's official title will be Vice President for Advanced Operations and Planning. Under the new structure, that will include the entire APG Division, the Marketing Division, and the TelCom Division."

"Jesus," Sanders said, sitting back in his chair. "That's pretty much everything."

Blackburn nodded slowly.

Sanders paused, thinking it over. "It sounds," he said finally, "like Meredith Johnson's going to be running this company."

"I wouldn't go *that* far," Blackburn said. "She won't have direct control over sales or finance or distribution in this new scheme. But I think there is no question Bob has placed her in direct line for succession, when he steps down as CEO sometime in the next two years." Blackburn shifted in his chair. "But that's the future. For the present—"

"Just a minute. She'll have four APG division managers reporting to her?" Sanders said.

"Yes."

"And who are those managers going to be? Has that been decided?"

"Well." Phil coughed. He ran his hands over his chest, and plucked at the handkerchief in his breast pocket. "Of course, the actual decision to name the division managers will be Meredith's."

"Meaning I might not have a job."

"Oh hell, Tom," Blackburn said. "Nothing of the sort. Bob wants everyone in the divisions to stay. Including you. He'd hate very much to lose you."

"But it's Meredith Johnson's decision whether I keep my job."

"Technically," Blackburn said, spreading his hands, "it has to be. But I think it's pretty much pro forma."

Sanders did not see it that way at all. Garvin could easily have named all the division managers at the same time he named Meredith Johnson to run the APG. If Garvin decided to turn the company over to some woman from Sales, that was certainly his choice. But Garvin could still make sure he kept his division heads in place—the heads who had served him and the company so well.

"Jesus," Sanders said. "I've been with this company twelve years."

"And I expect you will be with us many more," Blackburn said smoothly. "Look: it's in everybody's interest to keep the teams in place. Because as I said, she can't run them directly."

"Uh-huh."

Blackburn shot his cuffs and ran his hand through his hair. "Listen, Tom. I know you're disappointed that this appointment didn't come to you. But let's not make too much of Meredith appointing the division heads. Realistically speaking, she isn't going to make any changes. Your situation is secure." He paused. "You know the way Meredith is, Tom."

"I used to," Sanders said, nodding. "Hell, I lived with her for a while. But I haven't seen her in years."

Blackburn looked surprised. "You two haven't kept contact?"

"Not really, no. By the time Meredith joined the company, I was up here in Seattle, and she was based in Cupertino. I ran into her once, on a trip down there. Said hello. That's about it."

"Then you only know her from the old days," Blackburn said, as if it all suddenly made sense. "From six or seven years ago."

"It's longer than that," Sanders said. "I've been in Seattle eight years. So it must be . . ." Sanders thought back. "When I was going out with her, she worked for Novell in Mountain View. Selling Ethernet cards to small businesses for local area networks. When was that?" Although he remembered the relationship with Meredith Johnson vividly, Sanders was hazy about exactly when it had occurred. He tried to recall some memorable event—a birthday, a promotion, an apartment move—that would mark the date. Finally he remembered watching election returns with her on television: balloons rising up toward the ceiling, people cheering. She was drinking beer. That had been early in their relationship. "Jesus, Phil. It must be almost ten years ago."

"That long," Blackburn said.

When Sanders first met Meredith Johnson, she was one of the thousands of pretty saleswomen working in San Jose—young women in their twenties, not long out of college, who started out doing the product demos on the computer while a senior man stood beside her and did all the talking to the customer. Eventually, a lot of those women learned enough to do the selling themselves. At the time Sanders first knew Meredith, she had acquired enough jargon to rattle on about token rings and 10BaseT hubs. She didn't really have any deep knowledge, but she didn't need to. She was good-looking, sexy, and smart, and she had a kind of uncanny self-possession that carried her through awkward moments. Sanders had admired her, back in those days. But he never imagined that she had the ability to hold a major corporate position.

Blackburn shrugged. "A lot's happened in ten years, Tom," he said. "Meredith isn't just a sales exec. She went back to school, got an MBA. She worked at Symantec, then Conrad, and then she came to work with us. The last couple of years, she's been working very closely with Garvin. Sort of his protégé. He's been pleased with her work on a number of assignments."

Sanders shook his head. "And now she's my boss . . ."

"Is that a problem for you?"

"No. It just seems funny. An old girlfriend as my boss."

"The worm turns," Blackburn said. He was smiling, but Sanders sensed he was watching him closely. "You seem a little uneasy about this, Tom."

"It takes some getting used to."

"Is there a problem? Reporting to a woman?"

"Not at all. I worked for Eileen when she was head of HRI, and we got along great. It's not that. It's just funny to think of Meredith Johnson as my boss."

"She's an impressive and accomplished manager," Phil said. He stood up, smoothed his tie. "I think when you've had an opportunity to become reacquainted, you'll be very impressed. Give her a chance, Tom."

"Of course," Sanders said.

"I'm sure everything will work out. And keep your eye on the future. After all, you should be rich in a year or so."

"Does that mean we're still spinning off the APG Division?"

"Oh yes. Absolutely."

It was a much-discussed part of the merger plan that after Conley-White bought DigiCom, it would spin off the Advanced Products Division and take it public, as a separate company. That would mean enormous profits for everyone in the division. Because everyone would have the chance to buy cheap options before the stock was publicly sold.

"We're working out the final details now," Blackburn said. "But I expect that division managers like yourself will start with twenty thousand shares vested, and an initial option of fifty thousand shares at twenty-five cents a share, with the right to purchase another fifty thousand shares each year for the next five years."

"And the spin-off will go forward, even with Meredith running the divisions?"

"Trust me. The spin-off will happen within eighteen months. It's a formal part of the merger plan."

"There's no chance that she may decide to change her mind?"

"None at all, Tom." Blackburn smiled. "I'll tell you a little secret. Originally, this spin-off was Meredith's idea."

Blackburn left Sanders's office and went down the hall to an empty office and called Garvin. He heard the familiar sharp bark: "Garvin here."

"I talked to Tom Sanders."

"And?"

"I'd say he took it well. He was disappointed, of course. I think he'd already heard a rumor. But he took it well."

Garvin said, "And the new structure? How did he respond?"

"He's concerned," Blackburn said. "He expressed reservations."

"Why?"

"He doesn't feel she has the technical expertise to run the division."

Garvin snorted, "Technical expertise? That's the last goddamn thing I care about. Technical expertise is not an issue here."

"Of course not. But I think there was some uneasiness on the personal level. You know, they once had a relationship."

"Yes," Garvin said. "I know that. Have they talked?"

"He says, not for several years."

"Bad blood?"

"There didn't seem to be."

"Then what's he concerned about?"

"I think he's just getting used to the idea."

"He'll come around."

"I think so."

"Tell me if you hear otherwise," Garvin said, and hung up.

Alone in the office, Blackburn frowned. The conversation with Sanders left him vaguely uneasy. It had seemed to go well enough, and yet . . . Sanders, he felt sure, was not going to take this reorganization lying down. Sanders was popular in the Seattle division, and he could easily cause trouble. Sanders was too independent, he was not a team player, and the company needed team players now. The more Blackburn thought about it, the more certain he was that Sanders was going to be a problem.

Tom Sanders sat at his desk, staring forward, lost in thought. He was trying to put together his memory of a pretty young saleswoman in Silicon Valley with this new image of a corporate officer running company divisions, executing the complex groundwork required to take a division public. But his thoughts kept being interrupted by random images from the past: Meredith smiling, wearing one of his shirts, naked beneath it. An opened suitcase on the bed. White stockings and white garter belt. A bowl of popcorn on the blue couch in the living room. The television with the sound turned off.

And for some reason, the image of a flower, a purple iris, in stained glass. It was one of those hackneyed Northern California hippie images. Sanders knew where it came from: it was on the glass of the front door to the apartment where he had lived, back in Sunnyvale. Back in the days when he had known Meredith. He wasn't sure why he should keep thinking of it now, and he—

"Tom?"

He glanced up. Cindy was standing in the doorway, looking concerned.

"Tom, do you want coffee?"

"No, thanks."

"Don Cherry called again while you were with Phil. He wants you to come and look at the Corridor."

"They having problems?"

"I don't know. He sounded excited. You want to call him back?"

"Not right now. I'll go down and see him in a minute."

She lingered at the door. "You want a bagel? Have you had breakfast?"

"I'm fine."

"Sure?"

"I'm fine, Cindy. Really."

She went away. He turned to look at his monitor, and saw that the icon for his e-mail was blinking. But he was thinking again about Meredith Johnson.

Sanders had more or less lived with her for about six months. It had been quite an intense relationship for a while. And yet, although he kept having isolated, vivid images, he realized that in general his memories from that time were surprisingly vague. Had he really lived with Meredith for six months? When exactly had they first met, and when had they broken up? Sanders was surprised at how difficult it was for him to fix the chronology in his mind. Hoping for clarity, he considered other aspects of his life: what had been his position at DigiCom in those days? Was he still working in Marketing, or had he already moved to the technical divisions? He wasn't sure, now. He would have to look it up in the files.

He thought about Blackburn. Blackburn had left his wife and moved in with Sanders around the time Sanders was involved with Meredith. Or was it afterward, when things had gone bad? Maybe Phil had moved into his apartment around the time he was breaking up with Meredith. Sanders wasn't sure. As he considered it, he realized he wasn't sure about anything from that time. These events had all happened a decade ago, in another city, at another period in his life, and his memories were in disarray. Again, he was surprised at how confused he was.

He pushed the intercom. "Cindy? I've got a question for you."

"Sure, Tom."

"This is the third week of June. What were you doing the third week of June, ten years ago?"

She didn't even hesitate. "That's easy: graduating from college."

Of course that would be true. "Okay," he said. "Then how about June, nine years ago."

"Nine years ago?" Her voice sounded suddenly cautious, less certain. "Gee . . . Let's see, June . . . Nine years ago? . . . June . . . Uh . . . I think I was with my boyfriend in Europe."

"Not your present boyfriend?"

"No . . . This guy was a real jerk."

Sanders said, "How long did that last?"

"We were there for a month."

"I mean the relationship."

"With him? Oh, let's see, we broke up . . . oh, it must have been

... uh, December ... I think it was December, or maybe January, after the holidays ... Why?"

"Just trying to figure something out," Sanders said. Already he was relieved to hear the uncertain tone of her voice, as she tried to piece together the past. "By the way, how far back do we have office records? Correspondence, and call books?"

"I'd have to check. I know I have about three years."

"And what about earlier?"

"Earlier? How much earlier?"

"Ten years ago," he said.

"Gee, that'd be when you were in Cupertino. Do they have that stuff in storage down there? Did they put it on fiche, or was it just thrown out?"

"I don't know."

"You want me to check?"

"Not now," he said, and clicked off. He didn't want her making any inquiries in Cupertino now. Not right now.

Sanders rubbed his eyes with his fingertips. His thoughts drifting back over time. Again, he saw the stained-glass flower. It was oversize, bright, banal. Sanders had always been embarrassed by the banality of it. In those days, he had lived in one of the apartment complexes on Merano Drive. Twenty units clustered around a chilly little swimming pool. Everybody in the building worked for a high-tech company. Nobody ever went in the pool. And Sanders wasn't around much. Those were the days when he flew with Garvin to Korea twice a month. The days when they all flew coach. They couldn't even afford business class.

And he remembered how he would come home, exhausted from the long flight, and the first thing he would see when he got to his apartment was that damned stained-glass flower on the door.

And Meredith, in those days, was partial to white stockings, a white garter belt, little white flowers on the snaps with—

"Tom?" He looked up. Cindy was at the doorway. She said, "If you want to see Don Cherry, you'd better go now because you have a ten-thirty with Gary Bosak."

He felt as if she was treating him like an invalid. "Cindy, I'm fine."

"I know. Just a reminder."

"Okay, I'll go now."

As he hurried down the stairs to the third floor, he felt relieved at the distraction. Cindy was right to get him out of the office. And he was curious to see what Cherry's team had done with the Corridor.

The Corridor was what everyone at DigiCom called VIE: the Virtual Information Environment. VIE was the companion piece to Twinkle, the second major element in the emerging future of digital information as envisioned by DigiCom. In the future, information was going to be stored on disks, or made available in large databases that users would dial into over telephone lines. At the moment, users saw information displayed on flat screens—either televisions or computer screens. That had been the traditional way of handling information for the last thirty years. But soon, there would be new ways to present information. The most radical, and the most exciting, was virtual environments. Users wore special glasses to see computer-generated, three-dimensional environments which allowed them to feel as though they were literally moving through another world. Dozens of high-tech companies were racing to develop virtual environments. It was exciting, but very difficult, technology. At DigiCom, VIE was one of Garvin's pet projects; he had thrown a lot of money at it; he had had Don Cherry's programmers working on it around the clock for two years.

And so far, it had been nothing but trouble.

The sign on the door said "VIE" and underneath, "When Reality Is Not Enough." Sanders inserted his card in the slot, and the door clicked open. He passed through an anteroom, hearing a half-dozen voices shouting from the main equipment room beyond. Even in the anteroom, he noticed a distinctly nauseating odor in the air.

Entering the main room, he came upon a scene of utter chaos. The windows were thrown wide; there was the astringent smell of cleaning fluid. Most of the programmers were on the floor, working with disassembled equipment. The VIE units lay scattered in pieces, amid a tangle of multicolored cables. Even the black circular walker pads had been taken apart, the rubber bearings being cleaned one by one. Still more wires descended from the ceiling to the laser scanners which were broken open, their circuit boards exposed. Everyone seemed to be talking at once. And in the center of the room, looking like a teenage Buddha in an electric blue T-shirt that said "Reality Sucks," was Don Cherry, the head of Programming. Cherry was twenty-two years old, widely acknowledged to be indispensable, and famous for his impertinence.

When he saw Sanders he shouted: "Out! Out! Damned management! Out!"

"Why?" Sanders said. "I thought you wanted to see me."

"Too late! You had your chance!" Cherry said. "Now it's over!"

For a moment, Sanders thought Cherry was referring to the promotion he hadn't gotten. But Cherry was the most apolitical of the Digi-Com division heads, and he was grinning cheerfully as he walked toward Sanders, stepping over his prostrate programmers. "Sorry, Tom. You're too late. We're fine-tuning now."

"Fine-tuning? It looks like ground zero here. And what's that terrible smell?"

"I know." Cherry threw up his hands. "I ask the boys to wash every day, but what can I say. They're programmers. No better than dogs."

"Cindy said you called me several times."

"I did," Cherry said. "We had the Corridor up and running, and I wanted you to see it. But maybe it's just as well you didn't."

Sanders looked at the complex equipment scattered all around him. "You had it *up?*"

"That was then. This is now. Now, we're fine-tuning." Cherry nodded to the programmers on the floor, working on the walker pads. "We finally got the bug out of the main loop, last night at midnight. The refresh rate doubled. The system really rips now. So we have to adjust the walkers and the servos to update responsiveness. It's a *mechanical* problem," he said disdainfully. "But we'll take care of it anyway."

The programmers were always annoyed when they had to deal with mechanical problems. Living almost entirely in an abstract world of computer code, they felt that physical machinery was beneath them.

Sanders said, "What is the problem, exactly?"

"Well, look," Cherry said. "Here's our latest implementation. The user wears this headset," he said, pointing to what looked like thick silver sunglasses. "And he gets on the walker pad, here."

The walker pad was one of Cherry's innovations. The size of a small round trampoline, its surface was composed of tightly packed rubber balls. It functioned like a multidirectional treadmill; walking on the balls, users could move in any direction. "Once he's on the walker," Cherry said, "the user dials into a database. Then the computer, over there—" Cherry pointed to a stack of boxes in the corner, "takes the information coming from the database and constructs a virtual environment which is projected inside the headset. When the user walks on the pad, the projection changes, so you feel like you're walking down a corridor lined with drawers of data on all sides. The user can stop anywhere, open any file drawer with his hand, and thumb through data. Completely realistic simulation."

"How many users?"

"At the moment, the system can handle five at one time."

"And the Corridor looks like what?" Sanders said. "Wire-frame?" In the earlier versions, the Corridor was outlined in skeletal black-and-white outlines. Fewer lines made it faster for the computer to draw.

"Wire-frame?" Cherry sniffed. "*Please.* We dumped that two weeks ago. Now we are talking 3-D surfaces fully modeled in 24-bit color, with anti-alias texture maps. We're rendering true curved surfaces—no polygons. Looks completely real."

"And what're the laser scanners for? I thought you did position by infrared." The headsets had infrared sensors mounted above them, so that the system could detect where the user was looking and adjust the projected image inside the headset to match the direction of looking.

"We still do," Cherry said. "The scanners are for body representation."

"Body representation?"

"Yeah. Now, if you're walking down the Corridor with somebody else, you can turn and look at them and you'll see them. Because the scanners are capturing a three-dimensional texture map in real time: they read body and expression, and draw the virtual face of the virtual person standing beside you in the virtual room. You can't see the person's eyes, of course, because they're hidden by the headset they're wearing. But the system generates a face from the stored texture map. Pretty slick, huh?"

"You mean you can see other users?"

"That's right. See their faces, see their expressions. And that's not all. If other users in the system aren't wearing a headset, you can still see them, too. The program identifies other users, pulls their photo out of the personnel file, and pastes it onto a virtual body image. A little kludgey, but not bad." Cherry waved his hand in the air. "And that's not all. We've also built in virtual help."

"Virtual help?"

"Sure, users always need online help. So we've made an angel to help you. Floats alongside you, answers your questions." Cherry was grinning. "We thought of making it a blue fairy, but we didn't want to offend anybody."

Sanders stared thoughtfully at the room. Cherry was telling him about his successes. But something else was happening here: it was impossible to miss the tension, the frantic energy of the people as they worked.

"Hey, Don," one of the programmers shouted. "What's the Z-count supposed to be?"

"Over five," Cherry said.

"I got it to four-three."

"Four-three sucks. Get it above five, or you're fired." He turned to Sanders. "You've got to encourage the troops."

Sanders looked at Cherry. "All right," he said finally. "Now what's the *real* problem?"

Cherry shrugged. "Nothing. I told you: fine-tuning."

"Don."

Cherry sighed. "Well, when we jumped the refresh rate, we trashed the builder module. You see, the room is being built in real time by the box. With a faster refresh off the sensors, we have to build objects much faster. Otherwise the room seems to lag behind you. You feel like you're drunk. You move your head, and the room swooshes behind you, catching up."

"And?"

"And, it makes the users throw up."

Sanders sighed. "Great."

"We had to take the walker pads apart because Teddy barfed all over everything."

"Great, Don."

"What's the matter? It's no big deal. It cleans up." He shook his head. "Although I do wish Teddy hadn't eaten huevos rancheros for breakfast. That was unfortunate. Little bits of tortilla everywhere in the bearings."

"You know we have a demo tomorrow for the C-W people."

"No problem. We'll be ready."

"Don, I can't have their top executives throwing up."

"Trust me," Cherry said. "We'll be ready. They're going to love it. Whatever problems this company has, the Corridor is not one of them."

"That's a promise?"

"That," Cherry said, "is a guarantee."

Sanders was back in his office by ten-twenty, and was seated at his desk when Gary Bosak came in. Bosak was a tall man in his twenties, wearing jeans, running shoes, and a Terminator T-shirt. He carried a large fold-over leather briefcase, the kind that trial attorneys used.

"You look pale," Bosak said. "But everybody in the building is pale today. It's tense as hell around here, you know that?"

"I've noticed."

"Yeah, I bet. Okay to start?"

"Sure."

"Cindy? Mr. Sanders is going to be unavailable for a few minutes."

Bosak closed the office door and locked it. Whistling cheerfully, he unplugged Sanders's desk phone, and the phone beside the couch in the corner. From there, he went to the window and closed the blinds. There was a small television in the corner; he turned it on. He snapped the latches on his briefcase, took out a small plastic box, and flipped the switch on the side. The box began to blink, and emitted a low white noise hiss. Bosak set it in the middle of Sanders's desk. Bosak never gave information until the white noise scrambler was in place, since most of what he had to say implied illegal behavior.

"I have good news for you," Bosak said. "Your boy is clean." He pulled out a manila file, opened it up, and started handing over pages. "Peter John Nealy, twenty-three, DigiCom employee for sixteen months. Now working as a programmer in APG. Okay, here we go. His high school and college transcripts . . . Employment file from Data General, his last employer. All in order. Now, the recent stuff . . . Credit rating from TRW . . . Phone bills from his apartment . . . Phone bills for his cellular line . . . Bank statement . . . Savings account . . . Last two 1040s . . . Twelve months of credit card charges, VISA and Master . . . Travel records . . . E-mail messages inside the company, and off the Internet . . . Parking tickets . . . And this is the clincher . . . Ramada Inn in

Sunnyvale, last three visits, his phone charges there, the numbers he called . . . Last three car rentals with mileage . . . Rental car cellular phone, the numbers called . . . That's everything."

"And?"

"I ran down the numbers he called. Here's the breakdown. A lot of calls to Seattle Silicon, but Nealy's seeing a girl there. She's a secretary, works in sales, no conflict. He also calls his brother, a programmer at Boeing, does parallel processing stuff for wing design, no conflict. His other calls are to suppliers and code vendors, and they're all appropriate. No calls after hours. No calls to pay phones. No overseas calls. No suspicious pattern in the calls. No unexplained bank transfers, no sudden new purchases. No reason to think he's looking for a move. I'd say he's not talking to anybody you care about."

"Good," Sanders said. He glanced down at the sheets of paper, and paused. "Gary . . . Some of this stuff is from our company. Some of these reports."

"Yeah. So?"

"How'd you get them?"

Bosak grinned. "Hey. You don't ask and I don't tell you."

"How'd you get the Data General file?"

Bosak shook his head. "Isn't this why you pay me?"

"Yes it is, but—"

"Hey. You wanted a check on an employee, you got it. Your kid's clean. He's working only for you. Anything else you want to know about him?"

"No." Sanders shook his head.

"Great. I got to get some sleep." Bosak collected all the files and placed them back in his folder. "By the way, you're going to get a call from my parole officer."

"Uh-huh."

"Can I count on you?"

"Sure, Gary."

"I told him I was doing consulting for you. On telecommunications security."

"And so you are."

Bosak switched off the blinking box, put it in his briefcase, and reconnected the telephones. "Always a pleasure. Do I leave the bill with you, or Cindy?"

"I'll take it. See you, Gary."

"Hey. Anytime. You need more, you know where I am."

Sanders glanced at the bill, from NE Professional Services, Inc., of Bellevue, Washington. The name was Bosak's private joke: the letters NE stood for "Necessary Evil." Ordinarily, high-tech companies employed retired police officers and private investigators to do background checks, but occasionally they used hackers like Gary Bosak, who could gain access to electronic data banks, to get information on suspect employees. The advantage of using Bosak was that he could work quickly, often making a report in a matter of hours, or overnight. Bosak's methods were of course illegal; simply by hiring him, Sanders himself had broken a half-dozen laws. But background checks on employees were accepted as standard practice in high-tech firms, where a single document or product development plan might be worth hundreds of thousands of dollars to competitors.

And in the case of Pete Nealy, a check was particularly crucial. Nealy was developing hot new compression algorithms to pack and unpack video images onto CD-ROM laser disks. His work was vital to the new Twinkle technology. High-speed digital images coming off the disk were going to transform a sluggish technology and produce a revolution in education. But if Twinkle's algorithms became available to a competitor, then DigiCom's advantage would be greatly reduced, and that meant—

The intercom buzzed. "Tom," Cindy said. "It's eleven o'clock. Time for the APG meeting. You want the agenda on your way down?"

"Not today," he said. "I think I know what we'll be talking about."

In the third-floor conference room, the Advanced Products Group was already meeting. This was a weekly meeting in which the division heads discussed problems and brought everyone up to date. It was a meeting that Sanders ordinarily led. Around the table were Don Cherry, the chief of Programming; Mark Lewyn, the temperamental head of Product Design, all in black Armani; and Mary Anne Hunter, the head of Data Telecommunications. Petite and intense, Hunter was dressed in a sweatshirt, shorts, and Nike running tights; she never ate lunch, but ordinarily went on a five-mile run after each meeting.

Lewyn was in the middle of one of his storming rages: "It's insulting to everybody in the division. I have no idea why she got this position. I don't know what her qualifications could be for a job like this, and—"

Lewyn broke off as Sanders came into the room. There was an awkward moment. Everyone was silent, glancing at him, then looking away.

"I had a feeling," Sanders said, smiling, "you'd be talking about this."

The room remained silent. "Come on," he said, as he slipped into a chair. "It's not a funeral."

Mark Lewyn cleared his throat. "I'm sorry, Tom. I think it's an outrage."

Mary Anne Hunter said, "Everybody knows it should have been you."

Lewyn said, "It's a shock to all of us, Tom."

"Yeah," Cherry said, grinning. "We've been trying like hell to get you sacked, but we never really thought it would work."

"I appreciate all this," Sanders said, "but it's Garvin's company, and he can do what he wants with it. He's been right more often than not. And I'm a big boy. Nobody ever promised me anything."

Lewyn said, "You're really okay with this?"

"Believe me. I'm fine."

"You talked with Garvin?"

"I talked with Phil."

Lewyn shook his head. "That sanctimonious asshole."

"Listen," Cherry said, "did Phil say anything about the spin-off?"

"Yes," Sanders said. "The spin-off is still happening. Eighteen months after the merger, they'll structure the IPO, and take the division public."

There were little shrugs around the table. Sanders could see they were relieved. Going public meant a lot of money to all the people sitting in the room.

"And what did Phil say about Ms. Johnson?"

"Not much. Just that she's Garvin's choice to head up the technical side."

At that moment Stephanie Kaplan, DigiCom's Chief Financial Officer, came into the room. A tall woman with prematurely gray hair and a notably silent manner, she was known as Stephanie Stealth, or the Stealth Bomber—the latter a reference to her habit of quietly killing projects she did not consider profitable enough. Kaplan was based in Cupertino, but she generally sat in once a month on the Seattle division meetings. Lately, she had been up more often.

Lewyn said, "We're trying to cheer up Tom, Stephanie."

Kaplan took a seat, and gave Sanders a sympathetic smile. She didn't speak.

Lewyn said, "Did *you* know this Meredith Johnson appointment was coming?"

"No," Kaplan said. "It was a surprise to everybody. And not everybody's happy about it." Then, as if she had said too much, she opened her briefcase, and busied herself with her notes. As usual, she slid into the background; the others quickly ignored her.

"Well," Cherry said, "I hear Garvin's got a real thing for her. Johnson's only been with the company four years, and she hasn't been especially outstanding. But Garvin took her under his wing. Two years ago, he began moving her up, fast. For some reason, he just thinks Meredith Johnson is *great*."

Lewyn said, "Is Garvin fucking her?"

"No, he just likes her."

"She must be fucking somebody."

"Wait a minute," Mary Anne Hunter said, sitting up. "What's this? If Garvin brought in some guy from Microsoft to run this division, nobody'd say he must be fucking somebody."

Cherry laughed. "It'd depend on who he was."

"I'm serious. Why is it when a woman gets a promotion, she must be fucking somebody?"

Lewyn said, "Look: if they brought in Ellen Howard from Microsoft, we wouldn't be having this conversation because we all know Ellen's very competent. We wouldn't like it, but we'd accept it. But nobody even *knows* Meredith Johnson. I mean, does anybody here know her?"

"Actually," Sanders said, "I know her."

There was silence.

"I used to go out with her."

Cherry laughed. "So *you're* the one she's fucking."

Sanders shook his head. "It was years ago."

Hunter said, "What's she like?"

"Yeah," Cherry said, grinning lasciviously. "What's she like?"

"Shut up, Don."

"Lighten up, Mary Anne."

"She worked for Novell when I knew her," Sanders said. "She was about twenty-five. Smart and ambitious."

"Smart and ambitious," Lewyn said. "That's fine. The world's full of smart and ambitious. The question is, can she run a technical division? Or have we got another Screamer Freeling on our hands?"

Two years earlier, Garvin had put a sales manager named Howard Freeling in charge of the division. The idea was to bring product development in contact with customers at an earlier point, to develop new products more in line with the emerging market. Freeling instituted focus groups, and they all spent a lot of time watching potential customers play with new products behind one-way glass.

But Freeling was completely unfamiliar with technical issues. So when confronted with a problem, he screamed. He was like a tourist in a foreign country who didn't speak the language and thought he could make the locals understand by shouting at them. Freeling's tenure at APG was a disaster. The programmers loathed him; the designers rebelled at his idea for neon-colored product boxes; the manufacturing glitches at factories in Ireland and Texas didn't get solved. Finally, when the production line in Cork went down for eleven days, Freeling flew over and screamed. The Irish managers all quit, and Garvin fired him.

"So: is that what we have? Another Screamer?"

Stephanie Kaplan cleared her throat. "I think Garvin learned his lesson. He wouldn't make the same mistake twice."

"So you think Meredith Johnson is up to the job?"

"I couldn't say," Kaplan replied, speaking very deliberately.

"Not much of an endorsement," Lewyn said.

"But I think she'll be better than Freeling," Kaplan said.

Lewyn snorted. "This is the Taller Than Mickey Rooney Award. You can still be very short and win."

"No," Kaplan said, "I think she'll be better."

Cherry said, "Better-looking, at least, from what I hear."

"Sexist," Mary Anne Hunter said.

"What: I can't say she's good-looking?"

"We're talking about her competence, not her appearance."

"Wait a minute," Cherry said. "Coming over here to this meeting, I pass the women at the espresso bar, and what are they talking about? Whether Richard Gere has better buns than Mel Gibson. They're talking about the crack in the ass, lift and separate, all that stuff. I don't see why they can talk about—"

"We're drifting afield," Sanders said.

"It doesn't matter what you guys say," Hunter said, "the fact is, this company is dominated by males; there are almost no women except Stephanie in high executive positions. I think it's great that Bob has appointed a woman to run this division, and I for one think we should support her." She looked at Sanders. "We all love you, Tom, but you know what I mean."

"Yeah, we all love you," Cherry said. "At least, we did until we got our cute new boss."

Lewyn said, "I'll support Johnson—if she's any good."

"No you won't," Hunter said. "You'll sabotage her. You'll find a reason to get rid of her."

"Wait a minute—"

"No. What is this conversation *really* about? It's about the fact that you're all pissed off because now you have to report to a woman."

"Mary Anne . . ."

"I mean it."

Lewyn said, "I think Tom's pissed off because he didn't get the job."

"I'm not pissed off," Sanders said.

"Well, I'm pissed off," Cherry said, "because Meredith used to be Tom's girlfriend, so now he has a special in with the new boss."

"Maybe." Sanders frowned.

Lewyn said, "On the other hand, maybe she hates you. All my old girlfriends hate me."

"With good reason, I hear," Cherry said, laughing.

Sanders said, "Let's get back to the agenda, shall we?"

"What agenda?"

"Twinkle."

There were groans around the table. "Not again."

"Goddamn Twinkle."

"How bad is it?" Cherry said.

"They still can't get the seek times down, and they can't solve the hinge problems. The line's running at twenty-nine percent."

Lewyn said, "They better send us some units."

"We should have them today."

"Okay. Table it till then?"

"It's okay with me." Sanders looked around the table. "Anybody else have a problem? Mary Anne?"

"No, we're fine. We still expect prototype card-phones off our test line within two months."

The new generation of cellular telephones were not much larger than a credit card. They folded open for use. "How's the weight?"

"The weight's now four ounces, which is not great, but okay. The problem is power. The batteries only run 180 minutes in talk mode. And the keypad sticks when you dial. But that's Mark's headache. We're on schedule with the line."

"Good." He turned to Don Cherry. "And how's the Corridor?"

Cherry sat back in his chair, beaming. He crossed his hands over his belly. "I am pleased to report," he said, "that as of half an hour ago, the Corridor is fan-fucking-*tastic*."

"Really?"

"That's great news."

"Nobody's throwing up?"

"*Please*. Ancient history."

Mark Lewyn said, "Wait a minute. Somebody threw up?"

"A vile rumor. That was then. This is now. We got the last delay bug out half an hour ago, and all functions are now fully implemented. We can take any database and convert it into a 3-D 24-bit color environment that you can navigate in real time. You can walk through any database in the world."

"And it's stable?"

"It's a rock."

"You've tried it with naïve users?"

"Bulletproof."

"So you're ready to demo for Conley?"

"We'll blow 'em away," Cherry said. "They won't fucking believe their eyes."

Coming out of the conference room, Sanders ran into a group of Conley-White executives being taken on a tour by Bob Garvin.

Robert T. Garvin looked the way every CEO wanted to look in the pages of *Fortune* magazine. He was fifty-nine years old and handsome, with a craggy face and salt-and-pepper hair that always looked wind-blown, as if he'd just come in from a fly-fishing trip in Montana, or a weekend sailing in the San Juans. In the old days, like everyone else, he had worn jeans and denim work shirts in the office. But in recent years, he favored dark blue Caraceni suits. It was one of the many changes that people in the company had noticed since the death of his daughter, three years before.

Brusque and profane in private, Garvin was all charm in public. Leading the Conley-White executives, he said, "Here on the third floor, you have our tech divisions and advanced product laboratories. Oh, Tom. Good." He threw his arm around Sanders. "Meet Tom Sanders, our division manager for advanced products. One of the brilliant young men who's made our company what it is. Tom, say hello to Ed Nichols, the CFO for Conley-White . . ."

A thin, hawk-faced man in his late fifties, Nichols carried his head tilted back, so that he seemed to be pulling away from everything, as if there were a bad smell. He looked down his nose through half-frame glasses at Sanders, regarding him with a vaguely disapproving air, and shook hands formally.

"Mr. Sanders. How do you do."

"Mr. Nichols."

". . . and John Conley, nephew of the founder, and vice president of the firm . . ."

Sanders turned to a stocky, athletic man in his late twenties. Wire-frame spectacles. Armani suit. Firm handshake. Serious expression. Sanders had the impression of a wealthy and very determined man.

"Hi there, Tom."

"Hi, John."

". . . and Jim Daly, from Goldman, Sachs . . ."

A balding, thin, storklike man in a pinstripe suit. Daly seemed distracted, befuddled, and shook hands with a brief nod.

". . . and of course, Meredith Johnson, from Cupertino."

She was more beautiful than he had remembered. And different in some subtle way. Older, of course, crow's-feet at the corners of her eyes, and faint creases in her forehead. But she stood straighter now, and she had a vibrancy, a confidence, that he associated with power. Dark blue suit, blond hair, large eyes. Those incredibly long eyelashes. He had forgotten.

"Hello, Tom, nice to see you again." A warm smile. Her perfume.

"Meredith, nice to see you."

She released his hand, and the group swept on, as Garvin led them down the hall. "Now, just ahead is the VIE Unit. You'll be seeing that work tomorrow."

Mark Lewyn came out of the conference room and said, "You met the rogues' gallery?"

"I guess so."

Lewyn watched them go. "Hard to believe those guys are going to be running this company," he said. "I did a briefing this morning, and let me tell you, they don't know *anything*. It's scary."

As the group reached the end of the hallway, Meredith Johnson looked back over her shoulder at Sanders. She mouthed, "I'll call you." And she smiled radiantly. Then she was gone.

Lewyn sighed. "I'd say," he said, "that you have an in with top management there, Tom."

"Maybe so."

"I just wish I knew why Garvin thinks she's so great."

Sanders said, "Well, she certainly looks great."

Lewyn turned away. "We'll see," he said. "We'll see."

At twenty past twelve, Sanders left his office on the fourth floor and headed toward the stairs to go down to the main conference room for lunch. He passed a nurse in a starched white uniform. She was looking in one office after another. "Where is he? He was just here a minute ago." She shook her head.

"Who?" Sanders said.

"The professor," she replied, blowing a strand of hair out of her eyes. "I can't leave him alone for a minute."

"What professor?" Sanders said. But by then he heard the female giggles coming from a room farther down the hall, and he already knew the answer. "Professor Dorfman?"

"Yes. Professor Dorfman," the nurse said, nodding grimly, and she headed toward the source of the giggles.

Sanders trailed after her. Max Dorfman was a German management consultant, now very elderly. At one time or another, he had been a visiting professor at every major business school in America, and he had gained a particular reputation as a guru to high-tech companies. During most of the 1980s, he had served on the board of directors of DigiCom, lending prestige to Garvin's upstart company. And during that time, he had been a mentor to Sanders. In fact, it was Dorfman who had convinced Sanders to leave Cupertino eight years earlier and take the job in Seattle.

Sanders said, "I didn't know he was still alive."

"Very much so," the nurse said.

"He must be ninety."

"Well, he doesn't act a day over eighty-five."

As they approached the room, he saw Mary Anne Hunter coming out. She had changed into a skirt and blouse, and she was smiling broadly, as if she had just left her lover. "Tom, you'll never guess who's here."

"Max," he said.

"That's right. Oh, Tom, you should see him: he's exactly the same."

"I'll bet he is," Sanders said. Even from outside the room, he could smell the cigarette smoke.

The nurse said, "Now, Professor," in a severe tone, and strode into the room. Sanders looked in; it was one of the employee lounges. Max Dorfman's wheelchair was pulled up to the table in the center of the room. He was surrounded by pretty assistants. The women were making a fuss over him, and in their midst Dorfman, with his shock of white hair, was grinning happily, smoking a cigarette in a long holder.

"What's he doing here?" Sanders said.

"Garvin brought him in, to consult on the merger. Aren't you going to say hello?" Hunter said.

"Oh, Christ," Sanders said. "You know Max. He can drive you crazy." Dorfman liked to challenge conventional wisdom, but his method was indirect. He had an ironic way of speaking that was provocative and mocking at the same moment. He was fond of contradictions, and he did not hesitate to lie. If you caught him in a lie, he would immediately say, "Yes, that's true. I don't know what I was thinking of," and then resume talking in the same maddening, elliptical way. He never really said what he meant; he left it for you to put it together. His rambling sessions left executives confused and exhausted.

"But you were such friends," Hunter said, looking at him. "I'm sure he'd like you to say hello."

"He's busy now. Maybe later." Sanders looked at his watch. "Anyway, we're going to be late for lunch."

He started back down the hallway. Hunter fell into step with him, frowning. "He always got under your skin, didn't he?"

"He got under everybody's skin. It was what he did best."

She looked at him in a puzzled way, and seemed about to say more, then shrugged. "It's okay with me."

"I'm just not in the mood for one of those conversations," Sanders said. "Maybe later. But not right now." They headed down the stairs to the ground floor.

In keeping with the stripped-down functionality of modern high-tech firms, DigiCom maintained no corporate dining room. Instead, lunches and dinners were held at local restaurants, most often at the nearby Il Terrazzo. But the need for secrecy about the merger obliged DigiCom to cater a lunch in the large, wood-paneled conference room on the ground floor. At twelve-thirty, with the principal managers of the DigiCom technical divisions, the Conley-White executives, and the Goldman, Sachs bankers all present, the room was crowded. The egalitarian ethos of the company meant that there was no assigned seating, but the principal C-W executives ended up at one side of the table near the front of the room, clustered around Garvin. The power end of the table.

Sanders took a seat farther down on the opposite side, and was surprised when Stephanie Kaplan slid into the chair to his right. Kaplan usually sat much closer to Garvin; Sanders was distinctly further down the pecking order. To Sanders's left was Bill Everts, the head of Human Resources—a nice, slightly dull guy. As white-coated waiters served the meal, Sanders talked about fishing on Orcas Island, which was Everts's passion. As usual, Kaplan was quiet during most of the lunch, seeming to withdraw into herself.

Sanders began to feel he was neglecting her. Toward the end of the meal, he turned to her and said, "I notice you've been up here in Seattle more often the last few months, Stephanie. Is that because of the merger?"

"No." She smiled. "My son's a freshman at the university, so I like to come up because I get to see him."

"What's he studying?"

"Chemistry. He wants to go into materials chemistry. Apparently it's going to be a big field."

"I've heard that."

"Half the time I don't know what he's talking about. It's funny, when your child knows more than you do."

He nodded, trying to think of something else to ask her. It wasn't easy: although he had sat in meetings with Kaplan for years, he knew little about her personally. She was married to a professor at San Jose State, a jovial chubby man with a mustache, who taught economics. When they were together, he did all the talking while Stephanie stood silently by. She was a tall, bony, awkward woman who seemed resigned to her lack of social graces. She was said to be a very good golfer—at least, good enough that Garvin wouldn't play her anymore. No one who knew her was surprised that she had made the error of beating Garvin too often; wags said that she wasn't enough of a loser to be promoted.

Garvin didn't really like her, but he would never think of letting her go. Colorless, humorless, and tireless, her dedication to the company was legendary; she worked late every night and came in most weekends. When she had had a bout of cancer a few years back, she refused to take even a single day off. Apparently she was cured of the cancer; at least, Sanders hadn't heard anything more about it. But the episode seemed to have increased Kaplan's relentless focus on her impersonal domain, figures and spreadsheets, and heightened her natural inclination to work behind the scenes. More than one manager had come to work in the morning, only to find a pet project killed by the Stealth Bomber, with no lingering trace of how or why it had happened. Thus her tendency to remain aloof in social situations was more than a reflection of her own discomfort; it was also a reminder of the power she wielded within the company, and how she wielded it. In her own way, she was mysterious—and potentially dangerous.

While he was trying to think of something to say, Kaplan leaned toward him confidentially and lowered her voice, "In the meeting this morning, Tom, I didn't really feel I could say anything. But I hope you're okay. About this new reorganization."

Sanders concealed his surprise. In twelve years, Kaplan had never said anything so directly personal to him. He wondered why she would do so now. He was instantly wary, unsure of how to respond.

"Well, it was a shock," he said.

She looked at him with a steady gaze. "It was a shock to many of us," she said quietly. "There was an uproar in Cupertino. A lot of people questioned Garvin's judgment."

Sanders frowned. Kaplan never said anything even obliquely critical of Garvin. Never. But now this. Was she testing him? He said nothing, and poked at his food.

"I can imagine you're uneasy about the new appointment."

"Only because it was so unexpected. It seemed to come out of the blue."

Kaplan looked at him oddly for a moment, as if he had disappointed her. Then she nodded. "It's always that way with mergers," she said. Her tone was more open, less confidential. "I was at CompuSoft when it merged with Symantec, and it was exactly the same: last-minute announcements, switches in the organization charts. Jobs promised, jobs lost. Everybody up in the air for weeks. It's not easy to bring two organizations together—especially these two. There are big differences in corporate cultures. Garvin has to make them comfortable." She gestured toward the end of the table where Garvin was sitting. "Just look at them," she said. "All the Conley people are wearing suits. Nobody in our company wears suits, except lawyers."

"They're East Coast," Sanders said.

"But it goes deeper than that. Conley-White likes to present itself as a diversified communications company, but it's really not so grand. Its primary business is textbooks. That's a lucrative business, but you're selling to school boards in Texas and Ohio and Tennessee. Many of them are deeply conservative. So Conley's conservative, by instinct and experience. They want this merger because they need to acquire a high-tech capacity going into the next century. But they can't get used to the idea of a very young company, where the employees work in T-shirts and jeans, and everybody goes by first names. They're in shock. Besides," Kaplan added, lowering her voice again, "there are internal divisions within Conley-White. Garvin has to deal with that, too."

"What internal divisions?"

She nodded toward the head of the table. "You may have noticed that their CEO isn't here. The big man hasn't honored us with his presence. He won't show up until the end of the week. For now, he's only sent his minions. Their highest-ranking officer is Ed Nichols, the CFO."

Sanders glanced over at the suspicious, sharp-faced man he had met earlier. Kaplan said, "Nichols doesn't want to buy this company. He thinks we're overpriced and underpowered. Last year, he tried to form a strategic alliance with Microsoft, but Gates blew him off. Then Nichols tried to buy InterDisk, but that fell through: too many problems, and InterDisk had that bad publicity about the fired employee. So they ended up with us. But Ed isn't happy about where he landed."

"He certainly doesn't look happy," Sanders said.

"The main reason is he hates the Conley kid."

Seated beside Nichols was John Conley, the bespectacled young lawyer in his twenties. Distinctly younger than anyone around him, Conley was speaking energetically, jabbing his fork in the air as he made a point to Nichols.

"Ed Nichols thinks Conley's an asshole."

"But Conley's only a vice president," Sanders said. "He can't have that much power."

Kaplan shook her head. "He's the heir, remember?"

"So? What does that mean? His grandfather's picture is on some boardroom wall?"

"Conley owns four percent of C-W stock, and controls another twenty-six percent still held by the family or vested in trusts controlled by the family. John Conley has the largest voting block of Conley-White stock."

"And John Conley wants the deal?"

"Yes." Kaplan nodded. "Conley handpicked our company to acquire. And he's going forward fast, with the help of his friends like Jim Daly at Goldman, Sachs. Daly's very smart, but investment bankers always have big fees riding on a merger. They'll do their due diligence, I'm not saying they won't. But it'd take a lot to get them to back out of the deal now."

"Uh-huh."

"So Nichols feels he's lost control of the acquisition, and he's being rushed into a deal that's a lot richer than it should be. Nichols doesn't see why C-W should make us all wealthy. He'd pull out of this deal if he could—if only to screw Conley."

"But Conley's driving this deal."

"Yes. And Conley's abrasive. He likes to make little speeches about youth versus age, the coming digital era, a young vision for the future. It enrages Nichols. Ed Nichols feels he's doubled the net worth of the company in a decade, and now this little twerp is giving him lectures."

"And how does Meredith fit in?"

Kaplan hesitated. "Meredith is suitable."

"Meaning what?"

"She's Eastern. She grew up in Connecticut and went to Vassar. The Conley people like that. They're comfortable with that."

"That's all? She has the right accent?"

"You didn't hear it from me," Kaplan said. "But I think they also see her as weak. They think they can control her once the merger is completed."

"And Garvin's going along with that?"

Kaplan shrugged. "Bob's a realist," she said. "He needs capitalization. He's built his company skillfully, but we're going to require massive infusions of cash for the next phase, when we go head-to-head with Sony and Philips in product development. Conley-White's textbook operation is a cash cow. Bob looks at them and sees green—and he's inclined to do what they want, to get their money."

"And of course, Bob likes Meredith."

"Yes. That's true. Bob likes her."

Sanders waited while she poked at her food for a while. "And you, Stephanie? What do you think?"

Kaplan shrugged. "She's able."

"Able but weak?"

"No." Kaplan shook her head. "Meredith has ability. That's not in question. But I'm concerned about her experience. She's not as seasoned as she might be. She's being put in charge of four major technical units that are expected to grow rapidly. I just hope she's up to it."

There was the clink of a spoon on a glass, and Garvin stepped to the front of the room. "Even though you're still eating dessert, let's get started, so we can finish by two o'clock," he said. "Let me remind you of the new timetable. Assuming everything continues as planned, we expect to make the formal announcement of the acquisition at a press conference here on Friday noon. And now, let me introduce our new associates from Conley-White . . ."

As Garvin named the C-W people, and they stood up around the table, Kaplan leaned over and whispered to Tom, "This is all fluff and feathers. The real reason for this lunch is you-know-who."

". . . and finally," Garvin said, "let me introduce someone that many of you know, but some of you do not, the new Vice President for Advanced Operations and Planning, Meredith Johnson."

There was scattered, brief applause as Johnson got up from her seat and walked to a podium at the front of the room. In her dark blue suit, she looked the model of corporate correctness, but she was strikingly beautiful. At the podium, she put on horn-rimmed glasses and lowered the conference room lights.

"Bob has asked me to review the way the new structure will work," she said, "and to say something about what we see happening in the coming months." She bent over the podium, where a computer was set up for presentations. "Now, if I can just work this thing . . . let me see . . ."

In the darkened room, Don Cherry caught Sanders's eye and shook his head slowly.

"Ah, okay, here we are," Johnson said, at the podium. The screen behind her came to life. Animated images generated by the computer were projected onto the screen. The first image showed a red heart, which broke into four pieces. "The heart of DigiCom has always been its Advanced Products Group, which consists of four separate divisions as you see here. But as all information throughout the world becomes digital, these divisions will inevitably merge." On the screen, the pieces of the heart slid back together, and the heart transformed itself into a spinning globe. It began to throw off products. "For the customer in the near future, armed with cellular phone, built-in fax modem, and hand-held computer or PDA, it will be increasingly irrelevant where in the world he or she is and where the information is coming from. We are talking about the true globalization of information, and this implies an array of new products for our major markets in business and education." The globe expanded and dissolved, became classrooms on all continents, students at desks. "In particular, education will be a growing focus of this company as technology moves from print to digital displays to virtual environments. Now, let's review exactly what this means, and where I see it taking us."

And she proceeded to do it all—hypermedia, embedded video, authoring systems, work-group structures, academic sourcing, customer acceptance. She moved on to the cost structures—projected research outlays and revenues, five-year goals, offshore variables. Then to major product challenges—quality control, user feedback, shorter development cycles.

Meredith Johnson's presentation was flawless, the images blending and flowing across the screen, her voice confident, no hesitation, no pauses. As she continued, the room became quiet, the atmosphere distinctly respectful.

"Although this is not the time to go into technical matters," she said, "I want to mention that new CD-drive seek times under a hundred milliseconds, combined with new compression algorithms, should shift the industry standard for CD to full-res digitized video at sixty fields per second. And we are talking about platform-independent RISC processors supported by 32-bit color active-matrix displays and portable hard copy at 1200 DPI and wireless networking in both LAN and WAN configurations. Combine that with an autonomously generated virtual database—especially when ROM-based software agents for object definition and classification are in place—and I think we can agree we are looking at prospects for a very exciting future."

Sanders saw that Don Cherry's mouth was hanging open. Sanders leaned over to Kaplan. "Sounds like she knows her stuff."

"Yes," Kaplan said, nodding. "The demo queen. She started out doing demos. Appearance has always been her strongest point." Sanders glanced at Kaplan; she looked away.

But then the speech ended. There was applause as the lights came up, and Johnson went back to her seat. The room broke up, people heading back to work. Johnson left Garvin, and went directly to Don Cherry, said a few words to him. Cherry smiled: the charmed geek. Then Meredith went across the room to Mary Anne, spoke briefly to her, and then to Mark Lewyn.

"She's smart," Kaplan said, watching her, "touching base with all the division heads—especially since she didn't name them in her speech."

Sanders frowned. "You think that's significant?"

"Only if she's planning to make changes."

"Phil said she wasn't going to."

"But you never know, do you?" Kaplan said, standing up, dropping her napkin on the table. "I've got to go—and it looks like you're next on her list."

Kaplan moved discreetly away as Meredith came up to Sanders. She was smiling. "I wanted to apologize, Tom," Meredith said, "for not mentioning your name and the names of the other division heads in my presentation. I don't want anybody to get the wrong idea. It's just that Bob asked me to keep it short."

"Well," Sanders said, "it looks like you won everybody over. The reaction was very favorable."

"I hope so. Listen," she said, putting her hand on his arm, "we've got a slew of due diligence sessions tomorrow. I've been asking all the heads to meet with me today, if they can. I wonder if you're free to come to my office at the end of the day for a drink. We can go over things, and maybe catch up on old times, too."

"Sure," he said. He felt the warmth of her hand on his arm. She didn't take it away.

"They've given me an office on the fifth floor, and with any luck there should be furniture in by later today. Six o'clock work for you?"

"Fine," he said.

She smiled. "You still partial to dry chardonnay?"

Despite himself, he was flattered that she remembered. He smiled, "Yes, I still am."

"I'll see if I can get one. And we'll go over some of the immediate problems, like that hundred-millisecond drive."

"Okay, fine. About that drive—"

"I know," she said, her voice lower. "We'll deal with it." Behind her, the Conley-White executives were coming up. "Let's talk tonight."

"Good."

"See you then, Tom."

"See you then."

As the meeting broke up, Mark Lewyn drifted over to him. "So, let's hear it: what'd she say to you?"

"Meredith?"

"No, the Stealthy One. Kaplan was bending your ear all during lunch. What's up?"

Sanders shrugged. "Oh, you know. Just small talk."

"Come on. Stephanie doesn't do small talk. She doesn't know how. And Stephanie talked more to you than I've seen her talk in years."

Sanders was surprised to see how anxious Lewyn was. "Actually," he said, "we talked mostly about her son. He's a freshman at the university."

But Lewyn wasn't buying it. He frowned and said, "She's up to something, isn't she. She never talks without a reason. Is it about me? I know she's critical of the design team. She thinks we're wasteful. I've told her many times that it's not true—"

"Mark," Sanders said. "Your name didn't even come up. Honest."

To change the subject, Sanders asked, "What'd you think of Johnson? Pretty strong presentation, I thought."

"Yes. She's impressive. There was only one thing that bothered me," Lewyn said. He was still frowning, still uneasy. "Isn't she supposed to be a late-breaking curve, forced on us by management at Conley?"

"That's what I heard. Why?"

"Her presentation. To put together a graphic presentation like that takes two weeks, at a minimum," Lewyn said. "In my design group, I get the designers on it a month in advance, then we run it through for timing, then say a week for revisions and re-do's, then another week while they transfer to a drive. And that's my own in-house group, working fast. For an executive, it'd take longer. They pawn it off on some assistant, who tries to make it for them. Then the executive looks at it, wants it all done over again. And it takes more time. So if this was her presentation, I'd say she's known about her new job for a while. Months."

Sanders frowned.

"As usual," Lewyn said, "the poor bastards in the trenches are the last to know. I just wonder what *else* we don't know."

Sanders was back at his office by 2:15. He called his wife to tell her he would be home late, that he had a meeting at six.

"What's happening over there?" Susan said. "I got a call from Adele Lewyn. She says Garvin's screwing everybody, and they're changing the organization around."

"I don't know yet," he said cautiously. Cindy had just walked in the room.

"Are you still getting a promotion?"

"Basically," he said, "the answer is no."

"I can't believe it," Susan said. "Tom, I'm sorry. Are you okay? Are you upset?"

"I would say so, yes."

"Can't talk?"

"That's right."

"Okay. I'll leave soup on. I'll see you when you get here."

Cindy placed a stack of files on his desk. When Sanders hung up, she said, "She already knew?"

"She suspected."

Cindy nodded. "She called at lunchtime," she said. "I had the sense. The spouses are talking, I imagine."

"I'm sure everybody's talking."

Cindy went to the door, then paused. Cautiously, she said, "And how was the lunch meeting?"

"Meredith was introduced as the new head of all the tech divisions. She gave a presentation. She says she's going to keep all the division heads in place, all reporting to her."

"Then there's no change for us? Just another layer on top?"

"So far. That's what they're telling me. Why? What do you hear?"

"I hear the same."

He smiled. "Then it must be true."

"Should I go ahead and buy the condo?" She had been planning this for some time, a condo in Queen Anne's Hill for herself and her young daughter.

Sanders said, "When do you have to decide?"

"I have another fifteen days. End of the month."

"Then wait. You know, just to be safe."

She nodded, and went out. A moment later, she came back. "I almost forgot. Mark Lewyn's office just called. The Twinkle drives have arrived from KL. His designers are looking at them now. Do you want to see them?"

"I'm on my way."

The Design Group occupied the entire second floor of the Western Building. As always, the atmosphere there was chaotic; all the phones were ringing, but there was no receptionist in the little waiting area by the elevators, which was decorated with faded, taped-up posters for a 1929 Bauhaus Exhibition in Berlin and an old science-fiction movie called *The Forbin Project*. Two Japanese visitors sat at a corner table, speaking rapidly, beside the battered Coke machine and the junk food dispenser. Sanders nodded to them, used his card to open the locked door, and went inside.

The floor was a large open space, partitioned at unexpected angles by slanted walls painted to look like pastel-veined stone. Uncomfortable-looking wire chairs and tables were scattered in odd places. Rock-and-roll music blared. Everybody was casually dressed; most of the designers wore shorts and T-shirts. It was clearly A Creative Area.

Sanders went through to Foamland, the little display of the latest product designs the group had made. There were models of tiny CD-ROM drives and miniature cellular phones. Lewyn's teams were charged with creating product designs for the future, and many of these seemed absurdly small: a cellular phone no larger than a pencil, and another that looked like a postmodern version of Dick Tracy's wrist radio, in pale green and gray; a pager the size of a cigarette lighter; and a micro-CD player with a flip-up screen that could fit easily in the palm of the hand.

Although these devices looked outrageously tiny, Sanders had long since become accustomed to the idea that the designs were at most two years in the future. The hardware was shrinking fast; it was difficult for Sanders to remember that when he began working at DigiCom, a "portable" computer was a thirty-pound box the size of a carry-on suitcase—and cellular telephones didn't exist at all. The first cellular phones that DigiCom manufactured were fifteen-pound wonders that you lugged around on a shoulder strap. At the time, people thought they were a miracle. Now, customers complained if their phones weighed more than a few ounces.

Sanders walked past the big foam-cutting machine, all twisted tubes and knives behind Plexiglas shields, and found Mark Lewyn and his team bent over three dark blue CD-ROM players from Malaysia. One of the players already lay in pieces on the table; under bright halogen lights, the team was poking at its innards with tiny screwdrivers, glancing up from time to time to the scope screens.

"What've you found?" Sanders said.

"Ah, hell," Lewyn said, throwing up his hands in artistic exasperation. "Not good, Tom. Not good."

"Talk to me."

Lewyn pointed to the table. "There's a metal rod inside the hinge. These clips maintain contact with the rod as the case is opened; that's how you maintain power to the screen."

"Yes . . ."

"But power is intermittent. It looks like the rods are too small. They're supposed to be fifty-four millimeters. These seem to be fifty-two, fifty-three millimeters."

Lewyn was grim, his entire manner suggesting unspeakable consequences. The bars were a millimeter off, and the world was coming to an end. Sanders understood that he would have to calm Lewyn down. He'd done it many times before.

He said, "We can fix that, Mark. It'll mean opening all the cases and replacing the bars, but we can do that."

"Oh sure," Lewyn said. "But that still leaves the clips. Our specs call for 16/10 stainless, which has requisite tension to keep the clips springy and maintain contact with the bar. These clips seem to be something else, maybe 16/4. They're too stiff. So when you open the cases the clips bend, but they don't spring back."

"So we have to replace the clips, too. We can do that when we switch the bars."

"Unfortunately, it's not that easy. The clips are heat-pressed into the cases."

"Ah, hell."

"Right. They are integral to the case unit."

"You're telling me we have to build new housings just because we have bad clips?"

"Exactly."

Sanders shook his head. "We've run off thousands so far. Something like four thousand."

"Well, we've got to do 'em again."

"And what about the drive itself?"

"It's slow," Lewyn said. "No doubt about it. But I'm not sure why. It might be power problems. Or it might be the controller chip."

"If it's the controller chip . . ."

"We're in deep shit. If it's a primary design problem, we have to go back to the drawing board. If it's only a fabrication problem, we have to change the production lines, maybe remake the stencils. But it's months, either way."

"When will we know?"

"I've sent a drive and power supply to the Diagnostics guys," Lewyn said. "They should have a report by five. I'll get it to you. Does Meredith know about this yet?"

"I'm briefing her at six."

"Okay. Call me after you talk to her?"

"Sure."

"In a way, this is good," Lewyn said.

"How do you mean?"

"We're throwing her a big problem right away," Lewyn said. "We'll see how she handles it."

Sanders turned to go. Lewyn followed him out. "By the way," Lewyn said. "*Are* you pissed off that you didn't get the job?"

"Disappointed," Sanders said. "Not pissed. There's no point being pissed."

"Because if you ask me, Garvin screwed you. You put in the time, you've demonstrated you can run the division, and he put in someone else instead."

Sanders shrugged. "It's his company."

Lewyn threw his arm over Sanders's shoulder, and gave him a rough hug. "You know, Tom, sometimes you're too reasonable for your own good."

"I didn't know being reasonable was a defect," Sanders said.

"Being *too* reasonable is a defect," Lewyn said. "You end up getting pushed around."

"I'm just trying to get along," Sanders said. "I want to be here when the division goes public."

"Yeah, true. You got to stay." They came to the elevator. Lewyn said, "You think she got it because she's a woman?"

Sanders shook his head. "Who knows."

"Pale males eat it again. I tell you. Sometimes I get so sick of the constant pressure to appoint women," Lewyn said. "I mean, look at this design group. We've got forty percent women here, better than any other division, but they always say, why don't you have more. More women, more—"

"Mark," he said, interrupting. "It's a different world now."

"And not a better one," Lewyn said. "It's hurting everybody. Look: when I started in DigiCom, there was only one question. Are you good? If you were good, you got hired. If you could cut it, you stayed. No more. Now, ability is only one of the priorities. There's also the question of whether you're the right sex and skin color to fill out the company's HR profiles. And if you turn out to be incompetent, we can't fire you. Pretty soon, we start to get junk like this Twinkle drive. Because no one's accountable anymore. No one is responsible. You can't build products on a *theory*. Because the product you're making is real. And if it stinks, it stinks. And no one will buy it."

Coming back to his office, Sanders used his electronic passcard to open the door to the fourth floor. Then he slipped the card in his trouser pocket, and headed down the hallway. He was moving quickly, thinking about the meeting with Lewyn. He was especially bothered by one thing that Lewyn had said: that he was allowing himself to be pushed around by Garvin—that he was being too passive, too understanding.

But Sanders didn't see it that way. When Sanders had said it was Garvin's company, he meant it. Bob was the boss, and Bob could do what he wanted. Sanders was disappointed not to get the job, but no one had promised it to him. Ever. He and others in the Seattle divisions had come, over a period of weeks, to assume that Sanders would get the job. But Garvin had never mentioned it. Nor had Phil Blackburn.

As a result, Sanders felt he had no reason to gripe. If he was disappointed, it was only because he had done it to himself. It was classic: counting your chickens before they hatched.

And as for being too passive—what did Lewyn expect him to do? Make a fuss? Yell and scream? That wouldn't do any good. Because clearly Meredith Johnson had this job, whether Sanders liked it or not. Resign? That *really* wouldn't do any good. Because if he quit, he would lose the profits pending when the company went public. That would be a real disaster.

So on reflection, all he could do was accept Meredith Johnson in the new job, and get on with it. And he suspected that if the situation were reversed, Lewyn, for all his bluster, would do exactly the same thing: grin and bear it.

But the bigger problem, as he thought it over, was the Twinkle drive. Lewyn's team had torn up three drives that afternoon, and they still didn't have any idea why they were malfunctioning. They had found some non-spec components in the hinge, which Sanders could track down. He'd find out soon enough why they were getting non-spec materials. But the real problem—the slowness of the drives—remained a mystery to which they had no clue, and that meant that he was going to—

"Tom? You dropped your card."

"What?" He looked up absently. An area assistant was frowning, pointing back down the hall.

"You dropped your card."

"Oh." He saw the passcard lying there, white against the gray carpet. "Thanks."

He went back to retrieve it. Obviously, he must be more upset than he realized. You couldn't get anywhere in the DigiCom buildings without a passcard. Sanders bent over, picked it up, and slipped it in his pocket.

Then he felt the second card, already there. Frowning, he took both cards out and looked at them.

The card on the floor wasn't his card, it was someone else's. He paused for a moment, trying to decide which was his. By design, the passcards were featureless: just the blue DigiCom logo, a stamped serial number, and a magstripe on the back.

He ought to be able to remember his card number, but he couldn't. He hurried back to his office, to look it up on his computer. He glanced at his watch. It was four o'clock, two hours before his meeting with Meredith Johnson. He still had a lot to do to prepare for that meeting. He frowned as he walked along, staring at the carpet. He would have to get the production reports, and perhaps also the design detail specs. He wasn't sure she would understand them, but he should be prepared with them, anyway. And what else? He did not want to go into this first meeting having forgotten something.

Once again, his thoughts were disrupted by images from his past. An opened suitcase. The bowl of popcorn. The stained-glass window.

"So?" said a familiar voice. "You don't say hello to your old friends anymore?"

Sanders looked up. He was outside the glass-walled conference room. Inside the room, he saw a solitary figure hunched over in a wheelchair, staring at the Seattle skyline, his back to Sanders.

"Hello, Max," Sanders said.

Max Dorfman continued to stare out the window. "Hello, Thomas."

"How did you know it was me?"

Dorfman snorted. "It must be magic. What do you think? Magic?" His voice was sarcastic. "Thomas: I can *see* you."

"How? You have eyes in the back of your head?"

"No, Thomas. I have a reflection in *front* of my head. I see you in the glass, of course. Walking with your head down, like a defeated *putz*." Dorfman snorted again, and then wheeled his chair around. His eyes were bright, intense, mocking. "You were such a promising man. And now you are hanging your head?"

Sanders wasn't in the mood. "Let's just say this hasn't been one of my better days, Max."

"And you want everybody to know about it? You want sympathy?"

"No, Max." He remembered how Dorfman had ridiculed the idea of sympathy. Dorfman used to say that an executive who wanted sympathy was not an executive. He was a sponge, soaking up something useless.

Sanders said, "No, Max. I was thinking."

"Ah. *Thinking*. Oh, I like thinking. Thinking is good. And what were you thinking about, Thomas: the stained glass in your apartment?"

Despite himself, Sanders was startled: "How did you know that?"

"Maybe it's magic," Dorfman said, with a rasping laugh. "Or perhaps I can read minds. You think I can read minds, Thomas? Are you stupid enough to believe that?"

"Max, I'm not in the mood."

"Oh well, then I must stop. If you're not in the mood, I must stop. We must at all costs preserve your mood." He slapped the arm of his wheelchair irritably. "*You told me, Thomas*. That's how I knew what you were thinking."

"I told you? When?"

"Nine or ten years ago, it must have been."

"What did I tell you?"

"Oh, you don't remember? No wonder you have problems. Better stare at the floor some more. It may do you good. Yes. I think so. Keep staring at the floor, Thomas."

"Max, for Christ's sake."

Dorfman grinned at him. "Do I irritate you?"

"You always irritate me."

"Ah. Well. Then perhaps there is hope. Not for you, of course—for me. I am old, Thomas. Hope has a different meaning, at my age. You wouldn't understand. These days, I cannot even get around by myself. I must have someone *push* me. Preferably a pretty woman, but as a rule they do not like to do such things. So here I am, with no pretty woman to push me. *Unlike you.*"

Sanders sighed. "Max, do you suppose we can just have an ordinary conversation?"

"What a good idea," Dorfman said. "I would like that very much. What is an ordinary conversation?"

"I mean, can we just talk like normal people?"

"If it will not bore you, Thomas, yes. But I am worried. You know how old people are worried about being boring."

"Max. What did you mean about the stained glass?"

He shrugged. "I meant Meredith, of course. What else?"

"What about Meredith?"

"How am I to know?" Dorfman said irritably. "All I know of this is what you told me. And all you told me is that you used to take trips, to Korea or Japan, and when you came back, Meredith would—"

"Tom, I'm sorry to interrupt," Cindy said, leaning in the door to the conference room.

"Oh, don't be sorry," Max said. "Who is this beautiful creature, Thomas?"

"I'm Cindy Wolfe, Professor Dorfman," she said. "I work for Tom."

"Oh, what a lucky man he is!"

Cindy turned to Sanders. "I'm really sorry, Tom, but one of the executives from Conley-White is in your office, and I thought you would want to—"

"Yes, yes," Dorfman said immediately. "He must go. Conley-White, it sounds *very* important."

"In a minute," Sanders said. He turned to Cindy. "Max and I were in the middle of something."

"No, no, Thomas," Dorfman said. "We were just talking about old times. You better go."

"Max—"

"You want to talk more, you think it's important, you come visit me. I am at the Four Seasons. You know that hotel. It has a *wonderful* lobby, such high ceilings. Very grand, especially for an old man. So, you go right along, Thomas." His eyes narrowed. "And leave the beautiful Cindy with me."

Sanders hesitated. "Watch out for him," he said. "He's a dirty old man."

"As dirty as possible," Dorfman cackled.

Sanders headed down the hallway to his office. As he left, he heard Dorfman say, "Now beautiful Cindy, please take me to the lobby where I have a car waiting. And on the way, if you don't mind indulging an old man, I have a few little questions. So many *interesting* things are happening in this company. And the secretaries always know everything, don't they?"

Mr. Sanders." Jim Daly stood quickly, as Sanders came into the room. "I'm glad they found you."

They shook hands. Sanders gestured for Daly to sit down, and slid behind his own desk. Sanders was not surprised; he had been expecting a visit from Daly or one of the other investment bankers for several days. Members of the Goldman, Sachs team had been speaking individually with people in various departments, going over aspects of the merger. Most of the time they wanted background information; although high technology was central to the acquisition, none of the bankers understood it very well. Sanders expected Daly to ask about progress on the Twinkle drive, and perhaps the Corridor.

"I appreciate your taking the time," Daly said, rubbing his bald head. He was a very tall, thin man, and he seemed even taller sitting down, all knees and elbows. "I wanted to ask you some things, ah, off the record."

"Sure," Sanders said.

"It's to do with Meredith Johnson," Daly said, in an apologetic voice. "If you, ah, don't mind, I'd prefer we just keep this conversation between us."

"All right," Sanders said.

"I understand that you have been closely involved with setting up the plants in Ireland and Malaysia. And that there has been a little controversy inside the company about how that was carried out."

"Well." Sanders shrugged. "Phil Blackburn and I haven't always seen eye to eye."

"Showing your good sense, in my view," Daly said dryly. "But I gather that in these disputes you represent technical expertise, and others in the company represent, ah, various other concerns. Would that be fair?"

"Yes, I'd say so." What was he getting at?

"Well, it's along those lines that I'd like to hear your thoughts. Bob

Garvin has just appointed Ms. Johnson to a position of considerable authority, a step which many in Conley-White applaud. And certainly it would be unfair to prejudge how she will carry out her new duties within the company. But by the same token, it would be derelict of me not to inquire about her past duties. Do you get my drift?"

"Not exactly," Sanders said.

"I'm wondering," Daly said, "what you feel about Ms. Johnson's past performance with regard to the technical operations of the company. Specifically, her involvement in the foreign operations of DigiCom."

Sanders frowned, thinking back. "I'm not aware that she's had much involvement," he said. "We had a labor dispute two years ago in Cork. She was part of the team that went over to negotiate a settlement. She lobbied in Washington about flat-panel display tariffs. And I know she headed the Ops Review Team in Cupertino, which approved the plans for the new plant at Kuala Lumpur."

"Yes, exactly."

"But I don't know that her involvement goes beyond that."

"Ah. Well. Perhaps I was given wrong information," Daly said, shifting in his chair.

"What did you hear?"

"Without going into specifics, let me say a question of judgment was raised."

"I see," Sanders said. Who would have said anything to Daly about Meredith? Certainly not Garvin or Blackburn. Kaplan? It was impossible to know for sure. But Daly would be talking only to highly placed officers.

"I was wondering," Daly said, "if you had any thoughts on her technical judgment. Speaking privately, of course."

At that moment, Sanders's computer screen beeped three times. A message flashed:

ONE MINUTE TO DIRECT VIDEO LINKUP: DC/M-DC/S
SEN: A. KAHN
REC: T. SANDERS

Daly said, "Is something wrong?"

"No," Sanders said. "It looks like I have a video feed coming in from Malaysia."

"Then I'll be brief and leave you to it," Daly said. "Let me put it to you directly. Within your division, is there any concern whether Meredith Johnson is qualified for this post?"

Sanders shrugged. "She's the new boss. You know how organizations are. There's always concern with a new boss."

"You're very diplomatic. I mean to say, is there concern about her expertise? She's relatively young, after all. Geographic move, uprooting. New faces, new staffing, new problems. And up here, she won't be so directly under Bob Garvin's, ah, wing."

"I don't know what to say," Sanders said. "We'll all have to wait and see."

"And I gather that there was trouble in the past when a non-technical person headed the division . . . a man named, ah, Screamer Freeling?"

"Yes. He didn't work out."

"And there are similar concerns about Johnson?"

Sanders said, "I've heard them expressed."

"And her fiscal measures? These cost-containment plans of hers? That's the crux, isn't it?"

Sanders thought: what cost-containment plans?

The screen beeped again.

30 SECONDS TO DIRECT VIDEO LINKUP: DC/M-DC/S

"There goes your machine again," Daly said, unfolding himself from the chair. "I'll let you go. Thank you for your time, Mr. Sanders."

"Not at all."

They shook hands. Daly turned and walked out of the room. Sanders's computer beeped three times in rapid succession:

15 SECONDS TO DIRECT VIDEO LINKUP: DC/M-DC/S

He sat down in front of the monitor and twisted his desk lamp so that the light shone on his face. The numbers on the computer were counting backward. Sanders looked at his watch. It was five o'clock—eight o'clock in Malaysia. Arthur would probably be calling from the plant.

A small rectangle appeared in the center of the screen and grew outward in progressive jumps. He saw Arthur's face, and behind him, the brightly lit assembly line. Brand-new, it was the epitome of modern manufacturing: clean and quiet, the workers in street clothes, arranged on both sides of the green conveyor belt. At each workstation there was a bank of fluorescent lights, which flared a little in the camera.

Kahn coughed and rubbed his chin. "Hello, Tom. How are you?" When he spoke, his image blurred slightly. And his voice was out of sync, since the bounce to the satellite caused a slight delay in the video, but the voice was transmitted immediately. This unsynchronized quality was very distracting for the first few seconds; it gave the linkup a dreamy quality. It was a little like talking to someone under water. Then you got used to it.

"I'm fine, Arthur," he said.

"Well, good. I'm sorry about the new organization. You know how I feel personally."

"Thank you, Arthur." He wondered vaguely how Kahn in Malaysia would have heard already. But in any company, gossip traveled fast.

"Yeah. Well. Anyway, Tom, I'm standing here on the floor," Kahn said, gesturing behind him. "And as you can see, we're still running very slow. And the spot checks are unimproved. What do the designers say? Have they gotten the units yet?"

"They came today. I don't have any news yet. They're still working on it."

"Uh-huh. Okay. And have the units gone to Diagnostics?" Kahn asked.

"I think so. Just went."

"Yeah. Okay. Because we got a request from Diagnostics for ten more drive units to be sent in heat-sealed plastic bags. And they specified that they wanted them sealed inside the factory. Right as they came off the line. You know anything about that?"

"No, this is the first I heard of it. Let me find out, and I'll get back to you."

"Okay, because I have to tell you, it seemed strange to me. I mean, ten units is a lot. Customs is going to query it if we send them all together. And I don't know what this sealing is about. We send them wrapped in plastic anyway. But not sealed. Why do they want them sealed, Tom?" Kahn sounded worried.

"I don't know," Sanders said. "I'll get into it. All I can think is that it's a full-court press around here. People really want to know why the hell those drives don't work."

"Hey, us too," Kahn said. "Believe me. It's making us crazy."

"When will you send the drives?"

"Well, I've got to get a heat-sealer first. I hope I can ship Wednesday, you can have them Thursday."

"Not good enough," Sanders said. "You should ship today, or tomorrow at the latest. You want me to run down a sealer for you? I can probably get one from Apple." Apple had a factory in Kuala Lumpur.

"No. That's a good idea. I'll call over there and see if Ron can loan me one."

"Fine. Now what about Jafar?"

"Hell of a thing," Kahn said. "I just talked to the hospital, and apparently he's got cramps and vomiting. Won't eat anything. The abo doctors say they can't figure out anything except, you know, a spell."

"They believe in spells?"

"Damn right," Kahn said. "They've got laws against sorcery here. You can take people to court."

"So you don't know when he'll be back?"

"Nobody's saying. Apparently he's really sick."

"Okay, Arthur. Anything else?"

"No. I'll get the sealer. And let me know what you find out."

"I will," Sanders said, and the transmission ended. Kahn gave a final wave, and the screen went blank.

SAVE THIS TRANSMISSION TO DISK OR DAT?

He clicked DAT, and it was saved to digital tape. He got up from the desk. Whatever all this was about, he'd better be informed before he had his meeting with Johnson at six. He went to the outer area, to Cindy's desk.

Cindy was turned away, laughing on the phone. She looked back and saw Sanders, and stopped laughing. "Listen, I got to go."

Sanders said, "Would you mind pulling the production reports on Twinkle for the last two months? Better yet, just pull everything since they opened the line."

"Sure."

"And call Don Cherry for me. I need to know what his Diagnostics group is doing with the drives."

He went back into his office. He noticed his e-mail cursor was blinking, and pushed the key to read them. While he waited, he looked at the three faxes on his desk. Two were from Ireland, routine weekly production reports. The third was a requisition for a roof repair at the Austin plant; it had been held up in Operations in Cupertino, and Eddie had forwarded it to Sanders to try and get action.

The screen blinked. He looked up at the first of his e-mail messages.

OUT OF NOWHERE WE GOT A BEAN COUNTER FROM OPERA-
TIONS DOWN HERE IN AUSTIN. HE'S GOING OVER ALL THE
BOOKS, DRIVING PEOPLE MAD. AND THE WORD IS WE GOT
MORE COMING DOWN TOMORROW. WHAT GIVES? THE RU-
MORS ARE FLYING, AND SLOWING HELL OUT OF THE LINE.
TELL ME WHAT TO SAY. IS THIS COMPANY FOR SALE OR
NOT?

EDDIE

Sanders did not hesitate. He couldn't tell Eddie what was going on. Quickly, he typed his reply:

THE BEAN COUNTERS WERE IN IRELAND LAST WEEK, TOO.
GARVIN'S ORDERED A COMPANY-WIDE REVIEW, AND
THEY'RE LOOKING AT EVERYTHING. TELL EVERYBODY
DOWN THERE TO FORGET IT AND GO BACK TO WORK.

TOM

He pushed the SEND button. The message disappeared.

"You called?" Don Cherry walked into the room without knocking, and dropped into the chair. He put his hands behind his head. "Jesus, what a day. I've been putting out fires all afternoon."

"Tell me."

"I got some dweebs from Conley down there, asking my guys what the difference is between RAM and ROM. Like they have time for this. Pretty soon, one of the dweebs hears 'flash memory' and he goes, 'How often does it flash?' Like it was a flashlight or something. And my guys have to put up with this. I mean, this is high-priced talent. They shouldn't be doing remedial classes for lawyers. Can't you stop it?"

"Nobody can stop it," Sanders said.

"Maybe Meredith can stop it," Cherry said, grinning.

Sanders shrugged. "She's the boss."

"Yeah. So—what's on your mind?"

"Your Diagnostics group is working on the Twinkle drives."

"True. That is, we're working on the bits and pieces that're left after Lewyn's nimble-fingered *artistes* tore the hell out of them. Why did they go to design first? Never, *ever*, let a designer near an actual piece of electronic equipment, Tom. Designers should only be allowed to draw pictures on pieces of paper. And only give them one piece of paper at a time."

"What have you found?" Sanders said. "About the drives."

"Nothing yet," Cherry said. "But we got a few ideas we're kicking around."

"Is that why you asked Arthur Kahn to send you ten drives, heat-sealed from the factory?"

"You bet your ass."

"Kahn was wondering about that."

"So?" Cherry said. "Let him wonder. It'll do him good. Keep him from playing with himself."

"I'd like to know, too."

"Well look," Cherry said. "Maybe our ideas won't amount to anything. At the moment, all we have is one suspicious chip. That's all Lewyn's clowns left us. It's not very much to go on."

"The chip is bad?"

"No, the chip is fine."

"What's suspicious about it?"

"Look," Cherry said. "We've got enough rumors flying around as it is. I can report that we're working on it, and we don't know yet. That's all. We'll get the sealed drives tomorrow or Wednesday, and we should know within an hour. Okay?"

"You thinking big problem, or little problem? I've got to know," Sanders said. "It's going to come up in the meetings tomorrow."

"Well, at the moment, the answer is we don't know. It could be anything. We're working on it."

"Arthur thinks it might be serious."

"Arthur might be right. But we'll solve it. That's all I can tell you."

"Don . . ."

"I understand you want an answer," Cherry said. "Do you understand that I don't have one?"

Sanders stared at him. "You could have called. Why'd you come up in person?"

"Since you asked," Cherry said, "I've got a small problem. It's delicate. Sexual harassment thing."

"*Another* one? It seems like that's all we have around here."

"Us and everybody else," Cherry said. "I hear UniCom's got fourteen suits going right now. Digital Graphics has even more. And MicroSym, look out. They're all pigs over there, anyway. But I'd like your read on this."

Sanders sighed. "Okay."

"In one of my programming groups, the remote DB access group. The group's all pretty old: twenty-five to twenty-nine years old. The supervisor for the fax modem team, a woman, has been asking one of the guys out. She thinks he's cute. He keeps turning her down. Today she asks him again in the parking lot at lunch; he says no. She gets in her car, rams his car, drives off. Nobody hurt, and he doesn't want to make a complaint. But he's worried, thinks it's a little out of hand. Comes to me for advice. What should I do?"

Sanders frowned. "You think that's the whole story? She's just mad at him because he turned her down? Or did he do something to provoke this?"

"He says no. He's a pretty straight guy. A little geeky, not real sophisticated."

"And the woman?"

"She's got a temper, no question. She blows at the team sometimes. I've had to talk to her about that."

"What does she say about the incident in the parking lot?"

"Don't know. The guy's asked me not to talk to her. Says he's embarrassed and doesn't want to make it worse."

Sanders shrugged. "What can you do? People are upset but nobody will talk . . . I don't know, Don. If a woman rammed his car, I'd guess he must have done something. Chances are he slept with her once, and won't see her again, and now she's pissed. That's my guess."

"That would be my guess, too," Cherry said, "but of course, maybe not."

"Damage to the car?"

"Nothing serious. Broken taillight. He just doesn't want it to get any worse. So, do I drop it?"

"If he won't file charges, I'd drop it."

"Do I speak to her informally?"

"I wouldn't. You go accusing her of impropriety—even informally—and you're asking for trouble. Nobody's going to support you. Because the chances are, your guy *did* do something to provoke her."

"Even though he says he didn't."

Sanders sighed. "Listen, Don, they always say they didn't. I never heard of one who said, 'You know, I deserve this.' Never happens."

"So, drop it?"

"Put a note in the file that he told you the story, be sure you characterize the story as alleged, and forget it."

Cherry nodded, turned to leave. At the door, he stopped and looked back. "So tell me this. How come we're both so convinced this guy must have done something?"

"Just playing the odds," Sanders said. "Now fix that damned drive for me."

At six o'clock, he said good night to Cindy and took the Twinkle files up to Meredith's office on the fifth floor. The sun was still high in the sky, streaming through the windows. It seemed like late afternoon, not the end of the day.

Meredith had been given the big corner office, where Ron Goldman used to be. Meredith had a new assistant, too, a woman. Sanders guessed she had followed her boss up from Cupertino.

"I'm Tom Sanders," he said. "I have an appointment with Ms. Johnson."

"Betsy Ross, from Cupertino, Mr. Sanders," she said. She looked at him. "Don't say anything."

"Okay."

"Everybody says something. Something about the flag. I get really sick of it."

"Okay."

"My whole life."

"Okay. Fine."

"I'll tell Miss Johnson you're here."

T om." Meredith Johnson waved from behind her desk, her other hand holding the phone. "Come in, sit down."

Her office had a view north toward downtown Seattle: the Space Needle, the Arly towers, the SODO building. The city looked glorious in the afternoon sun.

"I'll just finish this up." She turned back to the phone. "Yes, Ed, I'm with Tom now, we'll go over all of that. Yes. He's brought the documentation with him."

Sanders held up the manila folder containing the drive data. She pointed to her briefcase, which was lying open on the corner of the desk, and gestured for him to put it inside.

She turned back to the phone. "Yes, Ed, I think the due diligence will go smoothly, and there certainly isn't any impulse to hold anything back . . . No, no . . . Well, we can do it first thing in the morning if you like."

Sanders put the folder in her briefcase.

Meredith was saying, "Right, Ed, right. Absolutely." She came toward Tom and sat with one hip on the edge of the desk, her navy blue skirt riding up her thigh. She wasn't wearing stockings. "Everybody agrees that this is important, Ed. Yes." She swung her foot, the high heel dangling from her toe. She smiled at Sanders. He felt uncomfortable, and moved back a little. "I promise you, Ed. Yes. Absolutely."

Meredith hung up the phone on the cradle behind her, leaning back across the desk, twisting her body, revealing her breasts beneath the silk blouse. "Well, that's done." She sat forward again, and sighed. "The Conley people heard there's trouble with Twinkle. That was Ed Nichols, flipping out. Actually, it's the third call I've had about Twinkle this afternoon. You'd think that was all there was to this company. How do you like the office?"

"Pretty good," he said. "Great view."

"Yes, the city's beautiful." She leaned on one arm and crossed her legs. She saw that he noticed, and said, "In the summer, I'd rather not wear stockings. I like the bare feeling. So much cooler on a hot day."

Sanders said, "From now to the end of summer, it will be pretty much this way."

"I have to tell you, I dread the weather," she said. "I mean, after California . . ." She uncrossed her legs again, and smiled. "But you like it here, don't you? You seem happy here."

"Yes." He shrugged. "You get used to rain." He pointed to her briefcase. "Do you want to go over the Twinkle stuff?"

"Absolutely," she said, sliding off the desk, coming close to him. She looked him directly in the eyes. "But I hope you don't mind if I impose on you first. Just a little?"

"Sure."

She stepped aside. "Pour the wine for us."

"Okay."

"See if it's chilled long enough." He went over to the bottle on the side table. "I remember you always liked it cold."

"That's true," he said, spinning the bottle in the ice. He didn't like it so cold anymore, but he did in those days.

"We had a lot of fun back then," she said.

"Yes," he said. "We did."

"I swear," she said. "Sometimes I think that back when we were both young and trying to make it, I think that was the best it ever was."

He hesitated, not sure how to answer her, what tone to take. He poured the wine.

"Yes," she said. "We had a good time. I think about it often."

Sanders thought: I never do.

She said, "What about you, Tom? Do you think about it?"

"Of course." He crossed the room carrying the glasses of wine to her, gave her one, clinked them. "Sure I do. All us married guys think of the old days. You know I'm married now."

"Yes," she said, nodding. "Very married, I hear. With how many kids? Three?"

"No, just two." He smiled. "Sometimes it seems like three."

"And your wife is an attorney?"

"Yes." He felt safer now. The talk of his wife and children made him feel safer somehow.

"I don't know how somebody can be married," Meredith said. "I tried it." She held up her hand. "Four more alimony payments to the son of a bitch and I'm free."

"Who did you marry?"

"Some account executive at CoStar. He was cute. Amusing. But it turned out he was a typical gold digger. I've been paying him off for three years. *And* he was a lousy lay." She waved her hand, dismissing the subject. She looked at her watch. "Now come and sit down, and tell me how bad it is with the Twinkle drive."

"You want the file? I put it in your briefcase."

"No." She patted the couch beside her. "You just tell me yourself."

He sat down beside her.

"You look good, Tom." She leaned back and kicked off her heels, wiggled her bare toes. "God, what a day."

"Lot of pressure?"

She sipped her wine and blew a strand of hair from her face. "A lot to keep track of. I'm glad we're working together, Tom. I feel as though you're the one friend I can count on in all this."

"Thanks. I'll try."

"So: how bad is it?"

"Well. It's hard to say."

"Just tell me."

He felt he had no choice but to lay it all out for her. "We've built very successful prototypes, but the drives coming off the line in KL are running nowhere near a hundred milliseconds."

Meredith sighed, and shook her head. "Do we know why?"

"Not yet. We're working on some ideas."

"That line's a start-up, isn't it?"

"Two months ago."

She shrugged. "Then we have problems on a new line. That's not so bad."

"But the thing is," he said, "Conley-White is buying this company for our technology, and especially for the CD-ROM drive. As of today, we may not be able to deliver as promised."

"You want to tell them *that?*"

"I'm concerned they'll pick it up in due diligence."

"Maybe, maybe not." She leaned back in the couch. "We have to remember what we're really looking at. Tom, we've all seen production problems loom large, only to vanish overnight. This may be one of those situations. We're shaking out the Twinkle line. We've identified some early problems. No big deal."

"Maybe. But we don't know that. In reality, there may be a problem with controller chips, which means changing our supplier in Singapore. Or there may be a more fundamental problem. A design problem, originating here."

"Perhaps," Meredith said, "but as you say, we don't know that. And I don't see any reason for us to speculate. At this critical time."

"But to be honest—"

"It's not a matter of honesty," she said. "It's a matter of the underlying reality. Let's go over it, point by point. We've told them we have a Twinkle drive."

"Yes."

"We've built a prototype and tested the hell out of it."

"Yes."

"And the prototype works like gangbusters. It's twice as fast as the most advanced drives coming out of Japan."

"Yes."

"We've told them we're in production on the drive."

"Yes."

"Well, then," Meredith said, "we've told them all that anybody knows for sure, at this point. I'd say we are acting in good faith."

"Well, maybe, but I don't know if we can—"

"Tom." Meredith placed her hand on his arm. "I always liked your directness. I want you to know how much I appreciate your expertise and your frank approach to problems. All the more reason why I'm sure the Twinkle drive will get ironed out. We know that fundamentally it's a good product that performs as we say it does. Personally, I have complete confidence in it, and in your ability to make it work as planned. And I have no problem saying that at the meeting tomorrow." She paused, and looked intently at him. "Do you?"

Her face was very close to him, her lips half-parted. "Do I what?"

"Have a problem saying that at the meeting?"

Her eyes were light blue, almost gray. He had forgotten that, as he had forgotten how long her lashes were. Her hair fell softly around her face. Her lips were full. She had a dreamy look in her eyes. "No," he said. "I don't have a problem."

"Good. Then at least *that's* settled." She smiled and held out her glass. "Do the honors again?"

"Sure."

He got up from the couch and went over to the wine. She watched him.

"I'm glad you haven't let yourself go, Tom. You work out?"

"Twice a week. How about you?"

"You always had a nice tush. Nice hard tush."

He turned. "Meredith . . ."

She giggled. "I'm sorry. I can't help it. We're old friends." She looked concerned. "I didn't offend you, did I?"

"No."

"I can't imagine you ever getting prudish, Tom."

"No, no."

"Not you." She laughed. "Remember the night we broke the bed?"

He poured the wine. "We didn't exactly break it."

"Sure we did. You had me bent over the bottom of the footboard and—"

"I remember—"

"And first we broke the footboard, and then the bottom of the bed crashed down—but you didn't want to stop so we moved up and then when I was grabbing the headboard it all came—"

"I remember," he said, wanting to interrupt her, to stop this. "Those days were great. Listen, Meredith—"

"And then the woman from downstairs called up? Remember her? The old Lithuanian lady? She vanted ta know if somebody had died or vhat?"

"Yeah. Listen. Going back to the drive . . ."

She took the wineglass. "I *am* making you uncomfortable. What—did you think I was coming on to you?"

"No, no. Nothing like that."

"Good, because I really wasn't. I promise." She gave him an amused glance, then tilted her head back, exposing her long neck, and sipped the wine. "In fact, I—ah! Ah!" She winced suddenly.

"What is it?" he said, leaning forward, concerned.

"My neck, it goes into spasm, it's right there . . ." With her eyes still squeezed shut in pain, she pointed to her shoulder, near the neck.

"What should I—"

"Just rub it, squeeze—there—"

He put down his wineglass and rubbed her shoulder. "There?"

"Yes, ah, harder—squeeze—"

He felt the muscles of her shoulder relax, and she sighed. Meredith turned her head back and forth slowly, then opened her eyes. "Oh . . . Much better . . . Don't stop rubbing."

He continued rubbing.

"Oh, thanks. That feels good. I get this nerve thing. Pinched something, but when it hits, it's really . . ." She turned her head back and forth. Testing. "You did that very well. But you were always good with your hands, Tom."

He kept rubbing. He wanted to stop. He felt everything was wrong, that he was sitting too close, that he didn't want to be touching her. But it also felt good to touch her. He was curious about it.

"Good hands," she said. "God, when I was married, I thought about you all the time."

"You did?"

"Sure," she said. "I told you, he was terrible in bed. I hate a man who doesn't know what he's doing." She closed her eyes. "That was never your problem, was it."

She sighed, relaxing more, and then she seemed to lean into him, melting toward his body, toward his hands. It was an unmistakable sensation. Immediately, he gave her shoulder a final friendly squeeze, and took his hands away.

She opened her eyes. She smiled knowingly. "Listen," she said, "don't worry."

He turned and sipped his wine. "I'm not worried."

"I mean, about the drive. If it turns out we really have problems and need agreement from higher management, we'll get it. But let's not jump the gun now."

"Okay, fine. I think that makes sense." He felt secretly relieved to be talking once again about the drive. Back on safe ground. "Who would you take it to? Directly to Garvin?"

"I think so. I prefer to deal informally." She looked at him. "You've changed, haven't you."

"No . . . I'm still the same."

"I think you've changed." She smiled. "You never would have stopped rubbing me before."

"Meredith," he said, "it's different. You run the division now. I work for you."

"Oh, don't be silly."

"It's true."

"We're colleagues." She pouted. "Nobody around here really believes I'm superior to you. They just gave me the administrative work, that's all. We're colleagues, Tom. And I just want us to have an open, friendly relationship."

"So do I."

"Good. I'm glad we agree on that." Quickly, she leaned forward and kissed him lightly on the lips. "There. Was that so terrible?"

"It wasn't terrible at all."

"Who knows? Maybe we'll have to go to Malaysia together, to check on the assembly lines. They have very nice beaches in Malaysia. You ever been to Kuantan?"

"No."

"You'd love it."

"I'm sure."

"I'll show it to you. We could take an extra day or two. Stop over. Get some sun."

"Meredith—"

"Nobody needs to know, Tom."

"I'm married."

"You're also a man."

"What does that mean?"

"Oh Tom," she said, with mock severity, "don't ask me to believe you never have a little adventure on the side. I *know* you, remember?"

"You knew me a long time ago, Meredith."

"People don't change. Not *that* way."

"Well, I think they do."

"Oh, come on. We're going to be working together, we might as well enjoy ourselves."

He didn't like the way any of this was going. He felt pushed into an awkward position. He felt stuffy and puritanical when he said: "I'm married now."

"Oh, I don't care about your personal life," she said lightly. "I'm only responsible for your on-the-job performance. All work and no play, Tom. It can be bad for you. Got to stay playful." She leaned forward. "Come on. Just one little kiss . . ."

The intercom buzzed. "Meredith," the assistant's voice said.

She looked up in annoyance. "I told you, no calls."

"I'm sorry. It's Mr. Garvin, Meredith."

"All right." She got off the couch and walked across the room to her desk, saying loudly, "But after this, Betsy, no more calls."

"All right, Meredith. I wanted to ask you, is it okay if I leave in about ten minutes? I have to see the landlord about my new apartment."

"Yes. Did you get me that package?"

"I have it right here."

"Bring it in, and then you can leave."

"Thank you, Meredith. Mr. Garvin is on two."

Meredith picked up the phone and poured more wine. "Bob," she said. "Hi. What's up?" It was impossible to miss the easy familiarity in her voice.

She spoke to Garvin, her back turned to Sanders. He sat on the couch, feeling stranded, foolishly passive and idle. The assistant entered the room carrying a small package in a brown paper bag. She gave the package to Meredith.

"Of course, Bob," Meredith was saying. "I couldn't agree more. We'll certainly deal with that."

The assistant, waiting for Meredith to dismiss her, smiled at Sanders. He felt uncomfortable just sitting there on the couch, so he got up, walked to the window, pulled his cellular phone out of his pocket, and dialed Mark Lewyn's number. He had promised to call Lewyn anyway.

Meredith was saying, "That's a very good thought, Bob. I think we should act on it."

Sanders heard his call dial, and then an answering machine picked up. A male voice said, "Leave your message at the beep." Then an electronic tone.

"Mark," he said, "it's Tom Sanders. I've talked about Twinkle with Meredith. Her view is that we're in early production and we are shaking out the lines. She takes the position that we can't say for sure that there are any significant problems to be flagged, and that we should treat the situation as standard procedure for the bankers and C-W people tomorrow . . ."

The assistant walked out of the room, smiling at Sanders as she passed him.

". . . and that if we have problems with the drive later on that we have to get management involved with, we'll face that later. I've given her your thoughts, and she's talking to Bob now, so presumably we'll go into the meeting tomorrow taking that position . . ."

The assistant came to the door to the office. She paused briefly to twist the lock in the doorknob, then left, closing the door behind her.

Sanders frowned. *She had locked the door on her way out.* It wasn't so much the fact that she had done it, but the fact that he seemed to be in the middle of an arrangement, a planned event in which everyone else understood what was going on and he did not.

". . . Well, anyway, Mark, if there is a significant change in all this, I'll contact you before the meeting tomorrow, and—"

"Forget that phone," Meredith said coming up suddenly, very close to him, pushing his hand down, and pressing her body against his. Her lips mashed against his mouth. He was vaguely aware of dropping the phone on the windowsill as they kissed and she twisted, turning away, and they tumbled over onto the couch.

"Meredith, wait—"

"Oh God, I've wanted you all day," she said intensely. She kissed him again, rolling on top of him, lifting one leg to hold him down. His position was awkward but he felt himself responding to her. His immediate thought was that someone might come in. He had a vision of himself, lying on his back on the couch with his boss half-straddling him in her businesslike navy suit, and he was anxious about what the person seeing them would think, and then he was truly responding.

She felt it too, and it aroused her more. She pulled back for a breath. "Oh God, you feel so *good*, I can't stand the bastard touching me. Those stupid glasses. Oh! I'm so *hot*, I haven't had a decent fuck—" and then she threw herself back on him, kissing him again, her mouth mashed on him. Her tongue was in his mouth and he thought, *Jesus, she's pushing it.* He smelled her perfume, and it immediately brought back memories.

She shifted her body so she could reach down and touch him, and she moaned when she felt him through his trousers. She fumbled at the zipper. He had suddenly conflicting images, his desire for her, his wife and his kids, memories of the past, of being with her in the apartment in Sunnyvale, of breaking the bed. Images of his wife.

"Meredith—"

"*Oooh.* Don't talk. No! No . . ." She was gasping in little breaths, her mouth puckering rhythmically like a goldfish. He remembered that she got that way. He had forgotten until now. He felt her hot panting breath on his face, saw her flushed cheeks. She got his trousers open. Her hot hand on him.

"Oh, Jesus," she said, squeezing him, and she slid down his body, running her hands over his shirt.

"Listen, Meredith."

"Just let me," she said hoarsely. "Just for a minute." And then her mouth was on him. She was always good at this. Images flooding back to him. The way she liked to do it in dangerous places. While he was driving on the freeway. In the men's room at a sales conference. On the beach at Napili at night. The secret impulsive nature, the secret heat. When he was first introduced to her, the exec at ConTech had said, *She's one of the great cocksuckers.*

Feeling her mouth on him, feeling his back arch as the tension ran through his body, he had the uneasy sense of pleasure and danger at once. So much had happened during the day, so many changes, everything was so sudden. He felt dominated, controlled, and at risk. He had the feeling as he lay on his back that he was somehow agreeing to a situation that he did not understand fully, that was not fully recognized. There would be trouble later. He did not want to go to Malaysia with her. He did not want an affair with his boss. He did not even want a one-night stand. Because what always happened was that people found out, gossip at the water cooler, meaningful looks in the hallway. And sooner or later the spouses found out. It always happened. Slammed doors, divorce lawyers, child custody.

And he didn't want any of that. His life was arranged now, he had things in place. He had commitments. This woman from his past understood none of that. She was free. He was not. He shifted his body.

"Meredith—"

"God, you taste good."

"Meredith—"

She reached up, and pressed her fingers over his lips. "Ssshhh. I know you like it."

"I do like it," he said, "but I—"

"Then let me."

As she sucked him, she was unbuttoning his shirt, pinching his nipples. He looked down and saw her straddling his legs, her head bent over him. Her blouse was open. Her breasts swung free. She reached up, took his hands, and pulled them down, placing them on her breasts.

She still had perfect breasts, the nipples hard under his touch. She moaned. Her body squirming as she straddled him. He felt her warmth. He began to hear a buzzing in his ears, a suffusing intoxicated flush in his face as sounds went dull, the room seemed distant, and there was nothing but this woman and her body and his desire for her.

In that moment he felt a burst of anger, a kind of male fury that he was pinned down, that she was dominating him, and he wanted to be in control, to take her. He sat up and grabbed her hair roughly, lifting her head and twisting his body. She looked in his eyes and saw instantly.

"Yes!" she said, and she moved sideways, so he could sit up beside her. He slipped his hand between her legs. He felt warmth, and lacy underpants. He tugged at them. She wriggled, helping him, and he slid them down to her knees; then she kicked them away. Her hands were caressing his hair, her lips at his ear. "Yes," she whispered fiercely. "Yes!"

Her blue skirt was bunched up around her waist. He kissed her hard, pulling her blouse wide, pressing her breasts to his bare chest. He felt her heat all along his body. He moved his fingers, probing between her lips. She gasped as they kissed, nodding her head *yes*. Then his fingers were in her.

For a moment he was startled: she was not very wet, and then he remembered that, too. The way she would start, her words and body immediately passionate, but this central part of her slower to respond, taking her eventual arousal from his. She was always turned on most by his desire for her, and always came after he did—sometimes within a few seconds, but sometimes he struggled to stay hard while she rocked against him, pushing to her own completion, lost in her own private world while he was fading. He always felt alone, always felt as if she were using him. Those memories gave him pause, and she sensed his hesitation and grabbed him fiercely, fumbling at his belt, moaning, sticking her hot tongue in his ear.

But reluctance was seeping back into him now, his angry heat was fading, and unbidden the thought flashed through his mind: *It's not worth it.*

All his feelings shifted again, and now he had a familiar sensation. Going back to see an old lover, being attracted over dinner, then getting involved again, feeling desire and, suddenly, in the heat of the moment, in the press of flesh, being reminded of all the things that had been wrong with the relationship, feeling old conflicts and angers and irritations rise up again, and wishing that he had never started. Suddenly thinking of how to get out of it, how to stop what was started. But usually there was no way to get out of it.

Still his fingers were inside her, and she was moving her body against his hand, shifting to be sure he would touch the right place. She was wetter, her lips were swelling. She opened her legs wider for him. She was breathing very hard, stroking him with her fingers. "Oh God, I love the way you feel," she said.

Usually there was no way to get out of it.

His body was tense and ready. Her hard nipples brushed against his chest. Her fingers caressed him. She licked the bottom of his earlobe with a quick dart of her tongue and instantly there was nothing but his desire, hot and angry, more intense for the fact that he didn't really want to be there, that he felt she had manipulated him to this place. Now he would fuck her. He wanted to fuck her. Hard.

She sensed his change and moaned, no longer kissing him, leaning back on the couch, waiting. She watched him through half-closed eyes, nodding her head. His fingers still touched her, rapidly, repeatedly, making her gasp, and he turned, pushed her down on her back on the couch. She hiked up her skirt and spread her legs for him. He crouched over her and she smiled at him, a knowing, victorious smile. It made him furious to see this sense that she had somehow won, this watchful detachment, and he wanted to catch her, to make her feel as out of control as he felt, to make her part of this, to wipe that smug detachment from her face. He spread her lips but did not enter her, he held back, his fingers moving, teasing her.

She arched her back, waiting for him. "No, no . . . please . . ."

Still he waited, looking at her. His anger was fading as quickly as it had come, his mind drifting away, the old reservations returning. In an instant of harsh clarity, he saw himself in the room, a panting middle-aged, married man with his trousers down around his knees, bent over a woman on an office couch that was too small. What the hell was he doing?

He looked at her face, saw the way the makeup cracked at the corners of her eyes. Around her mouth.

She had her hands on his shoulders, tugging him toward her. "Oh please . . . No . . . No . . ." And then she turned her head aside and coughed.

Something snapped in him. He sat back coldly. "You're right." He got off the couch, and pulled up his trousers. "We shouldn't do this."

She sat up. "What are you doing?" She seemed puzzled. "You want this as much as I do. You know you do."

"No," he said. "We shouldn't do this, Meredith." He was buckling his belt. Stepping back.

She stared at him in dazed disbelief, like someone awakened from sleep. "You're not serious . . ."

"This isn't a good idea. I don't feel good about it."

And then her eyes were suddenly furious. "You fucking *son of a bitch.*"

She got off the couch fast, rushing at him, hitting him hard with bunched fists. "You bastard! You prick! You fucking bastard!" He was trying to button his shirt, turning away from her blows. "You shit! You bastard!"

She moved around him as he turned away, grabbing his hands, tearing at his shirt to keep him from buttoning it.

"You can't! You can't do this to me!"

Buttons popped. She scratched him, long red welts running down his chest. He turned again, avoiding her, wanting only to get out of there. To get dressed and get out of there. She pounded his back.

"You fucker, you can't leave me like this!"

"Cut it out, Meredith," he said. "It's over."

"*Fuck* you!" She grabbed a handful of his hair, pulling him down with surprising strength, and she bit his ear hard. He felt an intense shooting pain and he pushed her away roughly. She toppled backward, off balance, crashing against the glass coffee table, sprawling on the ground.

She sat there, panting. "You fucking son of a bitch."

"Meredith, just leave me alone." He was buttoning his shirt again. All he could think was: *Get out of here.* Get your stuff and get out of here.

He reached for his jacket, then saw his cellular phone on the windowsill.

He moved around the couch and picked up the phone. The wine-glass crashed against the window near his head. He looked over and saw her standing in the middle of the room, reaching for something else to throw.

"I'll kill you!" she said. "I'll fucking kill you."

"That's enough, Meredith," he said.

"The hell." She threw a small paper bag at him. It thunked against the glass and dropped to the floor. A box of condoms fell out.

"I'm going home." He moved toward the door.

"That's right," she said. "You go home to your wife and your little fucking family."

Alarms went off in his head. He hesitated for a moment.

"Oh yes," she said, seeing him pause. "I know *all* about you, you asshole. Your wife isn't fucking you, so you come in here and lead me on, you set me up and then you walk out on me, you hostile violent fucking asshole. You think you can treat women this way? You asshole."

He reached for the doorknob.

"You walk out on me, you're dead!"

He looked back and saw her leaning unsteadily on the desk, and he thought, *She's drunk.*

"Good night, Meredith," he said. He twisted the knob, then remembered that the door had been locked. He unlocked the door and walked out, without looking back.

In the outer room, a cleaning woman was emptying trash baskets from the assistants' desks.

"I'll fucking kill you for this!" Meredith called after him.

The cleaning woman heard it, and stared at Sanders. He looked away from her, and walked straight to the elevator. He pushed the button. A moment later, he decided to take the stairs.

S anders stared at the setting sun from the deck of the ferry going back to Winslow. The evening was calm, with almost no breeze; the surface of the water was dark and still. He looked back at the lights of the city and tried to assess what had happened.

From the ferry, he could see the upper floors of the DigiCom buildings, rising behind the horizontal gray concrete of the viaduct that ran along the water's edge. He tried to pick out Meredith's office window, but he was already too far away.

Out here on the water, heading home to his family, slipping back into his familiar daily routine, the events of the previous hour had already begun to take on an unreal quality. He found it hard to believe that it had happened. He reviewed the events in his mind, trying to see just where he had gone wrong. He felt certain that it was all his fault, that he had misled Meredith in some important way. Otherwise, she would never have come on to him. The whole episode was an embarrassment for him, and probably for her, too. He felt guilty and miserable—and deeply uneasy about the future. What would happen now? What would she do?

He couldn't even guess. He realized then that he didn't really know her at all. They had once been lovers, but that was a long time ago. Now she was a new person, with new responsibilities. She was a stranger to him.

Although the evening was mild, he felt chilled. He went back inside the ferry. He sat in a booth and took out his phone to call Susan. He pushed the buttons, but the light didn't come on. The battery was dead. For a moment he was confused; the battery should last all day. But it was dead.

The perfect end to his day.

. . .

Feeling the throb of the ferry engines, he stood in the bathroom and stared at himself in the mirror. His hair was messed; there was a faint smear of lipstick on his lips, and another on his neck; two buttons of his shirt were missing, and his clothes were rumpled. He looked as if he had just gotten laid. He turned his head to see his ear. A tiny bruise marked where she had bitten him. He unbuttoned the shirt and looked at the deep red scratches running in parallel rows down his chest.

Christ.

How was he going to keep Susan from seeing this?

He dampened paper towels and scrubbed away the lipstick. He patted down his hair, and buttoned his sport coat, hiding most of his shirt. Then he went back outside, sat down at a booth by the window, and stared into space.

"Hey, Tom."

He looked up and saw John Perry, his neighbor on Bainbridge. Perry was a lawyer with Marlin, Howard, one of the oldest firms in Seattle. He was one of those irrepressibly enthusiastic people, and Sanders didn't much feel like talking to him. But Perry slipped into the seat opposite him.

"How's it going?" Perry asked cheerfully.

"Pretty good," Sanders said.

"I had a *great* day."

"Glad to hear it."

"Just *great*," Perry said. "We tried a case, and I tell you, we kicked ass."

"Great," Sanders said. He stared fixedly out the window, hoping Perry would take the hint and go away.

Perry didn't. "Yeah, and it was a damned tough case, too. Uphill all the way for us," he said. "Title VII, Federal Court. Client's a woman who worked at MicroTech, claimed she wasn't promoted because she was a female. Not a very strong case, to tell the truth. Because she drank, and so on. There were problems. But we have a gal in our firm, Louise Fernandez, a Hispanic gal, and she is just *lethal* on these discrimination cases. Lethal. Got the jury to award our client nearly half a million. That Fernandez can work the case law like nothing you've ever seen. She's won fourteen of her last sixteen cases. She acts so sweet and demure, and inside, she's just *ice*. I tell you, sometimes women scare the hell out of me."

Sanders said nothing.

He came home to a silent house, the kids already asleep. Susan always put the kids to bed early. He went upstairs. His wife was sitting up in bed, reading, with legal files and papers scattered across the bedcovers. When she saw him, she got out of bed and came over to hug him. Involuntarily, his body tensed.

"I'm really sorry, Tom," she said. "I'm sorry about this morning. And I'm sorry about what happened at work." She turned her face up and kissed him lightly on the lips. Awkwardly, he turned away. He was afraid she would smell Meredith's perfume, or—

"You mad about this morning?" she asked.

"No," he said. "Really, I'm not. It was just a long day."

"Lot of meetings on the merger?"

"Yes," he said. "And more tomorrow. It's pretty crazy."

Susan nodded. "It must be. You just got a call from the office. From a Meredith Johnson."

He tried to keep his voice casual. "Oh yes?"

"Uh-huh. About ten minutes ago." She got back in bed. "Who is she, anyway?" Susan was always suspicious when women from the office called.

Sanders said, "She's the new veep. They just brought her up from Cupertino."

"I wondered . . . She acted like she knew me."

"I don't think you've ever met." He waited, hoping he wouldn't have to say more.

"Well," she said, "she sounded very friendly. She said to tell you everything is fine for the due diligence meeting tomorrow morning at eight-thirty, and she'll see you then."

"Okay. Fine."

He kicked off his shoes, and started to unbutton his shirt, then stopped. He bent over and picked his shoes up.

"How old is she?" Susan asked.

"Meredith? I don't know. Thirty-five, something like that. Why?"

"Just wondered."

"I'm going to take a shower," he said.

"Okay." She picked up her legal briefs, and settled back in bed, adjusting the reading light.

He started to leave.

"Do you know her?" Susan asked.

"I've met her before. In Cupertino."

"What's she doing up here?"

"She's my new boss."

"*She's* the one."

"Yeah," he said. "She's the one."

"She's the woman that's close to Garvin?"

"Yeah. Who told you? Adele?" Adele Lewyn, Mark's wife, was one of Susan's best friends.

She nodded. "Mary Anne called, too. The phone never stopped ringing."

"I'll bet."

"So is Garvin fucking her or what?"

"Nobody knows," he said. "The general belief is that he's not."

"Why'd he bring her in, instead of giving the job to you?"

"I don't know, Sue."

"You didn't talk to Garvin?"

"He came around to see me in the morning, but I wasn't there."

She nodded. "You must be pissed. Or are you being your usual understanding self?"

"Well." He shrugged. "What can I do?"

"You can quit," she said.

"Not a chance."

"They passed over you. Don't you *have* to quit?"

"This isn't the best economy to find another job. And I'm forty-one. I don't feel like starting over. Besides, Phil insists they're going to spin off the technical division and take it public in a year. Even if I'm not running it, I'd still be a principal in that new company."

"And did he have details?"

He nodded. "They'll vest us each twenty thousand shares, and options for fifty thousand more. Then options for another fifty thousand shares each additional year."

"At?"

"Usually it's twenty-five cents a share."

"And the stock will be offered at what? Five dollars?"

"At least. The IPO market is getting stronger. Then, say it goes to ten. Maybe twenty, if we're hot."

There was a brief silence. He knew she was good with figures. "No," she said finally. "You can't possibly quit."

He had done the calculations many times. At a minimum, Sanders would realize enough on his stock options to pay off his mortgage in a single payment. But if the stock went through the roof, it could be truly fantastic—somewhere between five and fourteen million dollars. That was why going public was the dream of anyone who worked in a technical company.

He said, "As far as I'm concerned, they can bring in Godzilla to manage that division, and I'll still stay at least two more years."

"And is that what they've done? Brought in Godzilla?"

He shrugged. "I don't know."

"Do you get along with her?"

He hesitated. "I'm not sure. I'm going to take a shower."

"Okay," she said. He glanced back at her: she was reading her notes again.

After his shower, he plugged his phone into the charger unit on the sink, and put on a T-shirt and boxer shorts. He looked at himself in the mirror; the shirt covered his scratches. But he was still worried about the smell of Meredith's perfume. He splashed after-shave on his cheeks.

Then he went into his son's room to check on him. Matthew was snoring loudly, his thumb in his mouth. He had kicked down the covers. Sanders pulled them back up gently and kissed his forehead.

Then he went into Eliza's room. At first he could not see her; his daughter had lately taken to burrowing under a barricade of covers and pillows when she slept. He tiptoed in, and saw a small hand reach up, and wave to him. He came forward.

"Why aren't you asleep, Lize?" he whispered.

"I was having a dream," she said. But she didn't seem frightened.

He sat on the edge of the bed, and stroked her hair. "What kind of a dream?"

"About the beast."

"Uh-huh . . ."

"The beast was really a prince, but he was placed under a powerful spell by a 'chantress."

"That's right . . ." He stroked her hair.

"Who turned him into a hideous beast."

She was quoting the movie almost verbatim.

"That's right," he said.

"Why?"

"I don't know, Lize. That's the story."

"Because he didn't give her shelter from the bitter cold?" She was quoting again. "Why didn't he, Dad?"

"I don't know," he said.

"Because he had no love in his heart," she said.

"Lize, it's time for sleep."

"Give me a dream first, Dad."

"Okay. There's a beautiful silver cloud hanging over your bed, and—"

"That dream's no good, Dad." She was frowning at him.

"Okay. What kind of dream do you want?"

"With Kermit."

"Okay. Kermit is sitting right here by your head, and he is going to watch over you all night."

"And you, too."

"Yes. And me, too." He kissed her forehead, and she rolled away to face the wall. As he left the room he could hear her sucking her thumb loudly.

He went back to the bedroom and pushed aside his wife's legal briefs to get into bed.

"Was she still awake?" Susan asked.

"I think she'll go to sleep. She wanted a dream. About Kermit."

His wife nodded. "Kermit is a very big deal now."

She didn't comment on his T-shirt. He slipped under the covers and felt suddenly exhausted. He lay back against the pillow and closed his eyes. He felt Susan picking up the briefs on the bed, and a moment later she turned off the light.

"Mmm," she said. "You smell good."

She snuggled up against him, pressing her face against his neck, and threw her leg over his side. This was her invariable overture, and it always annoyed him. He felt pinned down by her heavy leg.

She stroked his cheek. "Is that after-shave for me?"

"Oh, Susan . . ." He sighed, exaggerating his fatigue.

"Because it works," she said, giggling. Beneath the covers, she put her hand on his chest. He felt it slide down, and slip under the T-shirt.

He had a burst of sudden anger. What was the matter with her? She never had any sense about these things. She was always coming on to him at inappropriate times and places. He reached down and grabbed her hand.

"Something wrong?"

"I'm really tired, Sue."

She stopped. "Bad day, huh?" she said sympathetically.

"Yeah. Pretty bad."

She got up on one elbow, and leaned over him. She stroked his lower lip with one finger. "You don't want me to cheer you up?"

"I really don't."

"Not even a little?"

He sighed again.

"You sure?" she asked, teasingly. "Really, *really* sure?" And then she started to slide beneath the covers.

He reached down and held her head with both hands. "Susan. Please. Come on."

She giggled. "It's only eight-thirty. You can't be *that* tired."

"I am."

"I bet you're not."

"Susan, damn it. I'm not in the mood."

"Okay, okay." She pulled away from him. "But I don't know why you put on the after-shave, if you're not interested."

"For Christ's sake."

"We hardly ever have sex anymore, as it is."

"That's because you're always traveling." It just slipped out.

"I'm not 'always traveling.' "

"You're gone a couple of nights a week."

"That's not 'always traveling.' And besides, it's my job. I thought you were going to be more supportive of my job."

"I am supportive."

"Complaining is not supportive."

"Look, for Christ's sake," he said, "I come home early whenever you're out of town, I feed the kids, I take care of things so you don't have to worry—"

"*Sometimes*," she said. "And sometimes you stay late at the office, and the kids are with Consuela until all hours—"

"Well, I have a job, too—"

"So don't give me this 'take care of things' crap," she said. "You're not home anywhere near as much as I am, I'm the one who has two jobs, and mostly you do exactly what you want, just like every other fucking man in the world."

"Susan . . ."

"Jesus, you come home early once in a while, and you act like a fucking martyr." She sat up, and turned on her bedside light. "Every woman I know works harder than any man."

"Susan, I don't want to fight."

"Sure, make it *my* fault. I'm the one with the problem. Fucking *men*."

He was tired, but he felt suddenly energized by anger. He felt suddenly strong, and got out of bed and started pacing. "What does being a man have to do with it? Am I going to hear how oppressed you are again now?"

"Listen," she said, sitting straighter. "Women are oppressed. It's a fact."

"Is it? How are you oppressed? You never wash a load of clothes. You never cook a meal. You never sweep a floor. Somebody does all that for you. You have somebody to do everything for you. You have somebody to take the kids to school and somebody to pick them up. You're a partner in a law firm, for Christ's sake. You're about as oppressed as Leona Helmsley."

She was staring at him in astonishment. He knew why: Susan had made her oppression speech many times before, and he had never contradicted her. Over time, with repetition it had become an accepted idea in their marriage. Now he was disagreeing. He was changing the rules.

"I can't believe you. I thought you were different." She squinted at him, her judicious look. "This is because a woman got your job, isn't it."

"What're we going to now, the fragile male ego?"

"It's true, isn't it? You're threatened."

"No it's not. It's crap. Who's got the fragile ego around here? Your ego's so fucking fragile, you can't even take a rejection in bed without picking a fight."

That stopped her. He saw it instantly: she had no comeback. She sat there frowning at him, her face tight.

"Jesus," he said, and turned to leave the room.

"You picked this fight," she said.

He turned back. "I did not."

"Yes, you did. You were the one who started in with the traveling."

"No. You were complaining about no sex."

"I was *commenting*."

"Christ. Never marry a lawyer."

"And your ego *is* fragile."

"Susan, you want to talk fragile? I mean, you're so fucking self-involved that you had a shitfit this morning because you wanted to look pretty for the *pediatrician*."

"Oh, there it is. Finally. You *are* still mad because I made you late. What is it? You think you didn't get the job because you were late?"

"No," he said, "I didn't—"

"You didn't get the job," she said, "because Garvin didn't give it to you. You didn't play the game well enough, and somebody else played it better. That's why. A woman played it better."

Furious, shaking, unable to speak, he turned on his heel and left the room.

"That's right, leave," she said. "Walk away. That's what you always do. Walk away. Don't stand up for yourself. You don't want to hear it, Tom. But it's the truth. If you didn't get the job, you have nobody to blame but yourself."

He slammed the door.

He sat in the kitchen in darkness. It was quiet all around him, except for the hum of the refrigerator. Through the kitchen window, he could see the moonlight on the bay, through the stand of fir trees.

He wondered if Susan would come down, but she didn't. He got up and walked around, pacing. After a while, it occurred to him that he hadn't eaten. He opened the refrigerator door, squinting in the light. It was stacked with baby food, juice containers, baby vitamins, bottles of formula. He poked among the stuff, looking for some cheese, or maybe a beer. He couldn't find anything except a can of Susan's Diet Coke.

Christ, he thought, not like the old days. When his refrigerator was full of frozen food and chips and salsa and lots of beer. His bachelor days.

He took out the Diet Coke. Now Eliza was starting to drink it, too. He'd told Susan a dozen times he didn't want the kids to get diet drinks. They ought to be getting healthy food. Real food. But Susan was busy, and Consuela indifferent. The kids ate all kinds of crap. It wasn't right. It wasn't the way he had been brought up.

Nothing to eat. Nothing in his own damned refrigerator. Hopeful, he lifted the lid of a Tupperware container and found a partially eaten peanut butter and jelly sandwich, with Eliza's small toothmarks in one side. He picked the sandwich up and turned it over, wondering how old it was. He didn't see any mold.

What the hell, he thought, and he ate the rest of Eliza's sandwich, standing there in his T-shirt, in the light of the refrigerator door. He was startled by his own reflection in the glass of the oven. "Another privileged member of the patriarchy, lording it over the manor."

Christ, he thought, where did women come up with this crap?

He finished the sandwich and rubbed the crumbs off his hands. The wall clock said 9:15. Susan went to sleep early. Apparently she wasn't coming down to make up. She usually didn't. It was his job to make up.

He was the peacemaker. He opened a carton of milk and drank from it, then put it back on the wire shelf. He closed the door. Darkness again.

He walked over to the sink, washed his hands, and dried them on a dish towel. Having eaten a little, he wasn't so angry anymore. Fatigue crept over him. He looked out the window and through the trees and saw the lights of a ferry, heading west toward Bremerton. One of the things he liked about this house was that it was relatively isolated. It had some land around it. It was good for the kids. Kids should grow up with a place to run and play.

He yawned. She definitely wasn't coming down. It'd have to wait until morning. He knew how it would go: he'd get up first, fix her a cup of coffee, and take it to her in bed. Then he'd say he was sorry, and she would reply that she was sorry, too. They'd hug, and he would go get dressed for work. And that would be it.

He went back up the dark stairs to the second floor, and opened the door to the bedroom. He could hear the quiet rhythms of Susan's breathing.

He slipped into bed, and rolled over on his side. And then he went to sleep.

TUESDAY

It rained in the morning, hard sheets of drumming downpour that slashed across the windows of the ferry. Sanders stood in line to get his coffee, thinking about the day to come. Out of the corner of his eye, he saw Dave Benedict coming toward him, and quickly turned away, but it was too late. Benedict waved, "Hey, guy." Sanders didn't want to talk about DigiCom this morning.

At the last moment, he was saved by a call: the phone in his pocket went off. He turned away to answer it.

"Fucking A, Tommy boy." It was Eddie Larson in Austin.

"What is it, Eddie?"

"You know that bean counter Cupertino sent down? Well, get this: there's *eight* of 'em here now. Independent accounting firm of Jenkins, McKay, out of Dallas. They're going over all the books, like a swarm of roaches. And I mean everything: receivables, payables, A and L's, year to date, *everything*. And now they're going back through every year to 'eighty-nine."

"Yeah? Disrupting everything?"

"Better believe it. The gals don't even have a place to sit down and answer the phone. Plus, everything from 'ninety-one back is in storage, downtown. We've got it on fiche here, but they say they want original documents. They want the damned paper. And they get all squinty and paranoid when they order us around. Treating us like we're thieves or something trying to pull a fast one. It's insulting."

"Well," Sanders said, "hang in there. You've got to do what they ask."

"The only thing that really bothers me," Eddie said, "is they got another seven more coming in this afternoon. Because they're also doing a complete inventory of the plant. Everything from the furniture in the offices to the air handlers and the heat stampers out on the line. We got a guy there now, making his way down the line, stopping at each work station. Says, 'What's this thing called? How do you spell it? Who makes it? What's the model number? How old is it? Where's the serial number?' You ask me, we might as well shut the line down for the rest of the day."

Sanders frowned. "They're doing an *inventory?*"

"Well, that's what they call it. But it's beyond any damn inventory I ever heard of. These guys have worked over at Texas Instruments or someplace, and I'll give 'em one thing: they know what they're talking about. This morning, one of the Jenkins guys came up and asked me what kind of glass we got in the ceiling skylights. I said, 'What kind of glass?' I thought he was shitting me. He says, 'Yeah, is it Corning two-forty-seven, or two-forty-seven slash nine.' Or some damned thing like that. They're different kinds of UV glass, because UV can affect chips on the production line. I never even heard that UV can affect chips. 'Oh yeah,' this guy says. 'Real problem if your ASDs get over two-twenty.' That's annual sunny days. Have you heard of that?"

Sanders wasn't really listening. He was thinking about what it meant that somebody—either Garvin, or the Conley-White people—would ask for an inventory of the plant. Ordinarily, you called for an inventory only if you were planning to sell a facility. Then you had to do it, to figure your writedowns at the time of transfer of assets, and—

"Tom, you there?"

"I'm here."

"So I say to this guy, I never heard that. About the UV and the chips. And we been putting chips in the phones for years, never any trouble. And then the guy says, 'Oh, not for installing chips. UV affects it if you're *manufacturing* chips.' And I say, we don't do that here. And he says, 'I know.' So, I'm wondering: what the hell does he care what kind of glass we have? Tommy boy? You with me? What's the story?" Larson said. "We're going to have fifteen guys crawling all over us by the end of the day. Now don't tell me this is *routine.*"

"It doesn't sound like it's routine, no."

"It sounds like they're going to sell the plant to somebody who makes chips, is what it sounds like. And that ain't us."

"I agree. That's what it sounds like."

"Fucking A," Eddie said. "I thought you told me this wasn't going to happen. Tom: people here are getting upset. And I'm one of 'em."

"I understand."

"I mean, I got people asking me. They just bought a house, their wife's pregnant, they got a baby coming, and they want to know. What do I tell 'em?"

"Eddie, I don't have any information."

"Jesus, Tom, you're the division head."

"I know. Let me check with Cork, see what the accountants did there. They were out there last week."

"I already talked to Colin an hour ago. Operations sent two people out there. For one day. Very polite. Not like this at all."

"No inventory?"

"No inventory."

"Okay," Sanders said. He sighed. "Let me get into it."

"Tommy boy," Eddie said. "I got to tell you right out. I'm concerned you don't already know."

"Me, too," Sanders said. "Me, too."

He hung up the phone. Sanders pushed K-A-P for Stephanie Kaplan. She would know what was going on in Austin, and he thought she would tell him. But her assistant said Kaplan was out of the office for the rest of the morning. He called Mary Anne, but she was gone, too. Then he dialed the Four Seasons Hotel, and asked for Max Dorfman. The operator said Mr. Dorfman's lines were busy. He made a mental note to see Max later in the day. Because if Eddie was right, then Sanders was out of the loop. And that wasn't good.

In the meantime, he could bring up the plant closing with Meredith at the conclusion of the morning meeting with Conley-White. That was the best he could do, for the moment. The prospect of talking to her made him uneasy. But he'd get through it somehow. He didn't really have a choice.

When he got to the fourth-floor conference room, nobody was there. At the far end, a wall board showed a cutaway of the Twinkle drive and a schematic for the Malaysia assembly line. There were notes scribbled on some of the pads, open briefcases beside some of the chairs.

The meeting was already under way.

Sanders had a sense of panic. He started to sweat.

At the far end of the room, an assistant came in, and began moving around the table, setting out glasses and water.

"Where is everybody?" he asked.

"Oh, they left about fifteen minutes ago," she said.

"Fifteen minutes ago? When did they start?"

"The meeting started at eight."

"Eight?" Sanders said. "I thought it was supposed to be eight-thirty."

"No, the meeting started at eight."

Damn.

"Where are they now?"

"Meredith took everybody down to VIE, to demo the Corridor."

Entering VIE, the first thing Sanders heard was laughter. When he walked into the equipment room, he saw that Don Cherry's team had two of the Conley-White executives up on the system. John Conley, the young lawyer, and Jim Daly, the investment banker, were both wearing headsets while they walked on the rolling walker pads. The two men were grinning wildly. Everyone else in the room was laughing too, including the normally sour-faced CFO of Conley-White, Ed Nichols, who was standing beside a monitor which showed an image of the virtual corridor that the users were seeing. Nichols had red marks on his forehead from wearing the headset.

Nichols looked over as Sanders came up. "This is *fantastic.*"

Sanders said, "Yes, it's pretty spectacular."

"Simply fantastic. It's going to wipe out all the criticism in New York, once they see this. We've been asking Don if he can run this on our own corporate database."

"No problem," Cherry said. "Just get us the programming hooks for your DB, and we'll plug you right in. Take us about an hour."

Nichols pointed to the headset. "And we can get one of these contraptions in New York?"

"Easy," Cherry said. "We can ship it out later today. It'll be there Thursday. I'll send one of our people to set it up for you."

"This is going to be a *great* selling point," Nichols said. "Just great." He took out his half-frame glasses. They were a complicated kind of glasses that folded up very small. Nichols unfolded them carefully and put them on his nose.

On the walker pad, John Conley was laughing. "Angel," he said. "How do I open this drawer?" Then he cocked his head, listening.

"He's talking to the help angel," Cherry said. "He hears the angel through his earphones."

"What's the angel telling him?" Nichols said.

"That's between him and his angel," Cherry laughed.

On the walker pad, Conley nodded as he listened, then reached forward into the air with his hand. He closed his fingers, as if gripping something, and pulled back, pantomiming someone opening a file drawer.

On the monitor, Sanders saw a virtual file drawer slide out from the wall of the corridor. Inside the drawer he saw neatly arranged files.

"Wow," Conley said. "This is amazing. Angel: can I see a file? . . . Oh. Okay."

Conley reached out and touched one of the file labels with his fingertip. Immediately the file popped out of the drawer and opened up, apparently hanging in midair.

"We have to break the physical metaphor sometimes," Cherry said. "Because users have only one hand. And you can't open a regular file with one hand."

Standing on the black walker pad, Conley moved his hand through the air in short arcs, mimicking someone turning pages with his hand. On the monitor, Sanders saw Conley was actually looking at a series of spreadsheets. "Hey," Conley said, "you people ought to be more careful. I have all your financial records here."

"Let me see that," Daly said, turning around on the walker pad to look.

"You guys look all you want," Cherry laughed. "Enjoy it while you can. In the final system, we'll have safeguards built in to control access. But for now, we bypass the entire system. Do you notice that some of the numbers are red? That means they have more detail stored away. Touch one."

Conley touched a red number. The number zoomed out, creating a new plane of information that hung in the air above the previous spreadsheet.

"Wow!"

"Kind of a hypertext thing," Cherry said, with a shrug. "Sort of neat, if I say so myself."

Conley and Daly were giggling, poking rapidly at numbers on the spreadsheet, zooming out dozens of detail sheets that now hung in the air all around them. "Hey, how do you get rid of all this stuff?"

"Can you find the original spreadsheet?"

"It's hidden behind all this other stuff."

"Bend over, and look. See if you can get it."

Conley bent at the waist, and appeared to look under something. He reached out and pinched air. "I got it."

"Okay, now you see a green arrow in the right corner. Touch it."

Conley touched it. All the papers zoomed back into the original spreadsheet.

"Fabulous!"

"I want to do it," Daly said.

"No, you can't. I'm going to do it."

"No, me!"

"Me!"

They were laughing like delighted kids.

Blackburn came up. "I know this is enjoyable for everyone," he said to Nichols, "but we're falling behind our schedule and perhaps we ought to go back to the conference room."

"All right," Nichols said, with obvious reluctance. He turned to Cherry. "You sure you can get us one of these things?"

"Count on it," Cherry said. "Count on it."

Walking back to the conference room, the Conley-White executives were in a giddy mood; they talked rapidly, laughing about the experience. The DigiCom people walked quietly beside them, not wanting to disrupt the good mood. It was at that point that Mark Lewyn fell into step alongside Sanders and whispered, "Hey, why didn't you call me last night?"

"I did," Sanders said.

Lewyn shook his head. "There wasn't any message when I got home," he said.

"I talked to your answering machine, about six-fifteen."

"I never got a message," Lewyn said. "And then when I came in this morning, you weren't here." He lowered his voice. "Christ. What a mess. I had to go into the meeting on Twinkle with no idea what the approach was going to be."

"I'm sorry," Sanders said. "I don't know what happened."

"Fortunately, Meredith took over the discussion," Lewyn said. "Otherwise I would have been in deepest shit. In fact, I— We'll do this later," he said, seeing Johnson drop back to talk to Sanders. Lewyn stepped away.

"Where the hell were you?" Johnson said.

"I thought the meeting was for eight-thirty."

"I called your house last night, specifically because it was changed to eight. They're trying to catch a plane to Austin for the afternoon. So we moved everything up."

"I didn't get that message."

"I talked to your wife. Didn't she tell you?"

"I thought it was eight-thirty."

Johnson shook her head, as if dismissing the whole thing. "Anyway," she said, "in the eight o'clock session, I had to take another approach to Twinkle, and it's very important that we have some coordination in the light of—"

"Meredith?" Up at the front of the group, Garvin was looking back at her. "Meredith, John has a question for you."

"Be right there," she said. With a final angry frown at Sanders, she hurried up to the head of the group.

B ack in the conference room, the mood was light. They were all still joking as they took their seats. Ed Nichols began the meeting by turning to Sanders. "Meredith's been bringing us up to date on the Twinkle drive. Now that you're here, we'd like your assessment as well."

I had to take another approach to Twinkle, Meredith had said. Sanders hesitated. "My assessment?"

"Yes," Nichols said. "You're in charge of Twinkle, aren't you?"

Sanders looked at the faces around the table, turned expectantly toward him. He glanced at Johnson, but she had opened her briefcase and was rummaging through her papers, taking out several bulging manila envelopes.

"Well," Sanders said. "We built several prototypes and tested them thoroughly. There's no doubt that the prototypes performed flawlessly. They're the best drives in the world."

"I understand that," Nichols said. "But now you are in production, isn't that right?"

"That's right."

"I think we're more interested in your assessment of the production."

Sanders hesitated. What had she told them? At the other end of the room, Meredith Johnson closed her briefcase, folded her hands under her chin, and stared steadily at him. He could not read her expression.

What had she told them?

"Mr. Sanders?"

"Well," Sanders began, "we've been shaking out the lines, dealing with the problems as they arise. It's a pretty standard start-up experience for us. We're still in the early stages."

"I'm sorry," Nichols said. "I thought you've been in production for two months."

"Yes, that's true."

"Two months doesn't sound like 'the early stages' to me."

"Well—"

"Some of your product cycles are as short as nine months, isn't that right?"

"Nine to eighteen months, yes."

"Then after two months, you must be in full production. How do you assess that, as the principal person in charge?"

"Well, I'd say the problems are of the order of magnitude we generally experience at this point."

"I'm interested to hear that," Nichols said, "because earlier today, Meredith indicated to us that the problems were actually quite serious. She said you might even have to go back to the drawing board."

Shit.

How should he play it now? He'd already said that the problems were not so bad. He couldn't back down. Sanders took a breath and said, "I hope I haven't conveyed the wrong impression to Meredith. Because I have full confidence in our ability to manufacture the Twinkle drive."

"I'm sure you do," Nichols said. "But we're looking down the barrel at competition from Sony and Philips, and I'm not sure that a simple expression of your confidence is adequate. How many of the drives coming off the line meet specifications?"

"I don't have that information."

"Just approximately."

"I wouldn't want to say, without precise figures."

"Are precise figures available?"

"Yes. I just don't have them at hand."

Nichols frowned. His expression said: why don't you have them when you knew this is what the meeting was about?

Conley cleared his throat. "Meredith indicated that the line is running at twenty-nine percent capacity, and that only five percent of the drives meet specifications. Is that your understanding?"

"That's more or less how it has been. Yes."

There was a brief silence around the table. Abruptly, Nichols sat forward. "I'm afraid I need some help here," he said. "With figures like that, on what do you base your confidence in the Twinkle drive?"

"The reason is that we've seen all this before," Sanders replied. "We've seen production problems that look insurmountable but then get resolved quickly."

"I see. So you think your past experience will hold true here."

"Yes, I do."

Nichols sat back in his seat and crossed his arms over his chest. He looked extremely dissatisfied.

Jim Daly, the thin investment banker, sat forward and said, "Please don't misunderstand, Tom. We're not trying to put you on the spot," he said. "We have long ago identified several reasons for acquisition of this company, irrespective of any specific problem with Twinkle. So I don't think Twinkle is a critical issue today. We just want to know where we stand on it. And we'd like you to be as frank as possible."

"Well, there *are* problems," Sanders said. "We're in the midst of assessing them now. We have some ideas. But some of the problems may go back to design."

Daly said, "Give us worst case."

"Worst case? We pull the line, rework the housings and perhaps the controller chips, and then go back on."

"Causing a delay of?"

Nine to twelve months. "Up to six months," Sanders said.

"Jesus," somebody whispered.

Daly said, "Johnson suggested that the maximum delay would be six *weeks.*"

"I hope that's right. But you asked for worst case."

"Do you really think it will take six months?"

"You asked for worst case. I think it's unlikely."

"But possible?"

"Yes, possible."

Nichols sat forward again and gave a big sigh. "Let me see if I understand this right. If there *are* design problems with the drive, they occurred under your stewardship, is that correct?"

"Yes, it is."

Nichols shook his head. "Well. Having gotten us into this mess, do you really think you're the person to clean it up?"

Sanders suppressed a surge of anger. "Yes I do," he said. "In fact, I think I'm the best possible person to do it. As I said, we've seen this kind of situation before. And we've handled it before. I'm close to all the people involved. And I am sure we can resolve it." He wondered how he could explain to these people in suits the reality of how products were made. "When you're working the cycles," he said, "it's sometimes not so serious to go back to the boards. Nobody likes to do it, but it may have advantages. In the old days, we made a complete generation of new products every year or so. Now, more and more, we also make incremental changes within generations. If we have to redo the chips, we may be able to code in the video compression algorithms, which weren't available when we started. That will enhance the end-user perception of speed by more than the simple drive specs. We won't go back to build a hundred-millisecond drive. We'll go back to build an eighty-millisecond drive."

"But," Nichols said, "in the meantime, you won't have entered the market."

"No, that's true."

"You won't have established your brand name, or established market share for your product stream. You won't have your dealerships, or your OEMs, or your ad campaign, because you won't have a product line to support it. You may have a better drive, but it'll be an unknown drive. You'll be starting from scratch."

"All true. But the market responds fast."

"And so does the competition. Where will Sony be by the time you get to market? Will they be at eighty milliseconds, too?"

"I don't know," Sanders said.

Nichols sighed. "I wish I had more confidence about where we are on this thing. To say nothing of whether we're properly staffed to fix it."

Meredith spoke for the first time. "I may be a little bit at fault here," she said. "When you and I spoke about Twinkle, Tom, I understood you to say that the problems were quite serious."

"They are, yes."

"Well, I don't think we want to be covering anything up here."

He said quickly, "I'm not covering anything up." The words came out almost before he realized it. He heard his voice, high-pitched, tight.

"No, no," Meredith said soothingly. "I didn't mean to suggest you were. It's just that these technical issues are hard for some of us to grasp. We're looking for a translation into layman's terms of just where we are. If you can do that for us."

"I've been trying to do that," he said. He knew he sounded defensive. But he couldn't help it.

"Yes, Tom, I know you have," Meredith said, her voice still soothing. "But for example: if the laser read-write heads are out of sync with the m-subset instructions off the controller chip, what is that going to mean for us, in terms of down time?"

She was just grandstanding, demonstrating her facility with tech-talk, but her words threw him off balance anyway. Because the laser heads were read-only, not read-write, and they had nothing to do with the m-subset off the controller chip. The position controls all came off the x-subset. And the x-subset was licensed code from Sony, part of the driver code that every company used in their CD drives.

To answer without embarrassing her, he had to move into fantasy, where nothing he could say was true. "Well," he said, "you raise a good point, Meredith. But I think the m-subset should be a relatively simple problem, assuming the laser heads are tracking to tolerance. Perhaps three or four days to fix."

He glanced quickly at Cherry and Lewyn, the only people in the room who would know that Sanders had just spoken gibberish. Both men nodded sagely as they listened. Cherry even rubbed his chin.

Johnson said, "And do you anticipate a problem with the asynchronous tracking signals from the mother board?"

Again, she was mixing everything up. The tracking signals came from the power source, and were regulated by the controller chip. There wasn't a mother board in the drive units. But by now he was in the swing. He answered quickly: "That's certainly a consideration, Meredith, and we should check it thoroughly. I expect we'll find that the asynchronous signals may be phase-shifted, but nothing more than that."

"A phase-shift is easy to repair?"

"Yes, I think so."

Nichols cleared his throat. "I feel this is an in-house technical issue," he said. "Perhaps we should move on to other matters. What's next on the agenda?"

Garvin said, "We've scheduled a demo of the video compression just down the hall."

"Fine. Let's do that."

Chairs scraped back. Everyone stood up, and they filed out of the room. Meredith was slower to close up her files. Sanders stayed behind for a moment, too.

When they were alone he said, "What the hell was all that about?"

"All what?"

"All that gobbledygook about controller chips and read heads. You don't know what you were talking about."

"Oh yes I do," she said angrily. "I was fixing the mess that you made." She leaned over the table and glared at him. "Look, Tom. I decided to take your advice last night, and tell the truth about the drive. This morning I said there were severe problems with it, that you were very knowledgeable, and you would tell them what the problems were. I set it up, for you to say what you told me you wanted to say. But then you came in and announced there were no problems of significance."

"But I thought we agreed last night—"

"These men aren't fools, and we're not going to be able to fool them." She snapped her briefcase shut. "I reported in good faith what you told me. And then you said I didn't know what I was talking about."

He bit his lip, trying to control his anger.

"I don't know what you think is going on here," she said. "These men don't care about technical details. They wouldn't know a drive head from a dildo. They're just looking to see if anybody's in charge, if anybody has a handle on the problems. They want reassurance. And you didn't reassure them. So I had to jump in and fix it with a lot of techno-bullshit. I had to clean up after you. I did the best I could. But let's face it: you didn't inspire confidence today, Tom. Not at all."

"Goddamn it," he said. "You're just talking about appearances. Corporate appearances in a corporate meeting. But in the end somebody has to actually build the damn drive—"

"I'll say—"

"And I've been running this division for eight years, and running it damn well—"

"Meredith." Garvin stuck his head in the door. They both stopped talking.

"We're waiting, Meredith," he said. He turned and looked coldly at Sanders.

She picked up her briefcase and swept out of the room.

Sanders went immediately downstairs to Blackburn's office. "I need to see Phil."

Sandra, his assistant, sighed. "He's pretty busy today."

"I need to see him now."

"Let me check, Tom." She buzzed the inner office. "Phil? It's Tom Sanders." She listened a moment. "He says go right in."

Sanders went into Blackburn's office and closed the door. Blackburn stood up behind his desk and ran his hands down his chest. "Tom. I'm glad you came down."

They shook hands briefly. "It isn't working out with Meredith," Sanders said at once. He was still angry from his encounter with her.

"Yes, I know."

"I don't think I can work with her."

Blackburn nodded. "I know. She already told me."

"Oh? What'd she tell you?"

"She told me about the meeting last night, Tom."

Sanders frowned. He couldn't imagine that she had discussed that meeting. "Last night?"

"She told me that you sexually harassed her."

"I *what?*"

"Now, Tom, don't get excited. Meredith's assured me she's not going to press charges. We can handle it quietly, in house. That will be best for everyone. In fact, I've just been going over the organization charts, and—"

"Wait a minute," Sanders said. "She's saying *I* harassed *her?*"

Blackburn stared at him. "Tom. We've been friends a long time. I can assure you, this doesn't have to be a problem. It doesn't have to get around the company. Your wife doesn't need to know. As I said, we can handle this quietly. To the satisfaction of everyone involved."

"Wait a minute, it's not true—"

"Tom, just give me a minute here, please. The most important thing now is for us to separate the two of you. So you aren't reporting to her. I think a lateral promotion for you would be ideal."

"Lateral promotion?"

"Yes. There's an opening for technical vice president in the cellular division in Austin. I want to transfer you there. You'll go with the same seniority, salary, and benefit package. Everything the same, except you'll be in Austin and you won't have to have any direct contact with her. How does that sound?"

"Austin."

"Yes."

"Cellular."

"Yes. Beautiful weather, nice working conditions . . . university town . . . chance to get your family out of this rain . . ."

Sanders said, "But Conley's going to sell off Austin."

Blackburn sat down behind his desk. "I can't imagine where you heard that, Tom," he said calmly. "It's completely untrue."

"You sure about that?"

"Absolutely. Believe me, selling Austin is the last thing they'd do. Why, it makes no sense at all."

"They why are they inventorying the plant?"

"I'm sure they're going over the whole operation with a fine-tooth comb. Look, Tom. Conley's worried about cash flow after the acquisition, and the Austin plant is, as you know, very profitable. We've given them the figures. Now they're verifying them, making sure they're real. But there's no chance they would sell it. Cellular is only going to grow, Tom. You know that. And that's why I think a vice presidency there in Austin is an excellent career opportunity for you to consider."

"But I'd be leaving the Advanced Products Division?"

"Well, yes. The whole point would be to move you out of this division."

"And then I wouldn't be in the new company when it spins off."

"That's true."

Sanders paced back and forth. "That's completely unacceptable."

"Well, let's not be hasty," Blackburn said. "Let's consider all the ramifications."

"Phil," he said. "I don't know what she told you, but—"

"She told me the whole story—"

"But I think you should know—"

"And I want you to know, Tom," Blackburn said, "that I don't have any judgment about what may have happened. That's not my concern or my interest. I'm just trying to solve a difficult problem for the company."

"Phil. Listen. I didn't do it."

"I understand that's probably how you feel, but—"

"I didn't harass her. She harassed me."

"I'm sure," Blackburn said, "it may have *seemed* like that to you at the time, but—"

"Phil, I'm telling you. She did everything but rape me." He paced angrily. "Phil: *she* harassed *me*."

Blackburn sighed and sat back in his chair. He tapped his pencil on the corner of his desk. "I have to tell you frankly, Tom. I find that difficult to believe."

"It's what happened."

"Meredith's a beautiful woman, Tom. A very vital, sexy woman. I think it's natural for a man to, uh, lose control."

"Phil, you aren't hearing me. She harassed me."

Blackburn gave a helpless shrug. "I hear you, Tom. I just . . . I find that difficult to picture."

"Well, she did. You want to hear what really happened last night?"

"Well." Blackburn shifted in his chair. "Of course I want to hear your version. But the thing is, Tom, Meredith Johnson is very well connected in this company. She has impressed a lot of extremely important people."

"You mean Garvin."

"Not only Garvin. Meredith has built a power base in several areas."

"Conley-White?"

Blackburn nodded. "Yes. There, too."

"You don't want to hear what I say happened?"

"Of course I do," Blackburn said, running his hands through his hair. "Absolutely, I do. And I want to be scrupulously fair. But I'm trying to tell you that no matter what, we're going to have to make some transfers here. And Meredith has important allies."

"So it doesn't matter what I say."

Blackburn frowned, watching him pace. "I understand that you are upset. I can see that. And you're a valued person in this company. But what I'm trying to do here, Tom, is to get you to look at the situation."

"What situation?" Sanders said.

Blackburn sighed. "Were there any witnesses, last night?"

"No."

"So it's your word against hers."

"I guess so."

"In other words, it's a pissing match."

"So? That's no reason to assume I'm wrong, and she's right."

"Of course not," Blackburn said. "But look at the situation. A man claiming sexual harassment against a woman is, well, pretty unlikely. I don't think there's ever been a case in this company. It doesn't mean it couldn't happen. But it does means that it'd be very uphill for you—even if Meredith wasn't so well connected." He paused. "I just don't want to see you get hurt in this."

"I've already been hurt."

"Again, we're talking about feelings here. Conflicting claims. And unfortunately, Tom, no witnesses." He rubbed his nose, tugged at his lapels.

"You move me out of the APD, and I'm hurt. Because I won't get to be part of the new company. The company I worked on for twelve years."

"That's an interesting legal position," Blackburn said.

"I'm not talking about a legal position. I'm talking about—"

"Look. Tom. Let me review this with Garvin. Meanwhile, why don't you go off and think this Austin offer over. Think about it carefully. Because no one wins in a pissing match. You may hurt Meredith, but you'll hurt yourself much more. That's my concern here, as your friend."

"If you were my friend—" Sanders began.

"I *am* your friend," Blackburn said. "Whether you know it at this moment, or not." He stood up behind his desk. "You don't need this splashed all over the papers. Your wife doesn't need to hear about this, or your kids. You don't need to be the gossip of Bainbridge for the rest of the summer. That isn't going to do you any good at all."

"I understand that, but—"

"But we have to face reality, Tom," Blackburn said. "The company is faced with conflicting claims. What's happened has happened. We have to go on from here. And all I'm saying is, I'd like to resolve this quickly. So think it over. Please. And get back to me."

After Sanders left, Blackburn called Garvin. "I just talked with him," he told Garvin.

"And?"

"He says it was the other way around. That she harassed him."

"Christ," Garvin said. "What a mess."

"Yes. But on the other hand, it's what you'd expect him to say," Blackburn said. "It's the usual response in these cases. The man always denies it."

"Yeah. Well. This is dangerous, Phil."

"I understand."

"I don't want this thing to blow up on us."

"No, no."

"There's nothing more important right now than getting this thing resolved."

"I understand, Bob."

"You made him the Austin offer?"

"Yes. He's thinking it over."

"Will he take it?"

"My guess is no."

"And did you push it?"

"Well, I tried to convey to him that we weren't going to back down on Meredith. That we were going to support her through this."

"Damn right we are," Garvin said.

"I think he was clear about that. So let's see what he says when he comes back to us."

"He wouldn't go off and file, would he?"

"He's too smart for that."

"We hope," Garvin said irritably, and hung up.

L*ook at the situation.*

Sanders stood in Pioneer Park and leaned against a pillar, staring at the light drizzle. He was replaying the meeting with Blackburn.

Blackburn hadn't even been willing to listen to Sanders's version. He hadn't let Sanders tell him. Blackburn already knew what had happened.

She's a very sexy woman. It's natural for a man to lose control.

That was what everyone at DigiCom would think. Every single person in the company would have that view of what had happened. Blackburn had said he found it difficult to believe that Sanders had been harassed. Others would find it difficult, too.

Blackburn had told him it didn't matter what happened. Blackburn was telling him that Johnson was well connected, and that nobody would believe a man had been harassed by a woman.

Look at the situation.

They were asking him to leave Seattle, leave the APG. No options, no big payoff. No return for his twelve long years of work. All that was gone.

Austin. Baking hot, dry, brand-new.

Susan would never accept it. Her practice in Seattle was successful; she had spent many years building it. They had just finished remodeling the house. The kids liked it here. If Sanders even suggested a move, Susan would be suspicious. She'd want to know what was behind it. And sooner or later, she would find out. If he accepted the transfer, he would be confirming his guilt to his wife.

No matter how he thought about it, how he tried to put it together in his mind, Sanders could see no good outcome. He was being screwed.

I'm your friend, Tom. Whether you know it right now or not.

He recalled the moment at his wedding when Blackburn, his best man, said he wanted to dip Susan's ring in olive oil because there was always a problem about getting it on the finger. Blackburn in a panic, in case some little moment in the ceremony went wrong. That was Phil: always worried about appearances.

Your wife doesn't need to hear about this.

But Phil was screwing him. Phil, and Garvin behind him. They were both screwing him. Sanders had worked hard for the company for many years, but now they didn't give a damn about him. They were taking Meredith's side, without any question. They didn't even want to hear his version of what had happened.

As Sanders stood in the rain, his sense of shock slowly faded. And with it, his sense of loyalty. He started to get angry.

He took out his phone and placed a call.

"Mr. Perry's office."

"It's Tom Sanders calling."

"I'm sorry, Mr. Perry is in court. Can I give him a message?"

"Maybe you could help me. The other day he mentioned that you have a woman there who handles sexual harassment cases."

"We have several attorneys who do that, Mr. Sanders."

"He mentioned a Hispanic woman." He was trying to remember what else Perry had said about her. Something about being sweet and demure? He couldn't recall for sure.

"That would be Ms. Fernandez."

"I wonder if you could connect me," Sanders said.

Fernandez's office was small, her desk stacked high with papers and legal briefs in neat piles, a computer terminal in the corner. She stood up as he came in. "You must be Mr. Sanders."

She was a tall woman in her thirties, with straight blond hair and a handsome, aquiline face. She was dressed in a pale, cream-colored suit. She had a direct manner and a firm handshake. "I'm Louise Fernandez. How can I help you?"

She wasn't at all what he had expected. She wasn't sweet and demure at all. And certainly not Hispanic. He was so startled that without thinking he said, "You're not what I—"

"Expected?" She raised an eyebrow. "My father's from Cuba. We left there when I was a child. Please sit down, Mr. Sanders." She turned and walked back around her desk.

He sat down, feeling embarrassed. "Anyway, thank you for seeing me so quickly."

"Not at all. You're John Perry's friend?"

"Yes. He mentioned the other day that you, uh, specialized in these cases."

"I do labor law, primarily constructive termination and Title VII suits."

"I see." He felt foolish that he had come. He was taken aback by her brisk manner and elegant appearance. In fact, she reminded him very much of Meredith. He felt certain that she would not be sympathetic to his case.

She put on horn-rimmed glasses and peered at him across the desk. "Have you eaten? I can get you a sandwich if you like."

"I'm not hungry, thanks."

She pushed a half-eaten sandwich to the side of her desk. "I'm afraid I have a court appearance in an hour. Sometimes things get a bit rushed." She got out a yellow legal pad and set it before her. Her movements were quick, decisive.

Sanders watched her, sure she was the wrong person. He should never have come here. It was all a mistake. He looked around the office. There was a neat stack of bar charts for a courtroom appearance.

Fernandez looked up from the pad, her pen poised. It was one of those expensive fountain pens. "Would you like to tell me the situation?"

"Uh . . . I'm not sure where to begin."

"We could start with your full name and address, and your age."

"Thomas Robert Sanders." He gave his address.

"And your age?"

"Forty-one."

"Occupation?"

"I'm a division manager at Digital Communications. The Advanced Products Division."

"How long have you been at that company?"

"Twelve years."

"Uh-huh. And in your present capacity?"

"Eight years."

"And why are you here today, Mr. Sanders?"

"I've been sexually harassed."

"Uh-huh." She showed no surprise. Her expression was completely neutral. "You want to tell me the circumstances?"

"My boss, ah, came on to me."

"And the name of your boss?"

"Meredith Johnson."

"Is that a man or a woman?"

"A woman."

"Uh-huh." Again, no sign of surprise. She continued making notes steadily, the pen scratching. "When did this happen?"

"Last night."

"What were the exact circumstances?"

He decided not to mention the merger. "She has just been appointed my new boss, and we had several things to go over. She asked if we could meet at the end of the day."

"She requested this meeting?"

"Yes."

"And where did the meeting take place?"

"In her office. At six o'clock."

"Anybody else present?"

"No. Her assistant came in briefly, at the start of the meeting, then left. Before anything happened."

"I see. Go on."

"We talked for a while, about business, and we had some wine. She had gotten some wine. And then she came on to me. I was over by the window and suddenly she started kissing me. Then pretty soon we were sitting on the couch. And then she started, uh . . ." He hesitated. "How much detail do you want?"

"Just the broad strokes for now." She bit her sandwich. "You say you were kissing."

"Yes."

"And she initiated this?"

"Yes."

"What was your reaction when she did that?"

"I was uncomfortable. I'm married."

"Uh-huh. What was the general atmosphere in the meeting, prior to this kiss?"

"It was a regular business meeting. We were talking about business. But all the time, she was making, uh, suggestive remarks."

"Like what."

"Oh, about how good I looked. How I was in shape. How glad she was to see me."

"How glad she was to see you," Fernandez repeated, with a puzzled look.

"Yes. We knew each other before."

"You had a prior relationship?"

"Yes."

"When was that?"

"Ten years ago."

"And were you married then?"

"No."

"Did you both work for the same company at that time?"

"No. I did, but she worked for another company."

"And how long did your relationship last?"

"About six months."

"And have you kept up contact?"

"No. Not really."

"Any contact at all?"

"Once."

"Intimate?"

"No. Just, you know, hello in the hallway. At the office."

"I see. In the last eight years, have you ever been to her house or apartment?"

"No."

"Dinners, drinks after work, anything?"

"No. I really haven't seen her at all. When she joined the company, she was in Cupertino, in Operations. I was in Seattle, in Advanced Products. We didn't have much contact."

"So during this time, she wasn't your superior?"

"No."

"Give me a picture of Ms. Johnson. How old?"

"Thirty-five."

"Would you characterize her as attractive?"

"Yes."

"Very attractive?"

"She was a Miss Teenage something as a kid."

"So you would say she's very attractive." The pen scratched on the legal pad.

"Yes."

"And how about other men—would you say they find her very attractive?"

"Yes."

"What about her manner with regard to sexual matters? Does she make jokes? Sexual jokes, innuendoes, ribald comments?"

"No, never."

"Body language? Flirtatious? Does she touch people?"

"Not really. She certainly knows she's good-looking, and she can play on that. But her manner is . . . kind of cool. She's the Grace Kelly type."

"They say Grace Kelly was very sexually active, that she had affairs with most of her leading men."

"I wouldn't know."

"Uh-huh. What about Ms. Johnson, does she have affairs inside the company?"

"I wouldn't know. I haven't heard anything."

Fernandez flipped to a new page on her pad. "All right. And how long has she been your supervisor? Or is she your supervisor?"

"Yes, she is. One day."

For the first time, Fernandez looked surprised. She glanced at him, and took another bite of her sandwich. "One day?"

"Yes. Yesterday was the first day of a new company organization. She had just been appointed."

"So the day she is appointed, she meets with you, in the evening."

"Yes."

"All right. You were telling me, you were sitting on the couch and she was kissing you. And what happened then?"

"She unzipped—well, first of all, she started rubbing me."

"Your genitals."

"Yes. And kissing me." He found himself sweating. He wiped his forehead with his hand.

"I understand this is difficult. I'll try to make this as brief as possible," Fernandez said. "And then?"

"Then, she unzipped my pants, and started rubbing me with her hand."

"Was your penis exposed?"

"Yes."

"Who exposed it?"

"She took it out."

"So she took your penis out of your pants, and then rubbed it with her hand, is that right?" She peered at him over her glasses, and for a moment he glanced away in embarrassment. But when he looked back at her, he saw that she was not the least embarrassed, that her manner was more than clinical, more than professional—that she was in some deep way detached, and very cold.

"Yes," he said. "That's what happened."

"And what was your reaction?"

"Well." He gave an embarrassed shrug. "It worked."

"You were sexually aroused."

"Yes."

"Did you say anything to her?"

"Like what?"

"I'm just asking whether you said anything to her."

"Like what? I don't know."

"Did you say anything at all?"

"I said something, I don't know. I was feeling very uncomfortable."

"Do you remember what you said?"

"I think I just kept saying 'Meredith,' trying to get her to stop, you know, but she kept interrupting me, or kissing me."

"Did you say anything else besides 'Meredith'?"

"I don't remember."

"How did you feel about what she was doing?"

"I felt uncomfortable."

"Why?"

"I was afraid of getting involved with her, because she was my boss now, and because I was married now and I didn't want any complications in my life. You know, an office affair."

"Why not?" Fernandez asked.

The question took him aback. "Why *not?*"

"Yes." She looked at him directly, her eyes cool, appraising. "After all, you're alone with a beautiful woman. Why not have an affair?"

"Jesus."

"It's a question most people would ask."

"I'm married."

"So what? Married people have affairs all the time."

"Well," he said. "For one thing, my wife is a lawyer and very suspicious."

"Do I know her?"

"Her name is Susan Handler. She's with Lyman, King."

Fernandez nodded. "I've heard of her. So. You were afraid that she would find out."

"Sure. I mean, you have an affair in the office, and everybody's going to know. There isn't any way to keep it quiet."

"So you were concerned about this becoming known."

"Yes. But that wasn't the main reason."

"What was the main reason?"

"She was my boss. I didn't like the position I was in. She was, you know . . . well, she had the right to fire me. If she wanted to. So it was like I *had* to do it. I was very uncomfortable."

"Did you tell her that?"

"I tried."

"How did you try?"

"Well, I just tried."

"Would you say that you indicated to her that her advances were not welcome?"

"Eventually, yes."

"How is that?"

"Well, eventually, we continued this . . . whatever you call it, foreplay or whatever, and she had her panties off, and—"

"I'm sorry. How did she come to have her panties off?"

"I took them off."

"Did she ask you to do that?"

"No. But I got pretty worked up at one point, I was going to do it, or at least I was thinking about doing it."

"You were going to have intercourse." Her voice again cool. The pen scratching.

"Yes."

"You were a willing participant."

"For a while there. Yes."

"In what way were you a willing participant?" she asked. "What I mean is, did you initiate touching her body or breast or genitals without her encouragement?"

"I don't know. She was pretty much encouraging everything."

"I am asking, did you volunteer. Did you do it on your own. Or did she, for example, take your hand and place it on her—"

"No. I did it on my own."

"What about your earlier reservations?"

"I was worked up. Excited. I didn't care at that point."

"All right. Go on."

He wiped his forehead. "I'm being very honest with you."

"That's exactly what you should be. It's the best thing all around. Please go on."

"And she was lying on the couch with her skirt pulled up, and she wanted me to enter her, to . . . and she was sort of moaning, you know, saying, 'No, no,' and suddenly I had this feeling again that I didn't want to do this, so I said, 'Okay, let's not,' and I got off the couch and started getting dressed."

"You broke off from the encounter yourself."

"Yes."

"Because she had said, 'no'?"

"No, that was just an excuse. Because I was feeling uneasy at that point."

"Uh-huh. So you got off the couch and started to get dressed . . ."

"Yes."

"And did you say anything at that time? To explain your actions?"

"Yes. I said that I didn't think this was a good idea, and I didn't feel good about it."

"And how did she respond?"

"She got very angry. She started throwing things at me. Then she started hitting me. And scratching me."

"Do you have any marks?"

"Yes."

"Where are they?"

"On my neck and chest."

"Have they been photographed yet?"

"No."

"All right. Now when she scratched you, how did you respond?"

"I just tried to get dressed and get out of there."

"You didn't respond directly to her attack?"

"Well, at one point I pushed her back, to get her away from me, and she tripped on a table and fell on the floor."

"You make it sound like pushing her was self-defense on your part."

"It was. She was ripping the buttons off my shirt. I had to go home, and I didn't want my wife to see my shirt, so I pushed her away."

"Did you ever do anything that was *not* self-defense?"

"No."

"Did you hit her at any time?"

"No."

"You're sure about that?"

"Yes."

"All right. What happened then?"

"She threw a wineglass at me. But by then I was pretty much dressed. I went and got my phone from her windowsill, and then I went—"

"I'm sorry. You got your phone? What phone is that?"

"I had a cellular phone." He took it out of his pocket and showed her. "We all carry them in the company, because we make them. And I had been using the phone to make a call from her office, when she started kissing me."

"Were you in the middle of a call when she started kissing you?"

"Yes."

"Whom were you talking to?"

"An answering machine."

"I see." She was clearly disappointed. "Go on, please."

"So I went and got my phone and got the hell out of there. She was screaming that I couldn't do this to her, that she would kill me."

"And you responded how?"

"Nothing. I just left."

"And this was at what time?"

"About six forty-five."

"Did anybody see you leave?"

"The cleaning lady."

"Do you happen to know her name?"

"No."

"Ever seen her before?"

"No."

"Do you think she worked for your company?"

"She had a company uniform on. You know, for the maintenance firm that cleans up our offices."

"Uh-huh. And then?"

He shrugged. "I went home."

"Did you tell your wife what happened?"

"No."

"Did you tell anybody what happened?"

"No."

"Why not?"

"I guess I was in shock."

She paused and looked back over her notes. "All right. You say you were sexually harassed. And you have described a very direct overture by this woman. Since she was your boss, I would have thought you'd feel yourself at some risk in turning her down."

"Well. I was concerned. Sure. But I mean, don't I have the right to turn her down? Isn't that what this is about?"

"Certainly you have that right. I'm asking about your state of mind."

"I was very upset."

"Yet you did not want to tell anybody what had happened? You did not want to share this upsetting experience with a colleague? A friend? A family member, perhaps a brother? Anybody at all?"

"No. It didn't even occur to me. I didn't know how to deal with what— I guess I was in shock. I just wanted it to go away. I wanted to think it had never happened."

"Did you make any notes?"

"No."

"All right. Now, you mentioned that you didn't tell your wife. Would you say you concealed it from your wife?"

He hesitated. "Yes."

"Do you often conceal things from her?"

"No. But in this instance, you know, involving an old girlfriend, I didn't think she would be sympathetic. I didn't want to deal with her about this."

"Have you had other affairs?"

"This wasn't an affair."

"I'm asking a general question. In terms of your relationship to your wife."

"No. I haven't had affairs."

"All right. I advise you to tell your wife at once. Make a full and complete disclosure. Because I promise you that she will find out, if she hasn't done so already. However difficult it may be to tell her, your best chance to preserve your relationship is to be completely honest with her."

"Okay."

"Now, going back to last night. What happened next?"

"Meredith Johnson called the house and spoke to my wife."

Fernandez's eyebrows went up. "I see. Did you expect that to happen?"

"God, no. It scared the hell out of me. But apparently she was friendly, and just called to say that the morning meeting was rescheduled for eight-thirty. Today."

"I see."

"But when I got to work today, I found that the meeting had actually been scheduled for eight."

"So you arrived late, and were embarrassed, and so on."

"Yes."

"And you believe that it was a setup."

"Yes."

Fernandez glanced at her watch. "I'm afraid I'm running out of time. Bring me up to date about what happened today quickly, if you can."

Without mentioning Conley-White, he described the morning meeting briefly and his subsequent humiliation. His argument with Meredith. His conversation with Phil Blackburn. The offer of a lateral transfer. The fact that the transfer would deny him the benefits of a possible spin-off. His decision to seek advice.

Fernandez asked few questions and wrote steadily. Finally, she pushed the yellow pad aside.

"All right. I think I have enough to get the picture. You're feeling slighted and ignored. And your question is, do you have a harassment case?"

"Yes," he said, nodding.

"Well. Arguably you do. It's a jury case, and we don't know what would happen if we went to trial. But based on what you have told me here, I have to advise you that your case is not strong."

Sanders felt stunned. "Jesus."

"I don't make the law. I'm just telling you frankly, so you can arrive at an informed decision. Your situation is not good, Mr. Sanders."

Fernandez pushed back from her desk and began to stuff papers into her briefcase. "I have five minutes, but let me review for you what sexual harassment actually is, under the law, because many clients aren't clear about it. Title VII of the Civil Rights Act of 1964 made sex discrimination in the workplace illegal, but as a practical matter what we call sexual harassment was not defined for many years. Since the middle nineteen-eighties, the Equal Employment Opportunities Commission has, under Title VII, produced guidelines to define sexual harassment. In the last few years, these EEOC guidelines have been further clarified by case law. So the definitions are quite explicit. According to the law, for a complaint to qualify as sexual harassment, the behavior must contain three elements. First, it must be sexual. That means, for example, that making a profane or scatological joke is not sexual harassment, even though a listener may find it offensive. The conduct must be sexual in nature. In your case, there's no doubt about the explicitly sexual element, from what you have told me."

"Okay."

"Second, the behavior must be unwelcome. The courts distinguish between behavior that is voluntary and behavior that is welcome. For example, a person may be having a sexual relationship with a superior and it's obviously voluntary—no one's holding a gun to the person's head. But the courts understand that the employee may feel that they have no choice but to comply, and therefore the sexual relationship was not freely entered into—it's not welcome.

"To determine if behavior is really unwelcome, the courts look at the surrounding behavior in broad terms. Did the employee make sexual jokes in the workplace, and thus indicate that such jokes from others were welcome? Did the employee routinely engage in sexual banter, or sexual teasing with other employees? If the employee engaged in an actual affair, did they allow the supervisor into their apartment, did they visit the supervisor in the hospital, or see them at times when they didn't strictly have to, or engage in other actions that would suggest that they were actively and willingly participating in the relationship. In addition, the courts look to see if the employee ever told the supervisor the behavior was unwelcome, if the employee complained to anyone else about the relationship or tried to take any action to evade the unwelcome situation. That consideration becomes more significant when the employee is highly placed, and presumably more free to act."

"But I didn't tell anybody."

"No. And you didn't tell her, either. At least, not explicitly, so far as I can determine."

"I didn't feel I could."

"I understand you didn't. But it's a problem for your case. Now, the third element in sexual harassment is discrimination on the basis of gender. The most common is quid pro quo—the exchange of sexual favors in return for keeping your job or getting a promotion. The threat of that may be explicit or implied. I believe you said it was your understanding that Ms. Johnson had the ability to fire you?"

"Yes."

"How did you gain that understanding?"

"Phil Blackburn told me."

"Explicitly?"

"Yes."

"And what about Ms. Johnson? Did she make any offer contingent on sex? Did she make any reference to her ability to fire you, in the course of the evening?"

"Not exactly, but it was there. It was always in the air."

"How did you know?"

"She said things like 'As long as we're working together, we might as well have a little fun.' And she talked about wanting to have an affair during company trips we would make together to Malaysia, and so on."

"You interpreted this as an implied threat to your job?"

"I interpreted it to mean that if I wanted to get along with her, I had better go along with her."

"And you didn't want to do that?"

"No."

"Did you say so?"

"I said I was married, and that things had changed between us."

"Well, under most circumstances, that exchange alone would probably serve to establish your case. If there were witnesses."

"But there weren't."

"No. Now, there is a final consideration, which we call hostile working environment. This is ordinarily invoked in situations where an individual is harassed in a pattern of incidents that may not in themselves be sexual but that cumulatively amount to harassment based on gender. I don't believe you can claim hostile work environment on this single incident."

"I see."

"Unfortunately, the incident you describe is simply not as clear-cut as it might be. We would then turn to ancillary evidence of harassment. For example, if you were fired."

"I think in effect I have been fired," Sanders said. "Because I'm being pulled out of the division, and I won't get to participate in the spin-off."

"I understand. But the company's offer to transfer you laterally makes things complicated. Because the company can argue—very successfully, I think—that it does not owe you anything more than a lateral transfer. That it has never promised you the golden egg of a spin-off. That such a spin-off is in any case speculative, intended to occur at some future time, and it might never happen. That the company is not required to compensate you for your hopes—for some vague expectation of a future that might never occur. And therefore the company will claim that a lateral transfer is fully acceptable, and that you are being unreasonable if you turn it down. That you are in effect quitting, not being fired. It will place the burden back on you."

"That's ridiculous."

"Actually, it's not. Suppose, for example, you found out that you had terminal cancer and were going to die in six months. Would the company be required to pay the proceeds of the spin-off to your survivors? Clearly, no. If you're working in the company when it spins off, you participate. If you're not, you don't. The company has no broader obligation."

"You're saying I might as well have cancer."

"No, I'm saying that you're angry and you feel the company owes you something that the court will not agree it does. In my experience, sexual harassment claims often have this quality. People come in feeling angry and wronged, and they think they have rights that they simply don't have."

He sighed. "Would it be different if I were a woman?"

"Basically, no. Even in the most clear-cut situations—the most extreme and outrageous situations—sexual harassment is notoriously difficult to prove. Most cases occur as yours has: behind closed doors, with no witnesses. It's one person's word against another's. In that circumstance, where there is no clear-cut corroborating evidence, there is often a prejudice against the man."

"Uh-huh."

"Even so, one-fourth of all sexual harassment cases are brought by men. Most of those are brought against male bosses, but one-fifth are brought against women. And the number is increasing all the time, as we have more women bosses in the workplace."

"I didn't know that."

"It isn't much discussed," she said, peering over her glasses. "But it's happening. And from my point of view, it's to be expected."

"Why do you say that?"

"Harassment is about power—the undue exercise of power by a superior over a subordinate. I know there's a fashionable point of view that says women are fundamentally different from men, and that women would never harass an employee. But from where I sit, I've seen it all. I've seen and heard everything that you can imagine—and a lot that you wouldn't believe if I told you. That gives me another perspective.

Personally, I don't deal much in theory. I have to deal with the facts. And on the basis of facts, I don't see much difference in the behavior of men and women. At least, nothing that you can rely on."

"Then you believe my story?"

"Whether I believe you is not at issue. What's at issue is whether you realistically have a case, and therefore what you should do in your circumstances. I can tell you that I've heard it all before. You're not the first man I've been asked to represent, you know."

"What do you advise me to do?"

"I can't advise you," Fernandez said briskly. "The decision you face is much too difficult. I can only lay out the situation." She pushed her intercom button. "Bob, tell Richard and Eileen to bring the car around. I'll meet them in front of the building." She turned back to Sanders.

"Let me review your problems," she said. She ticked them off on her fingers. "One: you claim that you got into an intimate situation with a younger, very attractive woman but you turned her down. In the absence of witnesses or corroborating evidence, that isn't going to be an easy story to sell to a jury.

"Two: if you bring a lawsuit, the company will fire you. You're looking at three years before you come to trial. You have to think about how you'll support yourself during that time, about how you'll make your house payments, and your other expenses. I might take you on a contingency basis, but you'll still have to pay all direct costs throughout the trial. That will be a minimum of one hundred thousand dollars. I don't know whether you'll want to mortgage your house to pay for it. But it has to be dealt with.

"Three: a lawsuit will bring all this out into the open. It'll be in the papers and on the evening news for years before the trial begins. I can't adequately describe how destructive an experience that is—for you, and for your wife and family. Many families don't survive the pre-trial period intact. There are divorces, suicides, illnesses. It's *very* difficult.

"Four: because of the offer of lateral transfer, it's not clear what we can claim as damages. The company will claim that you have no case, and we'll have to fight it. But even with a stunning victory, you may end up with only a couple of hundred thousand dollars after expenses and fees and three years of your life. And of course the company can appeal, delaying payment further.

"Five: if you bring a lawsuit, you'll never work in this industry again.

I know it's not supposed to work that way, but as a practical matter, you'll never be hired for another job. That's just how it goes. It would be one thing if you were fifty-five. But you're only forty-one. I don't know if you want to make that choice, at this point in your life."

"Jesus." He slumped back in the chair.

"I'm sorry, but these are the facts of litigation."

"But it's so unjust."

She put on her raincoat. "Unfortunately, the law has nothing to do with justice, Mr. Sanders," she said. "It's merely a method for dispute resolution." She snapped her briefcase shut and extended her hand. "I'm sorry, Mr. Sanders. I wish it were different. Please feel free to call me again if you have any further questions."

She hurried out of the office, leaving him sitting there. After a moment the assistant came in. "Can I do anything for you?"

"No," Sanders said, shaking his head slowly. "No, I was just leaving."

I n the car, driving to the courthouse, Louise Fernandez recounted
Sanders's story to the two junior lawyers traveling with her. One
lawyer, a woman, said, "You don't really believe him?"

"Who knows?" Fernandez said. "It was behind closed doors. There's
never a way to know."

The young woman shook her head. "I just can't believe a woman
would act that way. So aggressively."

"Why not?" Fernandez said. "Suppose this wasn't a case of harassment. Suppose this was a question of implied promise between a man
and a woman. The man claims that behind closed doors he was promised a big bonus, but the woman denies it. Would you assume that the
man was lying because a woman wouldn't act that way?"

"Not about that, no."

"In that situation, you'd think that anything was possible."

"But this isn't a contract," the woman said. "This is sexual behavior."

"So you think women are unpredictable in their contractual arrangements, but stereotypical in their sexual arrangements?"

The woman said, "I don't know if *stereotypical* is the word I'd use."

"You just said that you can't believe a woman would act aggressively
in sex. Isn't that a stereotype?"

"Well, no," the woman said. "It's not a stereotype, because it's true.
Women are different from men when it comes to sex."

"And black people have rhythm," Fernandez said. "Asians are work-
aholics. And Hispanics don't confront . . ."

"But this is different. I mean, there are studies about this. Men and
women don't even talk to each other the same way."

"Oh, you mean like the studies that show that women are less good
at business and strategic thinking?"

"No. Those studies are wrong."

"I see. Those studies are wrong. But the studies about sexual differences are right?"

"Well, sure. Because sex is fundamental. It's a primal drive."

"I don't see why. It's used for all sorts of purposes. As a way of relating, a way of placating, a way of provoking, as an offer, as a weapon, as a threat. It can be quite complicated, the ways sex is used. Haven't you found that to be true?"

The woman crossed her arms. "I don't think so."

Speaking for the first time, the young man said, "So what'd you tell this guy? Not to litigate?"

"No. But I told him his problems."

"What do you think he should do?"

"I don't know," Fernandez said. "But I know what he should have done."

"What?"

"It's terrible to say it," she said. "But in the real world? With no witnesses? Alone in the office with his boss? He probably should have shut up and fucked her. Because right now, that poor bastard has no options at all. If he's not careful, his life is over."

anders walked slowly back down the hill toward Pioneer Square. The rain had stopped, but the afternoon was still damp and gray. The wet pavement beneath his feet sloped steeply downward. Around him the tops of the skyscrapers disappeared into the low-hanging, chilly mist.

He was not sure what he had expected to hear from Louise Fernandez, but it was certainly not a detailed account of the possibility of his being fired, mortgaging his house, and never working again.

Sanders felt overwhelmed by the sudden turn that his life had taken, and by a realization of the precariousness of his existence. Two days ago, he was an established executive with a stable position and a promising future. Now he faced disgrace, humiliation, loss of his job. All sense of security had vanished.

He thought of all the questions Fernandez had asked him—questions that had never occurred to him before. Why hadn't he told anyone. Why hadn't he made notes. Why hadn't he told Meredith explicitly that her advances were unwelcome. Fernandez operated in a world of rules and distinctions that he did not understand, that had never crossed his mind. And now those distinctions turned out to be vitally important.

Your situation is not good, Mr. Sanders.

And yet . . . how could he have prevented this? What should he have done instead? He considered the possibilities.

Suppose he had called Blackburn right after the meeting with Meredith, and had told him in detail that Meredith had harassed him. He could have called from the ferry, lodged his complaint before she lodged hers. Would it have made a difference? What would Blackburn have done?

He shook his head, thinking about it. It seemed unlikely that anything would make a difference. Because in the end, Meredith was tied in to the power structure of the company in a way that Sanders was not. Meredith was a corporate player; she had power, allies. That was the message—the final message—of this situation. Sanders didn't count. He was just a technical guy, a cog in the corporate wheels. His job was to get along with his new boss, and he had failed to do that. Whatever he did now was just whining. Or worse: ratting on the boss. Whistle-blowing. And nobody liked a whistle-blower.

So what could he have done?

As he thought about it, he realized that he couldn't have called Blackburn right after the meeting because his cellular phone had gone dead, its power drained.

He had a sudden image of a car—*a man and a woman in a car, driving to a party*. Somebody had told him something once . . . a story about some people in a car.

It teased him. He couldn't quite get it.

There were plenty of reasons why the phone might be dead. The most likely explanation was nicad memory. The new phones used rechargeable nickel-cadmium batteries, and if they didn't completely discharge between uses, the batteries could reset themselves at a shorter duration. You never knew when it was going to show up. Sanders had had to throw out batteries before because they developed a short memory.

He took out his phone, turned it on. It glowed brightly. The battery was holding up fine today.

But there was something . . .

Driving in a car.

Something he wasn't thinking about.

Going to a party.

He frowned. He couldn't get it. It hung at the back of his memory, too dim to recover.

But it started him thinking: what else wasn't he getting? Because as he considered the whole situation, he began to have the nagging sense that there was something else that he was overlooking. And he had the feeling that Fernandez had overlooked it, too. Something hadn't come up in her questions to him. Something that everybody was taking for granted, even though—

Meredith.

Something about Meredith.

She had accused him of harassment. She had gone to Blackburn and accused him the next morning. Why would she do that? No doubt she felt guilty about what had happened at the meeting. And perhaps she was afraid Sanders would accuse her, so she decided to accuse him first. Her accusation was understandable in that light.

But if Meredith really had power, it didn't make sense to raise the sexual issue at all. She could just as easily have gone to Blackburn and said, Listen, it isn't working out with Tom. I can't deal with him. We have to make a change. And Blackburn would have done it.

Instead, she had accused him of harassment. And that must have been embarrassing to her. Because harassment implied a loss of control. It meant that she had not been able to control her subordinate in a meeting. Even if something unpleasant did happen, a boss would never mention it.

Harassment is about power.

It was one thing if you were a lowly female assistant fondled by a stronger, powerful man. But in this case Meredith was the boss. She had all the power. Why would she claim harassment by Sanders? Because the fact was, subordinates didn't harass their bosses. It just didn't happen. You'd have to be crazy to harass your boss.

Harassment is about power—the undue exercise of power by a superior over a subordinate.

For her to claim sexual harassment was, in an odd way, to admit that she was subordinate to Sanders. And she would never do that. Quite the contrary: Meredith was new to her job, eager to prove that she was in control of the situation. So her accusation made no sense—unless she was using it as a convenient way to destroy him. Sexual harassment had the advantage of being a charge that was difficult to recover from. You were presumed guilty until proven innocent—and it was hard to prove innocence. It tarnished any man, no matter how frivolous the accusation. In that sense, harassment was a very powerful accusation. The most powerful accusation she could make.

But then, she said that she wasn't going to press charges. And the question was—

Why not?

Sanders stopped on the street.

That was it.

She's assured me, she's not going to press charges.

Why wasn't Meredith going to press charges?

At the time that Blackburn said that, Sanders had never questioned it. Louise Fernandez had never questioned it. But the fact was, Meredith's refusal to press charges made no sense at all. She had already accused him. Why not press it? Why not carry it to its conclusion?

Maybe Blackburn had talked her out of it. Blackburn was always so concerned about appearances.

But Sanders didn't think that was what had happened. Because a formal accusation could still be handled quietly. It could be processed inside the company.

And from Meredith's standpoint, there were real advantages to a formal accusation. Sanders was popular at DigiCom. He had been with the company a long time. If her goal was to get rid of him, to banish him to Texas, why not defuse the inevitable corporate grumbling by letting the accusation work its way through the company grapevine? Why not make it official?

The more Sanders thought about it, the more it seemed that there was only one explanation: Meredith wasn't going to press charges because she couldn't.

She couldn't, because she had some other problem.

Some other consideration.

Something else was going on.

We can handle it quietly.

Slowly, Sanders began to see everything differently. In the meeting earlier that day, Blackburn hadn't been ignoring him or slighting him. Not at all: Blackburn was scrambling.

Blackburn was scared.

We can handle it quietly. It's best for everyone.

What did he mean, best for everyone?

What problem did Meredith have?

What problem *could* she have?

The more Sanders thought about it, the more it seemed that there could be only one possible reason why she wasn't pressing charges against him.

He took out his phone, called United Airlines, and booked three round-trip tickets to Phoenix.

And then he called his wife.

You goddamn son of a bitch," Susan said.

They were sitting in a corner table at Il Terrazzo. It was two o'clock; the restaurant was nearly deserted. Susan had listened to him for half an hour, without interruption or comment. He told her everything that had happened in his meeting with Meredith, and everything that had happened that morning. The Conley-White meeting. The conversation with Phil. The conversation with Fernandez. Now he had finished. She stared at him.

"I could really learn to despise you, you know that? You son of a bitch, why didn't you tell me she was your ex-girlfriend?"

"I don't know," he said. "I didn't want to go into it."

"You didn't want to go into it? Adele and Mary Anne are talking to me on the phone all day, and they know, but I don't? It's humiliating, Tom."

"Well," he said, "you know you've been upset a lot lately, and—"

"Cut the crap, Tom," she said. "This has nothing to do with me. You didn't tell me because you didn't want to."

"Susan, that's not—"

"Yes it is, Tom. I was asking you about her, last night. You could have told me if you wanted to. But you didn't." She shook her head. "Son of a bitch. I can't believe what an asshole you are. You've made a real mess of this. Do you realize what a mess this is?"

"Yes," he said, hanging his head.

"Don't act contrite with me, you asshole."

"I'm sorry," he said.

"You're sorry? Fuck you, you're sorry. Jesus Christ. I can't believe you. What an asshole. You spent the night with your goddamned *girlfriend*."

"I didn't spend the night. And she's not my girlfriend."

"What do you mean? She was your big heartthrob."

"She wasn't my 'big heartthrob.' "

"Oh yeah? Then why wouldn't you tell me?" She shook her head. "Just answer one question. Did you fuck her or not?"

"No. I didn't."

She stared at him intently, stirring her coffee. "You're telling me the truth?"

"Yes."

"Nothing left out? No inconvenient parts skipped?"

"No. Nothing."

"Then why would she accuse you?"

"What do you mean?" he said.

"I mean, there must be a reason she accused you. You must have done something."

"Well, I didn't. I turned her down."

"Uh-huh. Sure." She frowned at him. "You know, this is not just about you, Tom. This involves your whole family: me and the kids."

"I understand that."

"Why didn't you tell me? If you told me last night, I could have helped you."

"Then help me now."

"Well, there isn't much we can do now," Susan said, with heavy sarcasm. "Not after she's gone to Blackburn and made an accusation first. Now you're finished."

"I'm not so sure."

"Trust me, you haven't got a move," she said. "If you go to trial, it'll be living hell for at least three years, and I personally don't think you can win. You're a man bringing a charge of harassment against a woman. They'll laugh you out of court."

"Maybe."

"Trust me, they will. So you can't go to trial. What can you do? Move to Austin. Jesus."

"I keep thinking," Sanders said. "She accused me of harassment, but now she isn't pressing charges. And I keep thinking, Why isn't she pressing charges?"

"Who *cares*?" Susan said, with an irritable wave of the hand. "It could be any of a million reasons. Corporate politics. Or Phil talked her out of it. Or Garvin. It doesn't matter why. Tom, face the facts: *you have no move.* Not now, you stupid son of a bitch."

"Susan, will you settle down?"

"Fuck you, Tom. You're dishonest and irresponsible."

"Susan—"

"We've been married five years. I deserve better than this."

"Will you take it easy? I'm trying to tell you: I think I *do* have a move."

"Tom. You *don't.*"

"I think I do. Because this is a very dangerous situation," Sanders said. "It's dangerous for everybody."

"What does that mean?"

"Let's assume that Louise Fernandez told me the truth about my lawsuit."

"She did. She's a good lawyer."

"But she wasn't looking at it from the company's standpoint. She was looking at it from the plaintiff's standpoint."

"Yeah, well, you're a plaintiff."

"No, I'm not," he said. "I'm a *potential* plaintiff."

There was a moment of silence.

Susan stared at him. Her eyes scanned his face. She frowned. He watched her put it together. "You're kidding."

"No."

"You must be out of your mind."

"No. Look at the situation. DigiCom's in the middle of a merger with a very conservative East Coast company. A company that's already pulled out of one merger because an employee had a little bad publicity. Supposedly this employee used some rough language while firing a temp secretary, and then Conley-White pulled out. They're very skittish about publicity. Which means the last thing anybody at DigiCom wants is a sexual harassment suit against the new female vice president."

"Tom. Do you realize what you're saying?"

"Yes," he said.

"If you do this, they're going to go *crazy*. They're going to try to destroy you."

"I know."

"Have you talked to Max about this? Maybe you should."

"The hell with Max. He's a crazy old man."

"I'd ask him. Because this isn't really your thing, Tom. You were never a corporate infighter. I don't know if you can pull this off."

"I think I can."

"It'll be nasty. In a day or so, you're going to wish you had taken the Austin job."

"Fuck it."

"It'll get really mean, Tom. You'll lose your friends."

"Fuck it."

"Just so you're ready."

"I am." Sanders looked at his watch. "Susan, I want you to take the kids and visit your mother for a few days." Her mother lived in Phoenix. "If you go home now and pack, you can make the eight o'clock flight at Sea-Tac. I've booked three seats for you."

She stared at him, as if she were seeing a stranger. "You're really going to do this . . . ," she said slowly.

"Yes. I am."

"Oh boy." She bent over, picked up her purse from the floor, and pulled out her day organizer.

He said, "I don't want you or the kids to be involved. I don't want anybody pushing a news camera in their faces, Susan."

"Well, just a minute . . ." She ran her finger down her appointments. "I can move that . . . And . . . conference call . . . Yes." She looked up. "Yes. I can leave for a few days." She glanced at her watch. "I guess I better hurry and pack."

He stood up and walked outside the restaurant with her. It was raining; the light on the street was gray and bleak. She looked up at him and kissed him on the cheek. "Good luck, Tom. Be careful."

He could see that she was frightened. It made him frightened, too.

"I'll be okay."

"I love you," she said. And then she walked quickly away in the rain. He waited for a moment to see if she looked back at him, but she never did.

W alking back to his office, he suddenly realized how alone he felt. Susan was leaving with the kids. He was on his own now. He had imagined he would feel relieved, free to act without restraint, but instead he felt abandoned and at risk. Chilled, he thrust his hands into the pockets of his raincoat.

He hadn't handled the lunch with Susan well. And she would be going off, mulling over his answers.

Why didn't you tell me?

He hadn't answered that well. He hadn't been able to express the conflicting feelings he had experienced last night. The unclean feeling, and the guilt, and the sense that he had somehow done something wrong, even though he hadn't done anything wrong.

You could have told me.

He hadn't done anything wrong, he told himself. But then why hadn't he told her? He had no answer to that. He passed a graphics shop, and a plumbing supply store with white porcelain fixtures in a window display.

You didn't tell me because you didn't want to.

But that made no sense. Why wouldn't he want to tell her? Once again, his thoughts were interrupted by images from the past: the white garter belt . . . a bowl of popcorn. . . . the stained-glass flower on the door to his apartment.

Cut the crap, Tom. This has nothing to do with me.

Blood in the white bathroom sink, and Meredith laughing about it. Why was she laughing? He couldn't remember now; it was just an isolated image. A stewardess putting a tray of airline food in front of him. A suitcase on the bed. The television sound turned off. The stained-glass flower, in gaudy orange and purple.

Have you talked to Max?

She was right about that, he thought. He should talk to Max. And he would, right after he gave Blackburn the bad news.

Sanders was back at his office at two-thirty. He was surprised to find Blackburn there, standing behind Sanders's desk, talking on his phone. Blackburn hung up, looking a little guilty. "Oh, Tom. Good. I'm glad you're back." He walked back around Sanders's desk. "What have you decided?"

"I've thought this over very carefully," Sanders said, closing the door to the hallway.

"And?"

"I've decided to retain Louise Fernandez of Marin, Howard to represent me."

Blackburn looked puzzled. "To represent you?"

"Yes. In the event it becomes necessary to litigate."

"Litigate," Blackburn said. "On what basis would you litigate, Tom?"

"Sexual harassment under Title VII," Sanders said.

"Oh, Tom," Blackburn said, making a mournful face. "That would be unwise. That would be very unwise. I urge you to reconsider."

"I've reconsidered all day," Sanders said. "But the fact is, Meredith Johnson harassed me, she made advances to me and I turned her down. Now she's a woman scorned, and she is being vindictive toward me. I'm prepared to sue if it comes to that."

"Tom . . ."

"That's it, Phil. That's what'll happen if you transfer me out of the division."

Blackburn threw up his hands. "But what do you expect us to do? Transfer Meredith?"

"Yes," Sanders said. "Or fire her. That's the usual thing one does with a harassing supervisor."

"But you forget: she's accused you of harassment, too."

"She's lying," Sanders said.

"But there are no witnesses, Tom. No evidence either way. You and she are both our trusted employees. How do you expect us to decide who to believe?"

"That's your problem, Phil. All I have to say is, I'm innocent. And I'm prepared to sue."

Blackburn stood in the middle of the room, frowning. "Louise Fernandez is a smart attorney. I can't believe she recommended this course of action to you."

"No. This is my decision."

"Then it's very unwise," Blackburn said. "You are putting the company in a very difficult position."

"The company is putting me in a difficult position."

"I don't know what to say," Phil said. "I hope this doesn't force us to terminate you."

Sanders stared at him, meeting his gaze evenly. "I hope not, too," he said. "But I don't have confidence that the company has taken my complaint seriously. I'll fill out a formal charge of sexual harassment with Bill Everts in HR later today. And I'm asking Louise to draw up the necessary papers to file with the state Human Rights Commission."

"Christ."

"She should file first thing tomorrow morning."

"I don't see what the rush is."

"There's no rush. It's just a filing. To get the complaint on record. I'm required to do that."

"But this is very serious, Tom."

"I know it, Phil."

"I'd like to ask you to do me a favor, as your friend."

"What's that?"

"Hold off the formal complaint. At least, with the HRC. Give us a chance to conduct an in-house investigation before you take this outside."

"But you aren't conducting an in-house investigation, Phil."

"Yes, we are."

"You didn't even want to hear my side of the story this morning. You told me it didn't matter."

"That's not true," Blackburn said. "You misunderstood me entirely. Of course it matters. And I assure you, we will hear your story in detail as part of our investigation."

"I don't know, Phil," Sanders said. "I don't see how the company can be neutral on this issue. It seems everything is stacked against me. Everybody believes Meredith and not me."

"I assure you that is not the case."

"It certainly seems like it. You told me this morning how well connected she is. How many allies she has. You mentioned that several times."

"Our investigation will be scrupulous and impartial. But in any case it seems reasonable to ask you to wait for the outcome before filing with a state agency."

"How long do you want me to wait?"

"Thirty days."

Sanders laughed.

"But that's the standard time for a harassment investigation."

"You could do it in a day, if you wanted to."

"But you must agree, Tom, that we're very busy right now, with all the merger meetings."

"That's your problem, Phil. I have a different problem. I've been unjustly treated by my superior, and I feel I have a right, as a long-standing senior employee, to see my complaint resolved promptly."

Blackburn sighed. "All right. Let me get back to you," he said. He hurried out of the room.

Sanders slumped in his chair and stared into space.

It had begun.

Fifteen minutes later, Blackburn met with Garvin in the fifth-floor executive conference room. Also present at the meeting were Stephanie Kaplan and Bill Everts, the head of Human Resources at DigiCom.

Blackburn began the meeting by saying, "Tom Sanders has retained outside counsel and is threatening litigation over Meredith Johnson."

"Oh, Christ," Garvin said.

"He's claiming sexual harassment."

Garvin kicked the leg of the table. "That son of a bitch."

Kaplan said, "What does he say happened?"

"I don't have all the details yet," Blackburn said. "But in essence he claims that Meredith made sexual overtures to him in her office last night, that he turned her down, and that now she is being vindictive."

Garvin gave a long sigh. "Shit," he said. "This is just what I didn't want to happen. This could be a *disaster*."

"I know, Bob."

Stephanie Kaplan said, "Did she do it?"

"Christ," Garvin said. "Who knows in these situations. That's always the question." He turned to Everts. "Has Sanders come to you about this?"

"Not yet, no. I imagine he will."

"We have to keep it in-house," Garvin said. "That's essential."

"Essential," Kaplan said, nodding. "Phil has to make sure it stays in-house."

"I'm trying," Blackburn said. "But Sanders is talking about filing tomorrow with the HRC."

"That's a public filing?"

"Yes."

"How soon is it made public?"

"Probably within forty-eight hours. Depending on how fast HRC does the paperwork."

"Christ," Garvin said. "Forty-eight hours? What's the matter with him? Doesn't he realize what he's doing?"

Blackburn said, "I think he does. I think he knows exactly."

"Blackmail?"

"Well. Pressure."

Garvin said, "Have you talked to Meredith?"

"Not since this morning."

"Somebody's got to talk to her. I'll talk to her. But how are we going to stop Sanders?"

Blackburn said, "I asked him to hold off the HRC filing, pending our investigation, for thirty days. He said no. He said we should be able to conduct our investigation in one day."

"Well, he got that right," Garvin said. "For all kinds of reasons, we damn well better conduct the investigation in one day."

"Bob, I don't know if that's possible," Blackburn said. "We have significant exposure here. The corporation is required by law to conduct a thorough and impartial investigation. We can't appear to be rushed or—"

"Oh, for Christ's sake," Garvin said. "I don't want to hear this legal pissing and moaning. What are we talking about? Two people, right? And no witnesses, right? So there's just two people. How long does it take to interview two people?"

"Well, it may not be that simple," Blackburn said, with a significant look.

"I'll tell you what's simple," Garvin said. "This is what's simple. Conley-White is a company obsessed with its public image. They sell textbooks to school boards that believe in Noah's ark. They sell magazines for kids. They have a vitamin company. They have a health-food company that markets baby foods. Rainbow Mush or something. Now Conley-White's buying our company, and in the middle of the acquisition a high-profile female executive, the woman in line to become CEO within two years, is accused of seeking sexual favors from a married man. You know what they're going to do if that gets out? They're going to *bail*. You know that Nichols is looking for any excuse to weasel out of this thing. This is perfect for him. Christ."

"But Sanders has already questioned our impartiality," Blackburn said. "And I'm not sure how many people know about the, ah, prior questions that we—"

"Quite a few," Kaplan said. "And didn't it come up at an officers' meeting last year?"

"Check the minutes," Garvin said. "We have no legal problem with current corporate officers, is that right?"

"That's right," Blackburn said. "Current corporate officers cannot be questioned or deposed on these matters."

"And we haven't lost any corporate officers in the last year? Nobody retired or moved?"

"No."

"Okay. So fuck him." Garvin turned to Everts. "Bill, I want you to go back through the HR records, and look carefully at Sanders. See if he's dotted every *i* and crossed every *t*. If he hasn't, I want to know."

"Right," Everts said. "But my guess is he's clean."

"All right," Garvin said, "let's assume that he is. What's it going to take to make Sanders go away? What does he want?"

Blackburn said, "I think he wants his job, Bob."

"He can't have his job."

"Well, that's the problem," Blackburn said.

Garvin snorted. "What's our liability, assuming he ever got to trial?"

"I don't think he has a case, based on what happened in that office. Our biggest liability would come from any perceived failure to respect due process and conduct a thorough investigation. Sanders could win on that alone, if we're not careful. That's my point."

"So we'll be careful. Fine."

"Now, guys," Blackburn said. "I feel strongly obliged to insert a note of caution. The extreme delicacy of this situation means that we have to be mindful of the details. As Pascal once said, 'God is in the details.' And in this case, the competing balance of legitimate legal claims forces me to admit it's unclear precisely what our best—"

"Phil," Garvin said. "Cut the crap."

Kaplan said, "Mies."

Blackburn said, "What?"

"Mies van der Rohe said, 'God is in the details.' "

"Who gives a shit?" Garvin said, pounding the table. "The point is, Sanders has no case—he just has us by the balls. And he knows it."

Blackburn winced. "I wouldn't phrase it exactly that way, but—"

"But that's the fucking situation."

"Yes."

Kaplan said, "Tom's smart, you know. A little naïve, but smart."

"Very smart," Garvin said. "Remember, I trained him. Taught him all he knows. He's going to be a big problem." He turned to Blackburn. "Get to the bottom line. What're we dealing with? Impartiality, right?"

"Yes . . ."

"And we want to move him out."

"Right."

"Okay. Will he accept mediation?"

"I don't know. I doubt it."

"Why not?"

"Ordinarily, we only use mediation to resolve settlement packages for employees who are leaving."

"So?"

"I think that's how he'll view it."

"Let's try, anyway. Tell him it's nonbinding, and see if we can get him to accept it on that basis. Give him three names and let him pick one. Mediate it tomorrow. Do I need to talk to him?"

"Probably. Let me try first, and you back up."

"Okay."

Kaplan said, "Of course, if we go to an outside mediator, we introduce an unpredictable element."

"You mean the mediator could find against us? I'll take the risk," Garvin said. "The important thing is to get the thing resolved. Quietly—and fast. I don't want Ed Nichols backpedaling on me. We have a press conference scheduled for Friday noon. I want this issue dead and buried by then, and I want Meredith Johnson announced as the new head of the division on Friday. Everybody clear on what's going to happen?"

They said they were.

"Then do it," Garvin said, and walked out of the room. Blackburn hurried after him.

In the hallway outside, Garvin said to Blackburn, "Christ, what a mess. Let me tell you. I'm very unhappy."

"I know," Blackburn said mournfully. He was shaking his head sadly.

"You really screwed the pooch on this one, Phil. Christ. You could have handled this one better. A *lot* better."

"How? What could I have done? He says that she hustled him, Bob. It's a serious matter."

"Meredith Johnson is vital to the success of this merger," Garvin said flatly.

"Yes, Bob. Of course."

"We must keep her."

"Yes, Bob. But we both know that in the past she has—"

"She has proven herself an outstanding piece of executive talent," Garvin said, interrupting him. "I won't allow these ridiculous allegations to jeopardize her career."

Blackburn was aware of Garvin's unswerving support of Meredith. For years, Garvin had had a blind spot for Johnson. Whenever criticisms of Johnson arose, Garvin would somehow change the subject, shift to something else. There was no reasoning with him. But now Blackburn felt he had to try. "Bob," he said. "Meredith's only human. We know she has her limitations."

"Yes," Garvin said. "She has youth. Enthusiasm. Honesty. Unwillingness to play corporate games. And of course, she's a woman. That's a real limitation, being a woman."

"But Bob—"

"I tell you, I can't stomach the excuses anymore," Garvin said. "We don't have women in high corporate positions here. Nobody does. Corporate America is rooms full of men. And whenever I talk about putting a woman in, there's always a 'But Bob' that comes up. The hell with it, Phil. We've got to break the glass ceiling sometime."

Blackburn sighed. Garvin was shifting the subject again. He said, "Bob, nobody's disagreeing with—"

"Yes, they are. You're disagreeing, Phil. You're giving me excuses why Meredith isn't suitable. And I'm telling you that if I had named some other woman, there'd be other excuses why that other woman isn't suitable. And I tell you, I'm tired of it."

Blackburn said, "We've got Stephanie. We've got Mary Anne."

"Tokens," Garvin said, with a dismissing wave. "Sure, let the CFO be a woman. Let a couple of the midrange execs be women. Throw the broads a bone. The fact remains. You can't tell me that a bright, able young woman starting out in business isn't held back by a hundred little *reasons*, oh such good *reasons*, why she shouldn't be advanced, why she shouldn't attain a major position of power. But in the end, it's just prejudice. And it has to stop. We have to give these bright young women a decent opportunity."

Blackburn said, "Well, Bob. I just think it would be prudent for you to get Meredith's view of this situation."

"I will. I'll find out what the hell happened. I know she'll tell me. But this thing still has to be resolved."

"Yes, it does, Bob."

"And I want you to be clear. I expect you to do whatever is necessary to get it resolved."

"Okay, Bob."

"*Whatever* is necessary," Garvin said. "Put the pressure on Sanders. Make sure he feels it. Rattle his cage, Phil."

"Okay, Bob."

"I'll deal with Meredith. You just take care of Sanders. I want you to rattle his fucking cage until he's black and blue."

Bob." Meredith Johnson stood at one of the center tables in the Design Group laboratory, going over the torn-apart Twinkle drives with Mark Lewyn. She came over when she saw Garvin standing to one side. "I can't tell you how sorry I am about all this business with Sanders."

"We're having some problems with it," Garvin said.

"I keep going over what happened," she said. "Wondering what I should have done. But he was angry and out of control. He had too much to drink, and he behaved badly. Not that we all haven't done that at some time in our lives, but . . ." She shrugged. "Anyway, I'm sorry."

"Apparently, he's going to file a harassment charge."

"That's unfortunate," she said. "But I suppose it's part of the pattern—trying to humiliate me, to discredit me with the people in the division."

"I won't let that happen," Garvin said.

"He resented my getting the job, and he couldn't deal with having me as his superior. He had to try and put me in my place. Some men are like that." She shook her head sadly. "For all the talk about the new male sensibility, I'm afraid very few men are like you, Bob."

Garvin said, "My concern now, Meredith, is that his filing may interfere with the acquisition."

"I can't see why that would be a problem," she said. "I think we can keep it under control."

"It's a problem, if he files with the state HRC."

"You mean he's going to go *outside?*" she asked.

"Yes. That's exactly what I mean."

Meredith stared off into space. For the first time, she seemed to lose her composure. She bit her lip. "That could be very awkward."

"I'll say. I've sent Phil to see him, to ask if we can mediate. With an experienced outside person. Someone like Judge Murphy. I'm trying to arrange it for tomorrow."

"Fine," Meredith said. "I can clear my schedule for a couple of hours tomorrow. But I don't know what we can expect to come out of it. He won't admit what happened, I'm sure. And there isn't any record, or any witnesses."

"I wanted you to fill me in," Garvin said, "on exactly what did happen, last night."

"Oh, Bob," she sighed. "I blame myself, every time I go over it."

"You shouldn't."

"I know, but I do. If my assistant hadn't gone off to rent her apartment, I could have buzzed her in, and none of this would have happened."

"I think you better tell me, Meredith."

"Of course, Bob." She leaned toward him and spoke quietly, steadily, for the next several minutes. Garvin stood beside her, shaking his head angrily as he listened.

D on Cherry put his Nikes up on Lewyn's desk. "Yeah? So Garvin came in. Then what happened?"

"So Garvin's standing over there in the corner, hopping up and down from one foot to the other, the way he does. Waiting to be noticed. He won't come over, he's waiting to be noticed. And Meredith's talking to me about the Twinkle drive that I have spread all over the table, and I'm showing her what we've found is wrong with the laser heads—"

"She gets all that?"

"Yeah, she seems okay. She's not Sanders, but she's okay. Fast learner."

"And better perfume than Sanders," Cherry said.

"Yeah, I like her perfume," Lewyn said. "Anyway—"

"Sanders's perfume leaves a lot to be desired."

"Yeah. Anyway, pretty soon Garvin gets tired of hopping, and he gives a discreet little cough, and Meredith notices Garvin and she goes 'Oh,' with a little thrill in her voice, you know that little sharp intake of breath?"

"Uh-oh," Cherry said. "Are we talking humparoonie here or what?"

"Well, that's the *thing*," Lewyn said. "She goes running over to him, and he holds out his arms to her, and I tell you it looks like that ad where the two lovers run toward each other in slow motion."

"Uh-oh," Cherry said. "Garvin's wife is going to be pissed."

"But that's the thing," Lewyn said. "When they finally get together, standing there side by side, it isn't that way at all. They're talking, and she's sort of cooing and batting her eyes at him, and he's such a tough guy he doesn't acknowledge it, but it's working on him."

"She's seriously cute, that's why," Cherry said. "I mean face it, she's got an outstanding molded case, with superior fit and finish."

"But the thing is, it's not like lovers at all. I'm staring, trying not to stare, and I tell you, it's not lovers. It's something else. It's almost like father-daughter, Don."

"Hey. You can fuck your daughter. Millions do."

"No, you know what I think? I think Bob sees himself in her. He sees something that reminds him of himself when he was younger. Some kind of energy or something. And I tell you, she plays it, Don. He crosses his arms, she crosses hers. He leans against the wall, she leans against the wall. She matches him exactly. And from a distance, I'm telling you: *she looks like him*, Don."

"No . . ."

"Yes. Think about it."

"It'd have to be from a *very* long distance," Cherry said. He took his feet off the table, and got up to leave. "So what're we saying here? Nepotism in disguise?"

"I don't know. But Meredith's got some kind of rapport with him. It isn't pure business."

"Hey," Cherry said. "Nothing's pure business. I learned that one a long time ago."

Louise Fernandez came into her office, and dropped her briefcase on the floor. She thumbed through a stack of phone messages and turned to Sanders. "What's going on? I have three calls this afternoon from Phil Blackburn."

"That's because I told him I had retained you as my attorney, that I was prepared to litigate my claim. And I, uh, suggested that you were filing with the HRC in the morning."

"I couldn't possibly file tomorrow," she said. "And I wouldn't recommend that we do so now, in any event. Mr. Sanders, I take false statements very seriously. Don't ever characterize my actions again."

"I'm sorry," he said. "But things are happening very fast."

"Just so we are clear. I don't like it, and if it happens again, you'll be looking for new counsel." That coldness again, the sudden coldness. "Now. So you told Blackburn. What was his response?"

"He asked me if I would mediate."

"Absolutely not," Fernandez said.

"Why not?"

"Mediation is invariably to the benefit of the company."

"He said it would be nonbinding."

"Even so. It amounts to free discovery on their part. There's no reason to give it to them."

"And he said you could be present," he said.

"Of course I can be present, Mr. Sanders. That's no concession. You must have an attorney present at all times or the mediation will be invalid."

"Here are the three names he gave me, as possible mediators." Sanders passed her the list.

She glanced at it briefly. "The usual suspects. One of them is better than the other two. But I still don't—"

"He wants to do the mediation tomorrow."

"Tomorrow?" Fernandez stared at him, and sat back in her chair.

"Mr. Sanders, I'm all for a timely resolution, but this is ridiculous. We can't be ready by tomorrow. And as I said, I don't recommend that you agree to mediate under any circumstances. Is there something here I don't know?"

"Yes," he said.

"Let's have it."

He hesitated.

She said, "Any communication you make to me is privileged and confidential."

"All right. DigiCom is about to be acquired by a New York company called Conley-White."

"So the rumors are true."

"Yes," he said. "They intend to announce the merger at a press conference on Friday. And they intend to announce Meredith Johnson as the new vice president of the company, on Friday."

"I see," she said. "So that's Phil's urgency."

"Yes."

"And your complaint presents an immediate and serious problem for him."

He nodded. "Let's say it comes at a very sensitive time."

She was silent for a moment, peering at him over her reading glasses. "Mr. Sanders, I misjudged you. I had the impression you were a timid man."

"They're forcing me to do this."

"Are they." She gave him an appraising look. Then she pushed the intercom button. "Bob, let me see my calendar. I have to clear some things. And ask Herb and Alan to come in. Tell them to drop whatever they're doing. This is more important." She pushed the papers aside. "Are all the mediators on this list available?"

"I assume so."

"I'm going to request Barbara Murphy. Judge Murphy. You won't like her, but she'll do a better job than the others. I'll try and set it up for the afternoon if I can. We need the time. Otherwise, late morning. You realize the risk you're taking? I assume you do. This is a very dangerous game you've decided to play." She pushed the intercom. "Bob? Cancel Roger Rosenberg. Cancel Ellen at six. Remind me to call my husband and tell him I won't be home for dinner." She looked at Sanders. "Neither will you. Do you need to call home?"

"My wife and kids are leaving town tonight."

She raised her eyebrows. "You told her everything?"

"Yes."

"You *are* serious."

"Yes," he said. "I'm serious."

"Good," she said. "You're going to need to be. Let's be frank, Mr. Sanders. What you have embarked upon is not strictly a legal procedure. In essence, you're playing the pressure points."

"That's right."

"Between now and Friday, you're in a position to exert considerable pressure on your company."

"That's right."

"And they on you, Mr. Sanders. They on you."

He found himself in a conference room, facing five people, all taking notes. Seated on either side of Fernandez were two young lawyers, a woman named Eileen and a man named Richard. Then there were two investigators, Alan and Herb: one tall and handsome; the other chubby, with a pockmarked face and a camera hanging around his neck.

Fernandez made Sanders go over his story again, in greater detail. She paused frequently to ask questions, noting down times, names, and specific details. The two lawyers never said anything, although Sanders had the strong impression that the young woman was unsympathetic to him. The two investigators were also silent, except at specific points. After Sanders mentioned Meredith's assistant, Alan, the handsome one, said, "Her name again?"

"Betsy Ross. Like in the flag."

"She's on the fifth floor?"

"Yes."

"What time does she go home?"

"Last night, she left at six-fifteen."

"I may want to meet her casually. Can I go up to the fifth floor?"

"No. All visitors are stopped at reception in the downstairs lobby."

"What if I'm delivering a package? Would Betsy take delivery of a package?"

"No. Packages go to central receiving."

"Okay. What about flowers? Would they be delivered directly?"

"Yes, I guess so. You mean, like flowers for Meredith?"

"Yes," Alan said.

"I guess you could deliver those in person."

"Fine," Alan said, and made a note.

They stopped him a second time when he mentioned the cleaning woman he had seen on leaving Meredith's office.

"DigiCom uses a cleaning service?"

"Yes. AMS—American Management Services. They're over on—"

"We know them. On Boyle. What time do the cleaning crews enter the building?"

"Usually around seven."

"And this woman you didn't recognize. Describe her."

"About forty. Black. Very slender, gray hair, sort of curly."

"Tall? Short? What?"

He shrugged. "Medium."

Herb said, "That's not much. Can you give us anything else?"

Sanders hesitated. He thought about it. "No. I didn't really see her."

"Close your eyes," Fernandez said.

He closed them.

"Now take a deep breath, and put yourself back. It's yesterday evening. You have been in Meredith's office, the door has been closed for almost an hour, you have had your experience with her, now you are leaving the room, you are going out . . . How does the door open, in or out?"

"It opens in."

"So you pull the door open . . . you walk out . . . Fast or slow?"

"I'm walking fast."

"And you go into the outer room . . . What do you see?"

Through the door. Into the outer room, elevators directly ahead. Feeling disheveled, off balance, hoping there is no one to see him. Looking to the right at Betsy Ross's desk: clean, bare, chair pulled up to the edge of the desk. Notepad. Plastic cover on the computer. Desk light still burning.

Eyes swinging left, a cleaning woman at the other assistant's desk. Her big gray cleaning cart stands alongside her. The cleaning woman is lifting a trash basket to empty it into the plastic sack that hangs open from one end of the cart. The woman pauses in mid-lift, stares at him curiously. He is wondering how long she has been there, what she has heard from inside the room. A tinny radio on the cart is playing music.

"I'll fucking kill you for this!" Meredith calls after him.

The cleaning woman hears it. He looks away from her, embarrassed, and hurries toward the elevator. Feeling almost panic. He pushes the button.

"Do you see the woman?" Fernandez said.

"Yes. But it was so fast . . . And I didn't want to look at her." Sanders shook his head.

"Where are you now? At the elevator?"

"Yes."

"Can you see the woman?"

"No. I didn't want to look at her again."

"All right. Let's go back. No, no, keep your eyes closed. We'll do it again. Take a deep breath, and let it out slowly . . . Good . . . This time you're going to see everything in slow motion, like a movie. Now . . . come out through the door . . . and tell me when you see her for the first time."

Coming through the door. Everything slow. His head moving gently up and down with each footstep. Into the outer room. The desk to the right, tidy, lamp on. To the left, the other desk, the cleaning woman raising the—

"I see her."

"All right, now freeze what you see. Freeze it like a photograph."

"Okay."

"Now look at her. You can look at her now."

Standing with the trash basket in her hand. Staring at him, a bland expression. She's about forty. Short hair, curls. Blue uniform, like a hotel maid. A silver chain around her neck—no, hanging eyeglasses.

"She wears glasses around her neck, on a metal chain."

"Good. Just take your time. There's no rush. Look her up and down."

"I keep seeing her face . . ." *Staring at him. A bland expression.*

"Look away from her face. Look her up and down."

The uniform. Spray bottle clipped to her waist. Knee-length blue skirt. White shoes. Like a nurse. No. Sneakers. No. Thicker—running shoes. Thick soles. Dark laces. Something about the laces.

"She's got . . . sort of running shoes. Little old lady running shoes."

"Good."

"There's something funny about the laces."

"Can you see what's funny?"

"No. They're dark. Something funny. I . . . can't tell."

"All right. Open your eyes."

He looked at the five of them. He was back in the room. "That was weird," he said.

"If there was time," Fernandez said, "I would have a professional hypnotist take you through the entire evening. I've found it can be very useful. But there's no time. Boys? It's five o'clock. You better get started."

The two investigators collected their notes and left.

"What are they going to do?"

"If we were litigating this," Fernandez said, "we would have the right to depose potential witnesses—to question individuals within the company who might have knowledge bearing on the case. Under the present circumstances, we have no right to interrogate anybody, because you're entering into private mediation. But if one of the DigiCom assistants chooses to have a drink with a handsome delivery man after work, and if the conversation happens to turn to gossip about sex in the office, well, that's the way the cookie crumbles."

"We can use that information?"

Fernandez smiled. "Let's see what we find out first," she said. "Now, I want to go back over several points in your story, particularly starting at the time you decided not to have intercourse with Ms. Johnson."

"Again?"

"Yes. But I have a few things to do first. I need to call Phil Blackburn and arrange tomorrow's sessions. And I have some other things to check on. Let's break now and meet again in two hours. Meanwhile, have you cleaned out your office?"

"No," he said.

"You better clean it out. Anything personal or incriminating, get it out. From now on, expect your desk drawers to be gone through, your files to be searched, your mail to be read, your phone messages checked. Every aspect of your life is now public."

"Okay."

"So, go through your desk and your files. Remove anything of a personal nature."

"Okay."

"On your office computer, if you have any passwords, change them. Anything in electronic data files of a personal nature, get it out."

"Okay."

"Don't just remove it. Make sure you erase it, so it's unrecoverable."

"Okay."

"It's not a bad idea to do the same thing at home. Your drawers and files and computer."

"Okay." He was thinking: at home? Would they really break into his home?

"If you have any sensitive materials that you want to store, bring them to Richard here," she said, pointing to the young lawyer. "He'll have them taken to a safe-deposit box where they'll be kept for you. Don't tell me. I don't want to know anything about it."

"Okay."

"Now. Let's discuss the telephone. From now on, if you have any sensitive calls to make, don't use your office phone, your cellular phone, or your phone at home. Use a pay phone, and don't put it on a charge card, even your personal charge card. Get a roll of quarters and use them instead."

"You really think this is necessary?"

"I know it is necessary. Now. Is there anything in your past conduct with this company which might be said to be out of order?" She was peering at him over her glasses.

He shrugged. "I don't think so."

"Anything at all? Did you overstate your qualifications on your original job application? Did you abruptly terminate any employee? Have you had any kind of inquiry about your behavior or decisions? Were you ever the subject of an internal company investigation? And even if you weren't, did you ever, to your knowledge, do anything improper, however small or apparently minor?"

"Jesus," he said. "It's been twelve years."

"While you are cleaning out, think about it. I need to know anything that the company might drag up about you. Because if they can, they will."

"Okay."

"And one other point. I gather from what you've told me that nobody at your company is entirely clear why Johnson has enjoyed such a rapid rise among the executives."

"That's right."

"Find out."

"It won't be easy," Sanders said. "Everybody's talking about it, and nobody seems to know."

"But for everybody else," Fernandez said, "it's just gossip. For you, it's vital. We need to know where her connections are and why they exist. If we know that, we have a chance of pulling this thing off. But if we don't, Mr. Sanders, they're probably going to tear us apart."

He was back at DigiCom at six. Cindy was cleaning up her desk and was about to leave.

"Any calls?" he said, as he went into his office.

"Just one," she said. Her voice was tight.

"Who was that?"

"John Levin. He said it was important." Levin was an executive with a hard drive supplier. Whatever Levin wanted, it could wait.

Sanders looked at Cindy. She seemed tense, almost on the verge of tears.

"Something wrong?"

"No. Just a long day." A shrug: elaborate indifference.

"Anything I should know about?"

"No. It's been quiet. You didn't have any other calls." She hesitated. "Tom, I just want you to know, I don't believe what they are saying."

"What are they saying?" he asked.

"About Meredith Johnson."

"What about her?"

"That you sexually harassed her."

She blurted it out, and then waited. Watching him, her eyes moving across his face. He could see her uncertainty. Sanders felt uneasy in turn that this woman he had worked alongside for so many years would now be so openly unsure of him.

He said firmly, "It's not true, Cindy."

"Okay. I didn't think it was. It's just that everybody is—"

"There's no truth to it at all."

"Okay. Good." She nodded, put the call book in the desk drawer. She seemed eager to leave. "Did you need me to stay?"

"No."

"Good night, Tom."

"Good night, Cindy."

He went into his office and closed the door behind him. He sat behind his desk and looked at it a moment. Nothing seemed to have been touched. He flicked on his monitor, and began going through the drawers, rummaging through, trying to decide what to take out. He glanced up at the monitor, and saw that his e-mail icon was blinking. Idly, he clicked it on.

NUMBER OF PERSONAL MESSAGES: 3. DO YOU WANT TO READ THEM NOW?

He pressed the key. A moment later, the first message came up.

SEALED TWINKLE DRIVES ARE ON THEIR WAY TO YOU TODAY DHL. YOU SHOULD HAVE THEM TOMORROW. HOPE YOU FIND SOMETHING . . . JAFAR IS STILL SEVERELY ILL. THEY SAY HE MAY DIE.

ARTHUR KAHN

He pressed the key, and another message came up.

THE WEENIES ARE STILL SWARMING DOWN HERE. ANY NEWS YET?

EDDIE

Sanders couldn't worry about Eddie now. He pushed the key, and the third message came up.

I GUESS YOU HAVEN'T BEEN READING BACK ISSUES OF COMLINE. STARTING FOUR YEARS AGO.

AFRIEND

Sanders stared at the screen. ComLine was DigiCom's in-house newsletter—an eight-page monthly, filled with chatty accounts of hirings and promotions and babies born. The summer schedule for the softball team, things like that. Sanders never paid any attention to it and couldn't imagine why he should now.

And who was "Afriend"?

He clicked the REPLY button on the screen.

CAN'T REPLY - SENDER ADDRESS NOT AVAILABLE

He clicked the SENDER INFO button. It should give him the name and address of the person sending the e-mail message. But instead he saw dense rows of type:

```
FROM UU5.PSI.COM!UWA.PCM.COM.EDU!CHARON TUE JUN 16
04:43:31 REMOTE FROM DCCSYS
RECEIVED: FROM UUPSI5 BY DCCSYS.DCC.COM ID AA02599;
TUE, 16 JUN 4:42:19 PST
RECEIVED: FROM UWA.PCM.COM.EDU  BY  UU5.PSI.COM
(5.65B/4.0.071791-PSI/PSINET)
ID AA28153; TUE, 16 JUN 04:24:58 -0500
RECEIVED: FROM RIVERSTYX.PCM.COM.EDU BY UWA.PCM.
COM.EDU (4.1/SMI-4.1)
ID AA15969; TUE, 16 JUN 04:24:56 PST
RECEIVED: BY RIVERSTYX.PCM.COM.EDU (920330.SGI/5.6)
ID AA00448; TUE, 16 JUN 04:24:56 -0500
DATE: TUE, 16 JUN 04:24:56 -0500
FROM: CHARON@UWA.PCM.COM.EDU (AFRIEND)
MESSAGE-ID: < 9212220924.AA90448@RIVERSTYX.PCM.
COM.EDU >
TO: TSANDERS@DCC.COM
```

Sanders stared. The message hadn't come to him from inside the company at all. He was looking at an Internet routing. Internet was the vast worldwide computer network connecting universities, corporations, government agencies, and private users. Sanders wasn't knowledgeable about the Internet, but it appeared that the message from "Afriend," network name CHARON, had originated from UWA.PCM.COM.-EDU, wherever that was. Apparently some kind of educational institution. He pushed the PRINT SCREEN button, and made a mental note to turn this one over to Bosak. He needed to talk to Bosak anyway.

He went down the hall and got the sheet as it came out of the printer. Then he went back to his office and stared at the screen. He decided to try a reply to this person.

FROM: TSANDERS@DCC.COM
TO: CHARON@UWA.PCM.COM.EDU

ANY HELP GREATLY APPRECIATED.

SANDERS

He pushed the SEND button. Then he deleted both the original message and his own reply.

SORRY, YOU CANNOT DELETE THIS MAIL.

Sometimes e-mail was protected with a flag that prevented it from being deleted.
He typed: UNPROTECT MAIL.

THE MAIL IS UNPROTECTED.

He typed: DELETE MAIL.

SORRY, YOU CANNOT DELETE THIS MAIL.

What the hell is this? he thought. The system must be hanging up. Maybe it had been stymied by the Internet address. He decided to delete the message from the system at the control level.
He typed: SYSTEM.

WHAT LEVEL?

He typed: SYSOP.

SORRY, YOUR PRIVILEGES DO NOT INCLUDE SYSOP CONTROL.

"Christ," he said. They'd gone in and taken away his privileges. He couldn't believe it.
He typed: SHOW PRIVILEGES.

SANDERS, THOMAS L.
PRIOR USER LEVEL: 5 (SYSOP)
USER LEVEL CHANGE: TUE JUNE 16 4:50 PM PST
CURRENT USER LEVEL: 0 (ENTRY)
NO FURTHER MODIFICATIONS

There it was: they had locked him out of the system. User level zero was the level that assistants in the company were given.

Sanders slumped back in the chair. He felt as if he had been fired. For the first time, he began to realize what this was going to be like.

Clearly, there was no time to waste. He opened his desk drawer, and saw at once that the pens and pencils were neatly arranged. Someone had already been there. He pulled open the file drawer below. Only a half-dozen files were there; the others were all missing.

They had already gone through his desk.

Quickly, he got up and went out to the big filing cabinets behind Cindy's desk. These cabinets were locked, but he knew Cindy kept the key in her desk. He found the key, and unlocked the current year's files.

The cabinet was empty. There were no files there at all. They had taken everything.

He opened the cabinet for the previous year: empty.

The year before: empty.

All the others: empty.

Jesus, he thought. No wonder Cindy had been so cool. They must have had a gang of workmen up there with trolleys, cleaning everything out during the afternoon.

Sanders locked the cabinets again, replaced the key in Cindy's desk, and headed downstairs.

T he press office was on the third floor. It was deserted now except for a single assistant, who was closing up. "Oh. Mr. Sanders. I was just getting ready to leave."

"You don't have to stay. I just wanted to check some things. Where do you keep the back issues of ComLine?"

"They're all on that shelf over there." She pointed to a row of stacked issues. "Was there anything in particular?"

"No. You go ahead home."

The assistant seemed reluctant, but she picked up her purse and headed out the door. Sanders went to the shelf. The issues were arranged in six-month stacks. Just to be safe, he started ten stacks back—five years ago.

He began flipping through the pages, scanning the endless details of game scores and press releases on production figures. After a few minutes, he found it hard to pay attention. And of course he didn't know what he was looking for, although he assumed it was something about Meredith Johnson.

He went through two stacks before he found the first article.

NEW MARKETING ASSISTANT NAMED

Cupertino, May 10: DigiCom President Bob Garvin today announced the appointment of Meredith Johnson as Assistant Director of Marketing and Promotion for Telecommunications. She will report to Howard Gottfried in M and P. Ms. Johnson, 30, came to us from her position as Vice President for Marketing at Conrad Computer Systems of Sunnyvale. Before that, she was a senior administrative assistant at the Novell Network Division in Mountain View.

Ms. Johnson, who has degrees from Vassar College and Stanford Business School, was recently married to Gary Henley, a marketing executive at CoStar. Congratulations! As a new arrival to DigiCom, Ms. Johnson . . .

He skipped the rest of the article; it was all PR fluff. The accompanying photo was standard B-school graduate: against a gray background with light coming from behind one shoulder, it showed a young woman with shoulder-length hair in a pageboy style, a direct businesslike stare just shy of harsh, and a firm mouth. But she looked considerably younger than she did now.

Sanders continued to thumb through the issues. He glanced at his watch. It was almost seven, and he wanted to call Bosak. He came to the end of the year, and the pages were nothing but Christmas stuff. A picture of Garvin and his family ("Merry Christmas from the Boss! Ho Ho Ho!") caught his attention because it showed Bob with his former wife, along with his three college-age kids, standing around a big tree.

Had Garvin been going out with Emily yet? Nobody ever knew. Garvin was cagey. You never knew what he was up to.

Sanders went to the next stack, for the following year. January sales predictions. ("Let's get out and make it happen!") Opening of the Austin plant to manufacture cellular phones; a photo of Garvin in harsh sunlight, cutting the ribbon. A profile of Mary Anne Hunter that began, "Spunky, athletic Mary Anne Hunter knows what she wants out of life . . ." They had called her "Spunky" for weeks afterward, until she begged them to give it up.

Sanders flipped pages. Contract with the Irish government to break ground in Cork. Second-quarter sales figures. Basketball team scores against Aldus. Then a black box:

JENNIFER GARVIN

Jennifer Garvin, a third-year student at Boalt Hall School of Law in Berkeley, died on March 5 in an automobile accident in San Francisco. She was twenty-four years old. Jennifer had been accepted to the firm of Harley, Wayne and Myers following her graduation. A memorial service was held at the Presbyterian Church of Palo Alto for friends of the family and her many classmates. Those wishing to make memorial donations should send contributions to Mothers Against Drunk Drivers. All of us at Digital Communications extend our deepest sympathy to the Garvin family.

Sanders remembered that time as difficult for everyone. Garvin was snappish and withdrawn, drinking too much, and frequently absent from work. Not long afterward, his marital difficulties became public; within two years, he was divorced, and soon after that he married Emily Chen, a young executive in her twenties. But there were other changes, too. Everyone agreed: Garvin was no longer the same boss after the death of his daughter.

Garvin had always been a scrapper, but now he became protective, less ruthless. Some said that Garvin was stopping to smell the roses, but that wasn't it at all. He was newly aware of the arbitrariness of life, and it led him to control things, in a way that hadn't been true before. Garvin had always been Mr. Evolution: put it on the shore and see if it eats or dies. It made him a heartless administrator but a remarkably fair boss. If you did a good job, you were recognized. If you couldn't cut it, you were gone. Everybody understood the rules. But after Jennifer died, all that changed. Now he had overt favorites among staff and programs, and he nurtured those favorites and neglected others, despite the evidence in front of his face. More and more, he made business decisions arbitrarily. Garvin wanted events to turn out the way he intended them to. It gave him a new kind of fervor, a new sense of what the company should be. But it was also a more difficult place to work. A more political place.

It was a trend that Sanders had ignored. He continued to act as if he still worked at the old DigiCom—the company where all that mattered were results. But clearly, that company was gone.

Sanders continued thumbing through the magazines. Articles about early negotiations for a plant in Malaysia. A photo of Phil Blackburn in Ireland, signing an agreement with the city of Cork. New production figures for the Austin plant. Start of production of the A22 cellular model. Births and deaths and promotions. More DigiCom baseball scores.

JOHNSON TO TAKE OPERATIONS POST

Cupertino, October 20: Meredith Johnson has been named new Assistant Manager for Division Operations in Cupertino, replacing the very popular Harry Warner, who retired after fifteen years of service. The shift to Ops Manager takes Johnson out of marketing, where she has been very effective for the last year, since joining the company. In her new position, she will work closely with Bob Garvin on international operations for DigiCom.

But it was the accompanying picture that caught Sanders's attention. Once again, it was a formal head shot, but Johnson now looked completely different. Her hair was light blond. Gone was the neat business-school pageboy. She wore her hair short, in a curly, informal style. She was wearing much less makeup and smiling cheerfully. Overall, the effect was to make her appear much more youthful, open, innocent.

Sanders frowned. Quickly, he flipped back through the issues he had already looked at. Then he went back to the previous stack, with its year-end Christmas pictures: "Merry Christmas from the Boss! Ho Ho Ho!"

He looked at the family portrait. Garvin standing behind his three children, two sons and a daughter. That must be Jennifer. His wife, Harriet, stood to one side. In the picture, Garvin was smiling, his hand resting lightly on his daughter's shoulder, and she was tall and athletic-looking, with short, light blond, curly hair.

"I'll be damned," he said aloud.

He thumbed back quickly to the first article, to look at the original picture of Johnson. He compared it to the later one. There was no doubt about what she had done. He read the rest of the first article:

> As a new arrival to DigiCom, Ms. Johnson brings her considerable business acumen, her sparkling humor, and her sizzling softball pitch. She's a major addition to the DigiCom team! Welcome, Meredith!
>
> Her admiring friends are never surprised to learn that Meredith was once a finalist in the Miss Teen Connecticut contest. In her student days at Vassar, Meredith was a valued member of both the tennis team and the debating society. A member of Phi Beta Kappa, she took her major in psychology, with a minor in abnormal psych. Hope you won't be needing that around here, Meredith! At Stanford, she obtained her MBA with honors, graduating near the top of her class. Meredith told us, "I am delighted to join DigiCom and I look forward to an exciting career with this forward-looking company." We couldn't have said it better, Ms. Johnson!

"No shit," Sanders said. He had known almost none of this. From the start, Meredith had been based in Cupertino; Sanders never saw her. The one time he had run into her was soon after her arrival, before she changed her hair. Her hair—and what else?

He looked carefully at the two pictures. Something else was subtly different. Had she had plastic surgery? It was impossible to know. But her appearance was definitely changed between the two portraits.

He moved through the remaining issues of the magazine quickly now, convinced that he had learned what there was to know. Now he skimmed only the headlines:

GARVIN SENDS JOHNSON TO TEXAS
FOR AUSTIN PLANT OVERSIGHT

JOHNSON WILL HEAD NEW
OPERATIONS REVIEW UNIT

JOHNSON NAMED OPERATIONS VEEP
TO WORK DIRECTLY UNDER GARVIN

JOHNSON: TRIUMPH IN MALAYSIA
LABOR CONFLICT NOW RESOLVED

MEREDITH JOHNSON OUR RISING STAR
A SUPERB MANAGER; HER SKILL IN
TECHNICAL AREAS VERY STRONG

This final headline ran above a lengthy profile of Johnson, well placed on the second page of the magazine. It had appeared in ComLine only two issues ago. Seeing it now, Sanders realized that the article was intended for internal consumption—softening up the beachhead before the June landing. This article was a trial balloon that Cupertino had floated, to see if Meredith would be acceptable to run the technical divisions in Seattle. The only trouble was, Sanders never saw it. And nobody had ever mentioned it to him.

The article stressed the technical savvy that Johnson had acquired during her years with the company. She was quoted as saying, "I began my career working in technical areas, back with Novell. The technical fields have always been my first love; I'd love to go back to it. After all, strong technical innovation lies at the heart of a forward-looking company like DigiCom. Any good manager here must be able to run the technical divisions."

There it was.

He looked at the date: May 2. Published six weeks ago. Which meant that the article had been written at least two weeks before that.

As Mark Lewyn had suspected, Meredith Johnson knew she was going to be the head of the Advanced Products Division at least two months ago. Which meant, in turn, that Sanders had never been under consideration to become division head. He had never had a chance.

It was a done deal.

Months ago.

Sanders swore, took the articles over to the xerox machine and copied them, then put the stacks back on the shelf, and left the press office.

He got on the elevator. Mark Lewyn was there. Sanders said, "Hi, Mark." Lewyn didn't answer. Sanders pushed the button for the ground floor.

The doors closed.

"I just hope you know what the fuck you're doing," Lewyn said angrily.

"I think I do."

"Because you could fuck this thing up for everybody. You know that?"

"Fuck what up?"

"Just because you got your ass in the sling, it's not our problem."

"Nobody said it was."

"I don't know what's the matter with you," Lewyn said. "You're late for work, you don't call me when you say you will . . . What is it, trouble at home? More shit with Susan?"

"This has nothing to do with Susan."

"Yeah? I think it does. You've been late two days running and even when you're here, you walk around like you're dreaming. You're in fucking dreamland, Tom. I mean, what the hell were you doing, going to Meredith's office at night, anyway?"

"She asked me to come to her office. She's the boss. You're saying I shouldn't have gone?"

Lewyn shook his head in disgust. "This innocent act is a lot of crap. Don't you take any responsibility for anything?"

"What—"

"Look, Tom, everybody in the company knows that Meredith is a shark. Meredith Manmuncher, they call her. The Great White. Everybody knows she's protected by Garvin, that she can do what she wants. And what she wants is to play grabass with cute guys who show up in her office at the end of the day. She has a couple of glasses of wine, she gets a little flushed, and she wants service. A delivery boy, a trainee, a young account guy. Whatever. And nobody can say a word because Garvin thinks she walks on water. So, how come everybody else in the company knows it but you?"

Sanders was stunned. He did not know how to answer. He stared at Lewyn, who stood very close to him, his body hunched, hands in his pockets. He could feel Lewyn's breath on his face. But he could hardly hear Lewyn's words. It was as if they came to him from a great distance.

"Hey, Tom. You walk the same halls, you breathe the same air as the rest of us. You know who's doing what. You go marching up there to her office . . . and you know damned well what's coming. Meredith's done everything but announce to the world that she wants to suck your dick. All day long, she's touching your arm, giving you those *meaningful* little looks and squeezes. Oh, *Tom*. So *nice* to see you again. And now you tell me you didn't know what was coming in that office? *Fuck* you, Tom. You're an asshole."

The elevator doors opened. Before them, the ground-floor lobby was deserted, growing dark in the fading light of the June evening. A soft rain fell outside. Lewyn started toward the exit, then turned back. His voice echoed in the lobby.

"You realize," he said, "that you're acting like one of those women in all this. The way they always go, 'Who, me? I never intended that.' The way they go, 'Oh, it's not *my* responsibility. I never thought if I got drunk and kissed him and went to his room and lay down on his bed that he'd fuck me. Oh dear me no.' It's bullshit, Tom. Irresponsible bullshit. And you better think about what I'm saying, because there's a lot of us who have worked every bit as hard as you have in this company, and we don't want to see you screw up this merger and this spin-off for the rest of us. You want to pretend you can't tell when a woman's coming on to you, that's fine. You want to screw up your own life, it's your decision. But you screw up mine, and I'm going to fucking put you away."

Lewyn stalked off. The elevator doors started to close. Sanders stuck his hand out; the doors closed on his fingers. He jerked his hand, and the doors opened again. He hurried out into the lobby after Lewyn.

He grabbed Lewyn on the shoulder. "Mark, wait, listen—"

"I got nothing to say to you. I got kids, I got responsibilities. You're an asshole."

Lewyn shrugged Sanders's arm off, pushed open the door, and walked out. He strode quickly away, down the street.

As the glass doors closed, Sanders saw a flash of blond in the moving reflection. He turned.

"I thought that was a little unfair," Meredith Johnson said. She was standing about twenty feet behind him, near the elevators. She was wearing gym clothes—navy tights, and a sweatshirt—and she carried a gym bag in her hand. She looked beautiful, overtly sexual in a certain way. Sanders felt tense: there was no one else in the lobby. They were alone.

"Yes," Sanders said. "I thought it was unfair."

"I meant, to women," Meredith said. She swung the gym bag over her shoulder, the movement raising her sweatshirt and exposing her bare abdomen above her tights. She shook her head and pushed her hair back from her face. She paused a moment, and then she began to speak. "I want to tell you I'm sorry about all this," she said. She moved toward him in a steady, confident way, almost stalking. Her voice was low. "I never wanted any of this, Tom." She came a little closer, approaching slowly, as if he were an animal that might be frightened away. "I have only the warmest feelings for you." Still closer. "Only the warmest." Closer. "I can't help it, Tom, if I still want you." Closer. "If I did anything to offend you, I apologize." She was very close now, her body almost touching his, her breasts inches from his arm. "I'm truly sorry, Tom," she said softly. She seemed filled with emotion, her breasts rising and falling, her eyes moist and pleading as she looked up at him. "Can you forgive me? Please? You know how I feel about you."

He felt all the old sensations, the old stirrings. He clenched his jaw. "Meredith. The past is past. Cut it out, will you?"

She immediately changed her tone and gestured to the street. "Listen, I have a car here. Can I drop you somewhere?"

"No, thanks."

"It's raining. I thought you might want a lift."

"I don't think it's a good idea."

"Only because it's raining."

"This is Seattle," he said. "It rains all the time here."

She shrugged, walked to the door, and leaned her weight against it, thrusting out her hip. Then she looked back at him and smiled. "Remind me never to wear tights around you. It's embarrassing: you make me wet."

Then she turned away, pushed through the door, and walked quickly to the waiting car, getting in the back. She closed the door, looked back at him, and waved cheerfully. The car drove off.

Sanders unclenched his hands. He took a deep breath and let it out slowly. His whole body was tense. He waited until the car was gone, then went outside. He felt the rain on his face, the cool evening breeze.

He hailed a taxi. "The Four Seasons Hotel," he said to the driver.

Riding in the taxi, Sanders stared out the window, breathing deeply. He felt as though he couldn't get his breath. He had been badly unnerved by the meeting with Meredith. Especially coming so close after his conversation with Lewyn.

Sanders was distressed by what Lewyn had said, but you could never take Mark too seriously. Lewyn was an artistic hothead who handled his creative tensions by getting angry. He was angry about something most of the time. Lewyn liked being angry. Sanders had known him a long time. Personally, he had never understood how Adele, Mark's wife, put up with it. Adele was one of those wonderfully calm, almost phlegmatic women who could talk on the phone while her two kids crawled all over her, tugging at her, asking her questions. In a similar fashion, Adele just let Lewyn rage while she went on about her business. In fact, everyone just let Lewyn rage, because everyone knew that, in the end, it didn't mean anything.

Yet, it was also true that Lewyn had a kind of instinct for public perceptions and trends. That was the secret of his success as a designer. Lewyn would say, "Pastel colors," and everybody would groan and say that the new design colors looked like hell. But two years later, when the products were coming off the line, pastel colors would be just what everybody wanted. So Sanders was forced to admit that what Lewyn had said about him, others would soon be saying. Lewyn had said the company line: that Sanders was screwing up the chances for everybody else.

Well, screw them, he thought.

As for Meredith—he had had the distinct feeling that she had been toying with him in the lobby. Teasing him, playing with him. He could not understand why she was so confident. Sanders was making a very serious allegation against her. Yet she behaved as if there was no threat at all. She had a kind of imperviousness, an indifference, that made him deeply uneasy. It could only mean she knew that she had Garvin's backing.

The taxi pulled into the turnaround of the hotel. He saw Meredith's car up ahead. She was talking to the driver. She looked back and saw him.

There was nothing to do but get out and walk toward the entrance.

"Are you following me?" she said, smiling.

"No."

"Sure?"

"Yes, Meredith. I'm sure."

They went up the escalator from the street to the lobby. He stood behind her on the escalator. She looked back at him. "I wish you were."

"Yeah. Well, I'm not."

"It would have been nice," she said. She smiled invitingly.

He didn't know what to say; he just shook his head. They rode the rest of the way in silence until they came to the high ornate lobby. She said, "I'm in room 423. Come and see me anytime." She headed toward the elevators.

He waited until she was gone, then crossed the lobby and turned left to the dining room. Standing at the entrance, he saw Dorfman at a corner table, eating dinner with Garvin and Stephanie Kaplan. Max was holding forth, gesturing sharply as he spoke. Garvin and Kaplan both leaned forward, listening. Sanders was reminded that Dorfman had once been a director of the company—according to the stories, a very powerful director. It was Dorfman who had persuaded Garvin to expand beyond modems into cellular telephony and wireless communications, back in the days when nobody could see any link between computers and telephones. The link was obvious now but obscure in the early 1980s, when Dorfman had said, "Your business is not hardware. Your business is communications. Your business is access to information."

Dorfman had shaped company personnel as well. Supposedly, Kaplan owed her position to his glowing endorsement. Sanders had come to Seattle on Dorfman's recommendation. Mark Lewyn had been hired because of Dorfman. And any number of vice presidents had vanished over the years because Dorfman found them lacking in vision or stamina. He was a powerful ally or a lethal opponent.

And his position at the time of the merger was equally strong. Although Dorfman had resigned as a director years before, he still owned a good deal of DigiCom stock. He still had Garvin's ear. And he still had the contacts and prestige within the business and financial community that made a merger like this much simpler. If Dorfman approved the terms of the merger, his admirers at Goldman, Sachs and at First Boston would raise the money easily. But if Dorfman was dissatisfied, if he hinted that the merger of the two companies did not make sense, then the acquisition might unravel. Everyone knew it. Everyone understood very well the power he wielded—especially Dorfman himself.

Sanders hung back at the entrance to the restaurant, reluctant to come forward. After a while, Max glanced up and saw him. Still talking, he shook his head fractionally: *no.* Then, as he continued to talk, he made a subtle motion with his hand, tapping his watch. Sanders nodded, and went back into the lobby and sat down. He had the stack of ComLine photocopies on his lap. He browsed through them, studying again the way Meredith had changed her appearance.

A few minutes later, Dorfman rolled out in his wheelchair. "So, Thomas. I am glad you are not bored with your life."

"What does that mean?"

Dorfman laughed and gestured to the dining room. "They're talking of nothing else in there. The only topic this evening is you and Meredith. Everyone is so excited. So *worried.*"

"Including Bob?"

"Yes, of course. Including Bob." He wheeled closer to Sanders. "I cannot really speak to you now. Was there something in particular?"

"I think you ought to look at this," Sanders said, handing Dorfman the photocopies. He was thinking that Dorfman could take these pictures to Garvin. Dorfman could make Garvin understand what was really going on.

Dorfman examined them in silence a moment. "Such a lovely woman," he said. "So beautiful . . ."

"Look at the differences, Max. Look at what she did to herself."

Dorfman shrugged. "She changed her hair. Very flattering. So?"

"I think she had plastic surgery as well."

"It wouldn't surprise me," Dorfman said. "So many women do, these days. It is like brushing their teeth, to them."

"It gives me the creeps."

"Why?" Dorfman said.

"Because it's underhanded, that's why."

"What's underhanded?" Dorfman said, shrugging. "She is resource-ful. Good for her."

"I'll bet Garvin has no idea what she's doing to him," Sanders said.

Dorfman shook his head. "I'm not concerned about Garvin," he said. "I'm concerned about you, Thomas, and this outrage of yours—hmm?"

"I'll tell you why I'm outraged," Sanders said. "Because this is the kind of sneaky shit that a woman can pull but a man can't. She changes her appearance, she dresses and acts like Garvin's daughter, and that gives her an advantage. Because I sure as hell can't act like his daughter."

Dorfman sighed, shaking his head. "Thomas. Thomas."

"Well, I can't. Can I?"

"Are you enjoying this? You seem to be enjoying this outrage."

"I'm not."

"Then give it up," Dorfman said. He turned his wheelchair to face Sanders. "Stop talking this nonsense, and face what is true. Young people in organizations advance by alliances with powerful, senior people. True?"

"Yes."

"And it is always so. At one time, the alliance was formal—an apprentice and master, or a pupil and tutor. It was arranged, yes? But today, it is not formal. Today, we speak of mentors. Young people in business have mentors. True?"

"Okay . . ."

"So. How do young people attach themselves to a mentor? What is the process? First, by being agreeable, by being helpful to the senior person, doing jobs that need to be done. Second, by being attractive to the older person—imitating their attitudes and tastes. Third, by advo-cacy—adopting their agenda within the company."

"That's all fine," Sanders said. "What does it have to do with plastic surgery?"

"Do you remember when you joined DigiCom in Cupertino?"

"Yes, I remember."

"You came over from DEC. In 1980?"

"Yes."

"At DEC, you wore a coat and tie every day. But when you joined

DigiCom, you saw that Garvin wore jeans. And soon, you wore jeans, too."

"Sure. That was the style of the company."

"Garvin liked the Giants. You began to go to games in Candlestick Park."

"He was the boss, for Christ's sake."

"And Garvin liked golf. So you took up golf, even though you hated it. I remember you complained to me about how much you hated it. Chasing the stupid little white ball."

"Listen. I didn't have plastic surgery to make myself look like his kid."

"*Because you didn't have to*, Thomas," Dorfman said. He threw up his hands in exasperation. "Can you not see this point? Garvin liked brash, aggressive young men who drank beer, who swore, who chased women. And you did all those things in those days."

"I was young. That's what young men do."

"No, Thomas. That's what Garvin liked young men to do." Dorfman shook his head. "So much of this is unconscious. Rapport is unconscious, Thomas. But the task of building rapport is different, depending on whether you are the same sex as that person, or not. If your mentor is a man, you may act like his son, or brother, or father. Or you may act like that man when he was younger—you may remind him of himself. True? Yes, you see that. Good.

"But if you are a woman, everything is different. Now you must be your mentor's daughter, or lover, or wife. Or perhaps sister. In any case, very different."

Sanders frowned.

"I see this often, now that men are starting to work for women. Many times men cannot structure the relationship because they do not know how to act as the subordinate to a woman. Not with comfort. But in other cases, men slip easily into a role with a woman. They are the dutiful son, or the substitute lover or husband. And if they do it well, the women in the organization become angry, because they feel that they cannot compete as son or lover or husband to the boss. So they feel that the man has an advantage."

Sanders was silent.

"Do you understand?" Dorfman said.

"You're saying it happens both ways."

"Yes, Thomas. It is inevitable. It is the process."

"Come on, Max. There's nothing inevitable about it. When Garvin's daughter died, it was a personal tragedy. He was upset, and Meredith took advantage of—"

"*Stop,*" Dorfman said, annoyed. "Now you want to change human nature? There are always tragedies. And people always take advantage. This is nothing new. Meredith is intelligent. It is delightful to see such an intelligent, resourceful woman who is also beautiful. She is a gift from God. She is delightful. This is your trouble, Thomas. And it has been a long time coming."

"What does that—"

"And instead of dealing with your trouble, you waste your time with these . . . *trivialities.*" He handed back the pictures. "These are not important, Thomas."

"Max, will you—"

"You were never a good corporate player, Thomas. It was not your strength. Your strength was that you could take a technical problem and grind it down, push the technicians, encourage them and bully them, and finally get it solved. You could make it work. Is that not so?"

Sanders nodded.

"But now you abandon your strengths for a game that does not suit you."

"Meaning what?"

"You think that by threatening a lawsuit, you put pressure on her and on the company. In fact, you played into her hands. You have let her define the game, Thomas."

"I had to do something. She broke the law."

"She broke the law," Dorfman mimicked him, with a sarcastic whine. "Oh me, oh my. And you are so defenseless. I am filled with sorrow for your plight."

"It's not easy. She's well connected. She has strong supporters."

"Is that so? Every executive with strong supporters has also strong detractors. And Meredith has her share of detractors."

"I tell you, Max," Sanders said, "she's dangerous. She's one of those MBA image people, focused on image, everything image, never substance."

"Yes," Dorfman said, nodding approvingly. "Like so many young executives today. Very skilled with images. Very interested in manipulating that reality. A fascinating trend."

"I don't think she's competent to run this division."

"And what if she is not?" Dorfman snapped. "What difference does it make to you? If she's incompetent, Garvin will eventually acknowledge it and replace her. But by then, you will be long gone. Because you will lose this game with her, Thomas. She is better at politics than you. She always was."

Sanders nodded. "She's ruthless."

"Ruthless, schmoothless. She is *skilled*. She has an instinct. You lack it. You will lose everything if you persist this way. And you will deserve the fate that befalls you because you have behaved like a fool."

Sanders was silent. "What do you recommend I do?"

"Ah. So now you want advice?"

"Yes."

"Really?" He smiled. "I doubt it."

"Yes, Max. I do."

"All right. Here is my advice. Go back, apologize to Meredith, apologize to Garvin, and resume your job."

"I can't."

"Then you don't want advice."

"I can't do that, Max."

"Too much pride?"

"No, but—"

"You are infatuated with the anger. How dare this woman act this way. She has broken the law, she must be brought to justice. She is dangerous, she must be stopped. You are filled with *delicious*, righteous indignation. True?"

"Oh, hell, Max. I just can't do it, that's all."

"Of course you can do it. You mean you *won't*."

"All right. I won't."

Dorfman shrugged. "Then what do you want from me? You come to ask my advice in order not to take it? This is nothing special." He grinned. "I have a lot of other advice you won't take, either."

"Like what?"

"What do you care, since you won't take it?"

"Come on, Max."

"I'm serious. You won't take it. We are wasting our time here. Go away."

"Just tell me, will you?"

Dorfman sighed. "Only because I remember you from the days when you had sense. First point. Are you listening?"

"Yes, Max. I am."

"First point: you know everything you need to know about Meredith Johnson. So forget her now. She is not your concern."

"What does that mean?"

"Don't interrupt. Second point. Play your own game, not hers."

"Meaning what?"

"Meaning, solve the problem."

"Solve what problem? The lawsuit?"

Dorfman snorted and threw up his hands. "You are impossible. I am wasting my time."

"You mean drop the lawsuit?"

"Can you understand English? *Solve the problem.* Do what you do well. Do your job. Now go away."

"But Max—"

"Oh, I can't do anything for you," Dorfman said. "It's your life. You have your own mistakes to make. And I must return to my guests. But try to pay attention, Thomas. Do not sleep through this. And remember, all human behavior has a reason. All behavior is solving a problem. Even *your* behavior, Thomas."

And he spun in his wheelchair and went back to the dining room.

F ucking Max, he thought, walking down Third Street in the damp evening. It was infuriating, the way Max would never just say what he meant.

This is your trouble, Thomas. And it has been a long time coming.

What the hell was that supposed to mean?

Fucking Max. Infuriating and frustrating and exhausting, too. That was what Sanders remembered most about the sessions he used to have, when Max was on the DigiCom board. Sanders would come away exhausted. In those days, back in Cupertino, the junior execs had called Dorfman "The Riddler."

All human behavior is solving a problem. Even your behavior, Thomas.

Sanders shook his head. It made no sense at all. Meanwhile, he had things to do. At the end of the street, he stepped into a phone booth and dialed Gary Bosak's number. It was eight o'clock. Bosak would be home, just getting out of bed and having coffee, starting his working day. Right now, he would be yawning in front of a half-dozen modems and computer screens as he began to dial into all sorts of databases.

The phone rang, and a machine said, "You have reached NE Professional Services. Leave a message." And a beep.

"Gary, this is Tom Sanders. I know you're there, pick up."

A click, and then Bosak said, "Hey. The last person I thought I'd hear from. Where're you calling from?"

"Pay phone."

"Good. How's it going with you, Tom?"

"Gary, I need some things done. Some data looked up."

"Uh . . . Are we talking things for the company, or private things?"

"Private."

"Uh . . . Tom. I'm pretty busy these days. Can we talk about this next week?"

"That's too late."

"But the thing is, I'm pretty busy now."

"Gary, what is this?"

"Tom, come on. You know what this is."

"I need help, Gary."

"Hey. And I'd love to help you. But I just got a call from Blackburn who told me that if I had anything to do with you, anything at all, I could expect the FBI going through my apartment at six a.m. tomorrow morning."

"Christ. When was this?"

"About two hours ago."

Two hours ago. Blackburn was way ahead of him. "Gary . . ."

"Hey. You know I always liked you, Tom. But not this time. Okay? I got to go."

Click.

F rankly, none of this surprises me," Fernandez said, pushing aside a paper plate. She and Sanders had been eating sandwiches in her office. It was nine p.m., and the offices around them were dark, but her phone was still ringing, interrupting them frequently. Outside, it had begun to rain again. Thunder rumbled, and Sanders saw flashes of summer lightning through the windows.

Sitting in the deserted law offices, Sanders had the feeling that he was all alone in the world, with nobody but Fernandez and the encroaching darkness. Things were happening quickly; this person he had never met before today was fast becoming a kind of lifeline for him. He found himself hanging on every word she said.

"Before we go on, I want to emphasize one thing," Fernandez said. "You were right not to get in the car with Johnson. You are not to be alone with her ever again. Not even for a few moments. Not ever, under any circumstances. Is that clear?"

"Yes."

"If you do, it will destroy your case."

"I won't."

"All right," she said. "Now. I had a long talk with Blackburn. As you guessed, he's under tremendous pressure to get this matter resolved. I tried to move the mediation session to the afternoon. He implied that the company was ready to deal and wanted to get started right away. He's concerned about how long the negotiations will take. So we'll start at nine tomorrow."

"Okay."

"Herb and Alan have been making progress. I think they'll be able to help us tomorrow. And these articles about Johnson may be useful, too," she said, glancing at the photocopies of the ComLine pieces.

"Why? Dorfman says they're irrelevant."

"Yes, but they document her history in the company, and that gives us leads. It's something to work on. So is this e-mail from your friend." She frowned at the sheet of printout. "This is an Internet address."

"Yes," he said, surprised that she knew.

"We do a lot of work with high-technology companies. I'll have somebody check it out." She put it aside. "Now let's review where we are. You couldn't clean out your desk because they were already there."

"Right."

"And you would have cleaned out your computer files, but you've been shut out of the system."

"Yes."

"Which means that you can't change anything."

"That's right. I can't do anything. It's like I'm an assistant."

She said, "Were you going to change any files?"

He hesitated. "No. But I would have, you know, looked around."

"Nothing in particular you were aware of?"

"No."

"Mr. Sanders," she said, "I want to emphasize that I have no judgment here. I'm simply trying to prepare for what may happen tomorrow. I want to know what surprises they'll have for us."

He shook his head. "There isn't anything in the files that's embarrassing to me."

"You've thought it over carefully?"

"Yes."

"Okay," she said. "Then considering the early start, I think you better get some sleep. I want you sharp tomorrow. Will you be able to sleep?"

"Jeez, I don't know."

"Take a sleeping pill if you need to."

"I'll be okay."

"Then go home and go to bed, Mr. Sanders. I'll see you in the morning. Wear a coat and tie tomorrow. Do you have some kind of a blue coat?"

"A blazer."

"Fine. Wear a conservative tie and a white shirt. No after-shave."

"I never dress like that at the office."

"This is not the office, Mr. Sanders. That's just the point." She stood up and shook his hand. "Get some sleep. And try not to worry. I think everything is going to be fine."

"I bet you say that to all your clients."

"Yes, I do," she said. "But I'm usually right. Get some sleep, Tom. I'll see you tomorrow."

He came home to a dark, empty house. Eliza's Barbie dolls lay in an untidy heap on the kitchen counter. One of his son's bibs, streaked with green baby food, was on the counter beside the sink. He set up the coffeemaker for the morning and went upstairs. He walked past the answering machine but neglected to look at it, and failed to notice the blinking light.

Upstairs, when he undressed in the bathroom, he saw that Susan had taped a note to the mirror. "Sorry about lunch. I believe you. I love you. S."

It was just like Susan to be angry and then to apologize. But he was glad for the note and considered calling her now. But it was nearly midnight in Phoenix, which meant it was too late. She'd be asleep.

Anyway, as he thought about it, he realized that he didn't want to call her. As she had said at the restaurant, this had nothing to do with her. He was alone in this. He'd stay alone.

Wearing just shorts, he padded into his little office. There were no faxes. He switched on his computer and waited while it came up.

The e-mail icon was blinking. He clicked it.

TRUST NOBODY.

AFRIEND

Sanders shut off the computer and went to bed.

WEDNESDAY

n the morning, he took comfort in his routine, dressing quickly while listening to the television news, which he turned up loud, trying to fill the empty house with noise. He drove into town at 6:30, stopping at the Bainbridge Bakery to buy a pull-apart and a cup of cappuccino before going down to the ferry.

As the ferry pulled away from Winslow, he sat toward the stern, so he would not have to look at Seattle as it approached. Lost in his thoughts, he stared out the window at the gray clouds hanging low over the dark water of the bay. It looked like it would rain again today.

"Bad day, huh?" a woman said.

He looked up and saw Mary Anne Hunter, pretty and petite, standing with her hands on her hips, looking at him with concern. Mary Anne lived on Bainbridge, too. Her husband was a marine biologist at the university. She and Susan were good friends, and often jogged together. But he didn't often see Mary Anne on the ferry because she usually went in early.

"Morning, Mary Anne."

"What I can't understand is how they got it," she said.

"Got what?" Sanders said.

"You mean you haven't seen it? Jesus. You're in the papers, Tom." She handed him the newspaper under her arm.

"You're kidding."

"No. Connie Walsh strikes again."

Sanders looked at the front page, but saw nothing. He began flipping through quickly.

"It's in the Metro section," she said. "The first opinion column on the second page. Read it and weep. I'll get more coffee." She walked away.

Sanders opened the paper to the Metro section.

AS I SEE IT
by Constance Walsh

MR. PIGGY AT WORK

The power of the patriarchy has revealed itself again, this time in a local high-tech firm I'll call Company X. This company has appointed a brilliant, highly competent woman to a major executive position. But many men in the company are doing their damnedest to get rid of her.

One man in particular, let's call him Mr. Piggy, has been especially vindictive. Mr. Piggy can't tolerate a woman supervisor, and for weeks he has been running a bitter campaign of innuendo inside the company to keep it from happening. When that failed, Mr. Piggy claimed that his new boss sexually assaulted him, and nearly raped him, in her offices. The blatant hostility of this claim is matched only by its absurdity.

Some of you may wonder how a woman could rape a man. The answer is, of course, she can't. Rape is a crime of violence. It is exclusively a crime of males, who use rape with appalling frequency to keep women in their place. That is the deep truth of our society, and of all other societies before ours.

For their part, women simply do not oppress men. Women are powerless in the hands of men. And to claim that a woman committed rape is absurd. But that didn't stop Mr. Piggy, who is interested only in smearing his new supervisor. He's even bringing a formal charge of sexual harassment against her!

In short, Mr. Piggy has the nasty habits of a typical patriarch. As you might expect, they appear everywhere in his life. Although Mr. Piggy's wife is an outstanding attorney, he pressures her to give up her job and stay home with the kids. After all, Mr. Piggy doesn't want his wife out in the business world, where she might hear about his affairs with young women and his excessive drinking. He probably figures his new female supervisor wouldn't approve of that, either. Maybe she won't allow him to be late to work, as he so often is.

So Mr. Piggy has made his underhanded move, and another talented businesswoman sees her career unfairly jeopardized. Will she be able to keep the pigs in the pen at Company X? Stay tuned for updates.

"Christ," Sanders said. He read it through again.

Hunter came back with two cappuccinos in paper cups. She pushed one toward him. "Here. Looks like you need it."

"How did they get the story?" he said.

Hunter shook her head. "I don't know. It looks to me like there's a leak inside the company."

"But who?" Sanders was thinking that if the story made the paper, it must have been leaked by three or four p.m. the day before. Who in the company even knew that he was considering a harassment charge at that time?

"I can't imagine who it could be," Hunter said. "I'll ask around."

"And who's Constance Walsh?"

"You never read her? She's a regular columnist at the *Post-Intelligencer*," Hunter said. "Feminist perspectives, that kind of thing." She shook her head. "How is Susan? I tried to call her this morning, and there's no answer at your house."

"Susan's gone away for a few days. With the kids."

Hunter nodded slowly. "That's probably a good idea."

"We thought so."

"She knows about this?"

"Yes."

"And is it true? Are you charging harassment?"

"Yes."

"Jesus."

"Yes," he said, nodding.

She sat with him for a long time, not speaking. She just sat with him. Finally she said, "I've known you for a long time. I hope this turns out okay."

"Me, too."

There was another long silence. Finally, she pushed away from the table and got up.

"See you later, Tom."

"See you, Mary Anne."

He knew what she was feeling. He had felt it himself, when others in the company had been accused of harassment. There was suddenly a distance. It didn't matter how long you had known the person. It didn't matter if you were friends. Once an accusation was made, everybody pulled away. Because the truth was, you never knew what had happened. You couldn't afford to take sides—even with your friends.

He watched her walk away, a slender, compact figure in exercise clothes, carrying a leather briefcase. She was barely five feet tall. The men on the ferry were so much larger. He remembered that she had once told Susan that she took up running because of her fear of rape. "I'll just outrun them," she had said. Men didn't know anything about that. They didn't understand that fear.

But there was another kind of fear that only men felt. He looked at the newspaper column with deep and growing unease. Key words and phrases jumped out at him:

Vindictive . . . bitter . . . can't tolerate a woman . . . blatant hostility . . .

rape . . . crime of males . . . smearing his supervisor . . . affairs with young women . . . excessive drinking . . . late to work . . . unfairly jeopardized . . . pigs in the pen.

These characterizations were more than inaccurate, more than unpleasant. They were dangerous. And it was exemplified by what happened to John Masters—a story that had reverberated among many senior men in Seattle.

• • •

Masters was fifty, a marketing manager at MicroSym. A stable guy, solid citizen, married twenty-five years, two kids—the older girl in college, the younger girl a junior in high school. The younger girl starts to have trouble with school, her grades go down, so the parents send her to a child psychologist. The child psychologist listens to the daughter and then says, You know, this is the typical story of an abused child. Do you have anything like that in your past?

Gee, the girl says, I don't think so.

Think back, the psychologist says.

At first the girl resists, but the psychologist keeps at her: Think back. Try to remember. And after a while, the girl starts to recall some vague memories. Nothing specific, but now she thinks it's possible. Maybe Daddy did do something wrong, way back when.

The psychologist tells the wife what is suspected. After twenty-five years together, the wife and Masters have some anger between them. The wife goes to Masters and says, Admit what you did.

Masters is thunderstruck. He can't believe it. He denies everything. The wife says, You're lying, I don't want you around here. She makes him move out of the house.

The older daughter flies home from college. She says, What is this madness? You know Daddy didn't do anything. Come to your senses. But the wife is angry. The daughter is angry. And the process, once set in motion, can't be stopped.

The psychologist is required by state law to report any suspected abuse. She reports Masters to the state. The state is required by law to conduct an investigation. Now a social worker is talking to the daughter, the wife, and Masters. Then to the family doctor. The school nurse. Pretty soon, everybody knows.

Word of the accusation gets to MicroSym. The company suspends him from his job, pending the outcome. They say they don't want negative publicity.

Masters is seeing his life dissolve. His younger daughter won't talk to him. His wife won't talk to him. He's living alone in an apartment. He has money problems. Business associates avoid him. Everywhere he turns, he sees accusing faces. He is advised to get a lawyer. And he is so shattered, so uncertain, he starts going to a shrink himself.

His lawyer makes inquiries; disturbing details emerge. It turns out that the particular psychologist who made the accusation uncovers abuse in a high percentage of her cases. She has reported so many cases that the state agency has begun to suspect bias. But the agency can do nothing; the law requires that all cases be investigated. The social worker assigned to the case has been previously disciplined for her excessive zeal in pursuing questionable cases and is widely thought to be incompetent, but the state cannot fire her for the usual reasons.

The specific accusation—never formally presented—turns out to be that Masters molested his daughter in the summer of her third grade. Masters thinks back, has an idea. He gets his old canceled checks out of storage, digs up his old business calendars. It turns out that his daughter was at a camp in Montana that whole summer. When she came home in August, Masters was on a business trip in Germany. He did not return from Germany until after school had started again.

He had never even seen his daughter that summer.

Masters's shrink finds it significant that his daughter would locate the abuse at the one time when abuse was impossible. The shrink concludes that the daughter felt abandoned and has translated that into a memory of abuse. Masters confronts the wife and daughter. They listen to the evidence and admit that they must have the date wrong, but remain adamant that the abuse occurred.

Nevertheless, the facts about the summer schedule lead the state to drop its investigation, and MicroSym reinstates Masters. But Masters has missed a round of promotions, and a vague cloud of prejudice hangs over him. His career has been irrevocably damaged. His wife never reconciles, eventually filing for divorce. He never again sees his younger daughter. His older daughter, caught between warring family factions, sees less of him as time goes on. Masters lives alone, struggles to rebuild his life, and suffers a near-fatal heart attack. After his recovery, he sees a few friends, but now he is morose and drinks too much, a poor companion. Other men avoid him. No one has an answer to his constant question: What did I do wrong? What should I have done instead? How could I have prevented this?

Because, of course, he could not have prevented it. Not in a contemporary climate where men were assumed to be guilty of anything they were accused of.

Among themselves, men sometimes talked of suing women for false accusations. They talked of penalties for damage caused by those accusations. But that was just talk. Meanwhile, they all changed their behavior. There were new rules now, and every man knew them:

Don't smile at a child on the street, unless you're with your wife. Don't ever touch a strange child. Don't ever be alone with someone else's child, even for a moment. If a child invites you into his or her room, don't go unless another adult, preferably a woman, is also present. At a party, don't let a little girl sit on your lap. If she tries, gently push her aside. If you ever have occasion to see a naked boy or girl, look quickly away. Better yet, leave.

And it was prudent to be careful around your own children, too, because if your marriage went sour, your wife might accuse you. And then your past conduct would be reviewed in an unfavorable light: "Well, he was such an affectionate father—perhaps a little *too* affectionate." Or, "He spent so much time with the kids. He was always hanging around the house . . ."

This was a world of regulations and penalties entirely unknown to women. If Susan saw a child crying on the street, she picked the kid up. She did it automatically, without thinking. Sanders would never dare. Not these days.

And of course there were new rules for business, as well. Sanders knew men who would not take a business trip with a woman, who would not sit next to a female colleague on an airplane, who would not meet a woman for a drink in a bar unless someone else was also present. Sanders had always thought such caution was extreme, even paranoid. But now, he was not so sure.

The sound of the ferry horn roused Sanders from his thoughts. He looked up and saw the black pilings of the Colman Dock. The clouds were still dark, still threatening rain. He stood, belted his raincoat, and headed downstairs to his car.

On his way to the mediation center, he stopped by his office for a few minutes to pick up background documentation on the Twinkle drive. He thought it might be necessary in the morning's work. But he was surprised to see John Conley in his office, talking with Cindy. It was 8:15 in the morning.

"Oh, Tom," Conley said. "I was just trying to arrange an appointment with you. Cindy tells me that you have a very busy schedule and may be out of the office most of the day."

Sanders looked at Cindy. Her face was tight. "Yes," he said, "at least for the morning."

"Well, I only need a few minutes."

Sanders waved him into the office. Conley went in, and Sanders closed the door.

"I'm looking forward to the briefing tomorrow for John Marden, our CEO," Conley said. "I gather you'll be speaking then."

Sanders nodded vaguely. He had heard nothing about a briefing. And tomorrow seemed very far away. He was having trouble concentrating on what Conley was saying.

"But of course we'll all be asked to take a position on some of these agenda items," Conley said. "And I'm particularly concerned about Austin."

"Austin?"

"I mean, the sale of the Austin facility."

"I see," Sanders said. So it was true.

"As you know, Meredith Johnson has taken an early and strong position in favor of the sale," Conley said. "It was one of the first recommendations she gave us, in the early stages of shaping this deal. Marden's worried about cash flow after the acquisition; the deal's going to add debt, and he's worried about funding high-tech development. Johnson thought we could ease the debt load by selling off Austin. But

I don't feel myself competent to judge the pros and cons on this. I was wondering what your view was."

"On a sale of the Austin plant?"

"Yes. Apparently there's tentative interest from both Hitachi and Motorola. So it's quite possible that it could be liquidated quickly. I think that's what Meredith has in mind. Has she discussed it with you?"

"No," Sanders said.

"She probably has a lot of ground to cover, settling in to her new job," Conley said. He was watching Sanders carefully as he spoke. "What do you think about a sale?"

Sanders said, "I don't see a compelling reason for it."

"Apart from cash-flow issues, I think her argument is that manufacturing cellular phones has become a mature business," Conley said. "As a technology, it's gone through its exponential growth phase, and it's now approaching a commodity. The high profits are gone. From now on, there will be only incremental sales increases, against increasing severe foreign competition. So, telephones aren't likely to represent a major income source in the future. And of course there's the question of whether we should be manufacturing in the States at all. A lot of DigiCom's manufacturing is already offshore."

"That's all true," Sanders said. "But it's beside the point. First of all, cellular phones may be reaching market saturation, but the general field of wireless communications is still in its infancy. We're going to see more and more wireless office nets and wireless field links in the future. So the market is still expanding, even if telephony is not. Second, I would argue that wireless is a major part of our company's future interest, and one way to stay competitive is to continue to make products and sell them. That forces you to maintain contact with your customer base, to keep knowledgeable about their future interests. I wouldn't opt out now. If Motorola and Hitachi see a business there, why don't we? Third, I think that we have an obligation—a social obligation, if you will—to keep high-paying skilled jobs in the U.S. Other countries don't export good jobs. Why should we? Each of our offshore manufacturing decisions has been made for a specific reason, and, personally, I hope we start to move them back here. Because there are many hidden costs in offshore fabrication. But most important of all, even though we are primarily a development unit here—making new products—we need manufacturing. If there's anything that the last twenty years has shown us, it's that design and manufacturing are all one process. You start splitting off the design engineers from the manufacturing guys and you'll end up with bad design. You'll end up with General Motors."

He paused. There was a brief silence. Sanders hadn't intended to speak so strongly; it just came out. But Conley just nodded thoughtfully. "So you believe selling Austin would hurt the development unit."

"No question about it. In the end, manufacturing is a discipline."

Conley shifted in his seat. "How do you think Meredith Johnson feels on these issues?"

"I don't know."

"Because you see, all this raises a related question," Conley said. "Having to do with executive judgment. To be frank, I've heard some rumblings in the division about her appointment. In terms of whether she really has a good enough grasp of the issues to run a technical division."

Sanders spread his hands. "I don't feel I can say anything."

"I'm not asking you to," Conley said. "I gather she has Garvin's support."

"Yes, she does."

"And that's fine with us. But you know what I'm driving at," Conley said. "The classic problem in acquisitions is that the acquiring company doesn't really understand what they are buying, and they kill the goose that lays the golden egg. They don't intend to; but they do. They destroy the very thing they want to acquire. I'm concerned that Conley-White not make a mistake like that."

"Uh-huh."

"Just between us. If this issue comes up in the meeting tomorrow, would you take the position you just took?"

"Against Johnson?" Sanders shrugged. "That could be difficult." He was thinking that he probably wouldn't be at the meeting tomorrow. But he couldn't say that to Conley.

"Well." Conley extended his hand. "Thanks for your candor. I appreciate it." He turned to go. "One last thing. It'd be very helpful if we had a handle on the Twinkle drive problem by tomorrow."

"I know it," Sanders said. "Believe me, we're working on it."

"Good."

Conley turned, and left. Cindy came in. "How are you today?"

"Nervous."

"What do you need me to do?"

"Pull the data on the Twinkle drives. I want copies of everything I took Meredith Monday night."

"It's on your desk."

He scooped up a stack of folders. On top was a small DAT cartridge. "What's this?"

"That's your video link with Arthur from Monday."

He shrugged, and dropped it in his briefcase.

Cindy said, "Anything else?"

"No." He glanced at his watch. "I'm late."

"Good luck, Tom," she said.

He thanked her and left the office.

Driving in morning rush-hour traffic, Sanders realized that the only surprise in his encounter with Conley was how sharp the young lawyer was. As for Meredith, her behavior didn't surprise him at all. For years, Sanders had fought the B-school mentality that she exemplified. After watching these graduates come and go, Sanders had finally concluded that there was a fundamental flaw in their education. They had been trained to believe that they were equipped to manage anything. But there was no such thing as general managerial skills and tools. In the end, there were only specific problems, involving specific industries and specific workers. To apply general tools to specific problems was to fail. You needed to know the market, you needed to know the customers, you needed to know the limits of manufacturing and the limits of your own creative people. None of that was obvious. Meredith couldn't see that Don Cherry and Mark Lewyn needed a link to manufacturing. Yet time and again, Sanders had been shown a prototype and had asked the one significant question: It looks fine, but can you make it on a production line? Can you build it, reliably and quickly, for a price? Sometimes they could, and sometimes they couldn't. If you took away that question, you changed the entire organization. And not for the better.

Conley was smart enough to see that. And smart enough to keep his ear to the ground. Sanders wondered how much Conley knew of what he hadn't said in their meeting. Did he also know about the harassment suit? It was certainly possible.

Christ, Meredith wanted to sell Austin. Eddie had been right all along. He considered telling him, but he really couldn't. And in any case, he had more pressing things to worry about. He saw the sign for the Magnuson Mediation Center and turned right. Sanders tugged at the knot on his tie, and pulled into a space in the parking lot.

The Magnuson Mediation Center was located just outside Seattle, on a hill overlooking the city. It consisted of three low buildings arranged around a central courtyard where water splashed in fountains and pools. The entire atmosphere was designed to be peaceful and relaxing, but Sanders was tense when he walked up from the parking lot and found Fernandez pacing.

"You see the paper today?" she said.

"Yeah, I saw it."

"Don't let it upset you. This is a very bad tactical move on their part," she said. "You know Connie Walsh?"

"No."

"She's a bitch," Fernandez said briskly. "Very unpleasant and very capable. But I expect Judge Murphy to take a strong position on it in the sessions. Now, this is what I worked out with Phil Blackburn. We'll begin with your version of the events of Monday night. Then Johnson will tell hers."

"Wait a minute. Why should I go first?" Sanders said. "If I go first, she'll have the advantage of hearing—"

"You are the one bringing the claim so you are obligated to present your case first. I think it will be to our advantage," Fernandez said. "This way Johnson will testify last, before lunch." They started toward the center building. "Now, there are just two things you have to remember. First, always tell the truth. No matter what happens, just tell the truth. Exactly as you remember it even if you think it hurts your case. Okay?"

"Okay."

"Second, don't get mad. Her lawyer will try to make you angry and trap you. Don't fall for it. If you feel insulted or start to get mad, request a five-minute break to consult with me. You're entitled to that, whenever you want. We'll go outside and cool off. But whatever you do, keep cool, Mr. Sanders."

"Okay."

"Good." She swung open the door. "Now let's go do it."

The mediation room was wood-paneled and spare. He saw a polished wooden table with a pitcher of water and glasses and some notepads; in the corner, a sideboard with coffee and a plate of pastries. Windows opened out on a small atrium with a fountain. He heard the sound of soft gurgling water.

The DigiCom legal team was already there, ranged along one side of the table. Phil Blackburn, Meredith Johnson, an attorney named Ben Heller, and two other grim-faced female attorneys. Each woman had an imposing stack of xeroxed papers before her on the table.

Fernandez introduced herself to Meredith Johnson, and the two women shook hands. Then Ben Heller shook hands with Sanders. Heller was a florid, beefy man with silver hair, and a deep voice. Well connected in Seattle, he reminded Sanders of a politician. Heller introduced the other women, but Sanders immediately forgot their names.

Meredith said, "Hello, Tom."

"Meredith."

He was struck by how beautiful she looked. She wore a blue suit with a cream-colored blouse. With her glasses and her blond hair pulled back, she looked like a lovely but studious schoolgirl. Heller patted her hand reassuringly, as if speaking to Sanders had been a terrible ordeal.

Sanders and Fernandez sat down opposite Johnson and Heller. Everybody got out papers and notes. Then there was an awkward silence, until Heller said to Fernandez, "How'd that King Power thing turn out?"

"We were pleased," Fernandez said.

"They fixed an award yet?"

"Next week, Ben."

"What are you asking?"

"Two million."

"Two *million?*"

"Sexual harassment's serious business, Ben. Awards are going up fast.

Right now the average verdict is over a million dollars. Especially when the company behaves that badly."

At the far end of the room, a door opened and a woman in her mid-fifties entered. She was brisk and erect, and wore a dark blue suit not very different from Meredith's.

"Good morning," she said. "I'm Barbara Murphy. Please refer to me as Judge Murphy, or Ms. Murphy." She moved around the room, shaking hands with everyone, then took a seat at the head of the table. She opened her briefcase and took out her notes.

"Let me tell you the ground rules for our sessions here," Judge Murphy said. "This is not a court of law, and our proceedings won't be recorded. I encourage everyone to maintain a civil and courteous tone. We're not here to make wild accusations or to fix blame. Our goal is to define the nature of the dispute between the parties, and to determine how best to resolve that dispute.

"I want to remind everyone that the allegations made on both sides are extremely serious and may have legal consequences for all parties. I urge you to treat these sessions confidentially. I particularly caution you against discussing what is said here with any outside person or with the press. I have taken the liberty of speaking privately to Mr. Donadio, the editor of the *Post-Intelligencer,* about the article that appeared today by Ms. Walsh. I reminded Mr. Donadio that all parties in 'Company X' are private individuals and that Ms. Walsh is a paid employee of the paper. The risk of a defamation suit against the *P-I* is very real. Mr. Donadio seemed to take my point."

She leaned forward, resting her elbows on the table. "Now then. The parties have agreed that Mr. Sanders will speak first, and he will then be questioned by Mr. Heller. Ms. Johnson will speak next, and will be questioned by Ms. Fernandez. In the interest of time, I alone will have the right to ask questions during the testimony of the principals, and I will set limits on the questions of opposing attorneys. I'm open to some discussion, but I ask your cooperation in letting me exercise judgment and keep things moving. Before we begin, does anybody have any questions?"

Nobody did.

"All right. Then let's get started. Mr. Sanders, why don't you tell us what happened, from your point of view."

Sanders talked quietly for the next half hour. He began with his meeting with Blackburn, where he learned that Meredith was going to be the new vice president. He reported the conversation with Meredith after her speech, in which she suggested a meeting about the Twinkle drive. He told what happened in the six o'clock meeting in detail.

As he spoke, he realized why Fernandez had insisted he tell this story over and over, the day before. The flow of events came easily to him now; he found that he could talk about penises and vaginas without hesitation. Even so, it was an ordeal. He felt exhausted by the time he described leaving the room and seeing the cleaning woman outside.

He then told about the phone call to his wife, and the early meeting the next morning, his subsequent conversation with Blackburn, and his decision to press charges.

"That's about it," he finished.

Judge Murphy said, "I have some questions before we go on. Mr. Sanders, you mentioned that wine was drunk during the meeting."

"Yes."

"How much wine would you say you had?"

"Less than a glass."

"And Ms. Johnson? How much would you say?"

"At least three glasses."

"All right." She made a note. "Mr. Sanders, do you have an employment contract with the company?"

"Yes."

"What is your understanding of what the contract says about transferring you or firing you?"

"They can't fire me without cause," Sanders said. "I don't know what it says about transfers. But my point is that by transferring me, they might as well be firing me—"

"I understand your point," Murphy said, interrupting him. "I'm asking about your contract. Mr. Blackburn?"

Blackburn said, "The relevant clause refers to 'equivalent transfer.' "

"I see. So it is arguable. Fine. Let's go on. Mr. Heller? Your questions for Mr. Sanders, please."

Ben Heller shuffled his papers and cleared his throat. "Mr. Sanders, would you like a break?"

"No, I'm fine."

"All right. Now, Mr. Sanders. You mentioned that when Mr. Blackburn told you on Monday morning that Ms. Johnson was going to be the new head of the division, you were surprised."

"Yes."

"Who did you think the new head would be?"

"I didn't know. Actually, I thought I might be in line for it."

"Why did you think that?"

"I just assumed it."

"Did anybody in the company, Mr. Blackburn or anybody else, lead you to think you were going to get the job?"

"No."

"Was there anything in writing to suggest you would get the job?"

"No."

"So when you say you assumed it, you were drawing a conclusion based on the general situation at the company, as you saw it."

"Yes."

"But not based on any real evidence?"

"No."

"All right. Now, you've said that when Mr. Blackburn told you that Ms. Johnson was going to get the job, he also told you that she could choose new division heads if she wanted, and you told him you interpreted that to mean Ms. Johnson had the power to fire you?"

"Yes, that's what he said."

"Did he characterize it in any way? For example, did he say it was likely or unlikely?"

"He said it was unlikely."

"And did you believe him?"

"I wasn't sure what to believe, at that point."

"Is Mr. Blackburn's judgment on company matters reliable?"

"Ordinarily, yes."

"But in any case, Mr. Blackburn did say that Ms. Johnson had the right to fire you."

"Yes."

"Did Ms. Johnson ever say anything like that to you?"

"No."

"She never made any statement that could be interpreted as an offer contingent upon your performance, including sexual performance?"

"No."

"So when you say that during your meeting with her you felt that your job was at risk, that was not because of anything Ms. Johnson actually said or did?"

"No," Sanders said. "But it was in the situation."

"You *perceived* it as being in the situation."

"Yes."

"As you had earlier perceived that you were in line for a promotion, when in fact you were not? The very promotion that Ms. Johnson ended up getting?"

"I don't follow you."

"I'm merely observing," Heller said, "that perceptions are subjective, and do not have the weight of fact."

"Objection," Fernandez said. "Employee perceptions have been held valid in contexts where the reasonable expectation—"

"Ms. Fernandez," Murphy said, "Mr. Heller hasn't challenged the validity of your client's perceptions. He has questioned their accuracy."

"But surely they are accurate. Because Ms. Johnson was his superior, and she could fire him if she wanted to."

"That's not in dispute. But Mr. Heller is asking whether Mr. Sanders has a tendency to build up unjustified expectations. And that seems to me entirely relevant."

"But with all due respect, Your Honor—"

"Ms. Fernandez," Murphy said, "we're here to clarify this dispute. I'm going to let Mr. Heller continue. Mr. Heller?"

"Thank you, Your Honor. So to summarize, Mr. Sanders: Although you felt your job was on the line, you never got that sense from Ms. Johnson?"

"No, I didn't."

"Or from Mr. Blackburn?"

"No."

"Or, in fact, from anyone else?"

"No."

"All right. Let's turn to something else. How did it happen that there was wine at the six o'clock meeting?"

"Ms. Johnson said that she would get a bottle of wine."

"You didn't ask her to do that?"

"No. She volunteered to do it."

"And what was your reaction?"

"I don't know." He shrugged. "Nothing in particular."

"Were you pleased?"

"I didn't think about it one way or the other."

"Let me put it a different way, Mr. Sanders. When you heard that an attractive woman like Ms. Johnson was planning to have a drink with you after work, what went through your head?"

"I thought I better do it. She's my boss."

"That's all you thought?"

"Yes."

"Did you mention to anyone that you wanted to be alone with Ms. Johnson in a romantic setting?"

Sanders sat forward, surprised. "No."

"Are you sure about that?"

"Yes." Sanders shook his head. "I don't know what you're driving at."

"Isn't Ms. Johnson your former lover?"

"Yes."

"And didn't you want to resume your intimate relationship?"

"No, I did not. I was just hoping we would be able to find some way to be able to work together."

"Is that difficult? I would have thought it'd be quite easy to work together, since you knew each other so well in the past."

"Well, it's not. It's quite awkward."

"Is it? Why is that?"

"Well. It just is. I had never actually worked with her. I knew her in a totally different context, and I just felt awkward."

"How did your prior relationship with Ms. Johnson end, Mr. Sanders?"

"We just sort of . . . drifted apart."

"You had been living together at the time?"

"Yes. And we had our normal ups and downs. And finally, it just didn't work out. So we split up."

"No hard feelings?"

"No."

"Who left whom?"

"It was sort of mutual, as I recall."

"Whose idea was it to move out?"

"I guess . . . I don't really remember. I guess it was mine."

"So there was no awkwardness or tension about how the affair ended, ten years ago."

"No."

"And yet you felt there was awkwardness now?"

"Sure," Sanders said. "Because we had one kind of relationship in the past, and now we were going to have another kind of relationship."

"You mean, now Ms. Johnson was going to be your superior."

"Yes."

"Weren't you angry about that? About her appointment?"

"A little. I guess."

"Only a little? Or perhaps more than a little?"

Fernandez sat forward and started to protest. Murphy shot her a warning look. Fernandez put her fists under her chin and said nothing.

"I was a lot of things," Sanders said. "I was angry and disappointed and confused and worried."

"So in your mind, although you were feeling many different and confusing feelings, you're certain that you did not, under any circumstances, contemplate having sex with Ms. Johnson that night."

"No."

"It never crossed your mind?"

"No."

There was a pause. Heller shuffled his notes, then looked up. "You're married, are you not, Mr. Sanders?"

"Yes, I am."

"Did you call your wife to tell her you had a late meeting?"

"Yes."

"Did you tell her with whom?"

"No."

"Why not?"

"My wife is sometimes jealous about my past relationships. I didn't see any reason to cause her anxiety or make her upset."

"You mean, if you told her you were having a late meeting with Ms. Johnson, your wife might think that you would renew your sexual acquaintance."

"I don't know what she would think," Sanders said.

"But in any case, you didn't tell her about Ms. Johnson."

"No."

"What did you tell her?"

"I told her I had a meeting and I would be home late."

"How late?"

"I told her it might run to dinner or after."

"I see. Had Ms. Johnson suggested dinner to you?"

"No."

"So you presumed, when you called your wife, that your meeting with Ms. Johnson might be a long one?"

"No," Sanders said. "I didn't. But I didn't know exactly how long it would be. And my wife doesn't like me to call once and say I'll be an hour late, and then call again to say it'll be two hours. That annoys her. So it's easier for her if I just tell her I may be home after dinner. That way, she doesn't expect me and doesn't wait for me; and if I get home early, it's great."

"So this is your usual policy with your wife."

"Yes."

"Nothing unusual."

"No."

"In other words, your usual procedure is to lie to your wife about events at the office because in your view she can't take the truth."

"Objection," Fernandez said. "What's the relevance?"

"That's not it at all," Sanders continued, angrily.

"How is it, Mr. Sanders?"

"Look. Every marriage has its own way to work things out. This is ours. It makes things smoother, that's all. It's about scheduling at home, not about lying."

"But wouldn't you say that you lied when you failed to tell your wife you were seeing Ms. Johnson that night?"

"Objection," Fernandez said.

Murphy said, "I think this is *quite* enough, Mr. Heller."

"Your Honor, I'm trying to show that Mr. Sanders intended to consummate an encounter with Ms. Johnson, and that all his behavior is consistent with that. And in addition, to show that he routinely treats women with contempt."

"You haven't shown that, you haven't even laid a groundwork for that," Murphy said. "Mr. Sanders has explained his reasons, and in the absence of contrary evidence I accept them. Do you have contrary evidence?"

"No, Your Honor."

"Very well. Bear in mind that inflammatory and unsubstantiated characterizations do not assist our mutual efforts at resolution."

"Yes, Your Honor."

"I want everyone here to be clear: these proceedings are potentially damaging to all parties—not only in their outcome, but in the conduct of the proceedings themselves. Depending on the outcome, Ms. Johnson and Mr. Sanders may find themselves working together in some capacity in the future. I will not permit these proceedings to unnecessarily poison such future relationships. Any further unwarranted accusations will cause me to halt these proceedings. Does anyone have any questions about what I've just said?"

No one did.

"All right. Mr. Heller?"

Heller sat back. "No further questions, Your Honor."

"All right," Judge Murphy said. "We'll break for five minutes, and return to hear Ms. Johnson's version."

You're doing fine," Fernandez said. "You're doing very well. Your voice was strong. You were clear and even. Murphy was impressed. You're doing fine." They were standing outside, by the fountains in the courtyard. Sanders felt like a boxer between rounds, being worked over by his trainer. "How do you feel?" she asked. "Tired?"

"A little. Not too bad."

"You want coffee?"

"No, I'm okay."

"Good. Because the hard part is coming up. You're going to have to be very strong when she gives her version. You won't like what she says. But it's important that you stay calm."

"Okay."

She put her hand on his shoulder. "By the way, just between us: How *did* the relationship end?"

"To tell the truth, I can't remember exactly."

Fernandez looked skeptical. "But this was important, surely . . ."

"It was almost ten years ago," Sanders said. "To me, it feels like another lifetime."

She was still skeptical.

"Look," Sanders said. "This is the third week in June. What was going on in your love life the third week of June, ten years ago? Can you tell me?"

Fernandez was silent, frowning.

"Were you married?" Sanders prompted.

"No."

"Met your husband yet?"

"Uh, let's see . . . no . . . not until . . . I must have met my husband . . . about a year later."

"Okay. Do you remember who you were seeing before him?"

Fernandez was silent. Thinking.

"How about *anything* that happened between you and a lover in June, ten years ago?"

She was still silent.

"See what I mean?" Sanders said. "Ten years is a long time. I remember the affair with Meredith, but I'm not clear about the last few weeks of it. I don't remember the details of how it ended."

"What do you remember?"

He shrugged. "We had more fights, more yelling. We were still living together, but somehow, we began to arrange our schedules so that we never saw each other. You know how that happens. Because when we did run into each other, we fought.

"And finally one night, we had a big argument while we were getting dressed to go to a party. Some formal party for DigiCom. I remember I had to wear a tux. I threw my cuff links at her and then I couldn't find them. I had to get down on the floor and look. But once we were driving to the party, we sort of calmed down, and we started talking about breaking up. In this very ordinary way. Very reasonable way. It just came out. Both of us. Nobody shouted. And in the end, we decided it was best if we broke it off."

Fernandez was looking at him thoughtfully. "That's it?"

"Yeah." He shrugged. "Except we never got to the party."

Something at the back of his mind. *A couple in a car, going to a party. Something about a cellular phone. All dressed up, going to the party and they make a call, and—*

He couldn't get it. It hung in his memory, just beyond recollection.

The woman made a call on the cellular phone, and then . . . Something embarrassing afterward . . .

"Tom?" Fernandez said, shaking his shoulder. "Looks like our time is about up. Ready to go back?"

"I'm ready," he said.

As they were heading back to the mediation room, Heller came over. He gave Sanders an oily smile, then turned to Fernandez. "Counselor," he said. "I wonder if this is the time to talk about settlement."

"Settlement?" Fernandez said, showing elaborate surprise. "Why?"

"Well, things aren't going so well for your client, and—"

"Things are going fine for my client—"

"And this whole inquiry will only get more embarrassing and awkward for him, the longer it continues—"

"My client isn't embarrassed at all—"

"And perhaps it is to everyone's advantage to end it now."

Fernandez smiled. "I don't think that's my client's wish, Ben, but if you have an offer to make, we will of course entertain it."

"Yes. I have an offer."

"All right."

Heller cleared his throat. "Considering Tom's current compensation base and associated benefits package, and taking into consideration his lengthy service with the company, we're prepared to settle for an amount equal to several years of compensation. We'll add an allowance for your fees and other miscellaneous expenses of termination, the cost of a headhunter to relocate to a new position, and all direct costs that may be associated with moving his household, and all together make it four hundred thousand dollars. I think that's very generous."

"I'll see what my client says," Fernandez said. She took Sanders by the arm, and walked a short distance away. "Well?"

"No," Sanders said.

"Not so fast," she said. "That's a pretty reasonable offer. It's as much as you're likely to get in court, without the delay and expenses."

"No."

"Want to counter?"

"No. Fuck him."

"I think we should counter."

"Fuck him."

Fernandez shook her head. "Let's be smart, not angry. What do you hope to gain from all this, Tom? There must be a figure you would accept."

"I want what I'll get when they take the company public," Sanders said. "And that's somewhere between five and twelve million."

"You *think*. It's a speculative estimate for a future event."

"That's what it'll be, believe me."

Fernandez looked at him. "Would you take five million now?"

"Yes."

"Alternatively, would you take the compensation package he outlined, plus the stock options you would get at the time of the offering?"

Sanders considered that. "Yes."

"All right. I'll tell him."

She walked back across the courtyard to Heller. The two spoke briefly. After a moment, Heller turned on his heel and stalked away.

Fernandez came back, grinning. "He didn't go for it." They headed back inside. "But I'll tell you one thing: this is a good sign."

"It is?"

"Yes. If they want to settle before Johnson gives her testimony, it's a very good sign."

n view of the acquisition," Meredith Johnson said, "I felt it was important that I meet with all the division heads on Monday." She spoke calmly and slowly, looking at everyone seated around the table in turn. Sanders had the sense of an executive giving a presentation. "I met with Don Cherry, Mark Lewyn, and Mary Anne Hunter during the afternoon. But Tom Sanders said he had a very busy schedule, and asked if we could meet at the end of the day. At his request, I scheduled the meeting with Tom at six o'clock."

He was amazed at the cool way that she lied. He had expected her to be effective, but he was still astonished to see her in action.

"Tom suggested that we could have a drink as well, and go over old times. That wasn't really my style, but I agreed. I was especially concerned to establish good relations with Tom, because I knew he was disappointed he had not gotten the job, and because we had a past history. I wanted our working relationship to be cordial. For me to refuse a drink with him seemed . . . I don't know—standoffish, or stiff. So I said yes.

"Tom came to the office at six o'clock. We had a glass of wine, and talked about the problems with the Twinkle drive. However, from the outset he kept making comments of a personal nature that I considered inappropriate—for example, comments about my appearance, and about how often he thought about our past relationship. Reference to sexual incidents in the past, and so on."

Son of a bitch. Sanders's whole body was tense. His hands were clenched. His jaw was tight.

Fernandez leaned over and put her hand on his wrist.

Meredith Johnson was saying, ". . . had some calls from Garvin and others. I took them at my desk. Then my assistant came in and asked if she could leave early, to deal with some personal matters. I said she could. She left the room. That was when Tom came over and suddenly started kissing me."

She paused for a moment, looking around the room. She met Sanders's eyes with a steady gaze.

"I was taken aback by his sudden and unexpected overture," she said, staring evenly at him. "At first, I tried to protest, and to defuse the situation. But Tom is much larger than I am. Much stronger. He pulled me over onto the couch and started to disrobe, and to take my clothes off as well. As you can imagine, I was horrified and frightened. The situation was out of control, and the fact that it was happening made our future working relationship very difficult. To say nothing of how I felt personally, as a woman. I mean, to be assaulted in this way."

Sanders stared at her, trying desperately to control his anger. He heard Fernandez, at his ear. *"Breathe."* He took a deep breath and let it out slowly. He had not been aware until then that he was holding his breath.

"I kept trying to make light of it," Meredith continued, "to make jokes, to get free. I was trying to say to him, Oh, come on Tom, let's not do this. But he was determined. And when he tore my underwear off, when I heard the sound of the cloth ripping, I realized that I could not get out of this situation in any diplomatic way. I had to acknowledge that Mr. Sanders was raping me and I became very scared and very angry. When he moved away from me on the couch, to free his penis from his trousers, prior to penetration, I kneed him in the groin. He rolled off the couch, onto the floor. Then he got to his feet, and I got to my feet.

"Mr. Sanders was angry that I had refused his advances. He started shouting at me, and then he hit me, knocking me down onto the floor. But by then I was angry, too. I remember saying, 'You can't do this to me,' and swearing at him. But I can't say I remember everything that he said or that I said. He came back at me one more time, but by then I had my shoes in my hand, and I hit him in the chest with my high heels, trying to drive him away. I think I tore his shirt. I'm not sure. I was so angry by then, I wanted to kill him. I'm sure I scratched him. I remember I said I wanted to kill him. I was so angry. Here it was my first day in this new job, I was under so much pressure, I was trying to do a good job and this . . . this *thing* had happened that ruined our relationship and was going to cause a lot of trouble for everybody in the company. He went off in an angry rage. After he left, the question for me was how to handle it."

She paused, shaking her head, apparently lost in the emotions of that moment.

Heller said gently, "How did you decide to handle it?"

"Well, it's a problem. Tom's an important employee, and he is not an easy person to replace. Furthermore, in my judgment it would not be wise to make a replacement in the middle of the acquisition. My first impulse was to see if we could forget the whole thing. After all, we're both adults. I was personally embarrassed, but I thought that Tom would probably be embarrassed, too, when he sobered up and had a chance to think it over. And I thought that maybe we could just go on from there. After all, awkward things happen sometimes. People can overlook them.

"So when the meeting time changed, I called his house to tell him. He wasn't there, but I had a very pleasant conversation with his wife. It was clear from our conversation that she did not know that Tom had been meeting me, or that Tom and I knew each other from the past. Anyway, I gave his wife the new meeting time, and asked her to tell Tom.

"The next day, at the meeting, things did not go well. Tom showed up late, and changed his story about the Twinkle drive, minimizing the problems and contradicting me. He was clearly undercutting my authority in a corporate meeting and I could not permit that. I went directly to Phil Blackburn and told him everything that had happened. I said I did not want to press formal charges, but I made it clear that I could not work with Tom and that a change would have to be made. Phil said he would talk to Tom. And eventually it was decided that we would try to mediate a resolution."

She sat back, and placed her hands flat on the table. "That's all, I think. That's everything." She looked around at everyone, meeting their eyes in turn. Very cool, very controlled.

It was a spectacular performance, and in Sanders it produced a quite unexpected effect: he felt guilty. He felt as if he had done the things that she said he had done. He felt sudden shame, and looked down at the table, hanging his head.

Fernandez kicked him in the ankle, hard. He jerked his head up, wincing. She was frowning at him. He sat up.

Judge Murphy cleared her throat. "Evidently," she said, "we are presented with two entirely incompatible reports. Ms. Johnson, I have only a few questions before we go on."

"Yes, Your Honor?"

"You're an attractive woman. I'm sure you've had to fend off your share of unwanted approaches in the course of your business career."

Meredith smiled. "Yes, Your Honor."

"And I'm sure you have developed some skill at it."

"Yes, Your Honor."

"You've said you were aware of tensions from your past relationship with Mr. Sanders. Considering those tensions, I would have thought that a meeting held in the middle of the day, without wine, would have been more professional—would have set a better tone."

"I'm sure that's correct in hindsight," Meredith said. "But at the time, this was all in the context of the acquisition meetings. Everybody was busy. I was just trying to fit the meeting with Mr. Sanders in before the Conley-White sessions the next day. That's all I was thinking about. Schedules."

"I see. And after Mr. Sanders left your office, why didn't you call Mr. Blackburn, or someone else in the company, to report what had happened?"

"As I said, I was hoping it could all be overlooked."

"Yet the episode you describe," Murphy said, "is a serious breach of normal business behavior. As an experienced manager, you must have known the chance of a good working relationship with Mr. Sanders was nil. I would have thought you'd feel obliged to report what happened to a superior at once. And from a practical standpoint, I would have thought you'd want to go on record as soon as possible."

"As I said, I was still hoping." She frowned, thinking. "You know, I guess . . . I felt responsible for Tom. As an old friend, I didn't want to be the reason why he lost his job."

"On the other hand, you are the reason why he lost his job."

"Yes. Again, in hindsight."

"I see. All right. Ms. Fernandez?"

"Thank you, Your Honor." Fernandez turned in her chair to face Johnson. "Ms. Johnson, in a situation like this, when private behavior occurs behind closed doors, we need to look at surrounding events where we can. So I'll ask you a few questions about surrounding events."

"Fine."

"You've said that when you made the appointment with Mr. Sanders, he requested wine."

"Yes."

"Where did the wine come from, that you drank that night?"

"I asked my assistant to get it."

"This is Ms. Ross?"

"Yes."

"She's been with you a long time?"

"Yes."

"She came up with you from Cupertino?"

"Yes."

"She is a trusted employee?"

"Yes."

"How many bottles did you ask Ms. Ross to buy?"

"I don't remember if I specified a particular number."

"All right. How many bottles did she get?"

"Three, I think."

"Three. And did you ask your assistant to buy anything else?"

"Like what?"

"Did you ask her to buy condoms?"

"No."

"Do you know if she bought condoms?"

"No, I don't."

"In fact, she did. She bought condoms from the Second Avenue Drugstore."

"Well, if she bought condoms," Johnson said, "it must have been for herself."

"Do you know of any reason why your assistant would say she bought the condoms for you?"

"No," Johnson said, speaking slowly. She was thinking it over. "I can't imagine she would do that."

"Just a moment," Murphy said, interrupting. "Ms. Fernandez, are you alleging that the assistant *did* say that she bought the condoms for Ms. Johnson?"

"Yes, Your Honor. We are."

"You have a witness to that effect?"

"Yes, we do."

Sitting beside Johnson, Heller rubbed the bottom of his lip with one finger. Johnson showed no reaction at all. She didn't even blink. She just continued to gaze calmly at Fernandez, waiting for the next question.

"Ms. Johnson, did you instruct your assistant to lock the door to your office when Mr. Sanders was with you?"

"I most certainly did not."

"Do you know if she locked the door?"

"No, I don't."

"Do you know why she would tell someone that you ordered her to lock the door?"

"No."

"Ms. Johnson. Your meeting with Mr. Sanders was at six o'clock. Did you have any appointments later that day?"

"No. His was the last."

"Isn't it true that you had a seven o'clock appointment that you canceled?"

"Oh. Yes, that's true. I had one with Stephanie Kaplan. But I canceled it because I wasn't going to have the figures ready for her to go over. There wasn't time to prepare."

"Are you aware that your assistant told Ms. Kaplan that you were canceling because you had another meeting that was going to run late?"

"I don't know what my assistant said to her," Meredith replied, showing impatience for the first time. "We seem to be talking a great deal about my assistant. Perhaps you should be asking her these questions."

"Perhaps we should. I'm sure it can be arranged. All right. Let's turn to something else. Mr. Sanders said he saw a cleaning woman when he left your office. Did you also see her?"

"No. I stayed in my office after he had gone."

"The cleaning woman, Marian Walden, says she overheard a loud argument prior to Mr. Sanders's departure. She says she heard a man say, 'This isn't a good idea, I don't want to do this,' and she heard a woman say, 'You fucking bastard, you can't leave me like this.' Do you recall saying anything like that?"

"No. I recall saying, 'You can't do this to me.'"

"But you don't recall saying, 'You can't leave me like this.'"

"No, I do not."

"Ms. Walden is quite clear that was what you said."

"I don't know what Ms. Walden thought she heard," Johnson said. "The doors were closed the entire time."

"Weren't you speaking quite loudly?"

"I don't know. Possibly."

"Ms. Walden said you were shouting. And Mr. Sanders has said you were shouting."

"I don't know."

"All right. Now, Ms. Johnson, you said that you informed Mr. Blackburn that you could not work with Mr. Sanders after the unfortunate Tuesday morning meeting, is that right?"

"Yes. That's right."

Sanders sat forward. He suddenly realized that he had overlooked that, while Meredith was making her original statement. He had been so upset, he hadn't realized that she had lied about when she saw Blackburn. Because Sanders had gone to Blackburn's office right after the meeting—and Blackburn already knew.

"Ms. Johnson, what time would you say you went to see Mr. Blackburn?"

"I don't know. After the meeting."

"About what time?"

"Ten o'clock."

"Not earlier?"

"No."

Sanders glanced over at Blackburn, who sat rigidly at the end of the table. He looked tense, and bit his lip.

Fernandez said, "Shall I ask Mr. Blackburn to confirm that? I imagine his assistant has a log, if he has difficulty with exact memory."

There was a short silence. She looked over at Blackburn. "No," Meredith said. "No. I was confused. What I meant to say was I talked to Phil after the initial meeting, and before the second meeting."

"The initial meeting being the one at which Sanders was absent? The eight o'clock meeting."

"Yes."

"So Mr. Sanders's behavior at the second meeting, where he contradicted you, could not have been relevant to your decision to speak to Mr. Blackburn. Because you had already spoken to Mr. Blackburn by the time that meeting took place."

"As I say, I was confused."

"I have no more questions of this witness, Your Honor."

Judge Murphy closed her notepad. Her expression was bland and unreadable. She looked at her watch. "It's now eleven-thirty. We will break for lunch for two hours. I'm allowing extra time so that counsel can meet to review the situation and to decide how the parties wish to proceed." She stood up. "I am also available if counsel wish to meet with me for any reason. Otherwise, I'll see you all back here at one-thirty sharp. Have a pleasant and productive lunch." She turned and walked out of the room.

Blackburn stood and said, "Personally, I'd like to meet with opposing counsel, right now."

Sanders glanced over at Fernandez.

Fernandez gave the faintest of smiles. "I'm amenable to that, Mr. Blackburn," she said.

The three lawyers stood beside the fountain. Fernandez was talking animatedly to Heller, their heads close together. Blackburn was a few paces away, a cellular phone pressed to his ear. Across the courtyard, Meredith Johnson talked on another phone, gesturing angrily as she talked.

Sanders stood off to one side by himself, and watched. There was no question in his mind that Blackburn would seek a settlement. Piece by piece, Fernandez had torn Meredith Johnson's version apart: demonstrating that she had ordered her assistant to buy wine, to buy condoms, to lock the door when Sanders was there, and to cancel later appointments. Clearly, Meredith Johnson was not a supervisor surprised by a sexual overture. She had been planning it all afternoon. Her crucial reaction—her angry statement that "You can't leave me"—had been overheard by the cleaning woman. And she had lied about the timing and motivation of her report to Blackburn.

There could be no doubt in anyone's mind that Meredith was lying. The only question now was what Blackburn and DigiCom would do about it. Sanders had sat through enough management sensitivity seminars on sexual harassment to know what the company's obligation was. They really had no choice.

They would have to fire her.

But what would they do about Sanders? That was another question entirely. He had the strong intuition that by bringing this accusation, he had burned his bridges at the company; he would never be welcomed back. Sanders had shot down Garvin's pet bird, and Garvin would not forgive him for it.

So: they wouldn't let him back. They would have to pay him off.

"They're calling it quits already, huh?"

Sanders turned and saw Alan, one of the investigators, coming up from the parking lot. Alan had glanced over at the lawyers and quickly appraised the situation.

"I think so," Sanders said.

Alan squinted at the lawyers. "They should. Johnson has a problem. And a lot of people in the company know about it. Especially her assistant."

Sanders said, "You talked to her last night?"

"Yeah," he said. "Herb found the cleaning woman and got her taped. And I had a late night with Betsy Ross. She's a lonely lady, here in a new town. She drinks too much, and I taped it all."

"Did she know that?"

"She doesn't have to," Alan said. "It's still admissible." He watched the lawyers for a moment. "Blackburn must be shitting staples about now."

Louise Fernandez was stalking across the courtyard, grim-faced, hunched over. "God*damn* it," she said, as she came up.

"What happened?" Sanders said.

Fernandez shook her head. "They won't make a deal."

"They won't make a deal?"

"That's right. They just deny every point. Her assistant bought wine? That was for Sanders. Her assistant bought condoms? That was for the assistant. The assistant says she bought them for Johnson? The assistant is an unreliable drunk. The cleaning lady's report? She couldn't know what she heard, she had the radio on. And always the constant refrain, 'You know, Louise, this won't stand up in court.' And Bullet-proof Betty is on the phone, running the whole thing. Telling every-body what to do." Fernandez swore. "I have to tell you. This is the kind of shit male executives pull. They look you right in the eye and say, 'It never happened. It just isn't there. You have no case.' It burns my ass. *Damn* it!"

"Better get some lunch, Louise," Alan said. To Sanders he said, "She sometimes forgets to eat."

"Yeah, fine. Sure. Eat." They started toward the parking lot. She was walking fast, shaking her head. "I can't understand how they can take this position," she said. "Because I know—I could see it in Judge Murphy's eyes—that she didn't think there'd be an afternoon session at all. Judge Murphy heard the evidence and concluded it's all over. So did I. But it's not over. Blackburn and Heller aren't moving *one inch*. They're not going to settle. They're basically inviting us to sue."

"So we'll sue," Sanders said, shrugging.

"Not if we're smart," Fernandez said. "Not *now*. This is exactly what I was afraid would happen. They got a lot of free discovery, and we got nothing. We're back to square one. And they have the next three years to work on that assistant, and that cleaning lady, and anything else we come up with. And let me tell you: in three years we won't even be able to *find* that assistant."

"But we have her on tape . . ."

"She still has to appear in court. And believe me, she never will. Look, DigiCom has huge exposure. If we show that DigiCom didn't respond in a timely and adequate fashion to what they knew about Johnson, they could be liable for extremely large damages. There was a case on point last month in California: nineteen point four million dollars, found for the plaintiff. With exposure like that, take my word for it: the assistant will be unavailable. She'll be on vacation in Costa Rica for the rest of her life."

"So what do we do?" Sanders said.

"For better or worse, we're committed now. We've taken this line and we have to continue it. Somehow, we have to force them to come to terms," she said. "But we're going to need something else to do that. You got anything else?"

Sanders shook his head. "No, nothing."

"Hell," Fernandez said. "What's going on? I thought DigiCom was worried about this allegation becoming public before they finished the acquisition. I thought they had a publicity problem."

Sanders nodded. "I thought they did, too."

"Then there's something we don't understand. Because Heller and Blackburn both act like they couldn't care less what we do. Now why is that?"

A heavyset man with a mustache walked past them, carrying a sheaf of papers. He looked like a cop.

"Who's he?" Fernandez said.

"Never seen him before."

"They were calling on the phone for somebody. Trying to locate somebody. That's why I ask."

Sanders shrugged. "What do we do now?"

"We eat," Alan said.

"Right. Let's go eat," Fernandez said, "and forget it for a while."

In the same moment, a thought popped into his mind: *Forget that phone.* It seemed to come from nowhere, like a command:

Forget that phone.

Walking beside him, Fernandez sighed. "We still have things we can develop. It's not over yet. You've still got things, right, Alan?"

"Absolutely," Alan said. "We've hardly begun. We haven't gotten to Johnson's husband yet, or to her previous employer. There's lots of stones left to turn over and see what crawls out."

Forget that phone.

"I better check in with my office," Sanders said, and took out his cellular phone to dial Cindy.

A light rain began to fall. They came to the cars in the parking lot. Fernandez said, "Who's going to drive?"

"I will," Alan said.

They went to his car, a plain Ford sedan. Alan unlocked the doors, and Fernandez started to get in. "And I thought that at lunch today we would be going to have a party," she said.

Going to a party . . .

Sanders looked at Fernandez sitting in the front seat, behind the rain-spattered windshield. He held the phone up to his ear and waited while the call went through to Cindy. He was relieved that his phone was working correctly. Ever since Monday night when it went dead, he hadn't trusted it completely. But it seemed to be fine. Nothing wrong with it at all.

The couple was going to a party and she made a call on a cellular phone. From the car . . .

Forget that phone.

Cindy said, "Mr. Sanders's office."

And when she called, she got an answering machine. She left a message on the answering machine. And then she hung up.

"Hello? Mr. Sanders's office. Hello?"

"Cindy, it's me."

"Oh, hi, Tom." Still reserved.

"Any messages?" he said.

"Uh, yes, let me look at the book. You had a call from Arthur in KL, he wanted to know if the drives arrived. I checked with Don Cherry's team; they got them. They're working on them now. And you had a call from Eddie in Austin; he sounded worried. And you had another call from John Levin. He called you yesterday, too. And he said it was important."

Levin was the executive with a hard drive supplier. Whatever was on his mind, it could wait.

"Okay. Thanks, Cindy."

"Are you going to be back in the office today? A lot of people are asking."

"I don't know."

"John Conley from Conley-White called. He wanted to meet with you at four."

"I don't know. I'll see. I'll call you later."

"Okay." She hung up.

He heard a dial tone.

And then she had hung up.

The story tugged at the back of his mind. The two people in the car. Going to the party. Who had told him that story? How did it go?

On her way to the party, Adele had made a call from the car and then she had hung up.

Sanders snapped his fingers. Of course! Adele! The couple in the car had been Mark and Adele Lewyn. And they had had an embarrassing incident. It was starting to come back to him now.

Adele had called somebody and gotten the answering machine. She left a message, and hung up the phone. Then she and Mark talked in the car about the person Adele had just called. They made jokes and unflattering comments for about fifteen minutes. And later they were very embarrassed . . .

Fernandez said, "Are you just going to stand there in the rain?"

Sanders didn't answer. He took the cellular phone down from his ear. The keypad and screen glowed bright green. Plenty of power. He looked at the phone and waited. After five seconds, it clicked itself off; the screen went blank. That was because the new generation of phones had an auto-shutdown feature to conserve battery power. If you didn't use the phone or press the keypad for fifteen seconds, the phone shut itself off. So it wouldn't go dead.

But his phone had gone dead in Meredith's office.

Why?

Forget that phone.

Why had his cellular phone failed to shut itself off? What possible explanation could there be? Mechanical problems: one of the keys stuck, keeping the phone on. It had been damaged when he dropped it, when Meredith first kissed him. The battery was low because he forgot to charge it the night before.

No, he thought. The phone was reliable. There was no mechanical fault. And it was fully charged.

No.

The phone had worked correctly.

They made jokes and unflattering comments for about fifteen minutes.

His mind began to race, with scattered fragments of conversation coming back to him.

"Listen, why didn't you call me last night?"

"I did, Mark."

Sanders was certain that he had called Mark Lewyn from Meredith's office. Standing in the parking lot in the rain, he again pressed L-E-W on his keypad. The phone turned itself back on, the little screen flashing LEWYN and Mark's home number.

"There wasn't any message when I got home."

"I talked to your answering machine, about six-fifteen."

"I never got a message."

Sanders was sure that he had called Lewyn and had talked to his answering machine. He remembered a man's voice saying the standard message, "Leave a message when you hear the tone."

Standing there with the phone in his hand, staring at Lewyn's phone number, he pressed the SEND button. A moment later, the answering machine picked up. A woman's voice said, "Hi, you've reached Mark and Adele at home. We're not able to come to the phone right now, but if you leave a message, we'll call you back." *Beep.*

That was a different message.

He *hadn't* called Mark Lewyn that night.

Which could only mean he hadn't pressed L-E-W that night. Nervous in Meredith's office, he must have pressed something else. He had gotten somebody else's answering machine.

And his phone had gone dead.

Because . . .

Forget that phone.

"Jesus Christ," he said. He suddenly put it together. He knew exactly what had happened. And it meant that there was the chance that—

"Tom, are you all right?" Fernandez said.

"I'm fine," he said. "Just give me a minute. I think I've got something important."

He hadn't pressed L-E-W.

He had pressed something else. Something very close, probably one letter off. With fumbling fingers, Sanders pushed L-E-L. The screen stayed blank: he had no number stored for that combination. L-E-M. No number stored. L-E-S. No number stored. L-E-V.

Bingo.

Printed across the little screen was:

LEVIN

And a phone number for John Levin.

Sanders had called John Levin's answering machine that night.

John Levin called. He said it was important.

I'll bet he did, Sanders thought.

He remembered now, with sudden clarity, the exact sequence of events in Meredith's office. He had been talking on the phone and she said, "Forget that phone," and pushed his hand down as she started kissing him. He had dropped the phone on the windowsill as they kissed, and left it there.

Later on, when he left Meredith's office, buttoning his shirt, he had picked up the cellular phone from the sill, but by then it was dead. Which could only mean that it had remained constantly on for almost an hour. It had remained on during the entire incident with Meredith.

In the car, when Adele finished the call, she hung the phone back in the cradle, She didn't press the END *button, so the phone line stayed open, and their entire conversation was recorded on the person's answering machine. Fifteen minutes of jokes and personal commentary, all recorded on his answering machine.*

And Sanders's phone had been dead because the line stayed open. The whole conversation had been recorded.

Standing in the parking lot, he quickly dialed John Levin's number. Fernandez got out of the car and came over to him. "What's going on?" Fernandez said. "Are we going to lunch, or what?"

"Just a minute."

The call went through. A click of the pickup, then a man's voice: "John Levin."

"John, it's Tom Sanders."

"Well, hey there, Tom boy!" Levin burst out laughing. "My *man*! Are you having a red-hot sex life these days, or what? I tell you, Tom, my ears were burning."

Sanders said, "Was it recorded?"

"Jesus Christ, Tom, you better believe it. I came in Tuesday morning to check my messages, and I tell you, it went on for half an hour, I mean—"

"John—"

"Whoever said married life was dull—"

"John. Listen. *Did you keep it?*"

There was a pause. Levin stopped laughing. "Tom, what do you think I am, a pervert? Of *course* I kept it. I played it for the whole office. They loved it!"

"John. Seriously."

Levin sighed. "Yeah. I kept it. It sounded like you might be having a little trouble, and . . . I don't know. Anyway, I kept it."

"Good. Where is it?"

"Right here on my desk," Levin said.

"John, I want that tape. Now listen to me: this is what I want you to do."

Driving in the car, Fernandez said, "I'm waiting."

Sanders said, "There's a tape of the whole meeting with Meredith. It was all recorded."

"How?"

"It was an accident. I was talking to an answering machine," he said, "and when Meredith started kissing me, I put the phone down but didn't end the call. So the phone stayed connected to the answering machine. And everything we said went right onto the answering machine."

"Hot damn," Alan said, slapping the steering wheel as he drove.

"This is an audio tape?" Fernandez said.

"Yes."

"Good quality?"

"I don't know. We'll see. John's bringing it to lunch."

Fernandez rubbed her hands together. "I feel better already."

"Yes?"

"Yes," she said. "Because if it's any good at all, we can really draw blood."

John Levin, florid and jovial, pushed away his plate and drained the last of his beer. "Now that's what I call a meal. *Excellent* halibut." Levin weighed nearly three hundred pounds, and his belly pressed up against the edge of the table.

They were sitting in a booth in the back room of McCormick and Schmick's on First Avenue. The restaurant was noisy, filled with the lunchtime business crowd. Fernandez pressed the headphones to her ears as she listened to the tape on a Walkman. She had been listening intently for more than half an hour, making notes on a yellow legal pad, her food still uneaten. Finally she got up. "I have to make a call."

Levin glanced at Fernandez's plate. "Uh . . . do you want that?"

Fernandez shook her head, and walked away.

Levin grinned. "Waste not, want not," he said, and pulled the plate in front of him. He began to eat. "So Tom, are you in shit or what?"

"Deep shit," Sanders said. He stirred a cappuccino. He hadn't been able to eat lunch. He watched Levin wolf down great bites of mashed potatoes.

"I figured that," Levin said. "Jack Kerry over at Aldus called me this morning and said you were suing the company because you refused to jump some woman."

"Kerry is an asshole."

"The worst," Levin nodded. "The absolute worst. But what can you do? After Connie Walsh's column this morning, everybody's been trying to figure out who Mr. Piggy is." Levin took another huge bite of food. "But how'd she get the story in the first place? I mean, she's the one who broke it."

Sanders said, "Maybe you told her, John."

"Are you kidding?" Levin said.

"You had the tape."

Levin frowned. "You keep this up, Tom, you're going to piss me off." He shook his head. "No, you ask me, it was a woman who told her."

"What woman knew? Only Meredith, and she wouldn't tell."

"I'll bet you anything it'll turn out to be a woman," Levin said. "If you ever find out—which I doubt." He chewed thoughtfully. "Swordfish is a little rubbery. I think we should tell the waiter." He looked around the room. "Uh, Tom."

"Yes?"

"There's a guy standing over there, hopping from one foot to the other. I think maybe you know him."

Sanders looked over his shoulder. Bob Garvin was standing by the bar, looking at him expectantly. Phil Blackburn stood a few paces behind.

"Excuse me," Sanders said, and he got up from the table.

Garvin shook hands with Sanders. "Tom. Good to see you. How are you holding up with all this?"

"I'm okay," Sanders said.

"Good, good." Garvin placed his hand in a fatherly way on Sanders's shoulder. "It's nice to see you again."

"Nice to see you too, Bob."

Garvin said, "There's a quiet place in the corner over there. I asked them for a couple of cappuccinos. We can talk for a minute. Is that okay?"

"That's fine," Sanders said. He was well acquainted with the profane, angry Garvin. This cautious, polite Garvin made him uneasy.

They sat in the corner of the bar. Garvin settled into his chair and faced him.

"Well, Tom. We go way back, you and I."

"Yes, we do."

"Those damn trips to Seoul, eating that crappy food, and your ass hurting like hell. You remember all that."

"Yes, I do."

"Yeah, those were the days," Garvin said. He was watching Sanders carefully. "Anyway, Tom, we know each other, so I'm not going to bullshit you. Let me just put all the cards on the table," Garvin said. "We've got a problem here, and it's got to be solved before it turns into a real mess for everybody. I want to appeal to your better judgment about how we proceed from here."

"My better judgment?" Sanders said.

"Yes," Garvin said. "I'd like to look at this thing from all sides."

"How many sides are there?"

"There are at least two," Garvin said, with a smile. "Look, Tom. I'm sure it's no secret that I've supported Meredith inside our company. I've always believed that she's got talent and the kind of executive vision that we want for the future. I've never seen her do anything before that would suggest otherwise. I know she's only human, but she's very talented and I support her."

"Uh-huh . . ."

"Now perhaps in this case . . . perhaps it is true that she's made a mistake. I don't know."

Sanders said nothing. He just waited, staring at Garvin's face. Garvin was doing a convincing impression of an open-minded man. Sanders didn't buy it.

"In fact, let's say she has," Garvin said. "Let's say she did make a mistake."

"She did, Bob," Sanders said, firmly.

"All right. Let's say she did. An error of judgment, let's call it. An overstepping of bounds. The point is, Tom, faced with a situation like this, I still strongly support her."

"Why?"

"Because she's a woman."

"What does that have to do with it?"

"Well, women in business have traditionally been excluded from executive positions, Tom."

"Meredith hasn't been excluded," Sanders said.

"And after all," Garvin said, "she's young."

"She's not that young," Sanders said.

"Sure she is. She's practically a college kid. She just got her MBA a couple of years ago."

"Bob," Sanders said. "Meredith Johnson's thirty-five. She's not a kid at all."

Garvin did not seem to hear that. He looked at Sanders sympathetically. "Tom, I can understand that you were disappointed about the job," he said. "And I can understand that in your eyes, Meredith made a mistake in the way she approached you."

"She didn't approach me, Bob. She jumped me."

Garvin showed a flash of irritation. "You're no kid either, you know."

"That's right, I'm not," Sanders said. "But I *am* her employee."

"And I know she holds you in the highest regard," Garvin said, settling back in his chair. "As does everybody in the company, Tom. You're vital to our future. You know it, I know it. I want to keep our team together. And I keep coming back to the idea that we have to make allowances for women. We have to cut them a little slack."

"But we're not talking about women," Sanders said. "We're talking about one particular woman."

"Tom—"

"And if a man had done what she did, you wouldn't be talking about cutting him slack. You'd fire him, and throw him out on his ass."

"Possibly so."

"Well, that's the problem," Sanders said.

Garvin said, "I'm not sure I follow you there, Tom." His tone carried a warning: Garvin didn't like being disagreed with. Over the years, as his company grew in wealth and success, Garvin had grown accustomed to deference. Now, approaching retirement, he expected obedience and agreement. "We have an obligation to attain equality," Garvin said.

"Fine. But equality *means* no special breaks," Sanders said. "Equality means treating people the same. You're asking for *in*equality toward Meredith, because you won't do what you would do to a man—fire him."

Garvin sighed. "If it was a clear case, Tom, I would. But I understand this particular situation isn't so clear."

Sanders considered telling him about the tape. Something made him hold back. He said, "I think it is."

"But there are always differences of opinion on these matters," Garvin said, leaning across the bar. "That's a fact, isn't it? Always a difference of opinion. Tom. Look: what did she do that was so bad? I mean, really. She made a pass? Fine. You could have decided it was flattering. She's a beautiful woman, after all. There are worse things that could happen. A beautiful woman puts her hand on your knee. Or you could have just said, no thank you. You could have handled it any number of ways. You're a grown-up. But this . . . *vindictiveness*. Tom. I have to tell you. I'm surprised at you."

Sanders said, "Bob, she broke the law."

"That really remains to be seen, doesn't it?" Garvin said. "You can throw open your personal life for a jury to inspect, if that's what you want to do. I wouldn't want to do it, myself. And I don't see that it helps anybody to take this into court. It's a no-win situation, all around."

"What're you saying?"

"You don't want to go to court, Tom." Garvin's eyes were narrow, dangerous.

"Why not?"

"You just don't." Garvin took a deep breath. "Look. Let's stay on track here. I've talked to Meredith. She feels as I do, that this thing has gotten out of hand."

"Uh-huh . . ."

"And I'm talking to you now, too. Because my hope, Tom, is that we can put this to rest, and go back to the way things were—now hear me out, please—go back to the way things were, before this unfortunate misunderstanding happened. You stay at your job, Meredith stays at hers. You two continue to work together like civilized adults. You move forward and build the company, take it public, and everybody makes a pile of money a year down the line. What's wrong with that?"

Sanders felt something like relief, and a sense of normalcy returning. He longed to escape from the lawyers and from the tension of the last three days. To sink back into the way things were seemed as appealing as a warm bath.

"I mean, look at it this way, Tom. Right after this thing happened on Monday night, nobody blew the whistle. You didn't call anybody. Meredith didn't call anybody. I think you both wanted this thing to go away. Then there was an unfortunate mix-up the next day, and an argument that needn't have happened. If you'd been on time for the meeting, if you and Meredith had been in sync on the story, none of this would have happened. You two would still be working together, and whatever happened between you would remain your private business. Instead, we have this. It's all a big mistake, really. So why not just forget it and go forward? And get rich. Tom? What's wrong with that?"

"Nothing," Sanders said, finally.

"Good."

"Except it won't work," Sanders said.

"Why not?"

A dozen answers flashed through his mind: Because she's not competent. Because she's a snake. Because she's a corporate player, all image, and this is a technical division that has to get out the product. Because she's a liar. Because I have no respect for her. Because she'll do it again. Because she has no respect for me. Because you're not treating me fairly. Because she's your pet. Because you chose her over me. Because . . .

"Things have gone too far," he said.

Garvin stared at him. "Things can go back."

"No, Bob. They can't."

Garvin leaned forward. His voice dropped. "Listen you little *feringi* pissant. I know exactly what's going on here. I took you in when you didn't know *bulkogi* from bullshit. I gave you your start, I gave you help, I gave you opportunities, all along the line. Now you want to play rough? Fine. You want to see the shit come down? Just fucking wait, Tom." He stood up.

Sanders said, "Bob, you've never been willing to listen to reason on the subject of Meredith Johnson."

"Oh, you think *I* have a problem with Meredith?" Garvin laughed harshly. "Listen, Tom: she was your girlfriend, but she was smart and independent, and you couldn't handle her. You were pissed when she dropped you. And now, all these years later, you're going to pay her back. That's what this is about. It has nothing to do with business ethics or breaking the law or sexual harassment or any other damned thing. It's personal, and it's petty. And you're so full of shit your eyes are brown."

And he stalked out of the restaurant, pushing angrily past Blackburn. Blackburn remained behind for a moment, staring at Sanders, and then hurried after his boss.

As Sanders walked back to his table, he passed a booth with several guys from Microsoft, including two major assholes from systems programming. Someone made a snorting pig sound.

"Hey Mr. Piggy," said a low voice.

"Suwee! Suwee!"

"Couldn't get it up, huh?"

Sanders walked on a few paces, then turned back. "Hey, guys," he said. "At least I'm not bending over and grabbing my ankles in late-night meetings with—" and he named a Programming head at Microsoft.

They all roared with laughter.

"Whoa ho!"

"Mr. Piggy speaks!"

"Oink oink."

Sanders said, "What're you guys doing in town, anyway? They run short on K-Y jelly in Redmond?"

"Whoa!"

"The Piggy is pissed!"

They were doubled over, laughing like college kids. They had a big pitcher of beer on the table. One of them said, "If Meredith Johnson pulled off her pants for me, I sure wouldn't call the police about it."

"No way, Jose!"

"Service with a smile!"

"Hard charger!"

"Ladies *first!*"

"Ka-jung! Ka-jung!"

They pounded the table, laughing.

Sanders walked away.

O utside the restaurant, Garvin paced back and forth angrily on the pavement. Blackburn stood with the phone at his ear.

"Where is that fucking car?" Garvin said.

"I don't know, Bob."

"I told him to *wait*."

"I know, Bob. I'm trying to get him."

"Christ Almighty, the simplest things. Can't even get the fucking cars to work right."

"Maybe he had to go to the bathroom."

"So? How long does that take? Goddamn Sanders. Could you believe him?"

"No, I couldn't, Bob."

"I just don't understand. He won't deal with me on this. And I'm bending over backward here. I offer him his job back, I offer him his stock back, I offer him everything. And what does he do? Jesus."

"He's not a team player, Bob."

"You got that right. And he's not willing to meet us. We've got to get him to come to the table."

"Yes we do, Bob."

"He's not feeling it," Garvin said. "That's the problem."

"The story ran this morning. It can't have made him happy."

"Well, he's not feeling it."

Garvin paced again.

"There's the car," Blackburn said, pointing down the street. The Lincoln sedan was driving toward them.

"Finally," Garvin said. "Now look, Phil. I'm tired of wasting time on Sanders. We tried being nice, and it didn't work. That's the long and the short of it. So what are we going to do, to make him feel it?"

"I've been thinking about that," Phil said. "What's Sanders doing? I mean really doing? He's smearing Meredith, right?"

"Goddamn right."

"He didn't hesitate to smear her."

"He sure as hell didn't."

"And it's not true, what he's saying about her. But the thing about a smear is that it doesn't have to be true. It just has to be something people are willing to believe is true."

"So?"

"So maybe Sanders needs to see what that feels like."

"Like what feels like? What're you talking about?"

Blackburn stared thoughtfully at the approaching car. "I think that Tom's a violent man."

"Oh hell," Garvin said, "he's not. I've known him for years. He's a pussycat."

"No," Blackburn said, rubbing his nose. "I disagree. I think he's violent. He was a football player in college, he's a rough-and-tumble sort of guy. Plays football on the company team, knocks people around. He has a violent streak. Most men do, after all. Men are violent."

"What kind of shit is this?"

"And you have to admit, he was violent to Meredith," Blackburn continued. "Shouting. Yelling. Pushing her. Knocking her over. Sex and violence. A man out of control. He's much bigger than she is. Just stand them side by side, anybody can see the difference. He's much bigger. Much stronger. All you have to do is look, and you see he is a violent abusive man. That nice exterior is just a cover. Sanders is one of those men who take out their hostility by beating up defenseless women."

Garvin was silent. He squinted at Blackburn. "You'll never make this fly."

"I think I can."

"Nobody in their right mind'll buy it."

Blackburn said, "I think somebody will."

"Yeah? Who?"

"Somebody," Blackburn said.

The car pulled up to the curb. Garvin opened the door. "Well, all I know," he said, "is that we need to get him to negotiate. We need to apply pressure to bring him to the table."

Blackburn said, "I think that can be arranged."

Garvin nodded. "It's in your hands, Phil. Just make sure it happens." He got in the car. Blackburn got in the car after Garvin. Garvin said to the driver, "Where the fuck have you been?"

The door slammed shut. The car drove off.

anders drove with Fernandez in Alan's car back to the mediation center. Fernandez listened to Sanders's report of the conversation with Garvin, shaking her head. "You never should have seen him alone. He couldn't have behaved that way if I was there. Did he really say you have to make allowances for women?"

"Yes."

"That's noble of him. He's found a virtuous reason why we should protect a harasser. It's a nice touch. Everyone should sit back and allow her to break the law because she's a woman. Very nice."

Sanders felt stronger hearing her words. The conversation with Garvin had rattled him. He knew that Fernandez was working on him, building him back up, but it worked anyway.

"The whole conversation is ridiculous," Fernandez said. "And then he threatened you?"

Sanders nodded.

"Forget it. It's just bluster."

"You're sure?"

"Absolutely," she said. "Just talk. But at least now you know why they say men just don't get it. Garvin gave you the same lines that every corporate guy has been giving for years: Look at it from the harasser's point of view. What did they do that was so wrong. Let bygones be bygones. Everybody just go back to work. We'll be one big happy family again."

"Incredible," Alan said, driving the car.

"It is, in this day and age," Fernandez said. "You can't pull that stuff anymore. How old is Garvin, anyway?"

"Almost sixty."

"That helps explain it. But Blackburn should have told him it's completely unacceptable. According to the law, Garvin really doesn't have any choice. At a minimum, he has to transfer Johnson, not you. And almost certainly, he should fire her."

"I don't think he will," Sanders said.

"No, of course he won't."

"She's his favorite," Sanders said.

"More to the point, she's his vice president," Fernandez said. She stared out the window as they went up the hill toward the mediation center. "You have to realize, all these decisions are about power. Sexual harassment is about power, and so is the company's resistance to dealing with it. Power protects power. And once a woman gets up in the power structure, she'll be protected by the structure, the same as a man. It's like the way doctors won't testify against other doctors. It doesn't matter if the doctor is a man or a woman. Doctors just don't want to testify against other doctors. Period. And corporate executives don't want to investigate claims against other executives, male or female."

"So it's just that women haven't had these jobs?"

"Yes. But they're starting to get them now. And now they can be as unfair as any man ever was."

"Female chauvinist sows," Alan said.

"Don't you start," Fernandez said.

"Tell him the figures," Alan said.

"What figures?" Sanders said.

"About five percent of sexual harassment claims are brought by men against women. It's a relatively small figure. But then, only five percent of corporate supervisors are women. So the figures suggest that women executives harass men in the same proportion as men harass women. And as more women get corporate jobs, the percentage of claims by men is going up. Because the fact is, harassment is a power issue. And power is neither male nor female. Whoever is behind the desk has the opportunity to abuse power. And women will take advantage as often as men. A case in point being the delightful Ms. Johnson. And her boss isn't firing her."

"Garvin says it's because the situation isn't clear."

"I'd say that tape is pretty damn clear," Fernandez said. She frowned. "Did you tell him about the tape?"

"No."

"Good. Then I think we can wrap this case up in the next two hours."

Alan pulled into the parking lot and parked the car. They all got out.

"All right," Fernandez said. "Let's see where we are with her significant others. Alan. We've still got her previous employer—"

"Conrad Computer. Right. We're on it.

"And also the one before that."

"Symantec."

"Yes. And we have her husband—"

"I've got a call into CoStar for him."

"And the Internet business? 'Afriend'?"

"Working on it."

"And we have her B-school, and Vassar."

"Right."

"Recent history is the most important. Focus on Conrad and the husband."

"Okay," Alan said. "Conrad's a problem, because they supply systems to the government and the CIA. They gave me some song and dance about neutral reference policy and nondisclosure of prior employees."

"Then get Harry to call them. He's good on negligent referral. He can shake them up if they continue to stonewall."

"Okay. He may have to."

Alan got back in the car. Fernandez and Sanders started walking up to the mediation center. Sanders said, "You're checking her past companies?"

"Yes. Other companies don't like to give damaging information on prior employees. For years, they would never give anything at all except the dates of employment. But now there's something called compelled self-publication, and something called negligent referral. A company can be liable now for failing to reveal a problem with a past employee. So we can try to scare them. But in the end, they may not give us the damaging information we want."

"How do you know they have damaging information to give?"

Fernandez smiled. "Because Johnson is a harasser. And with harassers, there's always a pattern. It's never the first time."

"You think she's done this before?"

"Don't sound so disappointed," Fernandez said. "What did you think? That she did all this because she thought you were so cute? I guarantee you she has done it before." They walked past the fountains in the courtyard toward the door to the center building. "And now," Fernandez said, "let's go cut Ms. Johnson to shreds."

Precisely at one-thirty, Judge Murphy entered the mediation room. She looked at the seven silent people sitting around the table and frowned. "Has opposing counsel met?"

"We have," Heller said.

"With what result?" Murphy said.

"We have failed to reach a settlement," Heller said.

"Very well. Let's resume." She sat down and opened her notepad. "Is there further discussion relating to the morning session?"

"Yes, Your Honor," Fernandez said. "I have some additional questions for Ms. Johnson."

"Very well. Ms. Johnson?"

Meredith Johnson put on her glasses. "Actually, Your Honor, I would like to make a statement first."

"All right."

"I've been thinking about the morning session," Johnson said, speaking slowly and deliberately, "and Mr. Sanders's account of the events of Monday night. And I've begun to feel that there may be a genuine misunderstanding here."

"I see." Judge Murphy spoke absolutely without inflection. She stared at Meredith. "All right."

"When Tom first suggested a meeting at the end of the day, and when he suggested that we have some wine, and talk over old times, I'm afraid I may have unconsciously responded to him in a way that he might not have intended."

Judge Murphy didn't move. Nobody was moving. The room was completely still.

"I believe it is correct to say that I took him at his word, and began to imagine a, uh, romantic interlude. And to be frank, I was not opposed to that possibility. Mr. Sanders and I had a very special relationship some years ago, and I remembered it as a very exciting relationship. So I believe it is fair to say that I was looking forward to our meeting, and that perhaps I presumed that it would lead to an encounter. Which I was, unconsciously, quite willing to have occur."

Alongside Meredith, Heller and Blackburn sat completely stone-faced, showing no reaction at all. The two female attorneys showed no reaction. This had all been worked out in advance, Sanders realized. What was going on? Why was she changing her story?

Johnson cleared her throat, then continued in the same deliberate way. "I believe it is correct to say that I was a willing participant in all the events of the evening. And it may be that I was too forward, at one point, for Mr. Sanders's taste. In the heat of the moment, I may have overstepped the bounds of propriety and my position in the company. I think that's possible. After serious reflection, I find myself concluding that my own recollection of events and Mr. Sanders's recollection of events are in much closer agreement than I had earlier recognized."

There was a long silence. Judge Murphy said nothing. Meredith Johnson shifted in her chair, took her glasses off, then put them back on again.

"Ms. Johnson," Murphy said finally, "do I understand you to say that you are now agreeing to Mr. Sanders's version of the events on Monday night?"

"In many respects, yes. Perhaps in most respects."

Sanders suddenly realized what had happened: *they knew about the tape.*

But how could they know? Sanders himself had learned of it only two hours ago. And Levin had been out of his office, having lunch with them. So Levin couldn't have told them. How could they know?

"And, Ms. Johnson," Murphy said, "are you also agreeing to the charge of harassment by Mr. Sanders?"

"Not at all, Your Honor. No."

"Then I'm not sure I understand. You've changed your story. You say you now agree that Mr. Sanders's version of the events is correct in most respects. But you do not agree that he has a claim against you?"

"No, Your Honor. As I said, I think it was all a misunderstanding."

"A *misunderstanding*," Murphy repeated, with an incredulous look on her face.

"Yes, Your Honor. And one in which Mr. Sanders played a very active role."

"Ms. Johnson. According to Mr. Sanders, you initiated kissing over his protests; you pushed him down on the couch over his protests; you unzipped his trousers and removed his penis over his protests; and you removed your own clothing over his protests. Since Mr. Sanders is your employee, and dependent on you for employment, it is difficult for me to comprehend why this is not a clear-cut and indisputable case of sexual harassment on your part."

"I understand, Your Honor," Meredith Johnson said calmly. "And I realize I have changed my story. But the reason I say it is a misunderstanding is that from the beginning, I genuinely believed that Mr. Sanders was seeking a sexual encounter with me, and that belief guided my actions."

"You do not agree that you harassed him."

"No, Your Honor. Because I thought I had clear *physical* indications that Mr. Sanders was a willing participant. At times he certainly took the lead. So now, I have to ask myself why he would take the lead—and then so suddenly withdraw. I don't know why he did that. But I believe he shares responsibility for what happened. That is why I feel that, at the very least, we had a genuine misunderstanding. And I want to say that I am sorry—truly, deeply sorry—for my part in this misunderstanding."

"You're sorry." Murphy looked around the room in exasperation. "Can anyone explain to me what is going on? Mr. Heller?"

Heller spread his hands. "Your Honor, my client told me what she intended to do here. I consider it a very brave act. She is a true seeker after truth."

"Oh, spare me," Fernandez said.

Judge Murphy said, "Ms. Fernandez, considering this radically different statement from Ms. Johnson, would you like a recess before you proceed with your questions?"

"No, Your Honor. I am prepared to go forward now," Fernandez said.

"I see," Murphy said, puzzled. "All right. Fine." Judge Murphy clearly felt that there was something everyone else in the room knew that she didn't.

Sanders was still wondering how Meredith knew about the tape. He looked over at Phil Blackburn, who sat at one end of the table, his cellular phone before him. He was rubbing the phone nervously.

Phone records, Sanders thought. That must be it.

DigiCom would have had somebody—most probably Gary Bosak—going through all of Sanders's records, looking for things to use against him. Bosak would have checked all the calls made on Sanders's cellular phone. When he did that, he would have discovered a call that lasted forty-five minutes on Monday night. It would stand out: a whopping big duration and charge. And Bosak must have looked at the time of the call and figured out what had happened. He'd realize that Sanders hadn't been talking on the phone during that particular forty-five minutes on Monday night. Therefore, there could only be one explanation. The call was running to an answering machine, which meant there was a tape. And Johnson knew it, and had adjusted her story accordingly. That was what had made her change.

"Ms. Johnson," Fernandez said. "Let's clear up a few factual points first. Are you now saying that you *did* send your assistant to buy wine and condoms, that you *did* tell her to lock the door, and that you *did* cancel your seven o'clock appointment in anticipation of a sexual encounter with Mr. Sanders?"

"Yes, I did."

"In other words, you lied earlier."

"I presented my point of view."

"But we are not talking about a point of view. We are talking about facts. And given this set of facts, I'm curious to know why you feel that Mr. Sanders shares responsibility for what happened in that room Monday night."

"Because I felt . . . I felt that Mr. Sanders had come to my office with the clear intention of having sex with me, and he later denied any such intention. I felt he had set me up. He led me on, and then accused me, when I had done nothing more than simply respond to him."

"You feel he set you up?"

"Yes."

"And that's why you feel he shares responsibility?"

"Yes."

"In what way did he set you up?"

"Well, I think it's obvious. Things had gone very far along, when he suddenly got off the couch and said he was not going to proceed. I'd say that was a setup."

"Why?"

"Because you can't go so far and then just stop. That's obviously a hostile act, intended to embarrass and humiliate me. I mean . . . anyone can see that."

"All right. Let's review that particular moment in detail," Fernandez said. "As I understand it, we're talking about the time when you were on the couch with Mr. Sanders, with both of you in a state of partial undress. Mr. Sanders was crouched on his knees on the couch, his penis was exposed, and you were lying on your back with your panties removed and your legs spread, is that correct?"

"Basically. Yes." She shook her head. "You make it sound so . . . crude."

"But that was the situation at that moment, was it not?"

"Yes. It was."

"Now, at that moment, did you say, 'No, no, please,' and did Mr. Sanders reply, 'You're right, we shouldn't be doing this,' and then get off the couch?"

"Yes," she said. "That's what he said."

"Then what was the misunderstanding?"

"When I said, 'No, no,' I meant, 'No, don't wait.' Because he was waiting, sort of teasing me, and I wanted him to go ahead. Instead, he got off the couch, which made me very angry."

"Why?"

"Because I wanted him to do it."

"But Ms. Johnson, you said, 'No, no.' "

"I know what I said," she replied irritably, "but in that situation, it's perfectly clear what I was really saying to him."

"Is it?"

"Of course. He knew exactly what I was saying to him, but he chose to ignore it."

"Ms. Johnson, have you ever heard the phrase, 'No means no'?"

"Of course, but in this situation—"

"I'm sorry, Ms. Johnson. Does no mean no, or not?"

"Not in this case. Because at that time, lying on that couch, it was absolutely clear what I was really saying to him."

"You mean it was clear to you."

Johnson became openly angry. "It was clear to him, too," she snapped.

"Ms. Johnson. When men are told that 'no means no,' what does that mean?"

"I don't know." She threw up her hands in irritation. "I don't know what you're trying to say."

"I'm trying to say that men are being told that they must take women at their literal word. That no means no. That men cannot assume that no means maybe or yes."

"But in this particular situation, with all our clothes off, when things had gone so far—"

"What does that have to do with it?" Fernandez said.

"Oh, come off it," Johnson said. "When people are getting together, they begin with little touches, then little kisses, then a little petting, then some more petting. Then the clothes come off, and you're touching various private parts, and so on. And pretty soon you have an expectation about what's going to happen. And you don't turn back. To turn back is a hostile act. That's what he did. He set me up."

"Ms. Johnson. Isn't it true that women claim the right to turn back at any point, up to the moment of actual penetration? Don't women claim the unequivocal right to change their minds?"

"Yes, but in this instance—"

"Ms. Johnson. If women have the right to change their minds, don't men as well? Can't Mr. Sanders change his mind?"

"It was a hostile act." Her face had a fixed, stubborn look. "He set me up."

"I'm asking whether Mr. Sanders has the same rights as a woman in this situation. Whether he has the right to withdraw, even at the last moment."

"No."

"Why?"

"Because men are different."

"How are they different?"

"Oh, for Christ's sake," Johnson said angrily. "What are we talking about here? This is Alice in Wonderland. Men and women are *different*. Everybody knows that. Men can't control their impulses."

"Apparently Mr. Sanders could."

"Yes. As a hostile act. Out of his desire to humiliate me."

"But what Mr. Sanders actually said at the time was, 'I don't feel good about this.' Isn't that true?"

"I don't remember his exact words. But his behavior was very hostile and degrading toward me as a woman."

"Let's consider," Fernandez said, "who was hostile and degrading toward whom. Didn't Mr. Sanders protest the way things were going earlier in the evening?"

"Not really. No."

"I thought he had." Fernandez looked at her notes. "Early on, did you say to Mr. Sanders, 'You look good' and 'You always had a nice hard tush'?"

"I don't know. I might have. I don't remember."

"And what did he reply?"

"I don't remember."

Fernandez said, "Now, when Mr. Sanders was talking on the phone, did you come up, push it out of his hand, and say, 'Forget that phone'?"

"I might have. I don't really remember."

"And did you initiate kissing at that point?"

"I'm not really sure. I don't think so."

"Well, let's see. How else could it have occurred? Mr. Sanders was talking on his cellular phone, over by the window. You were on another phone at your desk. Did he interrupt his call, set down his phone, come over, and start kissing you?"

She paused for a moment. "No."

"Then who initiated the kissing?"

"I guess I did."

"And when he protested and said, 'Meredith,' did you ignore him, press on, and say, 'God, I've wanted you all day. I'm so hot, I haven't had a decent fuck'?" Fernandez repeated these statements in a flat uninflected monotone, as if reading from a transcript.

"I may have . . . I think that might be accurate. Yes."

Fernandez looked again at her notes. "And then, when he said, 'Meredith, wait,' again clearly speaking in a tone of protest, did you say, 'Oh, don't talk, no, no, oh Jesus'?"

"I think . . . possibly I did."

"On reflection, would you say these comments by Mr. Sanders were protests that you ignored?"

"If they were, they were not very clear protests. No."

"Ms. Johnson. Would you characterize Mr. Sanders as fully enthusiastic throughout the encounter?"

Johnson hesitated a moment. Sanders could almost see her thinking, trying to decide how much the tape would reveal. Finally she said, "He was enthusiastic sometimes, not so much at other times. That's my point."

"Would you say he was ambivalent?"

"Possibly. Somewhat."

"Is that a yes or a no, Ms. Johnson?"

"Yes."

"All right. So Mr. Sanders was ambivalent throughout the session. He's told us why: because he was being asked to embark on an office affair with an old girlfriend who was now his boss. And because he was now married. Would you consider those valid reasons for ambivalence?"

"I suppose so."

"And in this state of ambivalence, Mr. Sanders was overwhelmed at the last moment with the feeling that he didn't want to go forward. And he told you how he felt, simply and directly. So, why would you characterize that as a 'setup'? I think we have ample evidence that it is just the opposite—an uncalculated, rather desperate human response to a situation which you entirely controlled. This was not a reunion of old lovers, Ms. Johnson, though you prefer to think it was. This was not a meeting of equals at all. The fact is, you are his superior and you controlled every aspect of the meeting. You arranged the time, bought the wine, bought the condoms, locked the door—and then you blamed your employee when he failed to please you. That is how you continue to behave now."

"And you're trying to put his behavior in a good light," Johnson said. "But what I'm saying is that as a practical matter, waiting to the last minute to stop makes people very angry."

"Yes," Fernandez said. "That's how many men feel, when women withdraw at the last minute. But women say a man has no right to be angry, because a woman can withdraw at any time. Isn't that true?"

Johnson rapped her fingers on the table irritably. "Look," she said. "You're trying to make some kind of federal case here, by trying to obscure basic facts. What did I do that was so wrong? I made him an offer, that's all. If Mr. Sanders wasn't interested, all he had to do was say, 'No.' But he never said that. Not once. Because he intended to *set me up.* He's angry he didn't get the job and he's retaliating the only way he can—by smearing me. This is nothing but guerrilla warfare and character assassination. I'm a successful woman in business, and he resents my success and he's out to get me. You're saying all kinds of things to avoid that central and unavoidable fact."

"Ms. Johnson. The central and unavoidable fact is that you're Mr. Sanders's superior. And your behavior toward him was illegal. And it *is* in fact a federal case."

There was a short silence.

Blackburn's assistant came into the room and handed him a note. Blackburn read the note and passed it to Heller.

Murphy said, "Ms. Fernandez? Are you ready to explain what's going on to me now?"

"Yes, Your Honor. It turns out there is an audio tape of the meeting."

"Really? Have you heard it?"

"I have, Your Honor. It confirms Mr. Sanders's story."

"Are you aware of this tape, Ms. Johnson?"

"No, I am not."

"Perhaps Ms. Johnson and her attorney would like to hear it, too. Perhaps we should all hear it," Murphy said, looking directly at Blackburn.

Heller put the note in his pocket and said, "Your Honor, I'd like to request a ten-minute recess."

"Very well, Mr. Heller. I'd say this development warrants it."

Outside in the courtyard, black clouds hung low. It was threatening to rain again. Over by the fountains, Johnson huddled with Heller and Blackburn. Fernandez watched them. "I just don't understand this," she said. "There they all are, talking again. What is there to talk about? Their client lied, and then changed her story. There's no question that Johnson's guilty of sexual harassment. We have it recorded on tape. So what are they talking about?"

Fernandez stared for a moment, frowning. "You know, I have to admit it. Johnson's a hell of a smart woman," she said.

"Yes," Sanders said.

"She's quick and she's cool."

"Uh-huh."

"Moved up the corporate ladder fast."

"Yes."

"So . . . how'd she let herself get into this situation?"

"What do you mean?" Sanders said.

"I mean, what's she doing coming on to you the very first day at work? And coming on so strongly? Leaving herself open to all these problems? She's too smart for that."

Sanders shrugged.

"You think it's just because you're irresistible?" Fernandez said. "With all due respect, I doubt it."

He found himself thinking of the time he first knew Meredith, when she was doing demos, and the way she used to cross her legs whenever she was asked a question she couldn't answer. "She could always use sex to distract people. She's good at that."

"I believe it," Fernandez said. "So what is she distracting us from now?"

Sanders had no answer. But his instinct was that something else was going on. "Who knows how people really are in private?" he said. "I once knew this woman, she looked like an angel, but she liked bikers to beat her up."

"Uh-huh," Fernandez said. "That's fine. I'm not buying it for Johnson. Because Johnson strikes me as very controlled, and her behavior with you was not controlled."

"You said it yourself, there's a pattern."

"Yeah. Maybe. But why the first day? Why right away? I think she had another reason."

Sanders said, "And what about me? Do you think I had another reason?"

"I assume you did," she said, looking at him seriously. "But we'll talk about that later."

Alan came up from the parking lot, shaking his head.

"What've you got?" Fernandez said.

"Nothing good. We're striking out everywhere," he said. He flipped open his notepad. "Okay. Now, we've checked out that Internet address. The message originated in the 'U District.' And 'Afriend' turns out to be Dr. Arthur A. Friend. He's a professor of inorganic chemistry at the University of Washington. That name mean anything to you?"

"No," Sanders said.

"I'm not surprised. At the moment, Professor Friend is in northern Nepal on a consulting job for the Nepalese government. He's been there for three weeks. He's not expected back until late July. So it probably isn't him sending the messages anyway."

"Somebody's using his Internet address?"

"His assistant says that's impossible. His office is locked while he's away, and nobody goes in there except her. So nobody has access to his computer terminal. The assistant says she goes in once a day and answers Dr. Friend's e-mail, but otherwise the computer is off. And nobody knows the password but her. So I don't know."

"It's a message coming out of a locked office?" Sanders said, frowning.

"I don't know. We're still working on it. But for the moment, it's a mystery."

"All right, fine," Fernandez said. "What about Conrad Computer?"

"Conrad has taken a very hard position. They will only release information to the hiring company, meaning DigiCom. Nothing to us. And they say that the hiring company has not requested it. When we pushed, Conrad called DigiCom themselves, and DigiCom told them they weren't interested in any information Conrad might have."

"Hmmm."

"Next, the husband," Alan said. "I talked to someone who worked in his company, CoStar. Says the husband hates her, has lots of bad things to say about her. But he's in Mexico on vacation with his new girlfriend until next week."

"Too bad."

"Novell," Alan said. "They keep only the last five years current. Prior to that, records are in cold storage at headquarters in Utah. They have no idea what they'll show, but they're willing to get them out if we'll pay for it. It'll take two weeks."

Fernandez shook her head. "Not good."

"No."

"I have a strong feeling that Conrad Computer is sitting on something," Fernandez said.

"Maybe, but we'll have to sue to get it. And there's no time." Alan looked across the courtyard at the others. "What's happening now?"

"Nothing. They're hanging tough."

"Still?"

"Yeah."

"Jesus," Alan said. "Who's she got behind her?"

"I'd love to know," Fernandez said.

Sanders flipped open his cellular phone and checked in with his office. "Cindy, any messages?"

"Just two, Tom. Stephanie Kaplan asked if she could meet with you today."

"She say why?"

"No. But she said it wasn't important. And Mary Anne has come by twice, looking for you."

"Probably wants to skin me," Sanders said.

"I don't think so, Tom. She's about the only one who—she's very concerned about you, I think."

"Okay. I'll call her."

He started to dial Mary Anne's number when Fernandez nudged

him in the ribs. He looked over and saw a slender, middle-aged woman walking up from the parking lot toward them.

"Buckle up," Fernandez said.

"Why? Who's that?"

"That," Fernandez said, "is Connie Walsh."

Connie Walsh was about forty-five years old, with gray hair and a sour expression. "Are you Tom Sanders?"

"That's right."

She pulled out a tape recorder. "Connie Walsh, from the *Post-Intelligencer*. Can we talk for a moment?"

"Absolutely not," Fernandez said.

Walsh looked over at her.

"I'm Mr. Sanders's attorney."

"I know who you are," Walsh said, and turned back to Sanders. "Mr. Sanders, our paper's going with a story on this discrimination suit at DigiCom. My sources tell me that you are accusing Meredith Johnson of sex discrimination, is that correct?"

"He has no comment," Fernandez said, stepping between Walsh and Sanders.

Walsh looked past her shoulder and said, "Mr. Sanders, is it also true that you and she are old lovers, and that your accusation is a way to even the score?"

"He has no comment," Fernandez said.

"It looks to me like he does," Walsh said. "Mr. Sanders, you don't have to listen to her. You can say something if you want to. And I really think you should take this opportunity to defend yourself. Because my sources are also saying that you physically abused Ms. Johnson in the course of your meeting. These are very serious charges people are making against you, and I imagine you'll want to respond. What do you have to say to her allegations? Did you physically abuse her?"

Sanders started to speak, but Fernandez shot him a warning glance, and put her hand on his chest. She said to Walsh, "Has Ms. Johnson made these allegations to you? Because she was the only other one besides Mr. Sanders who was there."

"I'm not free to say. I have the story from very well-informed sources."

"Inside or outside the company?"

"I really can't say."

"Ms. Walsh," Fernandez said, "I am going to forbid Mr. Sanders to talk to you. And you better check with the *P-I* counsel before you run any of these unsubstantiated allegations."

"They're not unsubstantiated, I have very reliable—"

"If there is any question in your counsel's mind, you might have her call Mr. Blackburn and he will explain what your legal position is in this matter."

Walsh smiled bleakly. "Mr. Sanders, do you want to make a comment?"

Fernandez said, "Just check with your counsel, Ms. Walsh."

"I will, but it won't matter. You can't squash this. Mr. Blackburn can't squash this. And speaking personally, I have to say I don't know how you can defend a case like this."

Fernandez leaned close to her, smiled, and said, "Why don't you step over here with me, and I'll explain something to you."

She walked with Walsh a few yards away, across the courtyard.

Alan and Sanders remained where they were. Alan sighed. He said, "Wouldn't you give anything to know what they were saying right now?"

Connie Walsh said, "It doesn't matter what you say. I won't give you my source."

"I'm not asking for your source. I'm simply informing you that your story is wrong—"

"Of course you'd say that—"

"And that there's documentary evidence that it's wrong."

Connie Walsh paused. She frowned. "Documentary evidence?"

Fernandez nodded slowly. "That's right."

Walsh thought it over. "But there can't be," she said. "You said it yourself. They were alone in the room. It's his word against hers. There's no documentary evidence."

Fernandez shook her head, and said nothing.

"What is it? A tape?"

Fernandez smiled thinly. "I really can't say."

"Even if there is, what can it show? That she pinched his butt a little? She made a couple of jokes? What's the big deal? Men have been doing that for hundreds of years."

"That's not the issue in this—"

"Give me a break. So this guy gets a little pinch, and he starts screaming bloody murder. That's not normal behavior in a man. This guy obviously hates and demeans women. That's clear, just to look at him. And there's no question: he hit her, in that meeting. The company had to call a doctor to examine her for a concussion. And I have several reliable sources that tell me he's known to be physically abusive. He and his wife have had trouble for years. In fact, she's left town with the kids and is going to file for divorce." Walsh was watching Fernandez carefully as she said it.

Fernandez just shrugged.

"It's a fact. The wife has left town," Walsh said flatly. "Unexpectedly. She took the kids. And nobody knows where she went. Now, you tell me what that means."

Fernandez said, "Connie, all I can do is advise you in my capacity as Mr. Sanders's attorney that documentary evidence contradicts your sources about this harassment charge."

"Are you going to show me this evidence?"

"Absolutely not."

"Then how do I know it exists?"

"You don't. You only know I have informed you of its existence."

"And what if I don't believe you?"

Fernandez smiled. "These are the decisions a journalist must make."

"You're saying it'd be reckless disregard."

"If you go with your story, yes."

Walsh stepped back. "Look. Maybe you've got some kind of a technical legal case here, and maybe you don't. But as far as I'm concerned, you're just another minority woman trying to get ahead with the patriarchy by getting down on her knees. If you had any self-respect you wouldn't be doing their dirty work for them."

"Actually, Connie, the person who seems to be caught in the grip of the patriarchy is you."

"That's a lot of crap," Walsh said. "And let me tell you, you're not going to evade the facts here. He led her on, and then he beat her up. He's an ex-lover, he's resentful, and he's violent. He's a typical man. And let me tell you, before I'm through, he'll wish he had never been born."

Sanders said, "Is she going to run the story?"

"No," Fernandez said. She stared across the courtyard at Johnson, Heller, and Blackburn. Connie Walsh had gone over to Blackburn and was talking to him. "Don't get distracted by this," Fernandez said. "It's not important. The main issue is: what're they going to do about Johnson."

A moment later, Heller came toward them. He said, "We've been going over things on our side, Louise."

"And?"

"We've concluded that we see no purpose to further mediation and are withdrawing, as of now. I've informed Judge Murphy that we will not continue."

"Really. And what about the tape?"

"Neither Ms. Johnson nor Mr. Sanders knew they were being taped. Under law, one party must know the interaction is being recorded. Therefore the tape is inadmissible."

"But Ben—"

"We argue that the tape should be disallowed, both from this mediation and from any subsequent legal proceeding. We argue that Ms. Johnson's characterization of the meeting as a misunderstanding between consenting adults is the correct one, and that Mr. Sanders bears a responsibility for that misunderstanding. He was an active participant, Louise, no way around it. He took her panties off. Nobody held a gun to his head. But since there was fault on both sides, the proper thing is for the two parties to shake hands, let go of all animosity, and return to work. Apparently Mr. Garvin has already proposed this to Mr. Sanders, and Mr. Sanders has refused. We believe that under the circumstances Mr. Sanders is acting unreasonably and that if he does not reconsider in a timely manner, he should be fired for his refusal to show up for work."

"Son of a bitch," Sanders said.

Fernandez laid a restraining hand on his arm. "Ben," she said calmly. "Is this a formal offer of reconciliation and return to the company?"

"Yes, Louise."

"And what are the sweeteners?"

"No sweeteners. Everybody just goes back to work."

"The reason I ask," Fernandez said, "is that I believe I can successfully argue that Mr. Sanders was aware the tape was being made, and thus it is indeed admissible. I will argue further that it is admissible under discovery of public records over common carriers as defined in *Waller* v. *Herbst*. I will argue further that the company knew of Ms. Johnson's long history of harassment, and has failed to take proper steps to investigate her behavior, either prior to this incident, or now. And I will argue that the company was derelict in protecting Mr. Sanders's reputation when it leaked the story to Connie Walsh."

"Wait a minute here—"

"I will argue that the company had a clear reason for leaking it: they desired to cheat Mr. Sanders out of his well-deserved reward for more than a decade of service to the company. And you've got an employee in Ms. Johnson who has had some trouble before. I will claim defamation and ask for punitive damages of sufficient magnitude to send a message to corporate America. I'll ask for sixty million dollars, Ben. And you'll settle for forty million—the minute I get the judge to allow the jury to hear this tape. Because we both know that when the jury hears that tape, they will take about five seconds to find against Ms. Johnson and the company."

Heller shook his head. "You've got a lot of long shots there, Louise. I don't think they'll ever let that tape be played in court. And you're talking about three years from now."

Fernandez nodded slowly. "Yes," she said. "Three years is a long time."

"You're telling me, Louise. Anything can happen."

"Yes, and frankly, I'm worried about that tape. So many untoward things can happen with evidence that is so scandalous. I can't guarantee somebody hasn't made a copy already. It'd be terrible if one fell into the hands of KQEM, and they started playing it over the radio."

"Christ," Heller said. "Louise, I can't believe you said that."

"Said what? I'm merely expressing my legitimate fears," Fernandez said. "I'd be derelict if I did not let you know my concerns. Let's face facts here, Ben. The cat's out of the bag. The press already has this story. Somebody leaked it to Connie Walsh. And she printed a story that's very damaging to Mr. Sanders's reputation. And it seems that somebody is still leaking, because now Connie is planning to write some unfounded speculation about physical violence by my client. It's unfortunate that someone on your side should have chosen to talk about this case. But we both know how it is with a hot story in the press—you never know where the next leak will come from."

Heller was uneasy. He glanced back at the others by the fountain. "Louise, I don't think there's any movement over there."

"Well, just talk to them."

Heller shrugged, and walked back.

"What do we do now?" Sanders said.

"We go back to your office."

"We?"

"Yes," Fernandez said. "This isn't the end. More is going to happen today, and I want to be there when it does."

Driving back, Blackburn talked on the car phone with Garvin. "The mediation's over. We called it off."

"And?"

"We're pushing Sanders hard to go back to work. But he's not responding so far. He's hanging tough. Now he's threatening punitive damages of sixty million dollars."

"Christ," Garvin said. "Punitive damages on what basis?"

"Defamation from corporate negligence dealing with the fact that we supposedly knew that Johnson had a history of harassment."

"I never knew of any history," Garvin said. "Did you know of any history, Phil?"

"No," Blackburn said.

"Is there any documentary evidence of such a history?"

"No," Blackburn said. "I'm sure there isn't."

"Good. Then let him threaten. Where did you leave it with Sanders?"

"We gave him until tomorrow morning to rejoin the company at his old job or get out."

"All right," Garvin said. "Now let's get serious. What have we got on him?"

"We're working on that felony charge," Blackburn said. "It's early, but I think it's promising."

"What about women?"

"There isn't any record on women. I know Sanders was screwing one of his assistants a couple of years back. But we can't find the records in the computer. I think he went in and erased them."

"How could he? We blocked his access."

"He must have done it some time ago. He's a cagey guy."

"Why the hell would he do it some time ago, Phil? He had no reason to expect any of this."

"I know, but we can't find the records now." Blackburn paused. "Bob, I think we should move up the press conference."

"To when?"

"Late tomorrow."

"Good idea," Garvin said. "I'll arrange it. We could even do it noon tomorrow. John Marden is flying in in the morning," he said, referring to Conley-White's CEO. "That'll work out fine."

"Sanders is planning to string this out until Friday," Blackburn said. "Let's just beat him to the punch. We've got him blocked as it is. He can't get into the company files. He can't get access to Conrad or anything else. He's isolated. He can't possibly come up with anything damaging between now and tomorrow."

"Fine," Garvin said. "What about the reporter?"

"I think she'll break the story on Friday," Blackburn said. "She already has it, I don't know where from. But she won't be able to resist trashing Sanders. It's too good a story; she'll go with it. And he'll be dead meat when she does."

"That's fine," Garvin said.

Meredith Johnson came off the fifth-floor elevator at DigiCom and ran into Ed Nichols. "We missed you at the morning meetings," Nichols said.

"Yeah, I had some things to take care of," she said.

"Anything I should know?"

"No," she said. "It's boring. Just some technical matters about tax exemptions in Ireland. The Irish government wants to expand local content at the Cork plant and we're not sure we can. This has been going on for more than a year."

"You look a little tired," Nichols said, with concern. "A little pale."

"I'm okay. I'll be happy when this is all over."

"We all will," Nichols said. "You have time for dinner?"

"Maybe Friday night, if you're still in town," she said. She smiled. "But really, Ed. It's just tax stuff."

"Okay, I believe you."

He waved and went down the hallway. Johnson went into her office.

She found Stephanie Kaplan there, working at the computer terminal on Johnson's desk. Kaplan looked embarrassed. "Sorry to use your computer. I was just running over some accounts while I waited for you."

Johnson threw her purse on the couch. "Listen, Stephanie," she said. "Let's get something straight right now. I'm running this division, and nobody's going to change that. And as far as I'm concerned, this is the time when a new vice president decides who's on their side, and who isn't. Somebody supports me, I'll remember. Somebody doesn't, I'll deal with that, too. Do we understand each other?"

Kaplan came around the desk. "Yes, sure, Meredith."

"Don't fuck with me."

"Never entered my mind, Meredith."

"Good. Thank you, Stephanie."

"No problem, Meredith."

Kaplan left the office. Johnson closed the door behind her and went directly to her computer terminal and stared intently at the screen.

Sanders walked through the corridors of DigiCom with a sense of unreality. He felt like a stranger. The people who passed him in the halls looked away and brushed past him, saying nothing.

"I don't exist," he said to Fernandez.

"Never mind," she said.

They passed the main part of the floor, where people worked in chest-high cubicles. Several pig grunts were heard. One person sang softly, "Because I used to fuck her, but it's all over now . . ."

Sanders stopped and turned toward the singing. Fernandez grabbed his arm.

"Never mind," she said.

"But Christ . . ."

"Don't make it worse than it is."

They passed the coffee machine. Beside it, someone had taped up a picture of Sanders. They had used it for a dartboard.

"Jesus."

"Keep going."

As he came to the corridor leading to his office, he saw Don Cherry coming the other way.

"Hi, Don."

"You screwed up bad on this one, Tom." He shook his head and walked on.

Even Don Cherry.

Sanders sighed.

"You knew this was going to happen," Fernandez said.

"Maybe."

"You did. This is the way it works."

Outside his office, Cindy stood up when she saw him. She said, "Tom, Mary Anne asked you to call her as soon as you got in."

"Okay."

"And Stephanie said to say never mind, she found out whatever she needed to know. She said, uh, not to call her."

"Okay."

He went in the office and closed the door. He sat down behind his desk and Fernandez sat opposite him. She took her cellular phone out of her briefcase, and dialed. "Let's get one thing squared away—Ms. Vries's office please . . . Louise Fernandez calling."

She cupped her hand over the phone. "This shouldn't take— Oh, Eleanor? Hi, Louise Fernandez. I'm calling you about Connie Walsh. Uh-huh . . . I'm sure you've been going over it with her. Yes, I know she feels strongly. Eleanor, I just wanted to confirm to you that there is a tape of the event, and it substantiates Mr. Sanders's version rather than Ms. Johnson's. Actually, yes, I could do that. Entirely off the record? Yes, I could. Well, the problem with Walsh's source is that the company now has huge liability and if you print a story that's wrong—even if you got it from a source—I think they have an action against you. Oh yes, I think absolutely Mr. Blackburn would sue. He wouldn't have any choice. Why don't you—I see. Uh-huh. Well, that could change, Eleanor. Uh-huh. And don't forget that Mr. Sanders is considering defamation right now, based on the Mr. Piggy piece. Yes, why don't you do that. Thank you."

She hung up and turned to Sanders. "We went to law school together. Eleanor is very competent and very conservative. She'd never have allowed the story in the first place, and would never have considered it now, if she didn't place a lot of reliance on Connie's source."

"Meaning?"

"I'm pretty sure I know who gave her the story," Fernandez said. She was dialing again.

"Who?" Sanders said.

"Right now, the important thing is Meredith Johnson. We've got to document the pattern, to demonstrate that she has harassed employees before. Somehow we've got to break this deadlock with Conrad Computer." She turned away. "Harry? Louise. Did you talk to Conrad? Uh-huh. And?" A pause. She shook her head irritably. "Did you explain to them about their liabilities? Uh-huh. Hell. So what's our next move? Because we've got a time problem here, Harry, that's what I'm concerned about."

While she was talking, Sanders turned to his monitor. The e-mail light was flashing. He clicked it.

YOU HAVE 17 MESSAGES WAITING.

Christ. He could only imagine. He clicked the READ button. They flashed up in order.

FROM: DON CHERRY, CORRIDOR PROGRAMMING TEAM
TO: ALL SUBJECTS

WE HAVE DELIVERED THE VIE UNIT TO CONLEY-WHITE'S PEOPLE. THE UNIT IS NOW ACTIVE INTO THEIR COMPANY DB SINCE THEY GAVE US THE HOOKS TODAY. JOHN CONLEY ASKED THAT IT BE DELIVERED TO A SUITE AT THE FOUR SEASONS HOTEL BECAUSE THEIR CEO IS ARRIVING THURS-DAY MORNING AND WILL SEE IT THEN. ANOTHER PRO-GRAMMING TRIUMPH BROUGHT TO YOU BY THE SWELL FOLKS AT VIE.

DON THE MAGNIFICENT

Sanders flipped to the next one.

FROM: DIAGNOSTICS GROUP
TO: APG TEAM

ANALYSIS OF TWINKLE DRIVES. THE PROBLEM WITH THE CONTROLLER TIMING LOOP DOES NOT SEEM TO COME FROM THE CHIP ITSELF. WE VERIFIED MICRO-FLUCTUATIONS IN CURRENT FROM THE POWER UNIT WHICH WAS APPAR-ENTLY ETCHED WITH SUBSTANDARD OR INADEQUATE RE-SISTANCES ON THE BOARD BUT THIS IS MINOR AND DOES NOT EXPLAIN OUR FAILURE TO MEET SPECS. ANALYSIS IS CONTINUING.

Sanders viewed the message with a sense of detachment. It didn't really tell him anything. Just words that concealed the underlying truth: they still didn't know what the problem was. At another time, he'd be on his way down to the Diagnostics team, to ride them hard to get to the bottom of it. But now ... He shrugged and went to the next message.

FROM: BASEBALL CENTRAL
TO: ALL PLAYERS
RE: NEW SUMMER SOFTBALL SCHEDULE

DOWNLOAD FILE BB.72 TO GET THE NEW REVISED SUMMER
SCHEDULE. SEE YOU ON THE FIELD!

He heard Fernandez say on the phone, "Harry, we've got to crack
this one somehow. What time do they close their offices in Sunnyvale?"
Sanders went to the next message.

NO MORE GROUP MESSAGES. DO YOU WANT TO READ PER-
SONAL MESSAGES?

He clicked the icon.

WHY DON'T YOU JUST ADMIT YOU ARE GAY?
(UNSIGNED)

He didn't bother to see where it had come from. They would
probably have manually entered it as coming from Garvin's address, or
something like that. He could check the real address inside the system,
but not without the access privileges they had taken away. He went to
the next message.

SHE'S BETTER LOOKING THAN YOUR ASSISTANT, AND YOU
DIDN'T SEEM TO MIND SCREWING HER.
(UNSIGNED)

Sanders clicked to the next one.

YOU SLIMY WEASEL - GET OUT OF THIS COMPANY.
YOUR BEST ADVICE

Christ, he thought. The next one:

LITTLE TOMMY HAD A PECKER
HE PLAYED WITH EVERY DAY

BUT WHEN A LADY TRIED TO TOUCH IT
LITTLE TOMMY SAID GO AWAY.

The verses ran on, down to the bottom of the screen, but Sanders didn't read the rest. He clicked and went on.

IF YOU WEREN'T FUCKING YOUR DAUGHTER SO MUCH YOU
MIGHT BE ABLE TO

He clicked again. He was clicking faster and faster, going through the messages.

GUYS LIKE YOU GIVE MEN A BAD NAME YOU ASSHOLE.

BORIS

Click.

YOU FILTHY LYING MALE PIG

Click.

HIGH TIME SOMEBODY STUCK IT TO THE WHINING BITCHES.
I'M TIRED OF THE WAY THEY BLAME EVERYBODY BUT
THEIRSELVES. TITS AND BLAME ARE SEX-LINKED TRAITS.
THEY'RE BOTH ON THE X-CHROMOSOME.

KEEP ON TRUCKIN'

He went through them, no longer reading. Eventually he was going so fast he almost missed one of the later ones:

JUST RECEIVED WORD THAT MOHAMMED JAFAR IS DYING.
HE'S STILL IN THE HOSPITAL, AND NOT EXPECTED TO SUR-
VIVE UNTIL MORNING. I GUESS MAYBE THERE'S SOME-
THING TO THIS SORCERY BUSINESS, AFTER ALL.

ARTHUR KAHN

Sanders stared at the screen. A man dying of sorcery? He couldn't begin to imagine what had really happened. The very idea seemed to belong to another world, not his. He heard Fernandez say, "I don't care, Harry, but Conrad has information relevant to the pattern, and somehow we have to get it out of them."

Sanders clicked to the final message.

YOU'RE CHECKING THE WRONG COMPANY.

AFRIEND

Sanders twisted the monitor around so Fernandez could see it. She frowned as she talked on the phone. "Harry, I got to go. Do what you can." She hung up. "What does it mean, we're checking the wrong company? How does this friend even know what we're doing? When did this come in?"

Sanders looked at the message headers. "One-twenty this afternoon."

Fernandez made a note on her legal pad. "That was about the time Alan was talking to Conrad. And Conrad called DigiCom, remember? So this message has to be coming from inside DigiCom."

"But it's on the Internet."

"Wherever it appears to be coming from, it's actually from somebody inside the company trying to help you."

His immediate thought, out of nowhere, was *Max*. But that didn't make any sense. Dorfman was tricky, but not in this way. Besides, Max wasn't knowledgeable about the minute-to-minute workings of the company.

No, this was somebody who wanted to help Sanders but who didn't want the help to be traced back.

"You're checking the wrong company . . ." he repeated aloud.

Could it be someone at Conley-White? Hell, he thought, it could be anybody.

"What does it mean, we're checking the wrong company?" he said. "We're checking all her past employers, and we're having a very difficult—"

He stopped.

You're checking the wrong company.

"I must be an idiot," he said. He started typing at his computer.

"What is it?" Fernandez said.

"They've restricted my access, but I still should be able to get this," he said, typing quickly.

"Get what?" she said, puzzled.

"You say harassers have a pattern, right?"

"Right."

"It shows up again and again, right?"

"Right."

"And we're checking her past employers, to get information about past episodes of harassment."

"Right. And failing."

"Yes. But the thing is," Sanders said, "she's worked here for the last four years, Louise. We're checking the wrong company."

He watched as the computer terminal flashed:

SEARCHING DATABASE

And then, after a moment, he turned the screen so Fernandez could see:

Digital Communications Data Reference Search Report

DB 4: Human Resources (Sub 5/Employee Records)
Search Criteria:
1. Disposition: Terminated a/o Transferred a/o Resigned
2. Supervisor: Johnson, Meredith
3. Other Criteria: males only

Summary Search Results:

Michael Tate	5/9/89	Terminate	Drug Use	HR RefMed
Edwin Sheen	7/5/89	Resign	Alt Employment	D-Silicon
William Rogin	11/9/89	Transfer	Own Request	Austin
Frederic Cohen	4/2/90	Resign	Alt Employment	Squire Sx
Robert Ely	6/1/90	Transfer	Own Request	Seattle
Michael Backes	8/11/90	Transfer	Own Request	Malaysia
Peter Saltz	1/4/91	Resign	Alt Employment	Novell
Ross Wald	8/5/91	Transfer	Own Request	Cork
Richard Jackson	11/14/91	Resign	Alt Employment	Aldus
James French	2/2/92	Transfer	Own Request	Austin

Fernandez scanned the list. "Looks like working for Meredith Johnson can be hazardous to your job. You're looking at the classic pattern: people last only a few months, and then resign or ask to be transferred elsewhere. Everything voluntary. Nobody ever fired, because that might trigger a wrongful termination suit. Classic. You know any of these people?"

"No," Sanders said, shaking his head. "But three of them are in Seattle," he said.

"I only see one."

"No, Aldus is here. And Squire Systems is out in Bellevue. So Richard Jackson and Frederic Cohen are up here, too."

"You have any way to get details of termination packages on these people?" she said. "That would be helpful. Because if the company paid anybody off, then we have a de facto case."

"No." Sanders shook his head. "Financial data is beyond minimal access."

"Try anyway."

"But what's the point? The system won't let me."

"Do it," Fernandez said.

He frowned. "You think they're monitoring me?"

"I guarantee it."

"Okay." He typed in the parameters and pressed the search key. The answer came back:

FINANCIAL DATABASE SEARCH IS BEYOND
LEVEL (0) ACCESS

He shrugged. "Just as I thought. No cigar."

"But the point is, we asked the question," Fernandez said. "It'll wake them right up."

S anders was heading toward the bank of elevators when he saw
Meredith coming toward him with three Conley-White executives. He turned quickly, then went to the stairwell and started walking
down the four flights to the street level. The stairwell was deserted.

One flight below, the door opened and Stephanie Kaplan appeared
and started coming up the stairs. Sanders was reluctant to speak to her;
Kaplan was, after all, the chief financial officer and close to both Garvin
and Blackburn. In the end, he said casually, "How's it going, Stephanie."

"Hello, Tom." Her nod to him was cool, reserved.

Sanders continued past her, going down a few more steps, when he
heard her say, "I'm sorry this is so difficult for you."

He paused. Kaplan was one flight above him, looking down. There
was no one else in the stairwell.

He said, "I'm managing."

"I know you are. But still, it must be hard. So much going on at once,
and nobody giving you information. It must be confusing to try to figure
everything out."

Nobody giving you information?

"Well, yes," he said, speaking slowly. "It is hard to figure things out,
Stephanie."

She nodded. "I remember when I first started out in business," she
said. "I had a woman friend who got a very good job in a company that
didn't usually hire women executives. In her new position, she had a lot
of stress and crises. She was proud of the way she was dealing with the
problems. But it turned out she'd only been hired because there was a
financial scandal in her division, and from the beginning they were
setting her up to take the fall. Her job was never about any of the things
she thought it was. She was a patsy. And she was looking the wrong way
when they fired her."

Sanders stared at her. Why was she telling him this? He said, "That's
an interesting story."

Kaplan nodded. "I've never forgotten it," she said.

On the stairs above, a door clanged open, and they heard footsteps descending. Without another word, Kaplan turned and continued up.

Shaking his head, Sanders continued down.

In the newsroom of the Seattle *Post-Intelligencer*, Connie Walsh looked up from her computer terminal and said, "You've got to be kidding."

"No, I'm not," Eleanor Vries said, standing over her. "I'm killing this story." She dropped the printout back on Walsh's desk.

"But you know who my source is," Walsh said. "And you know Jake was listening in to the entire conversation. We have very good notes, Eleanor. Very complete notes."

"I know."

"So, given the source, how can the company possibly sue?" Walsh said. "Eleanor: *I have the fucking story.*"

"You have *a* story. And the paper faces a substantial exposure already."

"Already? From what?"

"The Mr. Piggy column."

"Oh, for Christ's sake. There's no way to claim identification from that column."

Vries pulled out a xerox of the column. She had marked several passages in yellow highlighter. "Company X is said to be a high-tech company in Seattle that just named a woman to a high position. Mr. Piggy is said to be her subordinate. He is said to have brought a sexual harassment action. Mr. Piggy's wife is an attorney with young children. You say Mr. Piggy's charge is without merit, that he is a drunk and a womanizer. I think Sanders can absolutely claim identification and sue for defamation."

"But this is a column. An opinion piece."

"This column alleges facts. And it alleges them in a sarcastic and wildly overstated manner."

"It's an opinion piece. Opinion is protected."

"I don't think that's certain in this case at all. I'm disturbed that I allowed this column to run in the first place. But the point is, we cannot claim to be absent malice if we allow further articles to go out."

Walsh said, "You have no guts."

"And you're very free with other people's guts," Vries said. "The story's killed and that's final. I'm putting it in writing, with copies to you, Marge, and Tom Donadio."

"Fucking lawyers. What a world we live in. This story needs to be told."

"Don't screw around with this, Connie. I'm telling you. Don't."

And she walked away.

Walsh thumbed through the pages of the story. She had been working on it all afternoon, polishing it, refining it. Getting it exactly right. And now she wanted the story to run. She had no patience with legal thinking. This whole idea of protecting rights was just a convenient fiction. Because when you got right down to it, legal thinking was just narrow-minded, petty, self-protective—the kind of thinking that kept the power structure firmly in place. And in the end, fear served the power structure. Fear served men in power. And if there was anything that Connie Walsh believed to be true of herself, it was that she was not afraid.

After a long time, she picked up the phone and dialed a number. "KSEA-TV, good afternoon."

"Ms. Henley, please."

Jean Henley was a bright young reporter at Seattle's newest independent TV station. Walsh had spent many evenings with Henley, discussing the problems of working in the male-dominated mass media. Henley knew the value of a hot story in building a reporter's career.

This story, Walsh told herself, would be told. One way or another, it would be told.

Robert Ely looked up at Sanders nervously. "What do you want?" he asked. Ely was young, not more than twenty-six, a tense man with a blond mustache. He was wearing a tie and was in his shirtsleeves. He worked in one of the partitioned cubicles at the back of DigiCom's Accounting Department in the Gower Building.

"I want to talk about Meredith," Sanders said. Ely was one of the three Seattle residents on his list.

"Oh God," Ely said. He glanced around nervously. His Adam's apple bobbed. "I don't—I don't have anything to say."

"I just want to talk," Sanders said.

"Not *here*," Ely said.

"Then let's go to the conference room," Sanders said. They walked down the hall to a small conference room, but a meeting was being held there. Sanders suggested they go to the little cafeteria in the corner of Accounting, but Ely told him that wouldn't be private. He was growing more nervous by the minute.

"Really, I have nothing to tell you," he kept saying. "There's nothing, really nothing."

Sanders knew he had better find a quiet place at once, before Ely bolted and ran. They ended up in the men's room—white tile, spotlessly clean. Ely leaned against a sink. "I don't know why you are talking to me. I don't have anything I can tell you."

"You worked for Meredith, in Cupertino."

"Yes."

"And you left there two years ago?"

"Yes."

"Why did you leave?"

"Why do you think?" Ely said, in a burst of anger. His voice echoed off the tiles. "You know why, for Christ's sake. Everybody knows why. She made my life hell."

"What happened?" Sanders asked.

"What happened." Ely shook his head, remembering. "Every day, every day. 'Robert, would you stay late, we have some things to go over.' After a while, I tried to make excuses. Then she would say, 'Robert, I'm not sure you're showing the proper dedication to this company.' And she would put little comments in my performance review. Subtle little negative things. Nothing that I could complain about. But they were there. Piling up. 'Robert, I think you need my help here. Why don't you see me after work.' 'Robert, why don't you drop by my apartment and we'll discuss it. I really think you should.' I was—it was terrible. The, uh, person I was living with did not, uh . . . I was in a real bind."

"Did you report her?"

Ely laughed harshly. "Are you kidding? She's practically a member of Garvin's *family*."

"So you just put up with it . . ."

Ely shrugged. "Finally, the person I was living with got another job. When he came up here, I transferred, too. I mean, of course I wanted to go. It just worked out all around."

"Would you make a statement about Meredith now?"

"Not a chance."

"You realize," Sanders said, "that the reason she gets away with it is that nobody reports her."

Ely pushed away from the sink. "I have enough problems in my life without going public on this." He went to the door, paused, and turned back. "Just so you're clear: I've got nothing to say on the subject of Meredith Johnson. If anybody asks, I'll say our working relationship was correct at all times. And I'll also say that I never met you."

Meredith Johnson? Of course I remember her," Richard Jackson said. "I worked for her for more than a year." Sanders was in Jackson's office on the second floor of the Aldus Building, on the south side of Pioneer Square. Jackson was a good-looking man of thirty, with the hearty manner of an ex-athlete. He was a marketing manager at Aldus; his office was friendly, cluttered with product boxes for graphics programs: Intellidraw, Freehand, SuperPaint, and Pagemaker.

"Beautiful and charming woman," Jackson said. "Very intelligent. Always a pleasure."

Sanders said, "I was wondering why you left."

"I was offered this job, that's why. And I've never regretted it. Wonderful job. Wonderful company. I've had a great experience here."

"Is that the only reason you left?"

Jackson laughed. "You mean, did Meredith Manmuncher come on to me?" he said. "Hey, is the Pope Catholic? Is Bill Gates rich? Of *course* she came on to me."

"Did that have anything to do with your leaving?"

"No, no," Jackson said. "Meredith came on to everybody. She's sort of an equal opportunity employer, in that respect. She chased *everybody*. When I first started in Cupertino, she had this little gay guy she used to chase around the table. Terrorized the poor bastard. Little skinny nervous guy. Christ, she used to make him tremble."

"And you?"

Jackson shrugged. "I was a single guy, just starting out. She was beautiful. It was okay with me."

"You never had any difficulties?"

"Never. Meredith was fabulous. Shitty lay, of course. But you can't have everything. She's a very intelligent, very beautiful woman. Always dressed great. And she liked me, so she took me to all these functions. I met people, made contacts. It was great."

"So you saw nothing wrong?"

"Not a damn thing," Jackson said. "She could get a little bossy. That got old. There were a couple of other women I was seeing, but I always had to be on call for her. Even at the last minute. That could be irritating sometimes. You begin to think your life is not your own. And she's got a mean temper sometimes. But what the hell. You do what you have to do. Now I'm assistant manager here at thirty. I'm doing great. Great company. Great town. Great future. And I owe it to her. She's great."

Sanders said, "You were an employee of the company at the time that you were having your relationship, isn't that right?"

"Yeah, sure."

"Isn't she required by company policy to report any relationship with an employee? Did she report her relationship with you?"

"Christ, no," Jackson said. He leaned across his desk. "Let's get one thing straight, just between you and me. I think Meredith is great. If you have a problem with her, it's your problem. I don't know what it could be. You used to live with her, for Christ's sake. So there can't be any surprises. Meredith likes to fuck guys. She likes to tell them to do this, do that. She likes to order them around. That's who she is. And I don't see anything wrong with it."

Sanders said, "I don't suppose you'd—"

"Make a statement?" Jackson said. "Get serious. Listen, there's a lot of bullshit around now. I hear things like, 'You can't go out with the people you work with.' Christ, if I couldn't go out with the people I worked with, I'd still be a virgin. That's all anybody can go out with—the people you work with. That's the only people you get to know. And sometimes those people are your superiors. Big deal. Women screw men and get ahead. Men screw women and get ahead. Everybody's going to screw everybody else anyway, if they can. Because they want to. I mean, women are just as hot as men. They want it just like we do. That's real life. But you get some people who are pissed off, so they file a complaint, and say, 'Oh no, you can't do that to me.' I'm telling you, it's all bullshit. Like these sensitivity training seminars we all have to go to. Everybody sits there with their hands in their laps like a fucking Red Guard meeting, learning the correct way to address your fellow workers. But afterward everybody goes out and fucks around, the same as they always did. The assistants go, 'Oh, Mr. Jackson, have *you* been to the *gym*? You look so *strong*.' Batting their eyelashes. So what am I supposed to do? You can't make rules about this. People get hungry, they eat. Doesn't matter how many meetings they attend. This is all a gigantic jerk off. And anybody who buys into it is an asshole."

"I guess you answered my question," Sanders said. He got up to leave. Obviously, Jackson wasn't going to help him.

"Look," Jackson said. "I'm sorry you've got a problem here. But everyone's too damned sensitive these days. I see people now, kids right out of college, and they really think they should never experience an unpleasant moment. Nobody should ever say anything they don't like, or tell a joke they don't like. But the thing is, nobody can make the world be the way they want it to be all the time. Things always happen that embarrass you or piss you off. That's life. I hear women telling jokes about men every day. Offensive jokes. Dirty jokes. I don't get bent out of shape. Life is great. Who has time for this crap? Not me."

Sanders came out of the Aldus Building at five o'clock. Tired and discouraged, he trudged back toward the Hazzard Building. The streets were wet, but the rain had stopped, and the afternoon sunlight was trying to break through the clouds.

He was back in his office ten minutes later. Cindy was not at her desk, and Fernandez was gone. He felt deserted and alone and hopeless. He sat down and dialed the final number on his list.

"Squire Electronic Data Systems, good evening."

Sanders said, "Frederic Cohen's office, please."

"I'm sorry, Mr. Cohen has gone for the day."

"Do you know how I could reach him?"

"I'm afraid I don't. Do you want to leave voice mail?"

Damn, he thought. What was the point? But he said, "Yes, please."

There was a click. Then, "Hi, this is Fred Cohen. Leave a message at the tone. If it's after hours, you can try me on my car phone at 502-8804 or my home at 505-9943."

Sanders jotted the numbers down. He dialed the car phone first. He heard a crackle of static, then:

"I know, honey, I'm sorry I'm late, but I'm on my way. I just got tied up."

"Mr. Cohen?"

"Oh." A pause. "Yes. This is Fred Cohen."

"My name is Tom Sanders. I work over at DigiCom, and—"

"I know who you are." The voice sounded tense.

"I understand you used to work for Meredith Johnson."

"Yes. I did."

"I wonder if I could talk to you."

"What about?"

"About your experiences. Working for her."

There was a long pause. Finally, Cohen said, "What would be the point of that?"

"Well, I'm in a sort of a dispute with Meredith now, and—"

"I know you are."

"Yes, and you see, I would like to—"

"Look. Tom. I left DigiCom two years ago. Whatever happened is ancient history now."

"Well, actually," Sanders said, "it's not, because I'm trying to establish a pattern of behavior and—"

"I know what you're trying to do. But this is very touchy stuff, Tom. I don't want to get into it."

"If we could just talk," Sanders said. "Just for a few minutes."

"Tom." Cohen's voice was flat. "Tom, I'm married now. I have a wife. She's pregnant. I don't have anything to say about Meredith Johnson. Nothing at all."

"But—"

"I'm sorry. I've got to go."

Click.

Cindy came back in as he was hanging up the phone. She pushed a cup of coffee in front of him. "Everything okay?"

"No," he said. "Everything is terrible." He was reluctant to admit, even to himself, that he had no more moves left. He had approached three men, and they had each refused to establish a pattern of behavior for him. He doubted that the other men on the list would behave differently. He found himself thinking of what his wife, Susan, had said two days before. *You have no moves.* Now, after all this effort, it turned out to be true. He was finished. "Where's Fernandez?"

"She's meeting with Blackburn."

"What?"

Cindy nodded. "In the small conference room. They've been there about fifteen minutes now."

"Oh, Christ."

He got up from his desk and went down the hall. He saw Fernandez sitting with Blackburn in the conference room. Fernandez was making notes on her legal pad, head bent deferentially. Blackburn was running his hands down his lapels and looking upward as he spoke. He seemed to be dictating to her.

Then Blackburn saw him, and waved him over. Sanders went into the conference room. "Tom," Blackburn said, with a smile. "I was just coming to see you. Good news: I think we've been able to resolve this situation. I mean, really resolve it. Once and for all."

"Uh-huh," Sanders said. He didn't believe a word of it. He turned to Fernandez.

Fernandez looked up from her legal pad slowly. She appeared dazed. "That's the way it looks."

Blackburn stood and faced Sanders. "I can't tell you how pleased I am, Tom. I've been working on Bob all afternoon. And he's finally come to face reality. The plain fact is, the company has a problem, Tom. And we owe you a debt of gratitude for bringing it so clearly to our attention. This can't go on. Bob knows he has to deal with it. And he will."

Sanders just stared. He couldn't believe what he was hearing. But there was Fernandez, nodding and smiling.

Blackburn smoothed his tie. "But as Frank Lloyd Wright once said, 'God is in the details.' You know, Tom, we have one small immediate problem, a political problem, having to do with the merger. We're asking your help with the briefing tomorrow for Marden, Conley's CEO. But after that . . . well, you've been badly wronged, Tom. This company has wronged you. And we recognize that we have an obligation to make it up to you, whatever way we can."

Still disbelieving it, Sanders said harshly, "What exactly are we talking about?"

Blackburn's voice was soothing. "Well, Tom, at this point, that's really up to you," he said. "I've given Louise the parameters of a potential deal, and all the options that we would agree to. You can discuss it with her and get back to us. We'll sign any interim papers you require, of course. All that we ask in return is that you attend the meeting tomorrow and help us to get through the merger. Fair enough?"

Blackburn extended his hand and held it there.

Sanders stared.

"From the bottom of my heart, Tom, I'm sorry for all that has happened."

Sanders shook his hand.

"Thank you, Tom," Blackburn said. "Thank you for your patience, and thank you on behalf of this company. Now, sit down and talk with Louise, and let us know what you decide."

And Blackburn left the room, closing the door softly behind him.

He turned to Fernandez. "What the hell is this all about?"

Fernandez gave a long sigh. "It's called capitulation," she said. "Total and complete capitulation. DigiCom just folded."

Sanders watched Blackburn walk down the hallway away from the conference room. He was filled with confused feelings. Suddenly, he was being told it was all over, and over without a fight. Without blood being spilled.

Watching Blackburn, he had a sudden image of blood in the bathroom sink of his old apartment. And this time, he remembered where it came from. A part of the chronology fell into place.

Blackburn was staying at his apartment during his divorce. He was on edge, and drinking too much. One day he cut himself so badly while shaving that the sink was spattered with blood. Later on, Meredith saw the blood in the sink and on the towels, and she said, "Did one of you guys fuck her while she was having her period?" Meredith was always blunt that way. She liked to startle people, to shock them.

And then, one Saturday afternoon, she walked around the apartment in white stockings and a garter belt and a bra while Phil was watching television. Sanders said to her, "What are you doing that for?"

"Just cheering him up," Meredith replied. She threw herself back on the bed. "Now why don't you cheer me up?" she said. And she pulled her legs back, opening—

"Tom? Are you listening to me?" Fernandez was saying. "Hello? Tom? Are you there?"

"I'm here," Sanders said.

But he was still watching Blackburn, thinking about Blackburn. Now he remembered another time, a few years later. Sanders had started dating Susan, and Phil had dinner with the two of them one night. Susan went to the bathroom. "She's great," Blackburn said. "She's terrific. She's beautiful and she's great."

"But?"

"But . . ." Blackburn had shrugged. "She's a lawyer."

"So?"

"You can never trust a lawyer," Blackburn had said, and laughed. One of his rueful, wise laughs.

You can never trust a lawyer.

Now, standing in the DigiCom conference room, Sanders watched as Blackburn disappeared around a corner. He turned back to Fernandez.

". . . really had no choice," Fernandez was saying. "The whole situation finally became untenable. The fact situation with Johnson is bad. And the tape is dangerous—they don't want it played, and they're afraid it will get out. They have a problem about prior sexual harassment by Johnson; she's done it before, and they know it. Even though none of the men you talked to has agreed to talk, one of them might in the future, and they know it. And of course they've got their chief counsel revealing company information to a reporter."

Sanders said, "What?"

She nodded. "Blackburn was the one who gave the story to Connie Walsh. He acted in flagrant violation of all rules of conduct for an employee of the company. He's a major problem for them. And it all just became too much. These things could bring down the entire company. Looking at it rationally, they had to make a deal with you."

"Yeah," Sanders said. "But none of this is rational, you know?"

"You're acting like you don't believe it," Fernandez said. "Believe it. It just got too big. They couldn't sit on it anymore."

"So what's the deal?"

Fernandez looked at her notes. "You got your whole shopping list. They'll fire Johnson. They'll give you her job, if you want that. Or they'll reinstate you at your present position. Or they'll give you another position in the company. They'll pay you a hundred thousand in pain and suffering and they'll pay my fees. Or they'll negotiate a termination agreement, if you want that. In any case, they'll give you full stock options if and when the division goes public. Whether you choose to remain with the company or not."

"Jesus Christ."

She nodded. "Total capitulation."

"You really believe Blackburn means it?"

You can never trust a lawyer.

"Yes," she said. "Frankly, it's the first thing that has made any sense to me all day. They had to do this, Tom. Their exposure is too great, and the stakes are too high."

"And what about this briefing?"

"They're worried about the merger—as you suspected when all this began. They don't want to blow it with any sudden changes now. So they want you to participate in the briefing tomorrow with Johnson, as if everything was normal. Then early next week, Johnson will have a physical exam as part of her insurance for the new job. The exam will uncover serious health problems, maybe even cancer, which will force a regrettable change in management."

"I see."

He went to the window and looked out at the city. The clouds were higher, and the evening sun was breaking through. He took a deep breath.

"And if I don't participate in the briefing?"

"It's up to you, but I would, if I were you," Fernandez said. "At this point, you really are in a position to bring down the company. And what good is that?"

He took another deep breath. He was feeling better all the time.

"You're saying this is over," he said, finally.

"Yes. It's over, and you've won. You pulled it off. Congratulations, Tom."

She shook his hand.

"Jesus Christ," he said.

She stood up. "I'm going to draw up an instrument outlining my conversation with Blackburn, specifying these options, and send it to him for his signature in an hour. I'll call you when I have it signed. Meanwhile, I recommend you do whatever preparation you need for this meeting tomorrow, and get some much deserved rest. I'll see you tomorrow."

"Okay."

It was slowly seeping into him, the realization that it was over. Really over. It had happened so suddenly and so completely, he was a little dazed.

"Congratulations again," Fernandez said. She folded her briefcase and left.

He was back in his office at about six. Cindy was leaving; she asked if he needed her, and he said he didn't. Sanders sat at his desk and stared out the window for a while, savoring the conclusion of the day. Through his open door, he watched as people left for the night, heading down the hall. Finally he called his wife in Phoenix to tell her the news, but her line was busy.

There was a knock at his door. He looked up and saw Blackburn standing there, looking apologetic. "Got a minute?"

"Sure."

"I just wanted to repeat to you, on a personal level, how sorry I am about all this. In the press of complex corporate problems like this, human values may get lost, despite the best of intentions. While we intend to be fair to everyone, sometimes we fail. And what is a corporation if not a human group, a group of human beings? We're all people, underneath it all. As Alexander Pope once said, 'We're all just human.' So recognizing your own graciousness through all this, I want to say to you . . ."

Sanders wasn't listening. He was tired; all he really heard was that Phil realized he had screwed up, and now was trying to repair things in his usual manner, by sucking up to someone he had earlier bullied.

Sanders interrupted, saying, "What about Bob?" Now that it was over, Sanders was having a lot of feelings about Garvin. Memories going back to his earliest days with the company. Garvin had been a kind of father to Sanders, and he wanted to hear from Garvin now. He wanted an apology. Or something.

"I imagine Bob's going to take a couple of days to come around," Blackburn said. "This was a very difficult decision for him to arrive at. I had to work very hard on him, on your behalf. And now he's got to figure out how to break it to Meredith. All that."

"Uh-huh."

"But he'll eventually talk to you. I know he will. Meanwhile, I wanted to go over a few things about the meeting tomorrow," Blackburn said. "It's for Marden, their CEO, and it's going to be a bit more formal than the way we usually do things. We'll be in the big conference room on the ground floor. It'll start at nine, and go to ten. Meredith will chair the meeting, and she'll call on all the division heads to give a summary of progress and problems in their divisions. Mary Anne first, then Don, then Mark, then you. Everyone will talk three to four minutes. Do it standing. Wear a jacket and tie. Use visuals if you have them, but stay away from technical details. Keep it an overview. In your case, they'll expect to hear mostly about Twinkle."

Sanders nodded. "All right. But there isn't really much new to report. We still haven't figured out what's wrong with the drives."

"That's fine. I don't think anybody expects a solution yet. Just emphasize the success of the prototypes, and the fact that we've overcome production problems before. Keep it upbeat, and keep it moving. If you have a prototype or a mock-up, you might want to bring it along."

"Okay."

"You know the stuff—bright rosy digital future, minor technical glitches won't stand in the way of progress."

"Meredith's okay with that?" he said. He was slightly disturbed to hear that she was chairing the meeting.

"Meredith is expecting all the heads to be upbeat and nontechnical. There won't be a problem."

"Okay," Sanders said.

"Call me tonight if you want to go over your presentation," Blackburn said. "Or in the morning, early. Let's just finesse this session, and then we can move on. Start making changes next week."

Sanders nodded.

"You're the kind of man this company needs," Blackburn said. "I appreciate your understanding. And again, Tom, I'm sorry."

He left.

Sanders called down to the Diagnostics Group, to see if they had any further word. But there was no answer. He went out to the closet behind Cindy's desk and took out the AV materials: the big schematic drawing of the Twinkle drive, and the schematic of the production line in Malaysia. He could prop these on easels while he talked.

But as he thought about it, it occurred to him that Blackburn was right. A mock-up or a prototype would be good to have. In fact, he should probably bring one of the drives that Arthur had sent from KL.

It reminded him that he should call Arthur in Malaysia. He dialed the number.

"Mr. Kahn's office."

"It's Tom Sanders calling."

The assistant sounded surprised. "Mr. Kahn is not here, Mr. Sanders."

"When is he expected back?"

"He's out of the office, Mr. Sanders. I don't know when he'll be back."

"I see." Sanders frowned. That was odd. With Mohammed Jafar missing, it was unlike Arthur to leave the plant without supervision.

The assistant said, "Can I give him a message?"

"No message, thanks."

He hung up, went down to the third floor to Cherry's programming group, and put his card in the slot to let himself in. The card popped back out, and the LED blinked oooo. It took him a moment to realize that they had cut off his access. Then he remembered the other card he had picked up earlier. He pushed it in the slot, and the door opened. Sanders went inside.

He was surprised to find the unit deserted. The programmers all kept strange hours; there was almost always somebody there, even at midnight.

He went to the Diagnostics room, where the drives were being studied. There were a series of benches, surrounded by electronic equipment and blackboards. The drives were set out on the benches, all covered in white cloth. The bright overhead quartz lights were off.

He heard rock-and-roll music from an adjacent room, and went there. A lone programmer in his early twenties was sitting at a console typing. Beside him, a portable radio blared.

Sanders said, "Where is everybody?"

The programmer looked up. "Third Wednesday of the month."

"So?"

"OOPS meets on the third Wednesday."

"Oh." The Object Oriented Programmer Support association, or OOPS, was an association of programmers in the Seattle area. It was started by Microsoft some years earlier, and was partly social and partly trade talk.

Sanders said, "You know anything about what the Diagnostics team found?"

"Sorry." The programmer shook his head. "I just came in."

Sanders went back to the Diagnostics room. He flicked on the lights and gently removed the white cloth that covered the drives. He saw that only three of the CD-ROM drives had been opened, their innards exposed to powerful magnifying glasses and electronic probes on the tables. The remaining seven drives were stacked to one side, still in plastic.

He looked up at the blackboards. One had a series of equations and hastily scribbled data points. The other had a flowchart list that read:

A. Contr. Incompat.
   VLSI?
   pwr?
B. Optic Dysfunct-? voltage reg?/arm?/servo?
C. Laser R/O (a,b,c)
D. Σ Mechanical √ √
E. Gremlins

It didn't mean much to Sanders. He turned his attention back to the tables, and peered at the test equipment. It looked fairly standard, except that there were a series of large-bore needles lying on the table, and several white circular wafers encased in plastic that looked like camera filters. There were also Polaroid pictures of the drives in various stages of disassembly; the team had documented their work. Three of the Polaroids were placed in a neat row, as if they might be significant, but Sanders couldn't see why. They just showed chips on a green circuit board.

He looked at the drives themselves, being careful not to disturb anything. Then he turned to the stack of drives that were still wrapped in plastic. But looking closely, he noticed fine, needle-point punctures in the plastic covering four of the drives.

Nearby was a medical syringe and an open notebook. The notebook showed a column of figures:

PPU

7

II (repeat II)

5

2

And at the bottom someone had scrawled, "Fucking Obvious!" But it wasn't obvious to Sanders. He decided that he'd better call Don Cherry later tonight, to have him explain it. In the meantime, he took one of the extra drives from the stack to use in the presentation the following morning.

He left the Diagnostics room carrying all his presentation materials, the easel boards flapping against his legs. He headed downstairs to the ground floor conference room, which had an AV closet where speakers stored visual material before a presentation. He could lock his material away there.

In the lobby, he passed the receptionist's desk, now manned by a black security guard, who watched a baseball game and nodded to Sanders. Sanders went back toward the rear of the floor, moving quietly on the plush carpeting. The hallway was dark, but the lights were on in the conference room; he could see them shining from around the corner.

As he came closer, he heard Meredith Johnson say, "And then what?" And a man's voice answered something indistinct.

Sanders paused.

He stood in the dark corridor and listened. From where he stood, he could see nothing of the room.

There was a moment of silence, and then Johnson said, "Okay, so will Mark talk about design?"

The man said, "Yes, he'll cover that."

"Okay," Johnson said. "Then what about the . . ."

Sanders couldn't hear the rest. He crept forward, moving silently on the carpet, and cautiously peered around the corner. He still could not see into the conference room itself, but there was a large chrome sculpture in the hallway outside the room, a sort of propeller shape, and in the reflection of its polished surface he saw Meredith moving in the room. The man with her was Blackburn.

Johnson said, "So what if Sanders doesn't bring it up?"

"He will," Blackburn said.

"You're sure he doesn't—that the—" Again, the rest was lost.

"No, he—no idea."

Sanders held his breath. Meredith was pacing, her image in the reflection, twisting and distorted. "So when he does—I will say that this is a—is that—you mean?"

"Exactly," Blackburn said.

"And if he—"

Blackburn put his hand on her shoulder. "Yes, you have to—"

"—So—want me to—"

Blackburn said something quiet in reply, and Sanders heard none of it, except the phrase "—must demolish him."

"—Can do that—"

"—Make sure—counting on you—"

There was the shrill sound of a telephone. Both Meredith and Blackburn reached for their pockets. Meredith answered the call, and the two began to move toward the exit. They were heading toward Sanders.

Panicked, Sanders looked around, and saw a men's room to his right. He slipped inside the door as they came out of the conference room and started down the hallway.

"Don't worry about this, Meredith," Blackburn said. "It'll go fine."

"I'm not worried," she said.

"It should be quite smooth and impersonal," Blackburn said. "There's no reason for rancor. After all, you have the facts on your side. He's clearly incompetent."

"He still can't get into the database?" she said.

"No. He's locked out of the system."

"And there's no way he can get into Conley-White's system?"

Blackburn laughed. "No way in hell, Meredith."

The voices faded, moving down the hallway. Sanders strained to listen, finally heard the click of a door closing. He stepped out of the bathroom into the hallway.

The hallway was deserted. He stared toward the far door.

His own telephone rang in his pocket, the sound so loud it made him jump. He answered it. "Sanders."

"Listen," Fernandez said. "I sent the draft of your contract to Black-burn's office, but it came back with a couple of added statements that I'm not sure about. I think we better meet to discuss them."

"In an hour," Sanders said.

"Why not now?"

"I have something to do first," he said.

A h, Thomas." Max Dorfman opened the door to his hotel room and immediately wheeled away, back toward the television set. "You have finally decided to come."

"You've heard?"

"Heard what?" Dorfman said. "I am an old man. No one bothers with me anymore. I'm cast by the wayside. By everyone—including you." He clicked off the television set and grinned.

Sanders said, "What have you heard?"

"Oh, just a few things. Rumors, idle talk. Why don't you tell me yourself?"

"I'm in trouble, Max."

"Of course you are in trouble," Dorfman snorted. "You have been in trouble all week. You only noticed now?"

"They're setting me up."

"They?"

"Blackburn and Meredith."

"Nonsense."

"It's true."

"You believe Blackburn can set you up? Philip Blackburn is a spineless fool. He has no principles and almost no brains. I told Garvin to fire him years ago. Blackburn is incapable of original thought."

"Then Meredith."

"Ah. Meredith. Yes. So beautiful. Such lovely breasts."

"Max, please."

"You thought so too, once."

"That was a long time ago," Sanders said.

Dorfman smiled. "Times have changed?" he said, with heavy irony.

"What does that mean?"

"You are looking pale, Thomas."

"I can't figure anything out. I'm scared."

"Oh, you're scared. A big man like you is scared of this beautiful woman with beautiful breasts."

"Max—"

"Of course, you are right to be scared. She has done all these many terrible things to you. She has tricked you and manipulated you and abused you, yes?"

"Yes," Sanders said.

"You have been victimized by her and Garvin."

"Yes."

"Then why were you mentioning to me the flower, hmm?"

He frowned. For a moment he didn't know what Dorfman was talking about. The old man was always so confusing and he liked to be—

"The *flower*," Dorfman said irritably, rapping his knuckles on the wheelchair arm. "The stained-glass flower in your apartment. We were speaking of it the other day. Don't tell me you have forgotten it?"

The truth was that he had, until that moment. Then he remembered the image of the stained-glass flower, the image that had come unbidden to his mind a few days earlier. "You're right. I forgot."

"You *forgot*." Dorfman's voice was heavy with sarcasm. "You expect me to believe that?"

"Max, I did, I—"

He snorted. "You are impossible. I cannot believe you will behave so transparently. You didn't forget, Thomas. You merely chose not to confront it."

"Confront what?"

In his mind, Sanders saw the stained-glass flower, in bright orange and purple and yellow. The flower mounted in the door of his apartment. Earlier in the week, he had been thinking about it constantly, almost obsessing about it, and yet today—

"I cannot bear this charade," Dorfman said. "Of course you remember it all. But you are *determined* not to think of it."

Sanders shook his head, confused.

"Thomas. You told it all to me, ten years ago," Dorfman said, waving his hand. "You *confided* in me. Blubbering. You were very upset at the time. It was the most important thing in your life, at the time. Now you say it is all forgotten?" He shook his head. "You told me that you would take trips with Garvin to Japan and Korea. And when you returned, she would be waiting for you in the apartment. In some erotic costume, or whatever. Some erotic pose. And you told me that sometimes, when you got home, you would see her first through the stained glass. Isn't that what you told me, Thomas? Or do I have it wrong?"

He had it wrong.

It came back to Sanders in a rush then, like a picture zooming large and bright before his eyes. He saw everything, almost as if he was there once again: the steps leading up to his apartment on the second floor, and the sounds he heard as he went up the steps in the middle of the afternoon, sounds he could not identify at first, but then he realized what he was hearing as he came to the landing and looked in through the stained glass and he saw—

"I came back a day early," Sanders said.

"Yes, that's right. You came back *unexpectedly.*"

The glass in patterns of yellow and orange and purple. And through it, her naked back, moving up and down. She was in the living room, on the couch, moving up and down.

"And what did you do?" Dorfman said. "When you saw her?"

"I rang the bell."

"That's right. Very civilized of you. Very nonconfrontational and polite. You rang the bell."

In his mind he saw Meredith turning, looking toward the door. Her tangled hair falling across her face. She brushed the hair away from her eyes. Her expression changed as she saw him. Her eyes widened.

Dorfman prodded: "And then what? What did you do?"

"I left," Sanders said. "I went back to the . . . I went to the garage and got in my car. I drove for a while. A couple of hours. Maybe more. It was dark when I got back."

"You were upset, naturally."

He came back up the stairs, and again looked in through the stained glass. The living room was empty. He unlocked the door and entered the living room. There was a bowl of popcorn on the couch. The couch was creased. The television was on, soundless. He looked away from the couch and went into the bedroom, calling her name. He found her packing, her open suitcase on the bed. He said, "What are you doing?"

"Leaving," she said. She turned to face him. Her body was rigid, tense. "Isn't that what you want me to do?"

"I don't know," he said.

And then she burst into tears. Sobbing, reaching for a kleenex, blowing her nose loudly, awkwardly, like a child. And somehow in her distress he held his arms out, and she hugged him and said she was sorry, repeating the words, again and again, through her tears. Looking up at him. Touching his face.

And then somehow . . .

Dorfman cackled. "Right on the suitcase, yes? Right there on the suitcase, on her clothes that were being packed, you made your reconciliation."

"Yes," Sanders said, remembering.

"She aroused you. You wanted her back. She excited you. She challenged you. You wanted to possess her."

"Yes . . ."

"Love is wonderful," Dorfman sighed, sarcastic again. "So pure, so innocent. And then you were together again, is that right?"

"Yes. For a while. But it didn't work out."

It was odd, how it had finally ended. He had been so angry with her at first, but he had forgiven her, and he thought that they could go on. They had talked about their feelings, they had expressed their love, and he had tried to go on with the best will in the world. But in the end, neither of them could; the incident had fatally ruptured the relationship, and something vital had been torn from it. It didn't matter how often they told themselves that they could go on. Something else now ruled. The core was dead. They fought more often, managing in this way to sustain the old energy for a while. But finally, it just ended.

"And when it was over," Dorfman said, "that was when you came and talked to me."

"Yes," Sanders said.

"And what did you come to talk to me about?" Dorfman asked. "Or have you 'forgotten' that, too?"

"No. I remember. I wanted your advice."

He had gone to Dorfman because he was considering leaving Cupertino. He was breaking up with Meredith, his life was confused, everything was in disarray, and he wanted to make a fresh start, to go somewhere else. So he was considering moving to Seattle to head the Advanced Projects Division. Garvin had offered him the job in passing one day, and Sanders was thinking about taking it. He had asked Dorfman's advice.

"You were quite upset," Dorfman said. "It was an unhappy ending to a love affair."

"Yes."

"So you might say that Meredith Johnson is the reason you are here in Seattle," Dorfman said. "Because of her, you changed your career, your life. You made a new life here. And many people knew this fact of your past. Garvin knew. And Blackburn knew. That is why he was so careful to ask you if you could work with her. Everyone was so worried about how it would be. But you reassured them, Thomas, didn't you?"

"Yes."

"And your reassurances were false."

Sanders hesitated. "I don't know, Max."

"Come, now. You know *exactly*. It must have been like a bad dream, a nightmare from your past, to hear that this person you had run away from was now coming to Seattle, pursuing you up here, and that she would be your superior in the company. Taking the job that you wanted. That you thought you deserved."

"I don't know . . ."

"Don't you? In your place, I would be angry. I would want to be rid of her, yes? She hurt you once very badly, and you would not want to be hurt again. But what choice did you have? She had the job, and she was Garvin's protégé. She was protected by Garvin's power, and he would not hear a word against her. True?"

"True."

"And for many years you had not been close to Garvin, because Garvin didn't really want you to take the Seattle job in the first place. He had offered it to you, expecting you to turn it down. Garvin likes protégés. He likes admirers at his feet. He does not like his admirers to pack up and leave for another city. So Garvin was disappointed with you. Things were never the same. And now suddenly here was this woman out of your past, a woman with Garvin's backing. So, what choice did you have? What could you do with your anger?"

His mind was spinning, confused. When he thought back to the events of that first day—the rumors, the announcement by Blackburn, the first meeting with her—he did not remember feeling anger. His feelings had been so complicated on that day, but he had not felt anger, he was sure of it . . .

"Thomas, Thomas. Stop dreaming. There is no time for it."

Sanders was shaking his head. He couldn't think clearly.

"Thomas, *you arranged all this*. Whether you admit it or not, whether you are aware of it or not. On some level, what has happened is exactly what you intended. And you made sure it would happen."

He found himself remembering Susan. What had she said at the restaurant?

Why didn't you tell me? I could have helped you.

And she was right, of course. She was an attorney; she could have advised him if he had told her what happened the first night. She would have told him what to do. She could have gotten him out of it. But he hadn't told her.

There's not much we can do now.

"You wanted this confrontation, Thomas."

And then Garvin: *She was your girlfriend, and you didn't like it when she dropped you. So now you want to pay her back.*

"You worked all week to ensure this confrontation."

"Max—"

"So don't tell me you are a victim here. You're not a victim. You call yourself a victim because you don't want to take responsibility for your life. Because you are sentimental and lazy and naïve. You think other people should take care of you."

"Jesus, Max," Sanders said.

"You deny your part in this. You pretend to forget. You pretend to be unaware. And now you pretend to be confused."

"Max—"

"Oh! I don't know why I bother with you. How many hours do you have until this meeting? Twelve hours? Ten? Yet you waste your time talking to a crazy old man." He spun in his wheelchair. "If I were you, I would get to work."

"Meaning what?"

"Well, we know what your intentions are, Thomas. But what are *her* intentions, hmmm? She is solving a problem, too. She has a purpose here. So: what is the problem she is solving?"

"I don't know," Sanders said.

"Clearly. But how will you find out?"

Lost in thought, he walked the five blocks to Il Terrazzo. Fernandez was waiting for him outside. They went in together.

"Oh Christ," Sanders said, as he looked around.

"All the usual suspects," Fernandez said.

In the far section straight ahead, Meredith Johnson was having dinner with Bob Garvin. Two tables away, Phil Blackburn was eating with his wife, Doris, a thin bespectacled woman who looked like an accountant. Near them, Stephanie Kaplan was having dinner with a young man in his twenties—probably her son at the university, Sanders thought. And over to the right, by the window, the Conley-White people were in the midst of a working dinner, their briefcases open at their feet, papers scattered all over the table. Ed Nichols sat with John Conley to his right, and Jim Daly to his left. Daly was speaking into a tiny dictating machine.

"Maybe we should go somewhere else," Sanders said.

"No," Fernandez said. "They've already seen us. We can sit in the corner over there."

Carmine came over. "Mr. Sanders," he said with a formal nod.

"We'd like a table in the corner, Carmine."

"Yes of course, Mr. Sanders."

They sat to one side. Fernandez was staring at Meredith and Garvin. "She could be his daughter," she said.

"Everybody says so."

"It's quite striking."

The waiter brought menus. Nothing on it appealed to Sanders, but they ordered anyway. Fernandez was looking steadily at Garvin. "He's a fighter, isn't he."

"Bob? Famous fighter. Famous tough guy."

"She knows how to play him." Fernandez turned away and pulled papers out of her briefcase. "This is the contract that Blackburn sent back. It is all in order, except for two clauses. First, they claim the right to terminate you if you are shown to have committed a felony on the job."

"Uh-huh." He wondered what they might mean.

"And this second clause claims the right to terminate you if you have 'failed to demonstrate satisfactory performance in the job as measured by industry standards.' What does that mean?"

He shook his head. "They must have something in mind." He told her about the conversation he had overheard in the conference room.

As usual, Fernandez showed no reaction. "Possible," she said.

"Possible? They're going to do it."

"I meant legally. It's possible that they intend something of this sort. And it would work."

"Why?"

"A harassment claim brings up the entire performance of an employee. If there is dereliction, even a very old or minor dereliction, it may be used to dismiss the claim. I had one client who worked for a company for ten years. But the company was able to demonstrate that the employee had lied on the original application form, and the case was dismissed. The employee was fired."

"So this comes down to my performance."

"It may. Yes."

He frowned. What did they have on him?

She is solving a problem, too. So: what is the problem she is solving?

Beside him, Fernandez pulled the tape recorder out of her pocket. "There's a couple of other things I want to go over," she said. "There's something that happens early on in the tape."

"Okay."

"I want you to listen."

She gave the player to him. He held it close to his ear.

He heard his own voice saying clearly, "... we'll face that later. I've given her your thoughts, and she's talking to Bob now, so presumably we'll go into the meeting tomorrow taking that position. Well, anyway, Mark, if there is a significant change in all this, I'll contact you before the meeting tomorrow, and—"

"Forget that phone," Meredith's voice said loudly, and then there was the sound of rustling, like fabric, and a sort of hissing sound, and a dull *thunk* as the phone was dropped. The momentary sharp crackle of static.

More rustling. Then silence.

A grunt. Rustling.

As he listened, he tried to imagine the action in the room. They must have moved over to the couch, because now the voices were lower, less distinct. He heard himself say, "Meredith, wait—"

"Oh God," she said, "I've wanted you all day."

More rustling. Heavy breathing. It was hard to be certain what was happening. A little moan from her. More rustling.

She said, "Oh God, you feel so *good*, I can't stand the bastard touching me. Those stupid glasses. Oh! I'm so *hot*, I haven't had a decent fuck—"

More rustling. Static crackle. Rustling. More rustling. Sanders listened with a sense of disappointment. He could not really create images for what was going on—and he had been there. This tape would not be persuasive to someone else. Most of it sounded like obscure noise. With long periods of silence.

"Meredith—"

"*Ooob*. Don't talk. No! No . . ." He heard her gasping, in little breaths. Then more silence.

Fernandez said, "That's enough."

Sanders put the player down and shut it off. He shook his head.

"You can't tell anything from this. About what was really going on."

"You can tell enough," Fernandez said. "And don't you start worrying about the evidence. That's my job. But you heard her first statements?" She consulted her notepad. "Where she says, 'I've wanted you all day'? And then she says, 'Oh God you feel so good, I can't stand the bastard touching me. Those stupid glasses, oh I'm so hot, I haven't had a decent fuck.' You heard that part?"

"Yes. I heard it."

"Okay. Who is she talking about?"

"Talking about?"

"Yes. Who is the bastard she can't stand touching her?"

"I assume her husband," Sanders said. "We were talking about him earlier. Before the tape."

"Tell me what was said earlier."

"Well, Meredith was complaining about having to pay alimony to her husband, and then she said her husband was terrible in bed. She said, 'I hate a man who doesn't know what he's doing.' "

"So you think 'I can't stand the bastard touching me' refers to her husband?"

"Yes."

"I don't," Fernandez said. "They were divorced months ago. The divorce was bitter. The husband hates her. He has a girlfriend now; he's taken her to Mexico. I don't think she means the husband."

"Then who?"

"I don't know."

Sanders said, "I suppose it could be anybody."

"I don't think it's just anybody. Listen again. Listen to how she sounds."

He rewound the tape, held the player to his ear. After a moment, he put the player down. "She sounds almost angry."

Fernandez nodded. "Resentful is the term I'd use. She's in the midst of this episode with you, and she's talking about someone else. 'The bastard.' It's as if she wants to pay somebody back. Right at that moment, she's getting even."

Sanders said, "I don't know. Meredith's a talker. She always talked about other people. Old boyfriends, that stuff. She's not what you'd call a romantic."

He remembered one time when they were lying on the bed in the apartment in Sunnyvale, feeling a sort of relaxed glow. A Sunday afternoon. Listening to kids laughing in the street outside. His hand resting on her thigh, feeling the sweat. And in this thoughtful way she said, "You know, I once went out with this Norwegian guy, and he had a curved dick. Curved like a sword, sort of bent over to the side, and he—"

"Jesus, Meredith."

"What's the matter? It's true. He really did."

"Not now."

Whenever this sort of thing happened, she'd sigh, as if she was obliged to put up with his excessive sensitivity. "Why is it that guys always want to think they're the only ones?"

"We don't. We know we're not. Just not now, okay?"

And she'd sigh again . . .

Sitting in the restaurant, Fernandez said, "Even if it's not unusual for her to talk during sex—even if she is indiscreet or distancing—who is she talking about here?"

Sanders shook his head. "I don't know, Louise."

"And she says she can't stand him touching her . . . as if she has no choice. And she mentions his silly glasses." She looked over at Meredith, who was eating quietly with Garvin. "Him?"

"I don't think so."

"Why not?"

"Everybody says no. Everybody says Bob isn't screwing her."

"Everybody could be wrong."

Sanders shook his head. "It'd be incest."

"You're probably right."

The food came. Sanders poked at his pasta *puttanesca*, picking out the olives. He wasn't feeling hungry. Beside him, Fernandez ate heartily. They had ordered the same thing.

Sanders looked over at the Conley-White people. Nichols was holding up a clear plastic sheet of 35-millimeter transparencies. Slides. Of what? he wondered. His half-frame glasses were perched on his nose. He seemed to be taking a long time. Beside him, Conley glanced at his watch and said something about the time. The others nodded. Conley glanced over at Johnson, then turned back to his papers.

Daly said something. ". . . have that figure?"

"It's here," Conley said, pointing to the sheet.

"This is really very good," Fernandez said. "You shouldn't let it get cold."

"Okay." He took a bite. It had no taste. He put the fork down.

She wiped her chin with her napkin. "You know, you never really told me why you stopped. At the end."

"My friend Max Dorfman says I set it all up."

"Uh-huh," Fernandez said.

"Do you think that, too?"

"I don't know. I was just asking what you were feeling, at the time. At the time you pulled away."

He shrugged. "I just didn't want to."

"Uh-huh. Didn't feel like it when you got there, huh?"

"No, I didn't." Then he said, "You really want to know what it was? She coughed."

"She coughed?" Fernandez said.

Sanders saw himself again in the room, his trousers down around his knees, bent over Meredith on the office couch. He remembered thinking, What the hell am I doing? And she had her hands on his shoulders, tugging him toward her. "Oh please . . . No . . . No . . ."

And then she turned her head aside and coughed.

That cough was what did it. That was when he sat back, and said, "You're right," and got off the couch.

Fernandez frowned. "I have to say," Fernandez said. "A cough doesn't seem like a big deal."

"It was." He pushed his plate away. "I mean, you can't cough at a time like that."

"Why? Is this some etiquette I don't know about?" Fernandez said. "No coughing in the clinch?"

"It's not that at all," Sanders said. "It's just what it means."

"I'm sorry, you've lost me. What does a cough mean?"

He hesitated. "You know, women always think that men don't know what's going on. There's this whole idea that men can't find the place, they don't know what to do, all that stuff. How men are stupid about sex."

"I don't think you're stupid. What does a cough mean?"

"A cough means you're not involved."

She raised her eyebrows. "That seems a little extreme."

"It's just a fact."

"I don't know. My husband has bronchitis. He coughs all the time."

"Not at the last moment, he doesn't."

She paused, thinking about it. "Well, he certainly does right afterward. He breaks out in a fit of coughing. We always laugh about how he does that."

"Right after is different. But at the moment, right in the intense moment, I'm telling you—nobody coughs."

More images flashed through his mind. Her cheeks turn red. Her neck is blotchy, or her upper chest. Nipples no longer hard. They were hard at first, but not now. The eyes get dark, sometimes purple below. Lips swollen. Breathing changes. Sudden surging heat. Shift in the hips, shifting rhythm, tension but something else, something liquid. Forehead frowning. Wincing. Biting. So many different ways, but—

"Nobody coughs," he said again.

And then he felt a kind of sudden embarrassment, and pulled his plate back, and took a bite of pasta. He wanted a reason not to say more, because he had the feeling that he had overstepped the rules, that there was still this area, this kind of knowledge, this awareness that everyone pretended didn't exist . . .

Fernandez was staring at him curiously. "Did you read about this somewhere?"

He shook his head, chewing.

"Do men discuss it? Things like this?"

He shook his head, no.

"Women do."

"I know." He swallowed. "But anyway, she coughed, and that was why I stopped. She wasn't involved, and I was very—angry about it, I guess. I mean she was lying there panting and moaning, but she was really uninvolved. And I felt . . ."

"Exploited?"

"Something like that. Manipulated. Sometimes I think maybe if she hadn't coughed right then . . ." Sanders shrugged.

"Maybe I should ask her," Fernandez said, nodding her head in Meredith's direction.

Sanders looked up and saw that she was coming over to their table. "Oh, hell."

"Calmly, calmly. Everything's fine."

Meredith came over, a big smile on her face. "Hello, Louise. Hello, Tom." Sanders started to get up. "Don't get up, Tom, please." She rested her hand on his shoulder, gave it a little squeeze. "I just came by for a moment." She was smiling radiantly. She looked exactly like the confident boss, stopping to say hello to a couple of colleagues. Back at her table, Sanders saw Garvin paying the bill. He wondered if he would come over, too.

"Louise, I just wanted to say no hard feelings," Meredith said. "Everybody had a job to do. I understand that. And I think it served a purpose, clearing the air. I just hope we can go on productively from here."

Meredith was standing behind Sanders's chair as she talked. He had to twist his head and crane his neck to look at her.

Fernandez said, "Don't you want to sit down?"

"Well, maybe for a minute."

Sanders stood to get her a chair. He was thinking that to the Conley people, all this would look exactly right. The boss not wanting to intrude, waiting to be pressed by her co-workers to join them. As he brought the chair, he glanced over and saw that Nichols was looking at them, peering over his glasses. So was young Conley.

Meredith sat down. Sanders pushed the chair in for her. "You want anything?" Fernandez said solicitously.

"I just finished, thanks."

"Coffee? Anything?"

"I'm fine, thanks."

Sanders sat down. Meredith leaned forward. "Bob's been telling me about his plans to take this division public. It's very exciting. It looks like full speed ahead."

Sanders watched her with astonishment.

"Now, Bob has a list of names for the new company. When we spin it off next year. See how these sound to you: SpeedCore, SpeedStar, PrimeCore, Talisan, and Tensor. I think SpeedCore makes racing parts for stock cars. SpeedStar is right on the money—but maybe too right on. PrimeCore sounds like a mutual fund. How about Talisan or Tensor?"

"Tensor is a lamp," Fernandez said.

"Okay. But Talisan is pretty good, I think."

"The Apple-IBM joint venture is called Taligent," Sanders said.

"Oh. You're right. Too close. How about MicroDyne? That's not bad. Or ADG, for Advanced Data Graphics? Do either of those work, do you think?"

"MicroDyne is okay."

"I thought so, too. And there was one more . . . AnoDyne."

"That's a painkiller," Fernandez said.

"What is?"

"An anodyne is a painkiller. A narcotic."

"Oh. Forget that. Last one, SynStar."

"Sounds like a drug company."

"Yeah, it does. But we've got a year to come up with a better one. And MicroDyne isn't bad, to start. Sort of combining micro with dynamo. Good images, don't you think?"

Before they could answer, she pushed her chair back. "I've got to go. But I thought you'd like to hear the thinking. Thanks for your input. Good night, Louise. And Tom, I'll see you tomorrow." She shook hands with them both and crossed the room to Garvin. Together she and Garvin went over to the Conley table to say hello.

Sanders stared at her. " 'Good images,' " he repeated. "Christ. She's talking about names for a company, but she doesn't even know what the company is."

"It was quite a show."

"Sure," Sanders said. "She's all show. But it had nothing to do with us. It's for them." He nodded toward the Conley-White people, sitting across the restaurant. Garvin was shaking hands all around, and Meredith was talking to Jim Daly. Daly made a joke and she laughed, throwing her head back, showing her long neck.

"The only reason she talked to us was so that when I get fired tomorrow, she won't be seen as having planned it."

Fernandez was paying the bill. "You want to go?" she said. "I still have some things to check."

"Really? What do you have to check?"

"Alan may have gotten something more for us. There's a possibility."

At the Conley table, Garvin was saying good-bye. He gave a final wave, then crossed the room to talk to Carmine.

Meredith remained at the Conley-White table. She was standing behind John Conley, with her hands resting on his shoulders while she talked to Daly and Ed Nichols. Ed Nichols said something, peering over his glasses, and Meredith laughed, and came around to look over his shoulder at a sheet of figures he was holding. Her head was very close to Nichols. She nodded, talked, pointed to the sheet.

You're checking the wrong company.

Sanders stared at Meredith, smiling and joking with the three men from Conley-White. What had Phil Blackburn said to him yesterday?

The thing is, Tom, Meredith Johnson is very well connected in this company. She has impressed a lot of important people.

Like Garvin.

Not only Garvin. Meredith has built a power base in several areas.

Conley-White?

Yes. There, too.

Alongside him, Fernandez stood up. Sanders stood and said, "You know what, Louise?"

"What?"

"We've been checking the wrong company."

Fernandez frowned, then looked over at the Conley-White table. Meredith was nodding with Ed Nichols and pointing with one hand, her other hand flat on the table for balance. Her fingers were touching Ed Nichols. He was peering at the sheets of data over his glasses.

"Stupid glasses . . ." Sanders said.

No wonder Meredith wouldn't press harassment charges against him. It would have been too embarrassing for her relationship with Ed Nichols. And no wonder Garvin wouldn't fire her. It made perfect sense. Nichols was already uneasy about the merger—his affair with Meredith might be all that was holding it in place.

Fernandez sighed. "You think so? Nichols?"

"Yeah. Why not?"

Fernandez shook her head. "Even if it's true, it doesn't help us. They can argue paramour preference, they can argue lots of things—if there's even an argument that needs to be made. This isn't the first merger made in the sack, you know. I say, forget it."

"You mean to tell me," he said, "that there's nothing improper with her having an affair with someone at Conley-White and being promoted as a result?"

"Nothing at all. At least, not in the strict legal sense. So forget it."

Suddenly he remembered what Kaplan had said. *She was looking in the wrong direction when they fired her.*

"I'm tired," he said.

"We all are. They look tired, too."

Across the room, the meeting was breaking up. Papers were being put back into briefcases. Meredith and Garvin were chatting with them. They all started leaving. Garvin shook hands with Carmine, who opened the front door for his departing guests.

And then it happened.

There was the sudden harsh glare of quartz lights, shining in from the street outside. The group huddled together, trapped in the light. They cast long shadows back into the restaurant.

"What's going on?" Fernandez said.

Sanders turned to look, but already the group was ducking back inside, closing the door. There was a moment of sudden chaos. They heard Garvin say, "God*damn* it," and spin to Blackburn.

Blackburn stood, a stricken look on his face, and rushed over to Garvin. Garvin was shifting from foot to foot. He was simultaneously trying to reassure the Conley-White people and chew out Blackburn.

Sanders went over. "Everything okay?"

"It's the goddamned press," Garvin said. "KSEA-TV is out there."

"This is an outrage," Meredith said.

"They're asking about some harassment suit," Garvin said, looking darkly at Sanders.

Sanders shrugged.

"I'll speak to them," Blackburn said. "This is just ridiculous."

"I'll say it's ridiculous," Garvin said. "It's an outrage, is what it is."

Everyone seemed to be talking at once, agreeing that it was an outrage. But Sanders saw that Nichols looked shaken. Now Meredith was leading them out of the restaurant the back way, onto the terrace. Blackburn went out the front, into the harsh lights. He held up his hands, like a man being arrested. Then the door closed.

Nichols was saying, "Not good, not good."

"Don't worry, I know the news director over there," Garvin was saying. "I'll put this one away."

Jim Daly said something about how the merger ought to be confidential.

"Don't worry," Garvin said grimly. "It's going to be confidential as hell by the time I get through."

Then they were gone, out the back door, into the night. Sanders went back to the table, where Fernandez was waiting.

"A little excitement," Fernandez said calmly.

"More than a little," Sanders said. He glanced across the room at Stephanie Kaplan, still having dinner with her son. The young man was talking, gesturing with his hands, but Kaplan was staring fixedly at the back door, where the Conley-White people had departed. She had a curious expression on her face. Then, after a moment, she turned back and resumed her conversation with her son.

The evening was black, damp, and unpleasant. He shivered as he walked back to his office with Fernandez.

"How did a television crew get the story?"

"Probably from Walsh," Fernandez said. "But maybe another way. It's really a small town. Anyway, never mind that. You've got to prepare for the meeting tomorrow."

"I've been trying to forget that."

"Yeah. Well, don't."

Ahead they saw Pioneer Square, with windows in the buildings still brightly lit. Many of the companies here had business with Japan, and stayed open to overlap with the first hours of the day in Tokyo.

"You know," Fernandez said, "watching her with those men, I noticed how cool she was."

"Yes. Meredith is cool."

"Very controlled."

"Yes. She is."

"So why did she approach you so overtly—and on her first day? What was the rush?"

What is the problem she is trying to solve? Max had said. Now Fernandez was asking the same thing. Everyone seemed to understand except Sanders.

You're not a victim.

So, solve it, he thought.

Get to work.

He remembered the conversation when Meredith and Blackburn were leaving the conference room.

It should be quite smooth and impersonal. After all, you have the facts on your side. He's clearly incompetent.

He still can't get into the database?

No. He's locked out of the system.

And there's no way he can get into Conley-White's system?

No way in hell, Meredith.

They were right, of course. He couldn't get into the system. But what difference would it make if he could?

Solve the problem, Max had said. *Do what you do best.*

Solve the problem.

"Hell," Sanders said.

"It'll come," Fernandez said.

It was nine-thirty. On the fourth floor, cleaning crews worked in the central partition area. Sanders went into his office with Fernandez. He didn't really know why they were going there. There wasn't anything he could think to do, now.

Fernandez said, "Let me talk to Alan. He might have something." She sat down and began to dial.

Sanders sat behind his desk, and stared at the monitor. On the screen, his e-mail message read:

YOU'RE STILL CHECKING THE WRONG COMPANY.

AFRIEND

"I don't see how," he said, looking at the screen. He felt irritable, playing with a puzzle that everyone could solve except him.

Fernandez said, "Alan? Louise. What have you got? Uh-huh. Uh-huh. Is that . . . Well, that's very disappointing, Alan. No, I don't know, now. If you can, yes. When would you be seeing her? All right. Whatever you can." She hung up. "No luck tonight."

"But we've only got tonight."

"Yes."

Sanders stared at the message on the computer screen. Somebody inside the company was trying to help him. Telling him he was checking the wrong company. The message seemed to imply that there was a way for him to check the other company. And presumably, whoever knew enough to send this message also knew that Sanders had been cut out of the DigiCom system, his privileges revoked.

What could he do?

Nothing.

Fernandez said, "Who do you think this 'Afriend' is?"

"I don't know."

"Suppose you had to guess."

"I don't know."

"What comes into your mind?" she said.

He considered the possibility that 'Afriend' was Mary Anne Hunter. But Mary Anne wasn't really a technical person; her strength was marketing. She wasn't likely to be sending routed messages over the Internet. She probably didn't know what the Internet was. So: not Mary Anne.

And not Mark Lewyn. Lewyn was furious at him.

Don Cherry? Sanders paused, considering that. In a way, this was just like Cherry. But the only time that Sanders had seen him since this began, Cherry had been distinctly unfriendly.

Not Cherry.

Then who else could it be? Those were the only people with executive sysop access in Seattle. Hunter, Lewyn, Cherry. A short list.

Stephanie Kaplan? Unlikely. At heart, Kaplan was plodding and unimaginative. And she didn't know enough about computers to do this.

Was it somebody outside the company? It could be Gary Bosak, he thought. Gary probably felt guilty about having turned his back on Sanders. And Gary had a hacker's devious instincts—and a hacker's sense of humor.

It might very well be Gary.

But it still didn't do Sanders any good.

You were always good at technical problems. That was always your strength.

He pulled out the Twinkle CD-ROM drive, still in plastic. Why would they want it wrapped that way?

Never mind, he thought. Stay focused.

There was something wrong with the drive. If he knew what, he would have the answer. Who would know?

Wrapped in plastic.

It was something to do with the production line. It must be. He fumbled with the material on his desk and found the DAT cartridge. He inserted it into the machine.

It came up, showing his conversation with Arthur Kahn. Kahn was on one side of the screen, Sanders on the other.

Behind Arthur, the brightly lit assembly line beneath banks of fluorescent lights. Kahn coughed, and rubbed his chin. "Hello, Tom. How are you?"

"I'm fine, Arthur," he said.

"Well, good. I'm sorry about the new organization."

But Sanders wasn't listening to the conversation. He was looking at Kahn. He noticed now that Kahn was standing very close to the camera, so close that his features were slightly blurred, out of focus. His face was large, and blocked any clear view of the production line behind him. "You know how I feel personally," Kahn was saying, on the screen.

His face was blocking the line.

Sanders watched a moment more, and then switched the tape off.

"Let's go downstairs," he said.

"You have an idea?"

"Call it a last-ditch hope," he said.

The lights clicked on, harsh lights shining on the tables of the Diagnostic team. Fernandez said, "What is this place?"

"This is where they check the drives."

"The drives that don't work?"

"Right."

Fernandez gave a little shrug. "I'm afraid I'm not—"

"Me neither," Sanders said. "I'm not a technical person. I can just read people."

She looked around the room. "Can you read this?"

He sighed. "No."

Fernandez said, "Are they finished?"

"I don't know," he said.

And then he saw it. They *were* finished. They had to be. Because otherwise the Diagnostics team would be working all night, trying to get ready for the meeting tomorrow. But they had covered the tables up and gone to their professional association meeting because they were finished.

The problem was solved.

Everybody knew it but him.

That was why they had only opened three drives. They didn't need to open the others. And they had asked for them to be sealed in plastic . . .

Because . . .

The punctures . . .

"Air," he said.

"Air?"

"They think it's the air."

"What air?" she said.

"The air in the plant."

"The plant in Malaysia?"

"Right."

"This is about air in Malaysia?"

"No. Air in the *plant*."

He looked again at the notebook on the table. "PPU" followed by a row of figures. PPU stood for "particulates per unit." It was the standard measure of air cleanliness in a plant. And these figures, ranging from two to eleven—they were way off. They should be running zero particulates . . . one, at most. These figures were unacceptable.

The air in the plant was bad.

That meant that they would be getting dirt in the split optics, dirt in the drive arms, dirt in the chip joins . . .

He looked at the chips attached to the board.

"Christ," he said.

"What is it?"

"Look."

"I don't see anything."

"There's a space between the chips and the boards. The chips aren't seated."

"It looks okay to me."

"It's not."

He turned to the stacked drives. He could see at a glance that all the chips were seated differently. Some were tight, some had a gap of a few millimeters, so you could see the metal contacts.

"This isn't right," Sanders said. "This should never happen." The fact was that the chips were inserted on the line by automated chip pressers. Every board, every chip should look exactly the same coming off the line. But they didn't. They were all different. Because of that, you could get voltage irregularities, memory allocation problems—all kinds of random stuff. Which was exactly what they were getting.

He looked at the blackboard, the list of the flowchart. One item caught his eye.

D. Σ Mechanical $\checkmark$ $\checkmark$

The Diagnostics team had put two checks beside "Mechanical." The problem with the CD-ROM drives was a mechanical problem. Which meant it was a problem in the production line.

And the production line was his responsibility.

He'd designed it, he'd set it up. He'd checked all the specs on that line, from beginning to end.

And now it wasn't working right.

He was sure that it wasn't his fault. Something must have happened after he had set up the line. Somehow it had been changed around, and it didn't work anymore. But what had happened?

To find out, he needed to get onto the databases.

But he was locked out.

There wasn't any way to get online.

Immediately, he thought of Bosak. Bosak could get him on. So, for that matter, could one of the programmers on Cherry's teams. These kids were hackers: they would break into a system for a moment of minor amusement the way ordinary people went out for coffee. But there weren't any programmers in the building now. And he didn't know when they would be back from their meeting. Those kids were so unreliable. Like the kid that had thrown up all over the walker pad. That was the problem. They were just kids, playing with toys like the walker pad. Bright creative kids, fooling around, no cares at all, and—

"Oh, Jesus." He sat forward. "Louise."

"Yes?"

"There's a way to do this."

"Do what?"

"Get into the database." He turned and hurried out of the room. He was rummaging through his pockets, looking for the second electronic passcard.

Fernandez said, "Are we going somewhere?"

"Yes, we are."

"Do you mind telling me where?"

"New York," Sanders said.

T he lights flicked on one after another, in long banks. Fernandez stared at the room. "What is this? The exercise room from hell?"

"It's a virtual reality simulator," Sanders said.

She looked at the round walker pads, and all the wires, the cables hanging from the ceiling. "This is how you're going to get to New York?"

"That's right."

Sanders went over to the hardware cabinets. There were large hand-painted signs reading, "Do Not Touch" and "Hands Off, You Little Wonk." He hesitated, looking for the control console.

"I hope you know what you're doing," Fernandez said. She stood by one of the walker pads, looking at the silver headset. "Because I think somebody could get electrocuted with this."

"Yeah, I know." Sanders lifted covers off monitors and put them back on again, moving quickly. He found the master switch. A moment later, the equipment hummed. One after another, the monitors began to glow. Sanders said, "Get up on the pad."

He came over and helped her stand on the walker pad. Fernandez moved her feet experimentally, feeling the balls roll. Immediately, there was a green flash from the lasers. "What was that?"

"The scanner. Mapping you. Don't worry about it. Here's the headset." He brought the headset down from the ceiling and started to place it over her eyes.

"Just a minute." She pulled away. "What is this?"

"The headset has two small display screens. They project images right in front of your eyes. Put it on. And be careful. These things are expensive."

"How expensive?"

"A quarter of a million dollars apiece." He fitted the headset over her eyes and put the headphones over her ears.

"I don't see any images. It's dark in here."

"That's because you're not plugged in, Louise." He plugged in her cables.

"Oh," she said, in a surprised voice. "What do you know . . . I can see a big blue screen, like a movie screen. Right in front of me. At the bottom of the screen there are two boxes. One says 'ON' and one says 'OFF.' "

"Just don't touch anything. Keep your hands on this bar," he said, putting her fingers on the walker handhold. "I'm going to mount up."

"This thing on my head feels funny."

Sanders stepped up onto the second walker pad and brought the headset down from the ceiling. He plugged in the cable. "I'll be right with you," he said.

He put on the headset.

Sanders saw the blue screen, surrounded by blackness. He looked to his left and saw Fernandez standing beside him. She looked entirely normal, dressed in her street clothes. The video was recording her appearance, and the computer eliminated the walker pad and the headset.

"I can see you," she said, in a surprised voice. She smiled. The part of her face covered by the headset was computer animated, giving her a slightly unreal, cartoonlike quality.

"Walk up to the screen."

"How?"

"Just walk, Louise." Sanders started forward on the walker pad. The blue screen became larger and larger, until it filled his field of vision. He went over to the ON button, and pushed it with his finger.

The blue screen flashed. In huge lettering, stretching wide in front of them, it said:

DIGITAL COMMUNICATIONS DATA SYSTEMS

Beneath that was listed a column of oversize menu items. The screen looked exactly like an ordinary DigiCom monitor screen, the kind on everybody's office desk, now blown up to enormous size.

"A gigantic computer terminal," Fernandez said. "Wonderful. Just what everybody has been hoping for."

"Just wait." Sanders poked at the screen, selecting menu items. There was a kind of *whoosh* and the lettering on the screen curved inward, pulling back and deepening until it formed a sort of funnel that stretched away from them into the distance. Fernandez was silent.

That shut her up, he thought.

Now, as they watched, the blue funnel began to distort. It widened, became rectangular. The lettering and the blue color faded. Beneath his feet, a floor emerged. It looked like veined marble. The walls on both sides became wood paneling. The ceiling was white.

"It's a corridor," she said, in a soft voice.

The Corridor continued to build itself, progressively adding more detail. Drawers and cabinets appeared in the walls. Pillars formed along its length. Other hallways opened up, leading down to other corridors. Large light fixtures emerged from the walls and turned themselves on. Now the pillars cast shadows on the marble floors.

"It's like a library," she said. "An old-fashioned library."

"This part is, yes."

"How many parts are there?"

"I'm not sure." He started walking forward.

She hurried to catch up to him. Through his earphones, he heard the sound of their feet clicking on the marble floor. Cherry had added that—a nice touch.

Fernandez asked, "Have you been here before?"

"Not for several weeks. Not since it was finished."

"Where are we going?"

"I'm not exactly sure. But somewhere in here there's a way to get into the Conley-White database."

She said, "Where are we now?"

"We're in data, Louise. This is all just data."

"This corridor is data?"

"There is no corridor. Everything you see is just a bunch of numbers. It's the DigiCom company database, exactly the same database that people access every day through their computer terminals. Except it's being represented for us as a place."

She walked alongside him. "I wonder who did the decorating."

"It's modeled on a real library. In Oxford, I think."

They came to the junction, with other corridors stretching away. Big signs hung overhead. One said "Accounting." Another said "Human Resources." A third said "Marketing."

"I see," Fernandez said. "We're inside your company database."

"That's right."

"This is amazing."

"Yeah. Except we don't want to be here. Somehow, we have to get into Conley-White."

"How do we do that?"

"I don't know," Sanders said. "I need help."

"Help is here," said a soft voice nearby. Sanders looked over and saw an angel, about a foot high. It was white, and hovered in the air near his head. It held a flickering candle in its hands.

"Goddamn," Louise said.

"I am sorry," the angel said. "Is that a command? I do not recognize 'Goddamn.' "

"No," Sanders said quickly. "It's not a command." He was thinking that he would have to be careful or they would crash the system.

"Very well. I await your command."

"Angel: I need help."

"Help is here."

"How do I enter the Conley-White database?"

"I do not recognize 'the Conley-White database.' "

That made sense, Sanders thought. Cherry's team wouldn't have programmed anything about Conley-White into the Help system. He would have to phrase the question more generally. Sanders said, "Angel: I am looking for a database."

"Very well. Database gateways are accessed with the keypad."

"Where is the keypad?" Sanders said.

"Make a fist with your hand."

Sanders made a fist and a gray pad formed in the air so that he appeared to be holding it. He pulled it toward him and looked at it.

"Pretty neat," Fernandez said.

"I also know jokes," the angel said. "Would you like to hear one?"

"No," Sanders said.

"Very well. I await your command."

Sanders stared at the pad. It had a long list of operator commands, with arrows and push buttons. Fernandez said, "What is that, the world's most complicated TV remote?"

"Just about."

He found a push button marked OTHER DB. That seemed likely. He pressed it.

Nothing happened.

He pressed it again.

"The gateway is opening," the angel announced.

"Where? I don't see anything."

"The gateway is opening."

Sanders waited. Then he realized that the DigiCom system would have to connect to any remote database. The connection was going through; that was causing a delay.

"Connecting . . . now," the angel said.

The wall of the Corridor began to dissolve. They saw a large gaping black hole, and nothing beyond it.

"That's creepy," Fernandez said.

White wire-frame lines began to appear, outlining a new corridor. The spaces filled, one by one, creating the appearance of solid shapes.

"This one looks different," Fernandez said.

"We're connecting over a T-1 high-speed data line," Sanders said. "But even so, it's much slower."

The Corridor rebuilt itself as they watched. This time the walls were gray. They faced a black-and-white world.

"No color?"

"The system's trying to generate a simpler environment. Color means more data to push around. So this is black and white."

The new corridor added lights, a ceiling, a floor. After a moment, Sanders said, "Shall we go in?"

"You mean, the Conley-White database is in there?"

"That's right," Sanders said.

"I don't know," she said. She pointed: "What about this?"

Directly in front of them was a kind of flowing river of black-and-white static. It ran along the floor, and also along the walls. It made a loud hissing sound.

"I think that's just static off the phone lines."

"You think it's okay to cross?"

"We have to."

He started forward. Immediately, there was a growl. A large dog blocked their path. It had three heads that floated above its body, looking in all directions.

"What's that?"

"Probably a representation of their system security." Cherry and his sense of humor, he thought.

"Can it hurt us?"

"For God's sake, Louise. It's just a cartoon." Somewhere, of course, there was an actual monitoring system running on the Conley-White database. Perhaps it was automatic, or perhaps there was a real person who actually watched users come and go on the system. But now it was nearly one o'clock in the morning in New York. The dog was most likely just an automatic device of some kind.

Sanders walked forward, stepping through the flowing river of static. The dog growled as he approached. The three heads swiveled, watching him as he passed with cartoon eyes. It was a strange sensation. But nothing happened.

He looked back at Fernandez. "Coming?"

She moved forward tentatively. The angel remained behind, hovering in the air.

"Angel, are you coming?"

It didn't answer.

"Probably can't cross a gateway," Sanders said. "Not programmed."

They walked down the gray corridor. It was lined with unmarked drawers on all sides.

"It looks like a morgue," Fernandez said.

"Well, at least we're here."

"This is their company database in New York?"

"Yes. I just hope we can find it."

"Find what?"

He didn't answer her. He walked over to one file cabinet at random and pulled it open. He scanned the folders.

"Building permits," he said. "For some warehouse in Maryland, looks like."

"Why aren't there labels?"

Even as she said it, Sanders saw that labels were slowly emerging out of the gray surfaces. "I guess it just takes time." Sanders turned and looked in all directions, scanning the other labels. "Okay. That's better. HR records are on this wall, over here."

He walked along the wall. He pulled open a drawer.

"Uh-oh," Fernandez said.

"What?"

"Somebody's coming," she said, in an odd voice.

At the far end of the corridor, a gray figure was approaching. It was still too distant to make out details. But it was striding directly toward them.

"What do we do?"

"I don't know," Sanders said.

"Can he see us?"

"I don't know. I don't think so."

"We can see him, but he can't see us?"

"I don't know." Sanders was trying to figure it out. Cherry had installed another virtual system in the hotel. If someone was on that system, then he or she could probably see them. But Cherry had said that his system represented other users as well, such as somebody accessing the database from a computer. And somebody using a computer wouldn't be able to see them. A computer user wouldn't know who else was in the system.

The figure continued to advance. It seemed to come forward in jerks, not smoothly. They saw more detail; they could start to see eyes, a nose, a mouth.

"This is really creepy," Fernandez said.

The figure was still closer. The details were filling in.

"No kidding," Sanders said.

It was Ed Nichols.

Up close, they saw that Nichols's face was represented by a black-and-white photograph wrapped crudely around an egg-shaped head, atop a gray moving body that had the appearance of a mannequin or a puppet. It was a computer-generated figure. Which meant that Nichols wasn't on the virtual system. He was probably using his notebook computer in his hotel room. Nichols walked up to them and continued steadily past them.

"He can't see us."

Fernandez said, "Why does his face look that way?"

"Cherry said that the system pulls a photo from the file and pastes it on users."

The Nichols-figure continued on walking down the corridor, away from them.

"What's he doing here?"

"Let's find out."

They followed him back down the corridor until Nichols stopped at one file cabinet. He pulled it open and began to go through the records. Sanders and Fernandez came up and stood by his shoulder, and watched what he was doing.

The computer-generated figure of Ed Nichols was thumbing through his notes and e-mail. He went back two months, then three months, then six months. Now he began to pull out sheets of paper, which seemed to hang in the air as he read them. Memos. Notations. Personal and Confidential. Copies to File.

Sanders said, "These are all about the acquisition."

More notes came out. Nichols was pulling them quickly, one after another.

"He's looking for something specific."

Nichols stopped. He had found what he was looking for. His gray computer image held it in his hand and looked at it. Sanders read it over his shoulder, and said certain phrases aloud to Fernandez: "Memo dated December 4, last year. 'Met yesterday and today with Garvin and Johnson in Cupertino re possible acquisition of DigiCom . . .' bla bla . . . 'Very favorable first impression . . . Excellent grounding in critical areas we seek to acquire . . .' bla bla . . . 'Highly capable and aggressive executive staff at all levels. Particularly impressed with competence of Ms. Johnson despite youth.' I'll bet you were impressed, Ed."

The computer-generated Nichols moved down the hall to another drawer and opened it. He didn't find what he wanted and closed it. He went on to another drawer.

Then he began reading again, and Sanders read this one, too: " 'Memo to John Marden. Cost issues re DigiCom acquisition' . . . bla bla . . . 'Concern for high-technology development costs in new company' . . . bla bla . . . Here we are. 'Ms. Johnson has undertaken to demonstrate her fiscal responsibility in new Malaysia operation . . . Suggests savings can be made . . . Expected cost savings . . .' How the hell could she do that?"

"Do what?" Fernandez said.

"Demonstrate fiscal responsibility in the Malaysia operation? That was my operation."

"Uh-oh," Fernandez said. "You're not going to believe this."

Sanders glanced over at her. Fernandez was staring down the corridor. He turned to look.

Someone else was coming toward them.

"Busy night," he said.

But even from a distance, he could see that this figure was different. The head was more lifelike, and the body was fully detailed. The figure walked smoothly, naturally. "This could be trouble," he said. Sanders recognized him, even from a distance.

"It's John Conley," Fernandez said.

"Right. And he's on the walker pad."

"Which means?"

Conley abruptly stopped in the middle of the corridor, and stared.

"He can see us," Sanders said.

"He can? How?"

"He's on the system we installed in the hotel. That's why he's so detailed. He's on the other virtual system, so he can see us, and we can see him."

"Uh-oh."

"You said it."

Conley moved forward, slowly. He was frowning. He looked from Sanders to Fernandez to Nichols and back to Sanders. He seemed uncertain what to do.

Then he held his finger to his lips, a gesture for silence.

"Can he hear us?" Fernandez whispered.

"No," Sanders said, in a normal voice.

"Can we talk to him?"

"No."

Conley seemed to make a decision. He walked over to Sanders and Fernandez, until he was standing very close. He looked from one to the other. They could see his expression perfectly.

Then he smiled. He extended his hand.

Sanders reached out, and shook it. He didn't feel anything, but through the headset he saw what looked like his hand gripping Conley's.

Then Conley shook Fernandez's hand.

"This is extremely weird," Fernandez said.

Conley pointed toward Nichols. Then he pointed to his own eyes. Then to Nichols again.

Sanders nodded. They all went over to stand beside Nichols as he went through records.

"You mean Conley's watching him, too?"

"Yes."

"So we can all see Nichols . . ."

"Yes."

"But Nichols can't see any of us."

"Right."

The gray computer figure of Ed Nichols was pulling files hastily out of a drawer.

"What's he up to now?" Sanders said. "Ah. Going through expense records. Now he's found one: 'Sunset Shores Lodge, Carmel. December 5 and 6.' Two days after his memo. And look at these expenses. A hundred and ten dollars for breakfast? Somehow I don't think our Ed was alone there."

He looked over at Conley.

Conley shook his head, frowning.

Suddenly, the record Nichols was holding vanished.

"What happened?"

"I think he just deleted it."

Nichols thumbed through other records. He found four more for the Sunset Shores, and deleted them all. They vanished in midair. Then he closed the drawer, turned, and walked away.

Conley remained behind. He looked at Sanders and drew a finger quickly under his throat.

Sanders nodded.

Conley again put his finger to his lips.

Sanders nodded. He would keep quiet. "Come on," Sanders said to Fernandez. "We're done in here." He started back toward the DigiCom Corridor.

She walked alongside him, then said, "I think we have company."

Sanders looked back: Conley was following them.

"It's okay," he said. "Let him come."

They crossed the gateway, past the barking dog, and came back into the Victorian library. Fernandez sighed. "It feels good to be home again, doesn't it?"

Conley was walking along, showing no surprise. But then, he had seen the Corridor before. Sanders walked quickly. The angel floated alongside them.

"But you realize," Fernandez said, "that none of this makes any sense. Because Nichols is the one who's been opposed to the acquisition, and Conley is the one pushing for it."

"That's right," Sanders said. "It's perfect. Nichols is having it off with Meredith. He promotes her behind the scenes as the new head of the division. And how does he hide that fact? By continuously bitching and moaning to anybody who will listen."

"You mean, it's a cover."

"Sure. That's why Meredith never answered his complaints in any of the meetings. She knew he wasn't a real threat."

"And Conley?" she said.

Conley was still walking alongside them.

"Conley genuinely wants the acquisition. And he wants it to work well. Conley's smart, and I think he realizes that Meredith isn't competent for the job. But Conley sees Meredith as the price of Nichols's support. So Conley has gone along with the choice of Meredith—at least for the time being."

"And what are we doing now?"

"Finding out about the last missing piece."

"Which is?"

Sanders was looking down the hallway marked OPERATIONS. This wasn't really his area of the database, except in specific places of overlap. The files were marked alphabetically. He went down the row until he found DIGICOM/MALAYSIA SA.

He opened it up and searched the file section marked STARTUP. He found his own memos, feasibility studies, site reports, government negotiations, first set specifications, memos from their Singapore suppliers, more government negotiations, all stretching back two years.

"What are you looking for?"

"Building plans."

He expected to see the thick sheets of blueprints and inspection summaries, but instead there was just a thin file. He opened the first sheet, and a three-dimensional image of the factory floated in the air in front of him. It was just an outline at first, but it rapidly filled in and became solid-looking. Sanders, Fernandez, and Conley stood on three sides of it, looking at it. It was like a very large, detailed doll's house. They peered in through the windows.

Sanders pushed a button. The model became transparent, then turned into a cutaway; now they could see the assembly line, the physical plant. A green line—the conveyor belt—started moving, and the machines and workers assembled the CD-ROM drives as the parts came down the line.

"What are you looking for?"

"Revisions." He shook his head. "This is the first set of plans."

The second sheet was marked "Revisions 1/First Set" with the date. He opened it up. The model of the plant seemed to shimmer for a moment, but it remained the same.

"Nothing happened."

The next sheet was marked "Revisions 2/Detail Only." Again, when he opened it, the plant shimmered briefly but was unchanged.

"According to these records, the plant was never revised," Sanders said. "But we know it was."

"What's he doing?" Fernandez said. She was looking at Conley.

Sanders saw that Conley was slowly mouthing words, his facial movements exaggerated.

"He's trying to tell us something," she said to Sanders. "Can you see what it is?"

"No." Sanders watched a moment, but the cartoonlike quality of Conley's face made it impossible to read his lips. Finally Sanders shook his head.

Conley nodded, and took the keypad out of Sanders's hand. He pushed a button marked RELATED and Sanders saw a list of related databases flash up in the air. It was an extensive list, including the permits from the Malay government, the architect's notes, the contractor agreements, health and medical inspections, and more. All together, there were about eighty items on the list. Sanders felt sure he would have overlooked the one in the middle of the list that Conley was now pointing to:

OPERATIONS REVIEW UNIT

"What's that?" Fernandez said.

Sanders pressed the name and a new sheet fluttered up. He pushed a button marked SUMMARY and read the sheet aloud: " 'The Operations Review Unit was formed four years ago in Cupertino by Philip Blackburn to address problems not normally within Operations Management purview. The mission of the Review Unit was to improve management efficiency within DigiCom. Over the years, the Operations Review Unit has successfully resolved a number of management problems at DigiCom.' "

"Uh-huh," Fernandez said.

" 'Nine months ago, the Operations Review Unit, then headed by Meredith Johnson of Cupertino Operations, undertook a review of the proposed manufacturing facility in Kuala Lumpur, Malaysia. The immediate stimulus for the review was a conflict with the Malay government over the number and ethnic composition of workers employed at the proposed facility.' "

"Uh-oh," Fernandez said.

" 'Led by Ms. Johnson, with legal assistance from Mr. Blackburn, the Operations Review Unit had outstanding success in resolving the many problems facing DigiCom's Malaysian operation.' "

"What is this, a press release?" Fernandez said.

"Looks like it," Sanders said. He read on: " 'Specific issues concerned the number and ethnic composition of workers employed at the facility. The original plans called for seventy workers to be employed. Responding to the requests of the Malay government, Operations Review was able to increase the number of workers to eighty-five by reducing the amount of automation at the plant, thus making the facility more suitable to the economy of a developing country.' " Sanders looked over at Fernandez. "And screwing us completely," he said.

"Why?"

He continued: " 'In addition, a cost-savings review generated important fiscal benefits in a number of areas. Costs were reduced with no detriment to product quality at the plant. Air-handling capacity was revised to more appropriate levels, and outsourcing supplier contracts were reallocated, with substantial savings benefit to the company.' " Sanders shook his head. "That's it," he said. "That's the whole ball game."

"I don't understand," Fernandez said. "This makes sense to you?"

"You're damned right it does."

He pushed the DETAIL button for more pages.

"I am sorry," the angel said, "there is no more detail."

"Angel, where are the supporting memos and files?" Sanders knew that there had to be massive paperwork behind these summary changes. The renegotiations with the Malay government alone would fill drawers of files.

The angel said, "I am sorry. There is no more detail available."

"Angel, show me the files."

"Very well."

After a moment, a sheet of pink paper flashed up:

THE DETAIL FILES ON
OPERATIONS REVIEW UNIT/MALAYSIA
HAVE BEEN DELETED
SUNDAY 6/14 AUTHORIZATION DC/C/5905

"Hell," Sanders said.

"What does that mean?"

"Somebody cleaned up," Sanders said. "Just a few days ago. Who knew all this was going to happen? Angel, show me all communications between Malaysia and DC for the past two weeks."

"Do you wish telephone or video links?"

"Video."

"Press V."

He pushed a button, and a sheet uncurled in the air:

Date	Linking	To	Duration	Auth
6/1	A. Kahn > M. Johnson		0812–0814	ACSS
6/1	A. Kahn > M. Johnson		1343–1346	ADSS
6/2	A. Kahn > M. Johnson		1801–1804	DCSC
6/2	A. Kahn > T. Sanders		1822–1826	DCSE
6/3	A. Kahn > M. Johnson		0922–0924	ADSC
6/4	A. Kahn > M. Johnson		0902–0912	ADSC
6/5	A. Kahn > M. Johnson		0832–0832	ADSC
6/7	A. Kahn > M. Johnson		0904–0905	ACSS
6/11	A. Kahn > M. Johnson		2002–2004	ADSC
6/13	A. Kahn > M. Johnson		0902–0932	ADSC
6/14	A. Kahn > M. Johnson		1124–1125	ACSS
6/15	A. Kahn > T. Sanders		1132–1134	DCSE

"Burning up the satellite links," Sanders said, staring at the list. "Arthur Kahn and Meredith Johnson talked almost every day until June fourteenth. Angel, show me these video links."

"The links are not available for viewing except for 6/15."

That had been his own transmission to Kahn, two days earlier. "Where are the others?"

A message flashed up:

THE VIDEO FILES ON
OPERATIONS REVIEW UNIT/MALAYSIA
HAVE BEEN DELETED
SUNDAY 6/14 AUTHORIZATION DC/C/5905

Scrubbed again. He was pretty sure who had done it, but he had to be sure. "Angel, how do I check deletion authorization?"

"Press the data you desire," the Angel said.

Sanders pressed the authorization number. A small sheet of paper came upward out of the top sheet and hung in the air:

AUTHORIZATION DC/C/5905 IS
DIGITAL COMMUNICATIONS
CUPERTINO/OPERATIONS EXECUTIVE
SPECIAL PRIVILEGES NOTED
(NO OPERATOR ID NECESSARY)

"It was done by somebody very high up in Operations in Cupertino, a few days ago."

"Meredith?"

"Probably. And it means I'm screwed."

"Why?"

"Because now I know what was done at the Malaysia plant. I know exactly what happened: Meredith went in and changed the specs. But she's erased the data, right down to her voice transmissions to Kahn. Which means I can't prove any of it."

Standing in the corridor, Sanders poked the sheet, and it fluttered back down, dissolving into the top sheet. He closed his file, put it back in the drawer, and watched the model dissolve and disappear.

He looked over at Conley. Conley gave a little resigned shrug. He seemed to understand the situation. Sanders shook his hand, gripping air, and waved good-bye. Conley nodded and turned to leave.

"Now what?" Fernandez said.

"It's time to go," Sanders said.

The angel began to sing: "It's time to go, so long again till next week's show—"

"Angel, be quiet." The angel stopped singing. He shook his head. "Just like Don Cherry."

"Who's Don Cherry?" Fernandez asked.

"Don Cherry is a living god," the angel said.

They walked back to the entrance to the Corridor and then climbed out of the blue screen.

Back in Cherry's lab, Sanders took off the headset and, after a moment of disorientation, stepped off the walker pad. He helped Fernandez remove her equipment. "Oh," she said, looking around. "We're back in the real world."

"If that's what you call it," he said. "I'm not sure it's that much more real." He hung up her headset and helped her down from the walker pad. Then he turned off the power switches around the room.

Fernandez yawned and looked at her watch. "It's eleven o'clock. What are you going to do now?"

There was only one thing he could think of. He picked up the receiver on one of Cherry's data modem lines and dialed Gary Bosak's number. Sanders couldn't retrieve any data, but perhaps Bosak could— if he could talk him into it. It wasn't much of a hope. But it was all he could think to do.

An answering machine said, "Hi, this is NE Professional Services. I'm out of town for a few days, but leave a message." And then a beep.

Sanders sighed. "Gary, it's eleven o'clock on Wednesday. I'm sorry I missed you. I'm going home." He hung up.

His last hope.

Gone.

Out of town for a few days.

"Shit," he said.

"Now what?" Fernandez said, yawning.

"I don't know," he said. "I've got half an hour to make the last ferry. I guess I'll go home and try to get some sleep."

"And the meeting tomorrow?" she asked. "You said you need documentation."

Sanders shrugged. "Louise, I've done all that I can do. I know what I'm up against. I'll manage somehow."

"Then I'll see you tomorrow?"

"Yeah," he said. "See you tomorrow."

He felt less sanguine on the ferry going home, looking back at the lights of the city in the rippling black water. Fernandez was right; he ought to be getting the documentation he needed. Max would criticize him, if he knew. He could almost hear the old man's voice: "Oh, so you're *tired*? That's a good reason, Thomas."

He wondered if Max would be at the meeting tomorrow. But he found he couldn't really think about it. He couldn't imagine the meeting. He was too tired to concentrate. The loudspeaker announced that they were five minutes from Winslow, and he went belowdecks to get into his car.

He unlocked the door and slipped behind the wheel. He looked in the rearview mirror and saw a dark silhouette in the backseat.

"Hey," Gary Bosak said.

Sanders started to turn.

"Just keep looking forward," Bosak said. "I'll get out in a minute. Now listen carefully. They're going to screw you tomorrow. They're going to pin the Malaysia fiasco on you."

"I know."

"And if that doesn't work, they're going to hit you with employing me. Invasion of privacy. Felonious activity. All that crap. They've talked to my parole officer. Maybe you've seen him—a fat guy with a mustache?"

Sanders vaguely remembered the man walking up to the mediation center the day before. "I think so, yes. Gary, listen, I need some documents—"

"Don't talk. There's no time. They pulled all the documents relating to the plant off the system. Nothing's there anymore. It's gone. I can't help you." They heard the sound of the ferry horn. All around them, drivers were starting their engines. "But I'm not going down for this felony crap. And you're not, either. Take this." He reached forward, and handed Sanders an envelope.

"What's this?"

"Summary of some work I did for another officer of your company. Garvin. You might want to fax it to him in the morning."

"Why don't you?"

"I'm crossing the border tonight. I have a cousin in BC, I'll stay there for a while. You can leave a message on my machine if it turns out okay."

"All right."

"Stay cool, guy. The shit's really going to hit the fan tomorrow. Lots of changes coming."

Up ahead, the ramp went down with a metallic clang. The traffic officers were directing cars off the ferry.

"Gary. You've been monitoring me?"

"Yeah. Sorry about that. They told me I had to."

"Then who's 'Afriend'?"

Bosak laughed. He opened the door and got out. "I'm surprised at you, Tom. Don't you know who your friends are?"

The cars were beginning to pull out. Sanders saw brake lights on the car ahead of him flash red, and the car began to move.

"Gary—" he said, turning. But Bosak was gone.

He put the car in gear and drove off the ferry.

At the top of the driveway, he stopped to pick up his mail. There was a lot of it; he hadn't checked the mailbox for two days. He drove down to the house and left the car outside the garage. He unlocked the front door and went in. The house seemed empty and cold. It had a lemony odor. Then he remembered that Consuela had probably cleaned up.

He went into the kitchen and set up the coffeemaker for the morning. The kitchen was clean and the children's toys had been picked up; Consuela had definitely been there. He looked at the answering machine.

A red numeral was blinking: 14.

Sanders replayed the calls. The first was from John Levin, asking him to call, saying it was urgent. Then Sally, asking if the kids could arrange a play date. But then the rest were all hang-ups. And as he listened, they all seemed to sound exactly the same—the thin hissing background static of an overseas call and then the abrupt click of disconnection. Again and again.

Someone was trying to call him.

One of the later calls was apparently placed by an operator, because a woman's lilting voice said, "I'm sorry, there is no answer. Do you wish to leave a message?" And then a man's voice replied, "No." And then disconnection.

Sanders played it back, listening to that "No."

He thought it sounded familiar. Foreign, but still familiar.

"No."

He listened several times but could not identify the speaker.

"No."

One time, he thought the man sounded hesitant. Or was it hurried? He couldn't tell.

"Do you wish to leave a message?"

"No."

Finally he gave up, rewound the machine, and went upstairs to his office. He'd had no faxes. His computer screen was blank. No further help from "Afriend" tonight.

He read through the paper that Bosak had given him in the car. It was a single sheet, a memo addressed to Garvin, containing a report summary on a Cupertino employee whose name was blanked out. There was also a xerox of a check made out to NE Professional Services signed by Garvin.

It was after one when Sanders went into the bathroom and took a shower. He turned the water up hot, held his face close to the nozzle, and felt the stinging spray on his skin. With the sound of the shower roaring in his ears, he almost missed hearing the telephone ringing. He grabbed a towel and ran into the bedroom.

"Hello?"

He heard the static hiss of an overseas connection. A man's voice said, "Mr. Sanders, please."

"This is Mr. Sanders speaking."

"Mr. Sanders, sir," the voice said, "I do not know if you will remember me. This is Mohammed Jafar."

THURSDAY

The morning was clear. Sanders took an early ferry to work and got to his office at eight. He passed the downstairs receptionist and saw a sign that said "Main Conference Room in Use." For a horrified moment he thought that he had again mistaken the time for his meeting, and hurried to look in. But it was Garvin, addressing the Conley-White executives. Garvin was speaking calmly, and the executives were nodding as they listened. Then as he watched, Garvin finished and introduced Stephanie Kaplan, who immediately launched into a financial review with slides. Garvin left the conference room, and immediately his expression turned grim as he walked down the hallway toward the espresso bar at the end of the corridor, ignoring Sanders.

Sanders was about to head upstairs when he heard Phil Blackburn say, "I really feel I have a right to protest the way this matter has been handled."

"Well, you don't," Garvin said angrily. "You don't have any rights at all."

Sanders moved forward, toward the espresso bar. From his position across the hallway, he was able to see into the bar. Blackburn and Garvin were talking by the coffee machines.

"But this is extremely unfair," Blackburn said.

"Fuck unfair," Garvin said. "She named you as the source, you stupid asshole."

"But Bob, you told me—"

"I told you what?" Garvin said, eyes narrowing.

"You told me to handle it. To put pressure on Sanders."

"That's right, Phil. And you told *me* that you were going to take care of it."

"But you knew I talked to—"

"I knew you had done something," Garvin said. "But I didn't know what. Now she's named you as a source."

Blackburn hung his head. "I just think it's extremely unfair."

"Really? But what do you expect me to do? You're the fucking lawyer, Phil. You're the one always sweating about how things look. You tell me. What do I do?"

Blackburn was silent for a moment. Finally he said, "I'll get John Robinson to represent me. He can work out the settlement agreement."

"Okay, fine." Garvin nodded. "That's fine."

"But I just want to say to you, on a personal level, Bob, that I feel my treatment in this matter has been very unfair."

"Goddamn it, Phil, don't talk to me about your feelings. Your feelings are for sale. Now listen with both ears: Don't go upstairs. Don't clean out your desk. Go right to the airport. I want you on a plane in the next half hour. I want you fucking out of here, right now. Is that clear?"

"I just think you should acknowledge my contribution to the company."

"I am, you asshole," Garvin said. "Now get the fuck out of here, before I lose my temper."

Sanders turned and hurried upstairs. It was hard for him to keep from cheering. Blackburn was fired! He wondered if he should tell anybody; perhaps Cindy, he thought.

But when he got to the fourth floor, the hallways were buzzing; everyone was out of their offices, talking in the corridors. Obviously, rumors of the firing had already leaked. Sanders was not surprised that staffers were in hallways. Even though Blackburn was disliked, his firing would cause widespread uneasiness. Such a sudden change, involving a person so close to Garvin, conveyed to everyone a sense of peril. Everything was at risk.

Outside his office, Cindy said, "Tom, can you believe it? They say Garvin is going to fire Phil."

"You're kidding," Sanders said.

Cindy nodded. "Nobody knows why, but apparently it had something to do with a news crew last night. Garvin's been downstairs explaining it to the Conley-White people."

Behind him, somebody shouted, "It's on the e-mail!" The hallway was instantly deserted; everyone vanished into their offices. Sanders stepped behind his desk and clicked the e-mail icon. But it was slow coming up, probably because every employee in the building was clicking at exactly the same time.

Fernandez came in and said, "Is it true about Blackburn?"

"I guess so," Sanders said. "It's just coming over the e-mail now."

FROM: ROBERT GARVIN, PRESIDENT AND CEO
TO: ALL THE DIGICOM FAMILY

IT IS WITH GREAT SADNESS AND A DEEP SENSE OF PER-
SONAL LOSS THAT I TODAY ANNOUNCE THE RESIGNATION
OF OUR VALUED AND TRUSTED CHIEF CORPORATE COUN-
SEL, PHILIP A. BLACKBURN. PHIL HAS BEEN AN OUTSTAND-
ING OFFICER OF THIS COMPANY FOR NEARLY FIFTEEN
YEARS, A WONDERFUL HUMAN BEING, AND A CLOSE PER-
SONAL FRIEND AND ADVISOR AS WELL. I KNOW THAT LIKE
ME, MANY OF YOU WILL MISS HIS WISE COUNSEL AND GOOD
HUMOR PROFOUNDLY IN THE DAYS AND WEEKS TO COME.
AND I AM SURE THAT YOU WILL ALL JOIN ME IN WISHING
HIM THE BEST OF GOOD FORTUNE IN HIS NEW ENDEAVORS.
A HEARTY THANK YOU, PHIL. AND GOOD LUCK.

THIS RESIGNATION IS EFFECTIVE IMMEDIATELY. HOWARD
EBERHARDT WILL SERVE AS ACTING COUNSEL UNTIL SUCH
TIME AS A NEW PERMANENT APPOINTMENT IS MADE.

ROBERT GARVIN

Fernandez said, "What does it say?"

"It says, 'I fired his sanctimonious ass.' "

"It had to happen," Fernandez said. "Especially since he was the source on the Connie Walsh story."

Sanders said, "How did you know that?"

"Eleanor Vries."

"She told you?"

"No. But Eleanor Vries is a very cautious attorney. All those media attorneys are. The safest way to keep your job is to refuse to let things run. When in doubt, throw it out. So I had to ask myself, why did she let the Mr. Piggy story run, when it's clearly defamatory. The only possible reason is that she felt Walsh had an unusually strong source inside the company—a source that understood the legal implications. A source that, in giving the story, was in essence also saying, we won't sue if you print it. Since high-ranking corporate officers never know anything about law, it means the source could only be a high-ranking lawyer."

"Phil."

"Yes."

"Jesus."

"Does this change your plans?" Fernandez said.

Sanders had been considering that. "I don't think so," he said. "I think Garvin would have fired him later in the day, anyway."

"You sound confident."

"Yeah. I got some ammunition last night. And I hope more today."

Cindy came in and said, "Are you expecting something from KL? A big file?"

"Yes."

"This one's been coming in since 7 a.m. It must be a monster." She put a DAT cartridge on his desk. It was exactly like the DAT cartridge that had recorded his video link with Arthur Kahn.

Fernandez looked at him. He shrugged.

At eight-thirty, he transmitted Bosak's memo to Garvin's private fax machine. Then he asked Cindy to make copies of all the faxes that Mohammed Jafar had sent him the previous night. Sanders had been up most of the night, reading the material that Jafar had sent him. And it made interesting reading.

Jafar of course was not ill; he had never been ill. That had been a little story that Kahn had contrived with Meredith.

He pushed the DAT videocassette into the machine, and turned to Fernandez.

"You going to explain?" she said.

"I hope it'll be self-explanatory," Sanders said.

On the monitor, the following appeared:

5 SECONDS TO DIRECT VIDEO LINKUP: DC/M-DC/C
SEN: A. KAHN
REC: M. JOHNSON

On the screen, he saw Kahn at the factory, and then a moment later the screen split and he saw Meredith at her office in Cupertino.

"What is this?" Fernandez said.

"A recorded video communication. From last Sunday."

"I thought the communications were all erased."

"They were, here. But there was still a record in KL. A friend of mine sent it to me."

On the screen, Arthur Kahn coughed. "Uh, Meredith. I'm a little concerned."

"Don't be," Meredith said.

"But we still aren't able to manufacture to specs. We have to replace the air handlers, at the very least. Put in better ones."

"Not now."

"But we have to, Meredith."

"Not yet."

"But those handlers are inadequate, Meredith. We both thought they'd be okay, but they aren't."

"Never mind."

Kahn was sweating. He rubbed his chin nervously. "It's only a matter of time before Tom figures it out, Meredith. He's not stupid, you know."

"He'll be distracted."

"So you say."

"And besides, he's going to quit."

Kahn looked startled. "He is? I don't think he—"

"Trust me. He'll quit. He's going to hate working for me."

Sitting in Sanders's office, Fernandez leaned forward, staring at the screen. She said, "No shit."

Kahn said, "Why will he hate it?"

Meredith said, "Believe me. He will. Tom Sanders will be out in my first forty-eight hours."

"But how can you be sure—"

"What choice does he have? Tom and I have a history. Everybody in the company knows that. If any problem comes up, nobody will believe him. He's smart enough to understand that. If he ever wants to work again, he'll have no choice but to take whatever settlement he's offered and leave."

Kahn nodded, wiping the sweat from his cheek. "And then we say Sanders made the changes at the plant? He'll deny that he did."

"He won't even know. Remember. He'll be gone by then, Arthur."

"And if he isn't?"

"Trust me. He'll be gone. He's married, has a family. He'll go."

"But if he calls me about the production line—"

"Just evade it, Arthur. Be mystified. You can do that, I'm sure. Now, who else does Sanders talk to there?"

"The foreman, sometimes. Jafar. Jafar knows everything, of course. And he's one of those honest sorts. I'm afraid if—"

"Make him take a vacation."

"He just took one."

"Make him take another one, Arthur. I only need a week here."

"Jesus," Kahn said. "I'm not sure—"

She cut in: "Arthur."

"Yes, Meredith."

"This is the time when a new vice president counts favors that will be repaid in the future."

"Yes, Meredith."

"That's all."

The screen went blank. There were white streaking video lines, and then the screen was dark.

"Pretty cut and dried," Fernandez said.

Sanders nodded. "Meredith didn't think the changes would matter, because she didn't know anything about production. She was just cutting costs. But she knew that the changes at the plant would eventually be traced back to her, so she thought she had a way to get rid of me, to make me quit the company. And then she would be able to blame me for the problems at the plant."

"And Kahn went along with it."

Sanders nodded.

"And they got rid of Jafar."

Sanders nodded. "Kahn told Jafar to go visit his cousin in Johore for a week—to get out of town. To make it impossible for me to reach Jafar. But he never thought that Jafar would call me." He glanced at his watch. "Now, where is it?"

"What?"

On the screen, there was a series of tones, and they saw a handsome, dark-skinned newscaster at a desk, facing a camera and speaking rapidly in a foreign language.

"What's this?" Fernandez said.

"The Channel Three evening news, from last December." Sanders got up and pushed a button on the tape machine. The cassette popped out.

"What does it show?"

Cindy came back from the copying machine with wide eyes. She carried a dozen stacks of paper, each neatly clipped. "What're you going to do with this?"

"Don't worry about it," he said.

"But this is outrageous, Tom. What she's done."

"I know," he said.

"Everybody is talking," she said. "The word is that the merger is off."

"We'll see," Sanders said.

With Cindy's help, he began arranging the piles of paper in identical manila folders.

Fernandez said, "What exactly are you going to do?"

"Meredith's problem is that she lies," Sanders said. "She's smooth, and she gets away with it. She's gotten away with it her whole life. I'm going to see if I can get her to make a single, very big lie."

He looked at his watch. It was eight forty-five.

The meeting would start in fifteen minutes.

The conference room was packed. There were fifteen Conley-White executives down one side of the table, with John Marden in the middle, and fifteen DigiCom executives down the other side, with Garvin in the middle.

Meredith Johnson stood at the head of the table and said, "Next, we'll hear from Tom Sanders. Tom, I wonder if you could review for us where we stand with the Twinkle drive. What is the status of our production there."

"Of course, Meredith." Sanders stood, his heart pounding. He walked to the front of the room. "By way of background, Twinkle is our code name for a stand-alone CD-ROM drive player which we expect to be revolutionary." He turned to the first of his charts. "CD-ROM is a small laser disk used to store data. It is cheap to manufacture, and can hold an enormous amount of information in any form—words, images, sound, video, and so on. You can put the equivalent of six hundred books on a single small disk, or, thanks to our research here, an hour and a half of video. And any combination. For example, you could make a textbook that combines text, pictures, short movie sequences, animated cartoons, and so on. Production costs will soon be at ten cents a unit."

He looked down the table. The Conley-White people were interested. Garvin was frowning. Meredith looked tense.

"But for CD-ROM to be effective, two things need to happen. First, we need a portable player. Like this." He held up the player, and then passed it down the Conley-White side.

"A five-hour battery, and an excellent screen. You can use it on a train, a bus, or in a classroom—anywhere you can use a book."

The executives looked at it, turned it over in their hands. Then they looked back at Sanders.

"The other problem with CD-ROM technology," Sanders said, "is that it's slow. It's sluggish getting to all that wonderful data. But the Twinkle drives that we have successfully made in prototype are twice as fast as any other drive in the world. And with added memory for our packing and unpacking images, it is as quick as a small computer. We expect to get the unit cost for these drives down to the price of a video-game unit within a year. And we are manufacturing the drives now. We have had some early problems, but we are solving them."

Meredith said, "Can you tell us more about that? I gather from talking to Arthur Kahn that we're still not clear on why the drives have problems."

"Actually, we are," Sanders said. "It turns out that the problems aren't serious at all. I expect them to be entirely resolved in a matter of days."

"Really." She raised her eyebrows. "Then we've found what the trouble is?"

"Yes, we have."

"That's wonderful news."

"Yes, it is."

"Very good news indeed," Ed Nichols said. "Was it a design problem?"

"No," Sanders said. "There's nothing wrong with the design we made here, just as there was nothing wrong with the prototypes. What we have is a fabrication problem involving the production line in Malaysia."

"What sort of problems?"

"It turns out," Sanders said, "that we don't have the proper equipment on the line. We should be using automatic chip installers to lock the controller chips and the RAM cache on the board, but the Malays on the line have been installing chips by hand. Literally pushing them in with their thumbs. And it turns out that the assembly line is dirty, so we're getting particulate matter in the split optics. We should have level-seven air handlers, but we only have level-five handlers installed. And it turns out that we should be ordering components like hinge rods and clips from one very reliable Singapore supplier, but the components are actually coming from another supplier. Less expensive, less reliable."

Meredith looked uneasy, but only for a moment. "Improper equipment, improper conditions, improper components . . ." She shook her head. "I'm sorry. Correct me if I'm wrong, but didn't you set up that line, Tom?"

"Yes, I did," Sanders said. "I went out to Kuala Lumpur last fall and set it up with Arthur Kahn and the local foreman, Mohammed Jafar."

"Then how is it that we have so many problems?"

"Unfortunately, there was a series of bad judgment calls in setting up the line."

Meredith looked concerned. "Tom, we all know that you're extremely competent. How could this have happened?"

Sanders hesitated.

This was the moment.

"It happened because the line was changed," he said. "The specifications were altered."

"Altered? How?"

"I think that's something for you to explain to this group, Meredith," he said. "Since you ordered the changes."

"I ordered them?"

"That's right, Meredith."

"Tom, you must be mistaken," she said coolly. "I haven't had anything to do with that Malaysia line."

"Actually, you have," Sanders said. "You made two trips there, in November and December of last year."

"Two trips to Kuala Lumpur, yes. Because you mishandled a labor dispute with the Malaysian government. I went there and resolved the dispute. But I had nothing to do with the actual production line."

"I'd say you're mistaken, Meredith."

"I assure you," she said coldly. "I am not. I had nothing to do with the line, and any so-called changes."

"Actually, you went there and inspected the changes you ordered."

"I'm sorry, Tom. I didn't. I've never even seen the actual line."

On the screen behind her, the videotape of the newscast began to play silently with the sound off. The newscaster in coat and tie speaking to the camera.

Sanders said, "You never went to the plant itself?"

"Absolutely not, Tom. I don't know who could have told you such a thing—or why you would say it now."

The screen behind the newscaster showed the DigiCom building in Malaysia, then the interior of the plant. The camera showed the production lines and an official inspection tour taking place. They saw Phil

Blackburn, and alongside him, Meredith Johnson. The camera moved in on her as she chatted with one of the workers.

There was a murmur in the room.

Meredith spun around and looked. "This is outrageous. This is out of context. I don't know where this could have come from—"

"Malaysia Channel Three. Their version of the BBC. I'm sorry, Meredith." The newscast segment finished and the screen went blank. Sanders made a gesture, and Cindy began moving around the table, handing a manila folder to each person.

Meredith said, "Wherever this so-called tape came from—"

Sanders said, "Ladies and gentlemen, if you will open your packets, you will find the first of a series of memos from the Operations Review Unit, which was under the direction of Ms. Johnson in the period in question. I direct your attention to the first memo, dated November eighteenth of last year. You will notice that it has been signed by Meredith Johnson, and it stipulates that the line will be changed to accommodate the labor demands of the Malay government. In particular, this first memo states that automated chip installers will not be included, but that this work will be done by hand. That made the Malay government happy, but it meant we couldn't manufacture the drives."

Johnson said, "But you see, what you are overlooking is that the Malays gave us no choice—"

"In that case, we should never have built the plant there," Sanders said, cutting her off. "Because we can't manufacture the intended product at those revised specifications. The tolerances are inadequate."

Johnson said, "Well, that may be your own opinion—"

"The second memo, dated December third, indicates that a cost-savings review diminished air-handling capacities on the line. Again, this is a variance in the specifications that I established. Again, it is critical—we can't manufacture high-performance drives under these conditions. The long and the short of it is that these decisions doomed the drives to failure."

"Now look," Johnson said. "If anybody believes that the failure of these drives is anything but your—"

"The third memo," Sanders said, "summarizes cost savings from the Operations Review Unit. You'll see that it claims an eleven percent reduction in operating costs. That savings has already been wiped out by fabrication delays, not counting our time-to-market delay costs. Even if we immediately restore the line, this eleven percent savings translates into a production cost increase, over the run, of nearly seventy percent. First year, it's a hundred and ninety percent increase.

"Now the next memo," Sanders said, "explains why this cost-cutting was adopted in the first place. During acquisition talks between Mr. Nichols and Ms. Johnson in the fall of last year, Ms. Johnson indicated she would demonstrate that it was possible to reduce high-technology development costs, which were a source of concern to Mr. Nichols when they were meeting at—"

"Oh *Christ*," Ed Nichols said, staring at the paper.

Meredith pushed forward, stepping in front of Sanders. "Excuse me, Tom," she said, speaking firmly, "but I really must interrupt you. I'm sorry to have to say this, but no one here is fooled by this little charade." She swept her arm wide, encompassing the room. "Or by your so-called evidence." She spoke more loudly. "You weren't present when these management decisions were carefully taken by the best minds in this company. You don't understand the thinking that lies behind them. And the false postures you are striking now, the so-called memos that you are holding up to convince us . . . No one here is persuaded." She gave him a pitying look. "It's all empty, Tom. Empty words, empty phrases. When it comes right down to it, you're all show and no substance. You think you can come in here and second-guess the management team? I'm here to tell you that you can't."

Garvin stood abruptly, and said, "Meredith—"

"Let me finish," Meredith said. She was flushed, angry. "Because this is important, Bob. This is the heart of what is wrong with this division. Yes, there were some decisions taken that may be questionable in retrospect. Yes, we tried innovative procedures which perhaps went too far. But that hardly excuses the behavior we see today. This calculated, manipulative attitude by an individual who will do anything—anything at all—to get ahead, to make a name for herself at the expense of others, who will savage the reputation of anyone who stands in her path— I mean, that stands in *his* path—this ruthless demeanor that we are seeing . . . No one is fooled by this, Tom. Not for a minute. We're being asked to accept the worst kind of fraudulence. And we simply won't do it. It's wrong. This is all wrong. And it is bound to catch up with you.

I'm sorry. You can't come here and do this. It simply won't work—it hasn't worked. That's all."

She stopped to catch her breath and looked around the table. Everyone was silent, motionless. Garvin was still standing; he appeared to be in shock. Slowly, Meredith seemed to realize that something was wrong. When she spoke again, her voice was quieter.

"I hope that I have . . . that I have accurately expressed the sentiments of everyone here. That's all I intended to do."

There was another silence. Then Garvin said, "Meredith, I wonder if you would leave the room for a few minutes."

Stunned, she stared at Garvin for a long moment. Then she said, "Of course, Bob."

"Thank you, Meredith."

Walking very erect, she left the room. The door clicked shut behind her.

John Marden sat forward and said, "Mr. Sanders, please continue with your presentation. In your view, how long will it be until the line is repaired and fully functioning?"

It was noon. Sanders sat in his office with his feet on his desk and stared out the window. The sun was shining brightly on the buildings around Pioneer Square. The sky was clear and cloudless. Mary Anne Hunter, wearing a business suit, came in and said, "I don't get it."

"Get what?"

"That news tape. Meredith must have known about it. Because she was there when they were shooting it."

"Oh, she knew about it, all right. But she never thought I'd get it. And she never thought she'd appear in it. She thought they'd only show Phil. You know—a Muslim country. In a story about executives, they usually just show the men."

"Uh-huh. So?"

"But Channel Three is the government station," Sanders said. "And the story that night was that the government had been only partially successful in negotiating changes in the DigiCom plant—that the foreign executives had been intransigent and uncooperative. It was a story intended to protect the reputation of Mr. Sayad, the finance minister. So the cameras focused on her."

"Because . . ."

"Because she was a woman."

"Foreign she-devil in a business suit? Can't make a deal with a *feringi* woman?"

"Something like that. Anyway, the story focused on her."

"And you got the tape."

"Yeah."

Hunter nodded. "Well," she said, "it's fine with me." She left the room, and Sanders was alone again, staring out the window.

After a while, Cindy came in and said, "The latest word is the acquisition is off."

Sanders shrugged. He was flat, drained. He didn't care.

Cindy said, "Are you hungry? I can get you some lunch."

"I'm not hungry. What are they doing now?"

"Garvin and Marden are talking."

"Still? It's been more than an hour."

"They just brought in Conley."

"Only Conley? Nobody else?"

"No. And Nichols has left the building."

"What about Meredith?"

"Nobody's seen her."

He leaned back in his chair. He stared out the window. His computer gave three beeps.

30 SECONDS TO DIRECT VIDEO LINKUP: DC/M-DC/S
SEN: A. KAHN
REC: T. SANDERS

Kahn was calling. Sanders smiled grimly. Cindy came in and said, "Arthur's going to call."

"I see that."

15 SECONDS TO DIRECT VIDEO LINKUP: DC/M-DC/S

Sanders adjusted his desk lamp and sat back. The screen blossomed, and he saw the shimmering image resolve. It was Arthur, in the plant.

"Oh, Tom. Good. I hope it's not too late," Arthur said.

"Too late for what?" Sanders said.

"I know there's a meeting today. There's something I have to tell you."

"What's that, Arthur?"

"Well, I'm afraid I haven't been entirely straightforward with you, Tom. It's about Meredith. She made changes in the line six or seven months ago, and I'm afraid she intends to blame that on you. Probably in the meeting today."

"I see."

"I feel terrible about this, Tom," Arthur said, hanging his head. "I don't know what to say."

"Don't say anything, Arthur," Sanders said.

Kahn smiled apologetically. "I wanted to tell you earlier. I really did. But Meredith kept saying that you would be out. I didn't know what to do. She said there was a battle coming, and I had better pick the winner."

"You picked wrong, Arthur," Sanders said. "You're fired." He reached up and snapped off the television camera in front of him.

"What're you talking about?"

"You're fired, Arthur."

"But you can't do this to me . . . ," Kahn said. His image faded, began to shrink. "You can't—"

The screen was blank.

Fifteen minutes later, Mark Lewyn came by the office. He tugged at the neck of his black Armani T-shirt. "I think I'm an asshole," he said.

"Yeah. You are."

"It's just . . . I didn't understand the situation," he said.

"That's right, you didn't."

"What're you going to do now?"

"I just fired Arthur."

"Jesus. And what else?"

"I don't know. We'll see how it shakes out."

Lewyn nodded and went away nervously. Sanders decided to let him be nervous for a while. In the end, their friendship would be repaired. Adele and Susan were good friends. And Mark was too talented to replace in the company. But Lewyn could sweat for a while; it'd do him good.

At one o'clock, Cindy came in and said, "The word is Max Dorfman just went into the conference with Garvin and Marden."

"What about John Conley?"

"He's gone. He's with the accountants now."

"Then that's a good sign."

"And the word is Nichols was fired."

"Why do they think that?"

"He flew home an hour ago."

Fifteen minutes later, Sanders saw Ed Nichols walking down the hallway. Sanders got up and went out to Cindy's desk. "I thought you said Nichols went home."

"Well, that's what I heard," she said. "It's crazy. You know what they're saying about Meredith now?"

"What?"

"They say she's staying on."

"I don't believe it," Sanders said.

"Bill Everts told Stephanie Kaplan's assistant that Meredith Johnson is not going to be fired, that Garvin is backing her one hundred percent. Phil is going to take the rap for what happened in Malaysia but Garvin still believes Meredith is young and this shouldn't be held against her. So she's staying in her job."

"I don't believe it."

Cindy shrugged. "That's what they say," she said.

He went back to his office and stared out the window. He told himself it was just a rumor. After a while, the intercom buzzed. "Tom? Meredith Johnson just called. She wants to see you in her office right away."

Bright sunlight streamed in through the big windows on the fifth floor. The assistant outside Meredith's office was away from her desk. The door was ajar. He knocked.

"Come in," Meredith Johnson said.

She was standing, leaning back against the edge of her desk, her arms folded across her chest. Waiting.

"Hello, Tom," she said.

"Meredith."

"Come in. I won't bite."

He came in, leaving the door open.

"I must say that you outdid yourself this morning, Tom. I was surprised at how much you were able to learn in a short time. And it was really quite resourceful, the approach you took in the meeting."

He said nothing.

"Yes, it was a really excellent effort. You feeling proud of yourself?" she said, staring hard at him.

"Meredith . . ."

"You think you've finally paid me back? Well, I have news for you, Tom. You don't know *anything* about what's really going on."

She pushed away from the desk, and as she moved away, he saw a cardboard packing box on the desktop beside the telephone. She walked around behind the desk, and began putting pictures and papers and a pen set into the box.

"This whole thing was Garvin's idea. For three years, Garvin's been looking for a buyer. He couldn't find one. Finally he sent me out, and I found him one. I went through twenty-seven different companies until I got to Conley-White. They were interested, and I sold them hard. I put in the hours. I did whatever I had to do to keep the deal moving forward. *Whatever* I had to do." She pushed more papers into the box angrily.

Sanders watched her.

"Garvin was happy as long as I was delivering Nichols to him on a platter," Johnson said. "He wasn't fussy about how I was doing it. He wasn't even interested. He just wanted it done. I busted my ass for him. Because the chance to get this job was a big break for me, a real career opportunity. Why shouldn't I have it? I did the work. I put the deal together. I *earned* this job. I beat you fairly."

Sanders said nothing.

"But that's not how it turns out, is it? Garvin won't support me when the going gets tough. Everybody said he was like a father to me. But he was just using me. He was just making a deal, any way he could. And that's all he's doing now. Just another fucking deal, and who cares who gets hurt. Everybody moves on. Now I've got to find an attorney to negotiate my severance package. Nobody gives a damn."

She closed the box and leaned on it. "But I beat you, fair and square, Tom. I don't deserve this. I've been screwed by the damned system."

"No you haven't," Sanders said, staring her straight in the eye. "You've been fucking your assistants for years. You've been taking every advantage of your position that you could. You've been cutting corners. You've been lazy. You've been living on image and every third word out of your mouth is a lie. Now you're feeling sorry for yourself. You think the system is what's wrong. But you know what, Meredith? The system didn't screw you. The system *revealed* you, and dumped you out. Because when you get right down to it, you're completely full of shit." He turned on his heel. "Have a nice trip. Wherever you're going."

He left the room, and slammed the door behind him.

He was back in his office five minutes later, still angry, pacing back and forth behind his desk.

Mary Anne Hunter came in, wearing a sweatshirt and exercise tights. She sat down, and put her running shoes up on Sanders's desk. "What're you all worked up about? The press conference?"

"What press conference?"

"They've scheduled a press conference for four o'clock."

"Who says?"

"Marian in PR. Swears it came from Garvin himself. And Marian's assistant has been calling the press and the stations."

Sanders shook his head. "It's too soon." Considering all that had happened, the press conference should not be held until the following day.

"I think so," Hunter said, nodding. "They must be going to announce that the merger has fallen through. You heard what they're saying about Blackburn?"

"No, what?"

"That Garvin made him a million-dollar settlement."

"I don't believe it."

"That's what they say."

"Ask Stephanie."

"Nobody's seen her. Supposedly she went back to Cupertino, to deal with finances now that the merger is off." Hunter got up and walked to the window. "At least it's a nice day."

"Yeah. Finally."

"I think I'll go for a run. I can't stand this waiting."

"I wouldn't leave the building."

She smiled. "Yeah, I guess not." She stood at the window for a while. Finally she said, "Well, what do you know . . ."

Sanders looked up. "What?"

Hunter pointed down toward the street. "Minivans. With antennas on the top. I guess there is going to be a press conference, after all."

They held the press conference at four, in the main downstairs conference room. Strobes flashed as Garvin stood before the microphone, at the end of the table.

"I have always believed," he said, "that women must be better represented in high corporate office. The women of America represent our nation's most important underutilized resource as we go into the twenty-first century. And this is true in high technology no less than in other industries. It is therefore with great pleasure that I announce, as part of our merger with Conley-White Communications, that the new Vice President at Digital Communications Seattle is a woman of great talent, drawn from within the ranks in our Cupertino headquarters. She has been a resourceful and dedicated member of the DigiCom team for many years, and I am sure she will be even more resourceful in the future. I am pleased to introduce now the new Vice President for Advanced Planning, Ms. Stephanie Kaplan."

There was applause, and Kaplan stepped to the microphone and brushed back her shock of gray hair. She wore a dark maroon suit and smiled quietly. "Thank you, Bob. And thanks to everyone who has worked so hard to make this division so great. I want to say particularly that I look forward to working with the outstanding division heads we have here, Mary Anne Hunter, Mark Lewyn, Don Cherry, and, of course, Tom Sanders. These talented people stand at the center of our company, and I intend to work hand in hand with them as we move into the future. As for myself, I have personal as well as professional ties here in Seattle, and I can say no more than that I am delighted, just delighted, to be here. And I look forward to a long and happy time in this wonderful city."

Back in his office, Sanders got a call from Fernandez. "I finally heard from Alan. Are you ready for this? Arthur A. Friend is on sabbatical in Nepal. Nobody goes into his office except his assistant and a couple of his most trusted students. In fact, there's only one student who has been there during the time he is away. A freshman in the chemistry department named Jonathan—"

"Kaplan," Sanders said.

"That's right. You know who he is?" Fernandez said.

"He's the boss's son. Stephanie Kaplan's just been named the new head of the division."

Fernandez was silent for a moment. "She must be a very remarkable woman," she said.

Garvin arranged a meeting with Fernandez at the Four Seasons Hotel. They sat in the small, dark bar off Fourth Avenue in the late afternoon.

"You did a hell of a job, Louise," he said. "But justice was not served, I can tell you that. An innocent woman took the fall for a clever, scheming man."

"Come on, Bob," she said. "Is that why you called me over here? To complain?"

"Honest to God, Louise, this harassment thing has gotten out of hand. Every company I know has at least a dozen of these cases now. Where will it end?"

"I'm not worried," she said. "It'll shake out."

"Eventually, maybe. But meanwhile innocent people—"

"I don't see many innocent people in my line of work," she said. "For example, it's come to my attention that DigiCom's board members were aware of Johnson's problem a year ago and did nothing to address it."

Garvin blinked. "Who told you that? It's completely untrue."

She said nothing.

"And you could never have proved it."

Fernandez raised her eyebrows and said nothing.

"Who said that?" Garvin said. "I want to know."

"Look, Bob," she said. "The fact is, there's a category of behavior that no one condones anymore. The supervisor who grabs genitals, who squeezes breasts in the elevator, who invites an assistant on a business trip but books only one hotel room. All that is ancient history. If you have an employee behaving like that, whether that employee is male or female, gay or straight, you are obliged to stop it."

"Okay, fine, but sometimes it's hard to know—"

"Yes," Fernandez said. "And there's the opposite extreme. An employee doesn't like a tasteless remark and files a complaint. Somebody has to tell her it's not harassment. By then, her boss has been accused, and everybody in the company knows. He won't work with her any-more; there's suspicion, and bad feelings, and it's all a big mess at the company. I see that a lot. That's unfortunate, too. You know, my husband works in the same firm I do."

"Uh-huh."

"After we first met, he asked me out five times. At first I said no, but finally I said yes. We're happily married now. And the other day he said to me that, given the climate now, if we met today, he probably wouldn't ask me out five times. He'd just drop it."

"See? That's what I'm talking about."

"I know. But those situations will settle out eventually. In a year or two, everybody will know what the new rules are."

"Yes, but—"

"But the problem is that there's that third category, somewhere in the middle, between the two extremes," Fernandez said. "Where the behavior is gray. It's not clear what happened. It's not clear who did what to whom. That's the largest category of complaints we see. So far, society's tended to focus on the problems of the victim, not the prob-lems of the accused. But the accused has problems, too. A harassment claim is a weapon, Bob, and there are no good defenses against it. Anybody can use the weapon—and lots of people have. It's going to continue for a while, I think."

Garvin sighed.

"It's like that virtual reality thing you have," Fernandez said. "Those environments that seem real but aren't really there. We all live every day in virtual environments, defined by our ideas. Those environments are changing. It's changed with regard to women, and it's going to start changing with regard to men. The men didn't like it when it changed before, and the women aren't going to like it changing now. And some people will take advantage. But in the final analysis, it'll all work out."

"When? When will it all end?" Garvin said, shaking his head.

"When women have fifty percent of the executive positions," she said. "That's when it will end."

"You know I favor that."

"Yes," Fernandez said, "and I gather you have just appointed an outstanding woman. Congratulations, Bob."

Mary Anne Hunter was assigned to drive Meredith Johnson to the airport, to take a plane back to Cupertino. The two women sat in silence for fifteen minutes, Meredith Johnson hunched down in her trench coat, staring out the window.

Finally, when they were driving past the Boeing plant, Johnson said, "I didn't like it here, anyway."

Choosing her words carefully, Hunter said, "It has its good and bad points."

There was another silence. Then Johnson asked, "Are you a friend of Sanders?"

"Yes."

"He's a nice guy," Johnson said. "Always was. You know, we used to have a relationship."

"I heard that," Hunter said.

"Tom didn't do anything wrong, really," Johnson said. "He just didn't know how to handle a passing remark."

"Uh-huh," Hunter said.

"Women in business have to be perfect all the time, or they just get murdered. One little slip and they're dead."

"Uh-huh."

"You know what I'm talking about."

"Yes," Hunter said. "I know."

There was another long silence. Johnson shifted in her seat. She stared out the window.

"The system," Johnson said. "That's the problem. I was raped by the fucking system."

Sanders was leaving the building, on his way to the airport to pick up Susan and the kids, when he ran into Stephanie Kaplan. He congratulated her on the appointment. She shook his hand and said without smiling, "Thank you for your support."

He said, "Thank you for yours. It's nice to have a friend."

"Yes," she said. "Friendships are nice. So is competence. I'm not going to keep this job very long, Tom. Nichols is out as CFO of Conley, and their number-two man is a modest talent at best. They'll be looking for someone in a year or so. And when I go over there, someone will have to take over the new company here. I imagine it should be you."

Sanders bowed slightly.

"But that's in the future," Kaplan said crisply. "In the meantime, we have to get the work here back on track. This division is a mess. Everyone's been distracted by this merger, and the product lines have been compromised by Cupertino's ineptitude. We've got a lot to do to turn this around. I've set the first production meeting with all the division heads for seven a.m. tomorrow morning. I'll see you then, Tom."

And she turned away.

Sanders stood at the arrivals gate at Sea-Tac and watched the passengers come off the Phoenix plane. Eliza came running up to him, shouting "Daddy!" as she leapt into his arms. She had a suntan.

"Did you have a nice time in Phoenix?"

"It was great, Dad! We rode horses and ate tacos, and guess what?"

"What?"

"I saw a snake."

"A real snake?"

"Uh-huh. A green one. It was *this* big," she said, stretching her hands.

"That's pretty big, Eliza."

"But you know what? Green snakes don't hurt you."

Susan came up, carrying Matthew. She had a suntan, too. He kissed her, and Eliza said, "I told Daddy about the snake."

"How are you?" Susan said, looking at his face.

"I'm fine. Tired."

"Is it finished?"

"Yes. It's finished."

They walked on. Susan slipped her arm around his waist. "I've been thinking. Maybe I'm traveling too much. We ought to spend more time together."

"That'd be nice," he said.

They walked toward the baggage claim. Carrying his daughter, feeling her small hands on his shoulder, he glanced over and saw Meredith Johnson standing at the check-in counter of one of the departure gates. She was wearing a trench coat. Her hair was pulled back. She didn't turn and see him.

Susan said, "Somebody you know?"

"No," he said. "It's nobody."

POSTSCRIPT

Constance Walsh was fired by the Seattle *Post-Intelligencer* and sued the paper for wrongful termination and sexual discrimination under Title VII of the Civil Rights Act of 1964. The paper settled out of court.

Philip Blackburn was named chief counsel at Silicon Holographics of Mountain View, California, a company twice as large as DigiCom. He was later elected Chairman of the Ethics Panel of the San Francisco Bar Association.

Edward Nichols took early retirement from Conley-White Communications and moved with his wife to Nassau, Bahamas, where he worked part-time as a consultant to offshore firms.

Elizabeth "Betsy" Ross was hired by Conrad Computers in Sunnyvale, California, and soon after joined Alcoholics Anonymous.

John Conley was named Vice President for Planning at Conley-White Communications. He died in an automobile accident in Patchogue, New York, six months later.

Mark Lewyn was charged with sexual harassment under Title VII by an employee of the Design Group. Although Lewyn was cleared of the charge, his wife filed for divorce not long after the investigation was concluded.

Arthur Kahn joined Bull Data Systems in Kuala Lumpur, Malaysia.

Richard Jackson of Aldus was charged with sexual harassment under Title VII by an employee of American DataHouse, a wholesale distributor for Aldus. After an investigation, Aldus fired Jackson.

Gary Bosak developed a data encryption algorithm, which he licensed to IBM, Microsoft, and Hitachi. He became a multi-millionaire.

Louise Fernandez was appointed to the federal bench. She delivered a lecture to the Seattle Bar Association in which she argued that sexual harassment suits had become increasingly used as a weapon to resolve corporate disputes. She suggested that in the future there might be a need to revise laws or to limit the involvement of attorneys in such matters. Her speech was received coolly.

Meredith Johnson was named Vice President for Operations and Planning at IBM's Paris office. She subsequently married the United States Ambassador to France, Edward Harmon, following his divorce. She has since retired from business.

AFTERWORD

The episode related here is based on a true story. Its appearance in a novel is not intended to deny the fact that the great majority of harassment claims are brought by women against men. On the contrary: the advantage of a role-reversal story is that it may enable us to examine aspects concealed by traditional responses and conventional rhetoric. However readers respond to this story, it is important to recognize that the behavior of the two antagonists mirrors each other, like a Rorschach inkblot. The value of a Rorschach test lies in what it tells us about ourselves.

It is also important to emphasize that the story in its present form is fiction. Because allegations of sexual harassment in the workplace involve multiple, conflicting legal rights, and because such claims now create substantial risk not only for the individuals but for corporations, it has been necessary to disguise the real event with care. All the principals in this case agreed to be interviewed with the understanding that their identities would be concealed. I am grateful to them for their willingness to help clarify the difficult issues inherent in investigations of sexual harassment.

In addition, I am indebted to a number of attorneys, human relations officers, individual employees, and corporate officials who provided valuable perspectives on this evolving issue. It is characteristic of the extreme sensitivity surrounding any discussion of sexual harassment that everyone I talked to asked to remain anonymous.

A NOTE ON THE TYPE

This book was set in a digitized version of Janson. The hot-metal version of Janson was a recutting made direct from type cast from matrices long thought to have been made by the Dutchman Anton Janson, who was a practicing type founder in Leipzig during the years 1668–1687. However, it has been conclusively demonstrated that these types are actually the work of Nicholas Kis (1650–1702), a Hungarian, who most probably learned his trade from the master Dutch type founder Dirk Voskens. The type is an excellent example of the influential and sturdy Dutch types that prevailed in England up to the time William Caslon (1692–1766) developed his own incomparable designs from them.